Lecture Notes in Computer Science   **9045**

Commenced Publication in 1973
Founding and Former Series Editors:
Gerhard Goos, Juris Hartmanis, and Jan van Leeuwen

More information about this series at http://www.springer.com/series/7407

Ivan Dimov · István Faragó
Lubin Vulkov (Eds.)

Finite Difference Methods, Theory and Applications

6th International Conference, FDM 2014
Lozenetz, Bulgaria, June 18–23, 2014
Revised Selected Papers

 Springer

Editors
Ivan Dimov
Institute of Information and Communication
    Technologies
Bulgarian Academy of Sciences
Sofia
Bulgaria

Lubin Vulkov
University of Rousse
Rousse
Bulgaria

István Faragó
Institute of Mathematics and MTA-ELTE
    Numerical Analysis and Large Networks
    Research Group
Eötvös Loránd University
Budapest
Hungary

ISSN 0302-9743   ISSN 1611-3349 (electronic)
Lecture Notes in Computer Science
ISBN 978-3-319-20238-9   ISBN 978-3-319-20239-6   (eBook)
DOI 10.1007/978-3-319-20239-6

Library of Congress Control Number: 2015941871

LNCS Sublibrary: SL1 – Theoretical Computer Science and General Issues

Springer Cham Heidelberg New York Dordrecht London

Printed on acid-free paper

Springer International Publishing AG Switzerland is part of Springer Science+Business Media
(www.springer.com)

Preface

The International Conference on Finite Difference Methods is a traditional forum for scientists from all over the world, providing opportunities to share ideas and establish fruitful cooperation. The papers in this volume were presented at the 6th International Conference on Finite Difference Methods: Theory and Applications. FDM: T&A was held in Lozenetz, Bulgaria, in June 2014. The conference was organized and sponsored by Rousse University. This conference continued the tradition of four previous meetings: 1997 in Rousse (Bulgaria), organized by the Division of Numerical Analysis and Statistics; 1998 in Minsk (Belarus), organized by the Institute of Mathematics, Belarus Academy of Science; 2000 in Palanga, (Lithuania), organized by the Institute of Mathematics and Informatics (Vilnus); and 2006 and 2010 in Lozenetz (Bulgaria), organized by the Division of Numerical Analysis and Statistics, Rousse University.

The purpose of the first three conferences (held in Bulgaria, Belarus, and Lithuania) was to bring together scientists from the East and West to exchange ideas and establish research cooperation. We have observed that the contacts among scientists have become more regular and we are proud of our contribution in this respect.

The general topic for FDM: T&A 2014 was finite difference and combined finite difference methods as well as finite element methods and their various applications in physics, chemistry, engineering, biology, and finance. A number of modern numerical techniques were discussed and presented during the conference: splitting techniques, Green's function method, multigrid methods, and immersed interface method, among others.

Some contemporary topics were the focus of the following minisymposia:

(1) Financial Mathematics; (2) Asymptotic-Numerical Treatment of Problems with Boundary and Internal Layer; (3) Numerical Methods for Fractional Differential Equations; (4) Computational Methods in Geophysics.

The success of the conference and the present volume are due to the joint efforts of many colleagues from various institutions and organizations. We thank our colleagues for their help in the organization of this conference. We especially thank M. Koleva for her help in the preparation of this volume. We are also grateful to the organizers of the minisymposia.

The 7th International FDM:T&A Conference will be organized in June 2018.

March 2015

Ivan Dimov
István Faragó
Lubin Vulkov

Organization

FDM:T&A 2014 was organized by the University of Rousse Angel Kanchev, Rousse, Bulgaria.

Scientific Committee

Ivan Dimov, Bulgaria
Matthias Ehrhardt, Germany
István Faragó, Hungary
Martin Gander, Switzerland
Bosko Jovanovic, Serbia
Abdul Khaliq, USA
Raytcho Lazarov, USA
Piotr Matus, Belarus
Peter Minev, Canada
Nikolay Nefedof, Russia
Grigorii Shishkin, Russia
Vladimir Shidurov, Russia
Petr Vabishchevich, Russia
Song Wang, Australia
Ivan Yotov, USA
Zlotnik Zlotnik, Russia

Local Organizers

Chair

Luben Vulkov
Tatiana Chernogorova
Jury Dimitrov
Miglena Koleva
Walter Mudzimbabwe
Radoslav Valkov

Contents

Contributed Papers

Invited Papers

Simulation of Flow in Fractured Poroelastic Media: A Comparison of Different Discretization Approaches

I. Ambartsumyan[1], E. Khattatov[1], I. Yotov[1][(✉)], and P. Zunino[2]

[1] Department of Mathematics, University of Pittsburgh, Pittsburgh, PA 15260, USA
{ILA6,ELK58}@pitt.edu, yotov@math.pitt.edu
[2] Department of Mechanical Engineering and Materials Science, University of Pittsburgh, Pittsburgh, PA 15261, USA
paz13@pitt.edu

Abstract. We study two finite element computational models for solving coupled problems involving flow in a fracture and flow in poroelastic media. The Brinkman equation is used in the fracture, while the Biot system of poroelasticity is employed in the surrounding media. Appropriate equilibrium and kinematic conditions are imposed on the interfaces. We focus on the approximation of the interface conditions, which in this context feature the interaction of different variables, such as velocities, displacements, stresses and pressures. The aim of this study is to compare the Lagrange multiplier and the Nitsche's methods applied to enforce these non standard interface conditions.

1 Problem Set up

We consider a multiphysics model problem for flow in fractured and deformable porous media, where the simulation domain $\Omega \subset \mathbf{R}^d$, $d = 2, 3$, is a union of non-overlapping and possibly non-connected regions Ω_f and Ω_p. Here Ω_f is a fluid region with flow governed by the Brinkman equations and Ω_p is a poroelasticity region governed by the Biot system for coupled Darcy and elasticity equations. Let $\Gamma_{fp} = \partial\Omega_f \cap \partial\Omega_p$. Let $(\mathbf{u}_\star, p_\star)$ be the velocity-pressure pairs in $\Omega_\star$, $\star = f$, p, and let $\boldsymbol{\eta}_p$ be the displacement in Ω_p. Let ν be the fluid viscosity, K the symmetric and uniformly positive definite rock permeability tensor, $\mathbf{f}_\star$ body force terms, and $q_\star$ external source or sink terms. Let $\mathbf{D}(\mathbf{u}_f)$ and $\boldsymbol{\sigma}_f(\mathbf{u}_f, p_f)$ denote, respectively, the deformation rate tensor and the stress tensor:

$$\mathbf{D}(\mathbf{u}_f) = \frac{1}{2}(\nabla\mathbf{u}_f + \nabla\mathbf{u}_f^T), \quad \boldsymbol{\sigma}_f(\mathbf{u}_f, p_f) = -p_f\mathbf{I} + 2\nu\mathbf{D}(\mathbf{u}_f).$$

In the fracture region Ω_f, $(\mathbf{u}_f, p_f)$ satisfy the Brinkman equations

$$-\nabla \cdot \boldsymbol{\sigma}_f(\mathbf{u}_f, p_f) + \nu K_f^{-1}\mathbf{u}_f = \mathbf{f}_f, \quad \nabla \cdot \mathbf{u}_f = q_f \quad \text{in } \Omega_f, \tag{1}$$

© Springer International Publishing Switzerland 2015
I. Dimov et al. (Eds.): FDM 2014, LNCS 9045, pp. 3–14, 2015.
DOI: 10.1007/978-3-319-20239-6_1

where K_f represents the fracture permeability. Let $\boldsymbol{\sigma}_e(\boldsymbol{\eta})$ and $\boldsymbol{\sigma}_p(\boldsymbol{\eta}, p)$ be the elasticity and poroelasticity stress tensor, respectively:

$$\boldsymbol{\sigma}_e(\boldsymbol{\eta}) = \lambda_p(\nabla \cdot \boldsymbol{\eta})\mathbf{I} + 2\mu_p\mathbf{D}(\boldsymbol{\eta}), \quad \boldsymbol{\sigma}_p(\boldsymbol{\eta}, p) = \boldsymbol{\sigma}_e(\boldsymbol{\eta}) - \alpha p\mathbf{I}, \tag{2}$$

where α is the Biot-Willis constant. The poroelasticity region Ω_p is governed by the quasi-static Biot system [3]

$$-\nabla \cdot \boldsymbol{\sigma}_p(\boldsymbol{\eta}_p, p_p) = \mathbf{f}_p \quad \text{in } \Omega_p, \tag{3}$$

$$\nu K^{-1}\mathbf{u}_p + \nabla p_p = 0, \quad \frac{\partial}{\partial t}(s_0 p_p + \alpha\nabla \cdot \boldsymbol{\eta}_p) + \nabla \cdot \mathbf{u}_p = q_p \quad \text{in } \Omega_p, \tag{4}$$

where s_0 is a storage coefficient. The *interface conditions* on the fluid-poroelasticity interface Γ_{fp} are *mass conservation, balance of normal stress*, and the Beavers-Joseph-Saffman (BJS) law [2,13] modeling *slip with friction* [1,14]:

$$\mathbf{u}_f \cdot \mathbf{n}_f + \left(\frac{\partial\boldsymbol{\eta}_p}{\partial t} + \mathbf{u}_p\right) \cdot \mathbf{n}_p = 0, \tag{5}$$

$$-(\boldsymbol{\sigma}_f\mathbf{n}_f) \cdot \mathbf{n}_f = p_p, \tag{6}$$

$$-(\boldsymbol{\sigma}_f\mathbf{n}_f) \cdot \boldsymbol{\tau}_{f,j} = \nu\alpha_{BJS}\sqrt{K_j^{-1}}(\mathbf{u}_f - \frac{\partial\boldsymbol{\eta}_p}{\partial t}) \cdot \boldsymbol{\tau}_{f,j} \quad \text{on } \Gamma_{fp}, \tag{7}$$

as well as *conservation of momentum*:

$$(\boldsymbol{\sigma}_f\mathbf{n}_f) \cdot \mathbf{n}_f = (\boldsymbol{\sigma}_p\mathbf{n}_p) \cdot \mathbf{n}_p, \quad (\boldsymbol{\sigma}_f\mathbf{n}_f) \cdot \boldsymbol{\tau}_{f,j} = (\boldsymbol{\sigma}_p\mathbf{n}_p) \cdot \boldsymbol{\tau}_{p,j} \quad \text{on } \Gamma_{fp}, \tag{8}$$

where $\mathbf{n}_f$ and $\mathbf{n}_p$ are the outward unit normal vectors to $\partial\Omega_f$, and $\partial\Omega_p$, respectively, $\boldsymbol{\tau}_{f,j}, 1 \leq j \leq d-1$, is an orthogonal system of unit tangent vectors on Γ_{fp}, $K_j = (K\boldsymbol{\tau}_{f,j}) \cdot \boldsymbol{\tau}_{f,j}$, and $\alpha_{BJS} > 0$ is an experimentally determined friction coefficient. We note that the continuity of flux takes into account the normal velocity of the solid skeleton, while the BJS condition accounts for its tangential velocity.

2   Numerical Approximation

Our discretization approach is based on the finite element method. For this reason, we consider the weak formulation of the Biot-Brinkman system (1), (3) and (4). It is obtained by multiplying the equations in each region by suitable test functions, integrating by parts certain terms, and utilizing the interface and boundary conditions. For simplicity the latter are assumed to be homogeneous: $\mathbf{u}_f = 0$, $\mathbf{u}_p \cdot \mathbf{n}_p = 0$ and $\boldsymbol{\eta}_p = 0$ on $\partial\Omega$. Let $(\cdot, \cdot)_S$, $S \subset \mathbf{R}^d$, be the $L^2(S)$ inner product and let $\langle\cdot, \cdot\rangle_F$, $F \subset \mathbf{R}^{d-1}$, be the $L^2(F)$ inner product or duality pairing. Let us define

$$\mathbf{V}_f = H^1(\Omega_f)^d, \quad W_f = L^2(\Omega_f),$$
$$\mathbf{V}_p = H(\text{div}; \Omega_p), \quad W_p = L^2(\Omega_p),$$
$$\mathbf{X}_p = H^1(\Omega_p)^d.$$

The global spaces are products of the subdomain spaces and satisfy the boundary conditions. For simplicity assume for the moment that each region consists of a single subdomain. Let $b_*(\mathbf{v}, w) = -(\nabla \cdot \mathbf{v}, w)_{\Omega_*}$ and let

$$a_f(\mathbf{u}_f, \mathbf{v}_f) = (2\nu \mathbf{D}(\mathbf{u}_f) : \mathbf{D}(\mathbf{v}_f)_{\Omega_f} + (\nu K_f^{-1}\mathbf{u}_f, \mathbf{v}_f),$$

$$a_p^d(\mathbf{u}_p, \mathbf{v}_p) = (\nu K^{-1}\mathbf{u}_p, \mathbf{v}_p)_{\Omega_p},$$
$$a_p^e(\boldsymbol{\eta}_p, \boldsymbol{\xi}_p) = (\boldsymbol{\sigma}_e(\boldsymbol{\eta}_p) : \mathbf{D}(\boldsymbol{\xi}_p))_{\Omega_p}.$$

be the bilinear forms related to Brinkman, Darcy and the elasticity operators, respectively.

To proceed with the discretization, we denote with $\mathbf{V}_{f,h}, W_{f,h}$ the finite element spaces for the velocity and pressure approximation on the fluid domain Ω_f, with $\mathbf{V}_{p,h}, W_{p,h}$ the spaces for velocity and pressure approximation on the porous matrix Ω_p and with $\mathbf{X}_{p,h}$ the approximation spaces for the structure displacement. We assume that all the finite element approximation spaces comply with the prescribed Dirichlet conditions on external boundaries $\partial\Omega_f, \partial\Omega_p$. For the time discretization, we denote with t_n the current time step and with d_τ the first order (backward) discrete time derivative $d_\tau u^n := \tau^{-1}(u^n - u^{n-1})$.

2.1 Approximation of Interface Conditions Using the Lagrange Multiplier Method

To impose the interface conditions on Γ_{fp} we introduce a Lagrange multiplier $\lambda_h \in \Lambda_h = (\mathbf{V}_{p,h} \cdot \mathbf{n}_p)'$ with a physical meaning $\lambda_h = -(\boldsymbol{\sigma}_{f,h}\mathbf{n}_f) \cdot \mathbf{n}_f$. Then, we seek $\mathbf{u}_{f,h} \in \mathbf{V}_{f,h}, p_{f,h} \in W_{f,h}, \mathbf{u}_{p,h} \in \mathbf{V}_{p,h}, p_{p,h} \in W_{p,h}, \boldsymbol{\eta}_{p,h} \in \mathbf{X}_{p,h}$, and $\lambda_h \in \Lambda_h$ such that for all $\mathbf{v}_{f,h} \in \mathbf{V}_{f,h}, w_{f,h} \in W_{f,h}, \mathbf{v}_{p,h} \in \mathbf{V}_{p,h}, w_{p,h} \in W_{p,h}, \boldsymbol{\xi}_{p,h} \in \mathbf{X}_{p,h}$, and $\mu_h \in \Lambda_h$,

$$a_f(\mathbf{u}_{f,h}, \mathbf{v}_{f,h}) + a_p^d(\mathbf{u}_{p,h}, \mathbf{v}_{p,h}) + a_p^e(\boldsymbol{\eta}_{p,h}, \boldsymbol{\xi}_{p,h}) + a_{BJS}(\mathbf{u}_{f,h}, \boldsymbol{\eta}_{p,h}; \mathbf{v}_{f,h}, \boldsymbol{\xi}_{p,h})$$
$$+ b_f(\mathbf{v}_{f,h}, p_{f,h}) + b_p(\mathbf{v}_{p,h}, p_{p,h}) + \alpha b_p(\boldsymbol{\xi}_{p,h}, p_{p,h}) + b_\Gamma(\mathbf{v}_{f,h}, \mathbf{v}_{p,h}, \boldsymbol{\xi}_{p,h}; \lambda_h)$$
$$= (\mathbf{f}_{f,h}, \mathbf{v}_{f,h})_{\Omega_f} + (\mathbf{f}_{p,h}, \boldsymbol{\xi}_{p,h})_{\Omega_p}, \tag{9}$$

$$(s_0 d_\tau p_{p,h}, w_{p,h}) - \alpha b_p(d_\tau \boldsymbol{\eta}_{p,h}, w_{p,h}) - b_p(\mathbf{u}_{p,h}, w_{p,h}) - b_f(\mathbf{u}_{f,h}, w_{f,h})$$
$$= (q_{f,h}, w_{f,h})_{\Omega_f} + (q_{p,h}, w_{p,h})_{\Omega_p}, \tag{10}$$

$$b_\Gamma(\mathbf{u}_{f,h}, \mathbf{u}_{p,h}, d_\tau \boldsymbol{\eta}_{p,h}; \mu_h) = 0, \tag{11}$$

where

$$a_{BJS}(\mathbf{u}_f, \boldsymbol{\eta}_p; \mathbf{v}_f, \boldsymbol{\xi}_p) = \sum_{j=1}^{d-1} \langle \nu \alpha_{BJS} \sqrt{K_j^{-1}}(\mathbf{u}_f - \frac{\partial \boldsymbol{\eta}_p}{\partial t}) \cdot \boldsymbol{\tau}_{f,j}, (\mathbf{v}_f - \boldsymbol{\xi}_p) \cdot \boldsymbol{\tau}_{f,j} \rangle_{\Gamma_{fp}},$$

$$b_\Gamma(\mathbf{v}_f, \mathbf{v}_p, \boldsymbol{\xi}_p; \mu) = \langle \mathbf{v}_f \cdot \mathbf{n}_f + (\boldsymbol{\xi} + \mathbf{v}_p) \cdot \mathbf{n}_p, \mu \rangle_{\Gamma_{fp}}.$$

We note that the balance of normal stress, BJS, and conservation of momentum interface conditions (7) and (8) are natural and have been utilized in the derivation of the weak formulation, while the conservation of mass condition in (7) is imposed weakly in (11).

We solve problem (9), (10) and (11) using piecewise linear finite elements for the approximation of all the variables. It is well known that the equal-order approximation is unstable for saddle point problems such as Darcy, Stokes or Brinkman [4]. For this reason, we complement the discretization of Brinkman problem, namely (1), with the Brezzi-Pitkaranta stabilization operator acting on the pressure [4,5]. Owing to the pressure time derivative in equation (4) the Biot system is not a saddle point problem. For this reason, pressure stabilization is not required. Particular attention should be also devoted to the discretization of the Lagrange multiplier space. For the piecewise linear approximation adopted here, we observe that the Lagrange multiplier space coincides with the normal components of the interface trace spaces of $\mathbf{V}_{f,h}$, $\mathbf{V}_{p,h}$, $\mathbf{X}_{p,h}$. This property has two important consequences. First, it is straightforward to show that equation (11) is exactly satisfied. Second, owing to the results obtained in [12] for general elliptic problems and the more recent analysis of Stokes-Darcy equations [9], it can be shown that this property entails the unique solvability of the discrete system.

Besides the well posedness of the finite element method, stability of the time discretization must be also addressed. To this purpose, by taking

$$(\mathbf{v}_{f,h}, w_{f,h}, \mathbf{v}_{p,h}, w_{p,h}, \boldsymbol{\xi}_{p,h}, \mu) = (\mathbf{u}_{f,h}^n, p_{f,h}^n, \mathbf{u}_{p,h}^n, p_{p,h}^n, d_\tau \boldsymbol{\eta}_{p,h}^n, \lambda_h)$$

in (9)–(11) we obtain an energy equality

$$a_f(\mathbf{u}_{f,h}^n, \mathbf{u}_{f,h}^n) + a_p^d(\mathbf{u}_{p,h}^n, \mathbf{u}_{p,h}^n) + a_p^e(\boldsymbol{\eta}_{p,h}^n, d_\tau \boldsymbol{\eta}_{p,h}^n)$$

$$+ \sum_{j=1}^{d-1} \nu \alpha_{BJS} \| K_j^{-1/4}(\mathbf{u}_{f,h}^n - d_\tau \boldsymbol{\eta}_{p,h}^n) \cdot \boldsymbol{\tau}_{f,j} \|_{L^2(\Gamma_{fp})}^2 + (d_\tau s_0 p_{p,h}^n, p_{p,h}^n) = \mathcal{F}(t_n; \mathbf{u}_{f,h})$$

Using the following equality

$$\int_\Omega u^n d_\tau u^n = \frac{1}{2} d_\tau \|u^n\|_\Omega^2 + \frac{1}{2} \tau \|d_\tau u^n\|_\Omega^2$$

the energy equality becomes

$$\frac{1}{2} d_\tau \left(s_0 \| p_{p,h}^n \|_{L^2(\Omega_p)}^2 + a_p^e(\boldsymbol{\eta}_{p,h}^n, \boldsymbol{\eta}_{p,h}^n) \right) + \frac{\tau}{2} \left(s_0 \| d_\tau p_{p,h}^n \|_{L^2(\Omega_p)}^2 + a_p^e(d_\tau \boldsymbol{\eta}_{p,h}^n, d_\tau \boldsymbol{\eta}_{p,h}^n) \right)$$

$$+ a_f(\mathbf{u}_{f,h}^n, \mathbf{u}_{f,h}^n) + a_p^d(\mathbf{u}_{p,h}^n, \mathbf{u}_{p,h}^n) + \sum_{j=1}^{d-1} \nu \alpha_{BJS} \| K_j^{-1/4}(\mathbf{u}_{f,h}^n - d_\tau \boldsymbol{\eta}_{p,h}^n) \cdot \boldsymbol{\tau}_{f,j} \|_{L^2(\Gamma_{fp})}^2$$

$$= \mathcal{F}(t_n; \mathbf{u}_{f,h})$$

Rearranging the following terms,

$$a_f(\mathbf{u}_{f,h}^n, \mathbf{u}_{f,h}^n) + a_p^d(\mathbf{u}_{p,h}^n, \mathbf{u}_{p,h}^n) + \frac{\tau}{2} a_p^e(d_\tau \boldsymbol{\eta}_{p,h}^n, d_\tau \boldsymbol{\eta}_{p,h}^n)$$

$$= 2\nu \| \mathbf{D}(\mathbf{u}_{f,h}^n) \|_{\Omega_f}^2 + \nu K^{-1} \| \mathbf{u}_{p,h}^n \|_{\Omega_p}^2 + \frac{\tau}{2} \| d_\tau \nabla \cdot \boldsymbol{\eta}_{p,h}^n \|_{\Omega_p}$$

using the bound on generic forcing term,

$$\mathcal{F}(t_n; \mathbf{u}_{f,h}^n) \leq (2\epsilon'\nu)^{-1} \|\mathcal{F}(t_n)\|^2 + \frac{\epsilon'}{2}\nu \|\mathbf{D}(\mathbf{u}_{f,h}^n)\|_{\Omega_f}^2,$$

and combining these results and summing up with respect to time index $n = 1, ..., N$, the following energy estimate is obtained.

Theorem 1. *For any $\epsilon' > 0$, the discrete problem (9), (10) and (11) satisfies the following energy estimate:*

$$\frac{1}{2}\left(s_0\|p_{p,h}^n\|_{L^2(\Omega_p)}^2 + a_p^e(\boldsymbol{\eta}_{p,h}^n, \boldsymbol{\eta}_{p,h}^n)\right) + \tau \sum_{n=1}^{N}\left[2\nu(1 - \frac{\epsilon'}{4})\|\mathbf{D}(\mathbf{u}_{f,h})\|_{\Omega_f}^2\right.$$

$$+\nu K_f^{-1}\|\mathbf{u}_{f,h}\|_{\Omega_f}^2 + \nu K^{-1}\|\mathbf{u}_{p,h}\|_{\Omega_p}^2 + \frac{\tau}{2}\left(\|d_\tau \nabla \cdot \boldsymbol{\eta}_{p,h}^n\|_{\Omega_p} + s_0\|d_\tau p_{p,h}^n\|\right)$$

$$+\sum_{j=1}^{d-1}\nu\alpha_{BJS}\|K_j^{-1/4}(\mathbf{u}_{f,h} - \frac{\partial\boldsymbol{\eta}_{p,h}}{\partial t})\cdot\boldsymbol{\tau}_{f,j}\|_{L^2(\Gamma_{fp})}^2\right]$$

$$\leq \frac{1}{2}\left(s_0\|p_{p,h}^0\|_{L^2(\Omega_p)}^2 + a_p^e(\boldsymbol{\eta}_{p,h}^0, \boldsymbol{\eta}_{p,h}^0)\right) + \tau \sum_{n=1}^{N}(2\epsilon'\nu)^{-1}\|\mathcal{F}(t_n)\|^2. \qquad (12)$$

2.2   Approximation of Interface Conditions by Nitsche's Method

The enforcement of interface conditions by means of Lagrange multipliers leads to an accurate but expensive problem at the discrete level. For this reason, some alternatives have been developed. The most straightforward strategy consists in the application of a penalty method. The idea is to enrich the variational formulation with new terms corresponding to additional quadratic functionals which are minimized when the Dirichlet boundary conditions are exactly satisfied. The penalty method, however, suffers from lack of consistency with respect to the continuous formulation of the problem. Among several interpretations, Nitsches method can be seen as a variant of the penalty method. Indeed, it allows to weakly enforce boundary and interface conditions and it restores the strong consistency of the discrete scheme with respect to the continuous form of the variational formulation. For an introduction to this technique applied to general boundary and interface conditions we refer to [10], while this method is applied to FSI in [7], and to the Biot-Stokes system in [6].

Applying Nitsche's method to equations (1), (3) and (4) with the corresponding interface and boundary conditions, we obtain the following discrete problem formulation, that consists of a system of three coupled problems: for any index $n > 0$, find $\boldsymbol{\eta}_h^n \in \mathbf{X}_{p,h}$, $\mathbf{u}_{p,h}^n \in \mathbf{V}_{p,h}$, $p_{p,h}^n \in W_{p,h}$ and $\mathbf{u}_{f,h}^n \in \mathbf{V}_{f,h}$, $p_{f,h}^n \in W_{f,h}$ such that for any $\forall \boldsymbol{\xi}_h \in \mathbf{X}_{p,h}$, $\mathbf{v}_{p,h} \in \mathbf{V}_{p,h}$, $w_{p,h} \in W_{p,h}$, $\mathbf{v}_{f,h} \in \mathbf{V}_{f,h}$, $w_{f,h} \in W_{f,h}$ the following equations are satisfied:

$$a_p^e(\boldsymbol{\eta}_h^n, \boldsymbol{\xi}_h) - b_p(p_{p,h}^n, \boldsymbol{\xi}_h) + \langle \mathbf{n}_p \cdot \boldsymbol{\sigma}_{f,h}\mathbf{n}_p, (-\boldsymbol{\xi}_h)\cdot\mathbf{n}_p\rangle_{\Gamma_{fp}} \qquad (13)$$

$$+ \langle \nu\alpha_{BJS}\sqrt{K_j^{-1}}d_\tau\boldsymbol{\eta}_h^n \cdot \mathbf{t}_p, \boldsymbol{\xi}_h \cdot \mathbf{t}_p\rangle_{\Gamma_{fp}} + \langle \gamma\nu h^{-1}d_\tau\boldsymbol{\eta}_h^n \cdot \mathbf{n}_p, \boldsymbol{\xi}_h \cdot \mathbf{n}_p\rangle_{\Gamma_{fp}}$$

$$- \langle \nu\alpha_{BJS}\sqrt{K_j^{-1}}\mathbf{u}_{f,h}^n \cdot \mathbf{t}_p, \boldsymbol{\xi}_h \cdot \mathbf{t}_p\rangle_{\Gamma_{fp}} - \langle \gamma\nu h^{-1}(\mathbf{u}_{f,h}^n - \mathbf{u}_{p,h}^n)\cdot\mathbf{n}_p, \boldsymbol{\xi}_h \cdot \mathbf{n}_p\rangle_{\Gamma_{fp}} = 0,$$

$$s_0(d_\tau p_{p,h}^n, w_{p,h})_{\Omega_p} + a_p(u_{p,h}^n, v_{p,h}) - b_p(p_{p,h}^n, v_{p,h}) + b_p(w_{p,h}, u_{p,h}^n) \tag{14}$$
$$+ \langle \gamma \nu h^{-1} u_{p,h}^n \cdot n_p, v_{p,h} \cdot n_p \rangle_{\Gamma_{fp}} + b_s(w_{p,h}, d_\tau \eta_h^n)$$
$$- \langle \gamma \nu h^{-1} (u_{f,h}^n - d_\tau \eta_h^n) \cdot n_p, v_{p,h} \cdot n_p \rangle_{\Gamma_{fp}} - \langle n_p \cdot \sigma_{f,h} n_p, v_{p,h} \cdot n_p \rangle_{\Gamma_{fp}} = 0,$$

$$\langle \rho_f d_\tau u_{f,h}^n, v_{f,h} \rangle_{\Omega_f} + a_f(u_{f,h}^n, v_{f,h}) - b_f(p_{f,h}^n, v_{f,h}) + b_f(w_{f,h}, u_{f,h}^n) \tag{15}$$
$$- \langle n_f \cdot \sigma_{f,h} n_f, v_{f,h} \cdot n_f \rangle_{\Gamma_{fp}} - \langle n_f \cdot \sigma_{f,h}(\varsigma v_{f,h}, -w_{f,h}) n_f, u_{f,h}^n \cdot n_f \rangle_{\Gamma_{fp}}$$
$$+ \langle n_f \cdot \sigma_{f,h}(\varsigma v_{f,h}, -w_{f,h}) n_f, (u_{p,h}^n + d_\tau \eta_h^n) \cdot n_f \rangle_{\Gamma_{fp}}$$
$$+ \langle \gamma \nu h^{-1} u_{f,h}^n \cdot n_f, v_{f,h} \cdot n_f \rangle_{\Gamma_{fp}} + \langle \nu \alpha_{BJS} \sqrt{K_j^{-1}} u_{f,h}^n \cdot t_f, v_{f,h} \cdot t_f \rangle_{\Gamma_{fp}}$$
$$- \langle \gamma \nu h^{-1}(u_{p,h}^n + d_\tau \eta_h^n) \cdot n_f, v_{f,h} \cdot n_f \rangle_{\Gamma_{fp}} - \langle \nu \alpha_{BJS} \sqrt{K_j^{-1}} d_\tau \eta_h^n \cdot t_f, v_{f,h} \cdot t_f \rangle_{\Gamma_{fp}}$$
$$= \mathcal{F}(t^n; v_{f,h}),$$

where γ is a positive penalty (or stabilization) parameter and $\varsigma \in \{-1, 0, 1\}$ is a symmetry parameter that allows us to switch from the so called *symmetric, incomplete and skew-symmetric* problem formulations respectively. The value of γ will be determined below, in order to guarantee the stability of the scheme. To study the stability of the Nitsche's method, we use the following inverse inequality,

$$h \| D(u_h) n \|_{\Gamma_{fp}}^2 \le C_{TI} \| D(u_h) \|_{\Omega_f}^2, \tag{16}$$

where C_{TI} is a positive constant uniformly upper bounded with respect to the mesh characteristic size h, for a family of shape-regular and quasi-uniform meshes [8]. Since we solve the FSI problem on fixed domains, the constant C_{TI} does not depend on the solution. The following result shows that the scheme (13), (14) and (15) is stable for any time step.

Theorem 2. [6] *For any $\hat{\epsilon}', \check{\epsilon}'$ that satisfy*

$$1 - \frac{(\varsigma + 1)}{2} \hat{\epsilon}' C_{TI} - \frac{\check{\epsilon}'}{2} > 0$$

where $\varsigma \in \{-1, 0, 1\}$ provided that $\gamma > (\varsigma + 1)(\hat{\epsilon}')^{-1}$, there exist constants $0 < c < 1$ and $C > 1$, uniformly independent of the mesh characteristic size h, such that

$$\frac{1}{2} \left(2\mu_p \| D(\eta_h^N) \|_{\Omega_p}^2 + \lambda_p \| \nabla \cdot \eta_h^N \|_{\Omega_p}^2 + s_0 \| p_{p,h}^N \|_{\Omega_p}^2 \right) \tag{17}$$
$$+ c\tau \sum_{n=1}^N \left[2\nu \| D(u_{f,h}^n) \|_{\Omega_f}^2 + \nu K_f^{-1} \| u_{f,h}^n \|_{\Omega_f}^2 + \nu K^{-1} \| u_{p,h}^n \|_{\Omega_p}^2 \right.$$
$$+ \frac{\tau}{2} \left(\rho_f \| d_\tau u_{f,h}^n \|_{\Omega_f}^2 + 2\mu_p \| d_\tau D(\eta_h^n) \|_{\Omega_p}^2 + s_0 \| d_\tau p_{p,h}^n \|_{\Omega_p}^2 + \lambda_p \| d_\tau \nabla \cdot \eta_h^n \|_{\Omega_p}^2 \right)$$
$$+ \nu h^{-1} \left(\| (u_{f,h}^n - u_{p,h}^n - d_\tau \eta_h^n) \cdot n \|_{\Gamma_{fp}}^2 + \| (u_{f,h}^n - d_\tau \eta_h^n) \cdot t \|_{\Gamma_{fp}}^2 \right) \right]$$
$$\le \frac{1}{2} \left(2\mu_p \| D(\eta_h^0) \|_{\Omega_p}^2 + \lambda_p \| \nabla \cdot \eta_h^0 \|_{\Omega_p}^2 + s_0 \| p_{p,h}^0 \|_{\Omega_p}^2 \right)$$
$$+ \tau \sum_{n=1}^N \frac{C}{\nu} \| \mathcal{F}(t_n) \|^2.$$

More precisely, we have

$$c < \min\{(1 - \frac{(\varsigma + 1)}{2}\check{\epsilon}'C_{TI} - \frac{\check{\epsilon}'}{2}), ((\gamma - (\varsigma + 1)(\hat{\epsilon}')^{-1})\},$$
$$C > (2\check{\epsilon}')^{-1}.$$

3   Computational Results

We consider the test case motivated by the example investigated by Lesinigo et al. in [11], Sect. 7.2. The computational domain consists of two unit squares separated by a fracture of width $\delta = 0.1$. The squares represent the poroelastic subdomains of Ω_p. We assume that there are no external forces or mass sources. On the left and right boundaries we impose the homogeneous Dirichlet pressure and the homogeneous Dirichlet displacement conditions, while on the remaining external boundaries we impose zero normal flux and zero normal poroelastic stress. In order to generate a nontrivial flow pattern, a uniform flow is enforced on the bottom side of the fracture $\mathbf{u}_f \cdot \mathbf{n} = 10$, while the upper side of Ω_f is impermeable to flow. We expect to observe a vertical flow in Ω_f, which progressively fades out moving upwards, because of a significant leakage of fluid into Ω_p. Since the friction term is active in the Brinkman equation, we also expect that the pressure decreases along the direction of the flow in Ω_f.

We first consider a test case (we refer to it as Case A) designed to verify the agreement with the example in [11]. Since there the tangential flow interface condition is free slip and the porous media is not deformable, we take $\alpha_{BJS} = 0$ and Young modulus $E_p = 10^{10}$ Pa (very hard material), see the parameter set in Table 1. Furthermore, we investigate the behavior of the model in a regime different from [11]. In particular, we choose $\alpha_{BJS} = 1$ (this configuration is called Case B) and finally we consider a softer material characterized by $E_p = 10^3$ Pa (this is denoted as Case C). All the problems were solved over the time interval $[0, 1]$ s with time step $\Delta t = 0.1s$.

The simulation results obtained with the Lagrange multiplier method are reported in Fig. 1. On the left we observe that the main expected features of the solution are correctly captured. For a more quantitative comparison, we plot on the right the variation of velocity modulus and pressure along the (vertical)

Table 1. Poroelasticity and fluid parameters that are used in the numerical experiments denoted as Case A.

Parameters	Values	Parameters	Values
Young modulus E_p (Pa)	10^{10}	poisson ratio	0.3
First Lamé param. μ_p(Pa)	$3.84\,10^9$	Second Lamé param. λ_p(Pa)	$5.76\,10^9$
Hydraulic conductivity νK^{-1}(m^3 s/Kg)	$\mathbf{I}$	Mass storativity s_0(Pa)	1
Biot-Willis constant α	1	BJS friction coef. α_{BJS}	0
Hydraulic conductivity K_f^{-1}(m^3 s/Kg)	$10\mathbf{I}$	Brinkman viscosity ν	1

Case A: $\alpha_{BJS} = 0$, $E_p = 10^{10}$ Pa

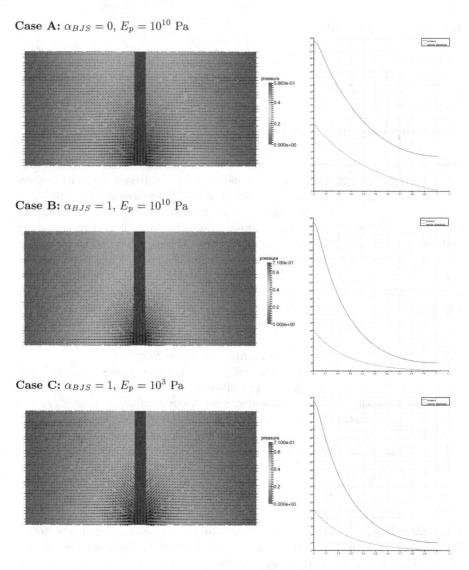

Case B: $\alpha_{BJS} = 1$, $E_p = 10^{10}$ Pa

Case C: $\alpha_{BJS} = 1$, $E_p = 10^3$ Pa

Fig. 1. Velocity glyphs at time $t = 1$. For visualization purposes, the velocity field on Ω_p has been magnified of a factor 10. The background color shows the pressure magnitude (left panel). Velocity and pressure plot along the vertical meanline of Ω_f (right panel).

meanline of Ω_f and we compare these results with the ones reported in Fig. 7a of [11]. An excellent agreement is observed. The analysis of the problem configurations A, B, C shows that the BJS friction factor α_{BJS} significantly affects the pressure profile in the fluid region Ω_f. Because the friction has increased, but the flow rare is prescribed at the boundary, we observe a significant increase

Case A: $\alpha_{BJS} = 0$, $E_p = 10^{10}$ Pa

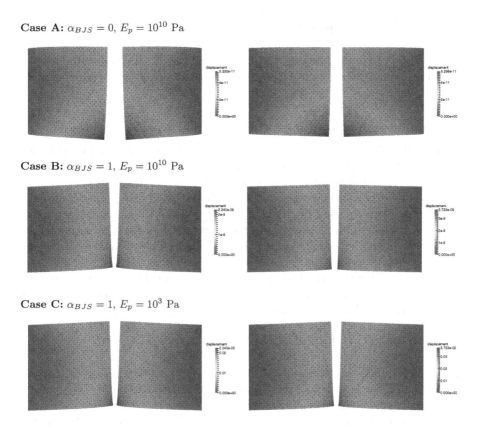

Case B: $\alpha_{BJS} = 1$, $E_p = 10^{10}$ Pa

Case C: $\alpha_{BJS} = 1$, $E_p = 10^3$ Pa

Fig. 2. Porous domain deformation at $t = 1$. The background color shows the fluid pressure magnitude. For visualization purposes, the displacement field has been magnified of a factor 10^7 for cases A and B. Results obtained with the Lagrange multiplier scheme are reported in the left, the ones corresponding to Nitsche's method are on the right.

of the pressure at the inlet. The pressure field in the porous medium is also sensitive to this variation, but with less intensity. We finally notice that cases B and C look very similar. This suggests that the entire flow field is almost unaffected by the stiffness of the material.

The sensitivity of the model with respect to the interface BJS friction coefficient also emerges in Fig. 2, where we show the displacement of the porous domain. In particular, we observe that the displacement field changes significantly when we move from $\alpha_{BJS} = 0$ to $\alpha_{BJS} = 1$. Furthermore, even though not revealed by the visualization because amplification factors are adopted, the displacement field is almost inversely proportional to the Young modulus of the solid material. We finally notice that, although $p_f > p_p$ along the fluid-solid interface, the fluid is inducing a traction on the solid in cases B and C. This effect happens in this specific problem configuration, because $p_f \simeq (2\nu \mathbf{D}(\mathbf{u}_f)\mathbf{n}_f) \cdot \mathbf{n}_f$ in Ω_f and as a result $(\boldsymbol{\sigma}_f(\mathbf{u}_f, p_f)\mathbf{n}_f) \cdot \mathbf{n}_f \simeq 0$. In the cases where $\alpha_{BJS} = 1$ the tangential interaction dominates over the normal one in the fluid-solid interaction. This justifies the differences between the displacement observed in cases A, B and C.

3.1 Comparison of Lagrange Multiplier and Nitsche's Methods

In this section we aim to accurately compare the two schemes proposed for the solution of the problem. First, from a preliminary inspection based on visualizations similar to the ones of Fig. 1, we observe that the two methods provide very similar results. They also provide equivalent accuracy and precision as confirmed by the analysis of the convergence rate with respect to the mesh characteristic size, h, reported in Table 2. As error indicators, we consider the dominating terms that appear in the left hand sides of the stability estimates, namely Eqs. (12) and (17) for the Lagrange and Nitsche's methods respectively. Finally, since the main distinction between the two proposed methods consists of the enforcement of interface conditions, and in particular the mass conservation condition (5), we analyze in Table 3 the numerical residual

$$\mathcal{R}_{\Gamma_i}(\mathbf{u}_f, \boldsymbol{\eta}_p, \mathbf{u}_p) := \int_{\Gamma_i} \left(\mathbf{u}_f \cdot \mathbf{n}_f + (\frac{\partial \boldsymbol{\eta}_p}{\partial t} + \mathbf{u}_p) \cdot \mathbf{n}_p \right), \Gamma_i = \Omega_f \cap \Omega_{p,i}, \; i = 1, 2,$$

for the approximate solutions provided by each method. The results of Table 3 confirm the higher accuracy of the Lagrange multiplier method for the approximation of interface conditions. Indeed, Table 3 shows that the Lagrange multiplier method is exactly enforcing the desired condition at every interface node.

Table 2. Convergence analysis with respect to the mesh characteristic size for the Lagrange multiplier method (top) and Nitsche's method (bottom). Since an analytical solution is not available for the considered problem, we calculate the error with respect to a numerical solution computed on a highly refined mesh with $h = 1/320$ and denoted by $\tilde{\mathbf{u}}_{f,h}, \tilde{p}_{p,h}, \mathbf{D}(\tilde{\boldsymbol{\eta}}_h), \tilde{\mathbf{u}}_{p,h}$. Then, the discretization error is given by $\mathbf{e}_{\mathbf{u},f,h}^N :=$ $\mathbf{u}_{f,h}^N - \tilde{\mathbf{u}}_{f,h}^N, e_{p,h}^N := p_{p,h}^N - \tilde{p}_{p,h}^N, \mathbf{D}(\mathbf{e}_{\eta,h}^N) := \mathbf{D}(\boldsymbol{\eta}_h^N) - \mathbf{D}(\tilde{\boldsymbol{\eta}}_h^N), \mathbf{e}_{\mathbf{u},p,h}^n := \mathbf{u}_{p,h}^n - \tilde{\mathbf{u}}_{p,h}^n$. To facilitate the interpretation of the results, error norms are normalized with respect to the corresponding norm of the solution.

h	$\|e_{p,h}^N\|_{\Omega_p}$	Rate	$\|\mathbf{D}(\mathbf{e}_{\eta,h}^N)\|_{\Omega_p}$	Rate	$\sqrt{\tau \sum_{n=1}^N \|\mathbf{e}_{\mathbf{u},f,h}^n\|_{\Omega_p}^2}$	Rate	$\sqrt{\tau \sum_{n=1}^N \|\mathbf{e}_{\mathbf{u},p,h}^n\|_{\Omega_p}^2}$	Rate
1/20	4.72E-02		4.03E-01		5.18E-02		1.77E-01	
1/40	1.35E-02	1.81	2.36E-01	0.77	1.94E-02	1.41	6.21E-02	1.51
1/80	3.38E-03	1.99	1.24E-01	0.93	5.97E-03	1.70	1.86E-02	1.74
1/160	6.93E-04	2.29	7.31E-02	0.76	1.40E-02	2.09	4.74E-03	1.98
h	$\|e_{p,h}^N\|_{\Omega_p}$	Rate	$\|\mathbf{D}(\mathbf{e}_{\eta,h}^N)\|_{\Omega_p}$	Rate	$\sqrt{\tau \sum_{n=1}^N \|\mathbf{e}_{\mathbf{u},f,h}^n\|_{\Omega_p}^2}$	Rate	$\sqrt{\tau \sum_{n=1}^N \|\mathbf{e}_{\mathbf{u},p,h}^n\|_{\Omega_p}^2}$	Rate
1/20	8.68E-02		4.35E-01		1.24E-01		2.23E-01	
1/40	3.19E-02	1.44	2.56E-01	0.76	3.39E-02	1.87	9.08E-02	1.30
1/80	1.11E-02	1.52	1.35E-01	0.93	1.06E-02	1.67	2.85E-02	1.67
1/160	3.30E-03	1.75	8.02E-02	0.75	2.49E-03	2.10	7.51E-03	1.92

Table 3. The behavior of the indicator $\mathcal{R}_{\Gamma_i}(\mathbf{u}_{f,h}, \boldsymbol{\eta}_{p,h}, \mathbf{u}_{p,h})$ when varying the mesh characteristic size.

	Lagrange multipliers		Nitsche	
h	$\mathcal{R}_{\Gamma_1}$	$\mathcal{R}_{\Gamma_2}$	$\mathcal{R}_{\Gamma_1}$	$\mathcal{R}_{\Gamma_2}$
1/20	4.4402E-12	3.8642E-12	2.7522E-01	2.7528E-01
1/40	1.9694E-12	1.9698E-12	4.8690E-03	4.8690E-03
1/80	4.2337E-13	4.2424E-13	1.5374E-03	1.5374E-03
1/160	1.0735E-13	1.0568E-13	3.8450E-04	3.8450E-04
1/320	2.5801E-13	2.5472E-13	9.3927E-05	9.3927E-05

Conversely, equation (5) is only approximately enforced by the Nitsche's scheme and as expected the residual decreases proportionally with h.

Acknowledgments. The first three authors have been partially supported by the NSF grants DMS 1115856 and DMS 1418947. The third and fourth authors have been partially supported by the DOE grant DE-FG02-04ER25618. The authors thank Martina Bukac and Rana Zakerzadeh for their contribution to the development of the software used in this paper.

References

1. Badia, S., Quaini, A., Quarteroni, A.: Coupling biot and navier-stokes equations for modelling fluid-poroelastic media interaction. J. Comput. Phys. **228**(21), 7986–8014 (2009). http://dx.doi.org/10.1016/j.jcp.2009.07.019
2. Beavers, G., Joseph, D.: Boundary conditions at a naturally impermeable wall. J. Fluid. Mech **30**, 197–207 (1967)
3. Biot, M.: General theory of three-dimensional consolidation. J. Appl. Phys. **12**, 155–164 (1941)
4. Boffi, D., Brezzi, F., Fortin, M.: Mixed Finite Element Methods and Applications. Springer Series in Computational Mathematics, vol. 44. Springer, Heidelberg (2013). http://dx.doi.org/10.1007/978-3-642-36519-5
5. Brezzi, F., Pitkäranta, J.: On the stabilization of finite element approximations of the stokes equations. In: Hackbusch, W. (ed.) Efficient solutions of elliptic systems (Kiel, 1984). Notes on Numerical Fluid Mechanics, vol. 10, pp. 11–19. Vieweg, Braunschweig (1984)
6. Bukac, M., Yotov, I., Zakerzadeh, R., Zunino, P.: Partitioning strategies for the interaction of a fluid with a poroelastic material based on a Nitsche's coupling approach. Comput. Methods Appl. Mech. Eng. **292**, 138–170 (2015). http://dx.doi.org/10.1016/j.cma.2014.10.047
7. Burman, E., Fernández, M.A.: Stabilization of explicit coupling in fluid-structure interaction involving fluid incompressibility. Comput. Methods Appl. Mech. Eng. **198**, 766–784 (2009)
8. Ern, A., Guermond, J.L.: Theory and Practice of Finite Elements. Applied Mathematical Sciences, vol. 159. Springer, New York (2004)

9. Girault, V., Vassilev, D., Yotov, I.: Mortar multiscale finite element methods for Stokes-Darcy flows. Numer. Math. **127**(1), 93–165 (2014). http://dx.doi.org/10.1007/s00211-013-0583-z
10. Hansbo, P.: Nitsche's method for interface problems in computational mechanics. GAMM-Mitt. **28**(2), 183–206 (2005)
11. Lesinigo, M., D'Angelo, C., Quarteroni, A.: A multiscale darcy-brinkman model for fluid flow in fractured porous media. Numer. Math. **117**(4), 717–752 (2011)
12. Pitkäranta, J.: Boundary subspaces for the finite element method with Lagrange multipliers. Numer. Math. **33**(3), 273–289 (1979). http://dx.doi.org/10.1007/BF01398644
13. Saffman, P.: On the boundary condition at the surface of a porous media. Stud. Appl. Math. **50**, 93–101 (1971)
14. Showalter, R.E.: Poroelastic filtration coupled to Stokes flow. In: Control theory of partial differential equations, Lectures Notes Pure Applied Mathematics, vol. 242, pp. 229–241. Chapman & Hall/CRC, Boca Raton (2005). http://dx.doi.org/10.1201/9781420028317.ch16

A Transparent Boundary Condition
for an Elastic Bottom in Underwater Acoustics

Anton Arnold[1] and Matthias Ehrhardt[2]([⊠])

[1] Institut für Analysis Und Scientific Computing, Technische Universität Wien,
Wiedner Hauptstr. 8, 1040 Wien, Austria
anton.arnold@tuwien.ac.at
http://www.asc.tuwien.ac.at/∼arnold/
[2] Lehrstuhl für Angewandte Mathematik und Numerische Analysis,
Fachbereich C – Mathematik und Naturwissenschaften,
Bergische Universität Wuppertal, Gaußstr. 20, 42119 Wuppertal, Germany
ehrhardt@math.uni-wuppertal.de
http://www-num.math.uni-wuppertal.de/∼ehrhardt/

Abstract. This work deals with the derivation of a novel transparent
boundary condition (TBC) for the coupling of the standard "parabolic"
equation (SPE) in underwater acoustics (assuming cylindrical symmetry)
with an elastic parabolic equation (EPE) for modelling the sea bottom
extending hereby the existing TBCs for a fluid model of the seabed.

Keywords: Transparent boundary condition · Elastic bottom ·
One-way Helmholtz equation · Standard "parabolic" equation · Seabed
interface

1 Introduction

"Parabolic" equation (PE) models appear in (underwater) acoustics as one-
way approximations to the Helmholtz equation in cylindrical coordinates with
azimuthal symmetry. These PE models have been widely used in the recent
past for wave propagation problems in various application areas, e.g. seismology,
optics and plasma physics but here we focus on their application to underwater
acoustics, where PEs have been introduced by Tappert [17]. For more details we
refer to [10].

In computational ocean acoustics one wants to determine the acoustic pressure $p(z, r)$ emerging from a time-harmonic point source situated in the water
at $(z_s, 0)$. The radial range variable is denoted by $r > 0$ and the depth variable
is $0 < z < z_b$. The water surface is located at $z = 0$, and the (horizontal) sea
bottom at $z = z_b$. We point out that irregular bottom surfaces and sub-bottom
layers can be included by simply extending the range of z. For an alternative
strategy based on transformation techniques, including proofs of well-posedness
in the case of upsloping and downsloping wedge-type domains in 2D and 3D
we refer to [2,6]. Further, the 3D treatment of a sloping sea bootom in a finite
element context was presented in [16].

© Springer International Publishing Switzerland 2015
I. Dimov et al. (Eds.): FDM 2014, LNCS 9045, pp. 15–24, 2015.
DOI: 10.1007/978-3-319-20239-6_2

In the sequel we denote the local sound speed by $c(z, r)$, the density by $\rho(z, r)$, and the attenuation by $\alpha(z, r) \geq 0$. $n(z, r) = c_0/c(z, r)$ is the refractive index, with a reference sound speed c_0. The reference wave number is $k_0 = 2\pi f/c_0$, where f denotes the (usually low) frequency of the emitted sound.

1.1   The Parabolic Approximations

The acoustic pressure $p(z, r)$ satisfies the *Helmholtz equation*

$$\frac{1}{r} \frac{\partial}{\partial r} \left(r \frac{\partial p}{\partial r} \right) + \rho \frac{\partial}{\partial z} \left(\rho^{-1} \frac{\partial p}{\partial z} \right) + k_0^2 N^2 p = 0, \qquad r > 0, \tag{1}$$

with the complex refractive index (where α accounts for damping in the medium)

$$N(z, r) = n(z, r) + i\alpha(z, r)/k_0. \tag{2}$$

In the far field approximation ($k_0 r \gg 1$) the (complex valued) outgoing acoustic field

$$\psi(z, r) = \sqrt{k_0 r}\, p(z, r)\, e^{-ik_0 r} \tag{3}$$

satisfies the *one-way Helmholtz equation*:

$$\psi_r = ik_0 \left(\sqrt{1 - L} - 1 \right) \psi, \qquad r > 0. \tag{4}$$

Here, $\sqrt{1 - L}$ is a pseudo-differential operator, and L the *Schrödinger operator*

$$L = -k_0^{-2} \rho\, \partial_z (\rho^{-1} \partial_z) + V(z, r) \tag{5}$$

with the complex valued "potential"

$$V(z, r) = 1 - N^2(z, r) = 1 - \left[n(z, r) + i\alpha(z, r)/k_0 \right]^2. \tag{6}$$

"Parabolic" approximations of (4) are formal approximations of the pseudo-differential operator $\sqrt{1 - L}$ by rational functions of L. This procedure yields a PDE that is easier to solve numerically than the pseudo-differential equation (4). For more details we refer to [17,18]. The linear approximation of $\sqrt{1 - \lambda}$ by $1 - \frac{\lambda}{2}$ gives the narrow angle or *standard "parabolic" equation* (SPE) of Tappert [17]

$$\psi_r = -\frac{ik_0}{2} L\psi, \qquad r > 0. \tag{7}$$

This Schrödinger equation (7) is a good description of waves with a propagation direction within about 15° of the horizontal. Rational approximations of the form

$$(1 - \lambda)^{\frac{1}{2}} \approx f(\lambda) = \frac{p_0 - p_1 \lambda}{1 - q_1 \lambda} \tag{8}$$

with real p_0, p_1, q_1 yield the *wide angle "parabolic" equations* (WAPE)

$$\psi_r = ik_0 \left(\frac{p_0 - p_1 L}{1 - q_1 L} - 1 \right) \psi, \qquad r > 0. \tag{9}$$

improving the description of the wave propagation up to angles of about $40°$.

Here we focus on a proper boundary condition (BC) at the sea bottom for the SPE (7) coupled to an elastic "parabolic" model for the sea bottom. At the water surface one usually employs a Dirichlet BC $\psi(z = 0, r) = 0$ and at the sea bottom one has to couple the wave propagation in the water to the wave propagation in the bottom.

1.2 The Coupling Condition

For the bottom $z > z_b$ one usually use a fluid model (i.e. assuming that (7) or (9) with possibly different rational approximation (8) also hold for $z > z_b$) with constant parameters c_b, ρ_b, α_b or with a linear squared refractive index [7,11].

In [4] we analyzed this coupling of WAPEs with different parameters p_0, p_1, q_1 and it turned out that the coupled model is well-defined (and the resulting evolution equation is conservative in $L^2(\mathbb{R}^+; (\sigma\rho)^{-1}dz)$) if the *coupling condition*

$$p_1(z)/q_1(z) =: \mu = \text{const} \tag{10}$$

is satisfied. Hence, it is not advisable to couple the WAPE and the SPE (where $p_1 = 1/2; q_1 = 0$) numerically; in this case the evolution is not conservative in the dissipation-free case ($\alpha \equiv 0$) [4]. If the parameters p_0, p_1, q_1 are fixed in one medium, condition (10) still leaves two free parameters to choose a different rational approximation model of $(1 - \lambda)^{\frac{1}{2}}$ in (8) for the second medium (cf. [8]). Hence, one can in fact obtain a better approximation in the second medium than with the originally intended "parabolic approximation".

1.3 Transparent Boundary Conditions

In practical simulations one is only interested in the acoustic field $\psi(z, r)$ in the water, i.e. for $0 < z < z_b$. While the physical problem is posed on the unbounded z-interval $(0, \infty)$, one wishes to restrict the computational domain in the z-direction by introducing an artificial boundary at or below the sea bottom. This artificial BC should of course change the model as little as possible, or ideally not at all.

In [13,15] Papadakis derived *impedance BCs* or *transparent boundary conditions* (TBC) for the SPE and the WAPE, which completely solves the problem of restricting the z–domain without changing the physical model: complementing the WAPE (9) with a TBC at z_b allows to recover — on the finite computational domain $(0, z_b)$ — the exact half-space solution on $0 < z < \infty$. As the SPE is a Schrödinger equation, similar strategies have been developed independently for quantum mechanical applications, cf. the review article [1].

Let us finally note, that Zhang and Tindle [20] proposed an alternative approach to the impedance BCs or TBCs of Papadakis. By minimizing the reflec-
tion coefficient at the water-bottom interface they derived in their *equivalent
fluid approximation* an expression for a complex fluid density that can be used
for modelling an elastic sea bottom in a classical fluid model. However, this
approach yields only satisfactory results for low shear wave speeds [20].

This work is organized as follows: In Sect. 2 we review the TBC for the SPE
coupled to an elastic bottom in the frequency domain and in Sect. 3 present in
detail the analytic inverse Laplace transformation to obtain this TBC in the
time domain. Finally, we draw a conclusion and summarize the basic inversion
rules previously used.

2   The Transparent Boundary Condition for a Fluid Bottom

The basic idea of the derivation is to explicitly solve the equation in the sea bottom, which is the exterior of the computational domain $(0, z_b)$. The TBC for the
SPE (or Schrödinger equation) was derived in [3,13,15] for various application
fields:

$$\psi(z_b, r) = -(2\pi k_0)^{-\frac{1}{2}} e^{\frac{\pi}{4} i} \frac{\rho_b}{\rho_w} \int_0^r \psi_z(z_b, r - \tau) e^{i\omega_b \tau} \tau^{-\frac{1}{2}} d\tau, \qquad (11)$$

with $\omega_b = k_0(N_b^2 - 1)/2$. This BC is nonlocal in the range variable r and involves
a mildly singular convolution kernel. Equivalently, it can be written as

$$\psi_z(z_b, r) = -\left(\frac{2k_0}{\pi}\right)^{\frac{1}{2}} e^{-\frac{\pi}{4} i} e^{i\omega_b r} \frac{\rho_w}{\rho_b} \frac{d}{dr} \int_0^r \psi(z_b, \tau) e^{-i\omega_b \tau} (r - \tau)^{-\frac{1}{2}} d\tau. \quad (12)$$

The r.h.s. of (12) can be expressed formally as a Riemann-Liouville fractional
derivative of order $\frac{1}{2}$, cf. [3]:

$$\psi_z(z_b, r) = -\sqrt{2k_0} e^{-\frac{\pi}{4} i} e^{i\omega_b r} \frac{\rho_w}{\rho_b} \partial_r^{1/2} [\psi(z_b, r) e^{-i\omega_b r}]. \qquad (13)$$

3   The Transparent Boundary Condition for an Elastic Bottom

The coupling of the SPE with an *elastic parabolic equation* (EPE) for the sea
bottom was described in [5,9,19]. Papadakis et al. [14,15] derived a TBC for
this coupling in the frequency regime. It reads for the Laplace transformed wave
field:

$$\hat{\psi}(z_b, s) = -\frac{\rho_b}{\rho_w} \frac{1}{k_0 N_s^4} \frac{1}{\sqrt[4]{M_p(s)}}$$

$$\times \left[(2M_s(s) + N_s^2)^2 - 4\sqrt[4]{M_p(s)} \sqrt[4]{M_s(s)} (M_s(s) + N_s^2) \right] \hat{\psi}_z(z_b, s), \quad (14)$$

with the notation

$$M_p(s) = 1 - N_p^2 - i\frac{2}{k_0}s, \qquad M_s(s) = 1 - N_s^2 - i\frac{2}{k_0}s. \tag{15}$$

Here, $N_p = n_p + i\alpha_p/k_0$ and $N_s = n_s + i\alpha_s/k_0$ denote the complex refractive indices for the compressional and shear waves in the bottom (cf. (2)).

4   The Transparent Boundary Condition in the Time Domain

In a tedious calculation the transformed TBC (14) can indeed be inverse Laplace transformed and it reads:

$$\psi(z_b, r) =$$
$$C\left[\int_0^r \psi_z(z_b, r - \tau)\, e^{i\omega_p \tau} g(\tau)\, d\tau - 2i\varphi \int_0^r \psi_{zr}(z_b, r - \tau)\, e^{i\omega_p \tau} \tau^{-\frac{1}{2}}\, d\tau\right], \tag{16}$$

with

$$C = -\frac{\rho_b}{\rho_w}\frac{2}{k_0^{5/2}N_s^4}\sqrt{\frac{2}{\pi}}\,e^{\frac{\pi}{4}i}, \quad \omega_p = \frac{k_0}{2}(N_p^2 - 1), \quad \varphi = -\frac{k_0}{2}(N_p^2 - N_s^2),$$

and the kernel $g(\tau)$ given by

$$g(\tau) = -3(1 - e^{i\varphi\tau})\tau^{-\frac{5}{2}} + i\frac{k_0}{2}(3N_p^2 - N_s^2 - 2N_s^2 e^{i\varphi\tau})\tau^{-\frac{3}{2}}$$
$$+ \frac{k_0^2}{2}(N_p^4 - N_p^2 N_s^2 + \tfrac{1}{2}N_s^4 + N_p^2 - N_s^2)\tau^{-\frac{1}{2}} = O(\tau^{-\frac{1}{2}}), \quad \text{for} \quad \tau \to \infty.$$

While this inverse transformation was carried out numerically in [14,15], our novel analytical TBC in the time regime may simplify both the analysis and the numerical solution of this coupled model. Let us remark that an asymptotic analysis of the elastic seabed was made by Makrakis [12].

4.1   Derivation of (16)

With the abbreviation $\hat{\psi}(s) := \hat{\psi}(z_b, s)$ and the notation

$$m_p(s) = \frac{k_0}{2}M_p(s) = -i\left[s - i\frac{k_0}{2}(N_p^2 - 1)\right], \tag{17}$$

$$m_s(s) = \frac{k_0}{2}M_s(s) = -i\left[s - i\frac{k_0}{2}(N_s^2 - 1)\right], \tag{18}$$

the transformed TBC (14) reads

$$\hat{\psi}(s)$$

$$= -\frac{\rho_b}{\rho_w}\frac{1}{k_0 N_s^4}\left[\frac{k_0}{2}\frac{\left(\frac{4}{k_0}m_s(s) + N_s^2\right)^2}{\sqrt[+]{m_p(s)}} - 4\sqrt{\frac{2}{k_0}}\sqrt[+]{m_s(s)}\left(\frac{2}{k_0}m_s(s) + N_s^2\right)\right]\hat{\psi}_z(s)$$

$$= -\frac{\rho_b}{\rho_w}\frac{8\sqrt{2}}{k_0^{5/2}N_s^4}\left[\frac{\left(m_s(s) + \frac{k_0}{4}N_s^2\right)^2}{\sqrt[+]{m_p(s)}} - 4\sqrt[+]{m_s(s)}\left(m_s(s) + \frac{k_0}{2}N_s^2\right)\right]\hat{\psi}_z(s), \quad (19)$$

where we denote the content of the square brackets by $f(s - \sigma)$ with

$$\sigma = i\frac{k_0}{2}\left(N_p^2 - 1\right). \quad (20)$$

We observe that we can write

$$m_p(s) = -i\left[s - \sigma\right], \qquad m_s(s) = -i\left[s - \sigma + i\frac{k_0}{2}(N_p^2 - N_s^2)\right].$$

The next step is a shift in the argument of $\hat{\psi}_z(s)$ in (19) by σ:

$$\hat{\psi}(s + \sigma) = -\frac{\rho_b}{\rho_w}\frac{8\sqrt{2}}{k_0^{5/2}N_s^4}\,f(s)\,\hat{\psi}_z(z_b, s + \sigma), \quad (21)$$

Taking the branch with positive real part $\sqrt[+]{-i} = e^{-\frac{\pi}{4}i}$ we get the kernel $f(s)$

$$f(s) = \frac{\left[-is + \frac{k_0}{2}(N_p^2 - N_s^2) + \frac{k_0}{4}N_s^2\right]^2}{e^{-\frac{\pi}{4}i}\sqrt[+]{s}}$$

$$- e^{-\frac{\pi}{4}i}\sqrt[+]{s + i\frac{k_0}{2}(N_p^2 - N_s^2)}(-i)\sqrt[+]{s + i\frac{k_0}{2}(N_p^2 - N_s^2)}\left[s + i\frac{k_0}{2}N_p^2\right],$$

where

$$\left[\cdots\right]^2 = -\left[s + i\frac{k_0}{2}N_p^2 - i\frac{k_0}{4}N_s^2\right]^2$$

$$= -\left[s + i\frac{k_0}{2}(N_p^2 - N_s^2)\right]\left[s + i\frac{k_0}{2}N_p^2\right] + \frac{k_0^2}{16}N_s^4,$$

i.e. we have

$$f(s) = e^{\frac{\pi}{4}i}\left\{\frac{1}{\sqrt[+]{s}}\left[\frac{k_0^2}{16}N_s^4 - \left[s + i\frac{k_0}{2}(N_p^2 - N_s^2)\right]\left[s + i\frac{k_0}{2}N_p^2\right]\right]\right.$$

$$\left. + \sqrt[+]{s + i\frac{k_0}{2}(N_p^2 - N_s^2)}\left[s + i\frac{k_0}{2}N_p^2\right]\right\}$$

$$= e^{\frac{\pi}{4}i}\left\{\left[\sqrt[+]{s + i\frac{k_0}{2}(N_p^2 - N_s^2)} - \sqrt[+]{s} - i\frac{k_0}{2}(N_p^2 - N_s^2)\frac{1}{\sqrt[+]{s}}\right]\cdot\right.$$

$$\left. \cdot\left[s + i\frac{k_0}{2}N_p^2\right] + \frac{k_0^2}{16}N_s^4\frac{1}{\sqrt[+]{s}}\right\}$$

$$= e^{\frac{\pi}{4}i}\left\{\left[\sqrt[+]{s - \gamma} - \sqrt[+]{s} + \gamma\frac{1}{\sqrt[+]{s}}\right]\left[s + i\frac{k_0}{2}N_p^2\right] + \frac{k_0^2}{16}N_s^4\frac{1}{\sqrt[+]{s}}\right\},$$

with
$$\gamma = -i\frac{k_0}{2}(N_p^2 - N_s^2).$$

Hence, inserting in (21) we obtain

$$\hat{\psi}(s+\sigma) = \tilde{C}\left\{i\frac{k_0}{2}N_p^2\left[\sqrt[4]{s-\gamma} - \sqrt[4]{s} + \gamma\frac{1}{\sqrt[4]{s}}\right] + \frac{k_0^2}{16}N_s^4\frac{1}{\sqrt[4]{s}}\right\}\hat{\psi}_z(s+\sigma)$$
$$+ \tilde{C}\left[\sqrt[4]{s-\gamma} - \sqrt[4]{s} + \gamma\frac{1}{\sqrt[4]{s}}\right]\left\{s\,\hat{\psi}_z(s+\sigma)\right\}, \tag{22}$$

where
$$\tilde{C} = -\frac{\rho_b}{\rho_w}\frac{8\sqrt{2}}{k_0^{5/2}N_s^4}\,e^{\frac{\pi}{4}i}. \tag{23}$$

Next, an inverse Laplace transformation of (22) yields the convolution integral

$$\psi(r)\,e^{-\sigma r} = \tilde{C}\int_0^r \psi_z(r-\tau)\,e^{-\sigma(r-\tau)}\,g_1(\tau)\,d\tau$$
$$+ \tilde{C}\int_0^r -\frac{\partial}{\partial\tau}\left[\psi_z(r-\tau)\,e^{-\sigma(r-\tau)}\right]g_2(\tau)\,d\tau, \tag{24}$$

$$g_1(\tau) = \mathcal{L}^{-1}\left\{i\frac{k_0}{2}N_p^2\left[\sqrt{s-\gamma} - \sqrt{s} + \gamma\frac{1}{\sqrt{s}}\right] + \frac{k_0^2}{16}N_s^4\frac{1}{\sqrt{s}}\right\}$$
$$= i\frac{k_0}{2}N_p^2\mathcal{L}^{-1}\left\{\sqrt{s-\gamma} - \sqrt{s}\right\} + \frac{k_0^2}{4}(N_p^2 - \tfrac{1}{2}N_s^2)^2\mathcal{L}^{-1}\left\{\frac{1}{\sqrt{s}}\right\}$$
$$= i\frac{k_0}{4\sqrt{\pi}}N_p^2(1 - e^{\gamma\tau})\,\tau^{-\frac{3}{2}} + \frac{k_0^2}{4\sqrt{\pi}}(N_p^2 - \tfrac{1}{2}N_s^2)^2\,\tau^{-\frac{1}{2}}, \tag{25}$$

$$g_2(\tau) = \mathcal{L}^{-1}\left\{\sqrt{s-\gamma} - \sqrt{s} + \gamma\frac{1}{\sqrt{s}}\right\} = \frac{1}{2\sqrt{\pi}}(1 - e^{\gamma\tau})\,\tau^{-\frac{3}{2}} + \frac{\gamma}{\sqrt{\pi}}\tau^{-\frac{1}{2}}$$
$$= \underbrace{\frac{1}{2\sqrt{\pi}}\left[(1 - e^{\gamma\tau})\,\tau^{-\frac{3}{2}} + \gamma\tau^{-\frac{1}{2}}\right]}_{=g_3(\tau)} + \underbrace{\frac{\gamma}{2\sqrt{\pi}}\tau^{-\frac{1}{2}}}_{=g_4(\tau)}, \tag{26}$$

$$I_3 = \int_0^r -\frac{\partial}{\partial\tau}\left[\psi_z(r-\tau)\,e^{-\sigma(r-\tau)}\right]g_3(\tau)\,d\tau$$
$$= \int_0^r \psi_z(r-\tau)\,e^{-\sigma(r-\tau)}g_3'(\tau)\,d\tau - \underbrace{\psi_z(r-\tau)\,e^{-\sigma(r-\tau)}g_3(\tau)\Big|_{\tau=0}^{\tau=r}}_{=0 \quad (\text{with } \psi_z(0)=0)}, \tag{27}$$

$$\psi(r)\,e^{-\sigma r} = \tilde{C} \int_0^r \psi_z(r-\tau)\,e^{-\sigma(r-\tau)}\big[g_1(\tau) + g_3'(\tau)\big]\,d\tau$$

$$+ \tilde{C} \int_0^r -\frac{\partial}{\partial r}\left[\psi_z(r-\tau)\,e^{-\sigma(r-\tau)}\right] g_4(\tau)\,d\tau$$

$$= \tilde{C} \int_0^r \psi_z(r-\tau)\,e^{-\sigma(r-\tau)}\big[g_1(\tau) + g_3'(\tau)\big]\,d\tau$$

$$+ \tilde{C} \int_0^r \left[\psi_{zr}(r-\tau)\,e^{-\sigma(r-\tau)} - \sigma\psi_z(r-\tau)\,e^{-\sigma(r-\tau)}\right] g_4(\tau)\,d\tau, \qquad (28)$$

i.e.

$$\psi(r) = \tilde{C}\left[\int_0^r \psi_z(r-\tau)\,e^{\sigma\tau}\,\tilde{g}(\tau)\,d\tau + \int_0^r \psi_{zr}(r-\tau)\,e^{\sigma\tau} g_4(\tau)\,d\tau\right], \qquad (29)$$

where

$$\tilde{g}(\tau) := \big[g_1(\tau) + g_3'(\tau) - \sigma g_4(\tau)\big]. \qquad (30)$$

We calculate

$$g_3'(\tau) = \frac{1}{2\sqrt{\pi}}\left[-\frac{3}{2}(1 - e^{\gamma\tau})\,\tau^{-\frac{5}{2}} - \gamma\,e^{\gamma\tau}\,\tau^{-\frac{3}{2}} - \frac{\gamma}{2}\,\tau^{-\frac{3}{2}}\right]$$

$$= \frac{1}{4\sqrt{\pi}}\left[-3(1 - e^{\gamma\tau})\,\tau^{-\frac{5}{2}} - 2\gamma(\tfrac{1}{2} + e^{\gamma\tau})\,\tau^{-\frac{3}{2}}\right], \qquad (31)$$

and

$$\sigma g_4(\tau) = \frac{\sigma\gamma}{2\sqrt{\pi}}\,\tau^{-\frac{1}{2}} = \frac{1}{4\sqrt{\pi}}\frac{k_0^2}{2}(N_p^2 - 1)(N_p^2 - N_s^2)\,\tau^{-\frac{1}{2}}, \qquad (32)$$

i.e. (30) gives finally

$$\tilde{g}(\tau) = \frac{1}{4\sqrt{\pi}}\left[ik_0 N_p^2(1 - e^{\gamma\tau})\,\tau^{-\frac{3}{2}} + k_0^2\big(N_p^2 - \tfrac{1}{2}N_s^2\big)^2\,\tau^{-\frac{1}{2}} - 3(1 - e^{\gamma\tau})\,\tau^{-\frac{5}{2}}\right.$$

$$\left.+ ik_0\big(N_p^2 - N_s^2\big)\big(\tfrac{1}{2} + e^{\gamma\tau}\big)\,\tau^{-\frac{3}{2}} - \frac{k_0^2}{2}\big(N_p^2 - 1\big)\big(N_p^2 - N_s^2\big)\,\tau^{-\frac{1}{2}}\right]$$

$$= \frac{1}{4\sqrt{\pi}}\left[\frac{k_0^2}{2}\big(N_p^4 - N_p^2 N_s^2 + \tfrac{1}{2}N_s^4 + N_p^2 - N_s^2\big)\,\tau^{-\frac{1}{2}}\right.$$

$$\left.+ i\frac{k_0}{2}\big(3N_p^2 - N_s^2 - 2N_s^2\,e^{\gamma\tau}\big)\,\tau^{-\frac{3}{2}} - 3(1 - e^{\gamma\tau})\tau^{-\frac{5}{2}}\right]$$

$$= O(\tau^{-\frac{1}{2}}), \quad \tau \to \infty. \qquad (33)$$

Finally, we define φ, ω, by setting $\gamma =: i\varphi$, $\sigma =: i\omega$ and

$$g(\tau) = 4\sqrt{\pi}\,\tilde{g}(\tau), \qquad C = \frac{\tilde{C}}{4\sqrt{\pi}}. \qquad (34)$$

This completes the calculation of (16).

5   Conclusion and Outlook

First, we will make first numerical investigations for these new TBCs and investigate their superiority compared to using their formulation in transformed space. Next, instead of using an ad-hoc discretization of the analytic transparent BC we will construct discrete TBCs of the fully discretized half-space problem in the spirit of [4].

Acknowledgments. The first author was supported by the FWF (project I 395-N16 and the doctoral school "Dissipation and dispersion in non-linear partial differential equations").

Appendix: Laplace–Transformations

$$\mathcal{L}^{-1}\left\{\sqrt{s-\gamma}-\sqrt{s}\right\} = \frac{1}{2\sqrt{\pi}}(1-e^{\gamma t})\,t^{-\frac{3}{2}}, \tag{L.1}$$

$$\mathcal{L}^{-1}\left\{\frac{1}{\sqrt{s}}\right\} = \frac{1}{\sqrt{\pi}}\,t^{-\frac{1}{2}}, \tag{L.2}$$

$$\mathcal{L}^{-1}\left\{\hat{\psi}(s+\sigma)\right\} = \psi(t)\,e^{-\sigma t}, \tag{L.3}$$

$$\mathcal{L}^{-1}\left\{s\,\hat{\psi}(s+\sigma)\right\} = \frac{d}{dt}\left\{\psi(t)\,e^{-\sigma t}\right\} \quad \text{if } \psi(0)=0. \tag{L.4}$$

References

1. Antoine, X., Arnold, A., Besse, C., Ehrhardt, M., Schädle, A.: A review of transparent and artificial boundary conditions techniques for linear and nonlinear Schrödinger equations. Commun. Comput. Phys. **4**, 729–796 (2008)
2. Antonopoulou, D.C., Dougalis, V.A., Zouraris, G.E.: Galerkin methods for parabolic and Schrödinger equations with dynamical boundary conditions and applications to underwater acoustics. SIAM J. Numer. Anal. **47**, 2752–2781 (2009)
3. Arnold, A.: Numerically absorbing boundary conditions for quantum evolution equations. VLSI Des. **6**, 313–319 (1998)
4. Arnold, A., Ehrhardt, M.: Discrete transparent boundary conditions for wide angle parabolic equations in underwater acoustics. J. Comp. Phys. **145**, 611–638 (1998)
5. Collins, M.D.: A higher-order parabolic equation for wave propagation in an ocean overlying an elastic bottom. J. Acoust. Soc. Am. **86**, 1459–1464 (1989)
6. Dougalis, V.A., Sturm, F., Zouraris, G.E.: On an initial-boundary value problem for a wide-angle parabolic equation in a waveguide with a variable bottom. Math. Meth. Appl. Sci. **32**, 1519–1540 (2009)
7. Ehrhardt, M., Mickens, R.E.: Solutions to the discrete airy equation: application to parabolic equation calculations. J. Comput. Appl. Math. **172**, 183–206 (2004)
8. Greene, R.R.: The rational approximation to the acoustic wave equation with bottom interaction. J. Acoust. Soc. Am. **76**, 1764–1773 (1984)

9. Greene, R.R.: A high-angle one-way wave equation for seismic wave propagation along rough and sloping interfaces. J. Acoust. Soc. Am. **77**, 1991–1998 (1985)
10. Jensen, F.B., Kuperman, W.A., Porter, M.B., Schmidt, H.: Computational Ocean Acoustics. AIP Press, New York (1994)
11. Levy, M.F.: Transparent boundary conditions for parabolic equation solutions of radiowave propagation problems. IEEE Trans. Antennas Propag. **45**, 66–72 (1997)
12. Makrakis, G.N.: Asymptotic study of the elastic seabed effects in ocean acoustics. Appl. Anal. **66**, 357–375 (1997)
13. Papadakis, J.S.: Impedance formulation of the bottom boundary condition for the parabolic equation model in underwater acoustics. NORDA Parabolic Equation Workshop, NORDA Technical note 143 (1982)
14. Papadakis, J.S., Taroudakis, M.I., Papadakis, P.J., Mayfield, B.: A new method for a realistic treatment of the sea bottom in the parabolic approximation. J. Acoust. Soc. Am. **92**, 2030–2038 (1992)
15. Papadakis, J.S.: Impedance bottom boundary conditions for the parabolic-type approximations in underwater acoustics. In: Vichnevetsky, R., Knight, D., Richter, G. (eds.) Advances in Computer Methods for Partial Differential Equations VII, pp. 585–590. IMACS, New Brunswick (1992)
16. Sturm, F., Kampanis, N.A.: Accurate treatment of a general sloping interface in a finite-element 3D narrow-angle PE model. J. Comput. Acoust. **15**, 285–318 (2007)
17. Tappert, F.D.: The parabolic approximation method. In: Keller, J.B., Papadakis, J.S. (eds.) Wave Propagation and Underwater Acoustics. Lecture Notes in Physics, vol. 70, pp. 224–287. Springer, New York (1977)
18. Thomson, D.J.: Wide-angle parabolic equation solutions to two range-dependent benchmark problems. J. Acoust. Soc. Am. **87**, 1514–1520 (1990)
19. Wetton, B.T., Brooke, G.H.: One-way wave equations for seismoacoustic propagation in elastic waveguides. J. Acoust. Soc. Am. **87**, 624–632 (1990)
20. Zhang, Z.Y., Tindle, C.T.: Improved equivalent fluid approximations for a low shear speed ocean bottom. J. Acoust. Soc. Am. **98**, 3391–3396 (1995)

Well-Posedness in Hölder Spaces of Elliptic Differential and Difference Equations

Allaberen Ashyralyev[1,2]([⊠])

[1] Department of Mathematics, Fatih University, Istanbul, Turkey
aashyr@fatih.edu.tr
[2] ITTU, Ashgabat, Turkmenistan

Abstract. In the present paper the well-posedness of the elliptic differential equation

$$-u''(t) + Au(t) = f(t)(-\infty < t < \infty)$$

in an arbitrary Banach space E with the general positive operator in Hö lder spaces $C^\beta(\mathbb{R}, E_\alpha)$ is established. The exact estimates in Hölder norms for the solution of the problem for elliptic equations are obtained. The high order of accuracy two-step difference schemes generated by an exact difference scheme or by Taylor's decomposition on three points for the approximate solutions of this differential equation are studied. The well-posedness of the these difference schemes in the difference analogy of Hölder spaces $C^\beta(\mathbb{R}_\tau, E_\alpha)$ are obtained. The almost coercive inequality for solutions in $C(\mathbb{R}_\tau, E)$ of these difference schemes is established.

Keywords: Abstract elliptic equation · Banach spaces · Exact estimates · Fractional spaces · Well-posedness

The role played by coercive inequalities (well-posedness) in the study of local boundary-value problems for elliptic and parabolic differential equations is well known (see, e.g., [1,2]). Well-posedness of local and nonlocal boundary value problems for elliptic differential equations have been studied extensively by many researchers (see [3–8, 11–13, 15–17] and the references therein). In the paper [10] the elliptic differential equation

$$- u''(t) + Au(t) = f(t), -\infty < t < \infty \tag{1}$$

with A positive operator in E was considered. A function $u(t)$ is called a solution of the problem (1) if the following conditions are satisfied:

(i) $u(t)$ is twice continuously differentiable function on $\mathbb{R} = (-\infty, \infty)$.
(ii) The element $u(t)$ belongs to $D(A)$ for all $t \in \mathbb{R}$, and the function $Au(t)$ is continuous on $\mathbb{R}$.
(iii) $u(t)$ satisfies the Eq. (1).

© Springer International Publishing Switzerland 2015
I. Dimov et al. (Eds.): FDM 2014, LNCS 9045, pp. 25–37, 2015.
DOI: 10.1007/978-3-319-20239-6_3

A solution of problem (1) defined in this manner will from now on be referred to as a solution of problem (1) in the space $C(E) = C(\mathbb{R}, E)$ of all continuous functions $\varphi(t)$ defined on $\mathbb{R}$ with values in E, equipped with the norm

$$||\varphi||_{C(E)} = \sup_{t \in \mathbb{R}} ||\varphi(t)||_E.$$

The well-posedness in $C(E)$ of the boundary value problem (1) means that coercive inequality

$$||u''||_{C(E)} + ||Au||_{C(E)} \le M||f||_{C(E)} \qquad (2)$$

is true for its solution $u(t) \in C(E)$ with some M, does not depend on $f(t) \in C(E)$. In this paper, positive constants, which can differ in time will be indicated with an M. On the other hand $M(\alpha, \beta, \cdots)$ is used to focus on the fact that the constant depends only on $\alpha, \beta, \cdots$. It is known that from the coercive inequality (2) the positivity of the operator A in the Banach space E follows under the assumption that the operator has bounded inverse $(I\lambda + A)^{-1}$ for any $\lambda \ge 0$, in E and estimate

$$||(\lambda I + A)^{-1}||_{E \to E} \le \frac{M}{1 + \lambda}$$

holds. It turns out that this positivity property of the operator A in E is necessary condition of well-posedness of the differential equation (1) in $C(E)$. The positivity of the operator A in E is not a sufficient condition for the well-posedness of the differential equation (1). As it turns out, problem (1) is not well-posed for all such operators. The counterexample given by Sobolevskii in [10]. It is known (see, for example [8]) that the operator $A^{1/2}$ has better spectral properties than the positive operator A. In particular, the operator $\lambda I + A^{1/2}$ has a bounded inverse for any complex number λ with $\text{Re}\lambda \ge 0$, and the estimate

$$||(\lambda I + A^{1/2})^{-1}||_{E \to E} \le M(|\lambda| + 1)^{-1}$$

is true for some $M \ge 1$. Thus, $A^{1/2}$ is a strongly positive operator in E, i.e. the following estimates hold:

$$||e^{-tA^{1/2}}||_{E \to E} \le Me^{-\delta t}, t||A^{1/2}e^{-tA^{1/2}}||_{E \to E} \le M, \quad t > 0, \quad \delta > 0 \qquad (3)$$

It is easy to see that formula

$$u(t) = \frac{1}{2}A^{-\frac{1}{2}} \int\limits_{-\infty}^{\infty} e^{-|t-s|A^{\frac{1}{2}}} f(s)ds \qquad (4)$$

defines the unique solution in $C(E)$ of differential equation (1) if, for example and $Af(t) \in C(E)$ or $f''(t) \in C(E)$. It turns out formula formula (4) defines the unique solution in $C(E)$ of differential equation (1) under essentially less restrictions on the smoothness of function $f(t)$. We introduce the Hölder space

$C^\beta(E) = C^\beta(\mathbb{R}, E)$, $\beta \in [0,1]$, of all E-valued abstract functions $\varphi(t)$ defined on $\mathbb{R}$ with the norm

$$\|\varphi\|_{C^\beta(E)} = \|\varphi\|_{C(E)} + \sup_{-\infty < t < t+\tau < \infty} \frac{\|\varphi(t+\tau) - \varphi(t)\|_E}{\tau^\beta}.$$

We call $u(t)$ a *solution* of problem (1) in $C^\beta(E)$, if it is a solution of this problem in $C(E)$, and $u''(t)$, $Au(t) \in C^\beta(E)$. Problem (1) is called *well-posed* in $C^\beta(E)$, if its solutions $u(t)$ in $C^\beta(E)$ satisfy the following *coercivity inequality*

$$\|u''\|_{C^\beta(E)} + \|Au\|_{C^\beta(E)} \le M(\beta)\|f\|_{C^\beta(E)}.$$

Theorem 1.1. *Equation (1) is well-posed in Banach spaces $C^\beta(E)$ for all $\beta \in (0,1)$, iff A is positive operator in a Banach space E. For solution problem $u(t)$ in $C^\beta(E)$ of equation (1) coercivity inequality* [10]

$$\|u''\|_{C^\beta(E)} + \|Au\|_{C^\beta(E)} \le \frac{M}{\beta(1-\beta)}\|f\|_{C^\beta(E)} \tag{5}$$

takes place.

For $\alpha \in (0,1)$, let $E_\alpha = E_{\alpha,\infty}(E, A^{1/2})$ be the *fractional space* consisting of all $v \in E$ for which the norm

$$\|v\|_{E_\alpha} = \|v\|_E + \sup_{\lambda > 0} \|\lambda^{1-\alpha} A^{1/2} e^{-\lambda A^{1/2}} v\|_E$$

is finite [18]. From (3) it follows that

$$\|e^{-tA^{1/2}}\|_{E_\alpha \to E_\alpha} \le M e^{-\delta t}, t\|A^{1/2} e^{-tA^{1/2}}\|_{E_\alpha \to E_\alpha} \le M, \quad t > 0, \quad \delta > 0 \tag{6}$$

Then, using (5), we get

$$\|u''\|_{C^\beta(E_\alpha)} + \|Au\|_{C^\beta(E_\alpha)} \le \frac{M}{\beta(1-\beta)}\|f\|_{C^\beta(E_\alpha)}. \tag{7}$$

Note that one can not put $\beta = 0$ and $\beta = 1$ in (5). In present paper the well-posedness of (1) in Hölder spaces $C^\beta(E_\alpha) = C^\beta(\mathbb{R}, E_\alpha)$, $(\alpha, \beta \in [0,1], \alpha + \beta \neq 0, \alpha + \beta \neq 2)$ is established. The exact estimates in Hölder norms for the solution of the problem for elliptic equations are obtained. The high order of accuracy two-step difference schemes generated by an exact difference scheme or by Taylor's decomposition on three points for the approximate solutions of this differential equation are studied. The well-posedness of these difference schemes in the difference analogy of Hölder spaces $C^\beta(\mathbb{R}_\tau, E_\alpha)$, $(0 \le \beta \le 1, 0 < \alpha < 1)$ are obtained. The almost coercive inequality for solutions in $C(\mathbb{R}_\tau, E)$ of these difference schemes is established.

1　Well-Posedness of Elliptic Differential Equations

Theorem 2.1. *For any α, $\beta \in [0,1]$, $\alpha + \beta \neq 0, \alpha + \beta \neq 2$, problem (1) is well-posed in $C^\beta(E_\alpha)$ and the estimate*

$$\|u''\|_{C^\beta(E_\alpha)} + \|Au\|_{C^\beta(E_\alpha)} \leq M \min\left\{\frac{1}{\beta(1-\beta)}, \frac{1}{\alpha(1-\alpha)}\right\} \|f\|_{C^\beta(E_\alpha)} \quad (8)$$

is valid.

The proof of Theorem 2.1 is based on the formula (4), estimate (3), and the definition of E_α−norm.

Now, we will consider the application of Theorem 2.1 For formulation it we need the following theorem

Theorem 2.2. *If A is positive operator in a Banach space E. Then $E_\alpha(A, E) = E_{2\alpha}(A^{\frac{1}{2}}, E)$ for all $0 < \alpha < \frac{1}{2}$ [9].*

We will consider $2m$-th order multidimensional elliptic equations

$$\begin{cases} -\dfrac{\partial^2 u}{\partial y^2} + \displaystyle\sum_{|r|=2m} a_r(x)\dfrac{\partial^{|r|}u}{\partial x_1^{r_1}\cdots\partial x_n^{r_n}} + \delta u(y,x) = f(y,x), \\[4mm] y \in \mathbb{R}, \quad x, r \in \mathbb{R}^n, \quad |r| = r_1 + \cdots + r_n, \end{cases} \quad (9)$$

where $a_r(x)$ and $f(y,x)$ are given sufficiently smooth functions and $\delta > 0$ is the sufficiently large number. It is assumed that the symbol

$$B^x(\xi) = \sum_{|r|=2m} a_r(x)\,(i\xi_1)^{r_1}\cdots(i\xi_n)^{r_n}, \xi = (\xi_1,\cdots,\xi_n) \in \mathbb{R}^n$$

of the differential operator of the form

$$B^x = \sum_{|r|=2m} a_r(x)\frac{\partial^{|r|}}{\partial x_1^{r_1}\cdots\partial x_n^{r_n}} \quad (10)$$

acting on functions defined on the space $\mathbb{R}^n$, satisfies the inequalities

$$0 < M_1|\xi|^{2m} \leq (-1)^m B^x(\xi) \leq M_2|\xi|^{2m} < \infty \quad (11)$$

for $\xi \neq 0$. The equation (9) has a unique smooth solution. This allows us to reduce the equation (9) to the equation (1) in a Banach space $E = C^\mu(\mathbb{R}^n)$ of all continuous bounded functions defined on $\mathbb{R}^n$ satisfying a Hölder condition with the indicator $\mu \in (0,1)$ with a strongly positive operator $A^x = B^x + \sigma I$ defined by (10) (see, [20]).

Theorem 2.3. *For the solution of the equation (9) the following estimate is satisfied:*

$$\|u\|_{C^{2+\beta}(C^{2\alpha m}(\mathbb{R}^n))} + \sum_{|\tau|=2m}\left\|\frac{\partial^{|r|}u}{\partial x_1^{r_1}\cdots\partial x_n^{r_n}}\right\|_{C^\beta(C^{2\alpha m}(\mathbb{R}^n))} \leq M(\delta, \alpha, \beta)\|f\|_{C^\beta(C^{2\alpha m}(\mathbb{R}^n))}.$$

The proof of Theorem 2.5 is based on the abstract Theorems 2.1 and 2.2 and on the theorem on the structure of the fractional spaces $E_\alpha(A^x, C^\mu(\mathbb{R}^n))$ and on the theorem about the well-posedness of elliptic equations.

Theorem 2.4. $E_\alpha(A^x, C^\mu(\mathbb{R}^n)) = C^{2m\alpha+\mu}(\mathbb{R}^n)$ *for all* $0 < \alpha < \frac{1}{2m}, 0 < \mu < 1$ [18].

Theorem 2.5. *Suppose that assumption (11) for the operator* A^x *holds. Then, for the solutions of the differential equation*

$$A^x u(x) = \omega(x), x \in \mathbb{R}^n$$

the coercive solvability estimate [19]

$$\sum_{|\tau|=2m} \left\| \frac{\partial^{|r|} u}{\partial x_1^{r_1} \cdots \partial x_n^{r_n}} \right\|_{C^{2m\alpha}(\mathbb{R}^n)} \leq M(\delta, \alpha) \|\omega\|_{C^{2m\alpha}(\mathbb{R}^n)}, 0 < \alpha < \frac{1}{2m}$$

is valid.

2 Difference Schemes Generated by an Exact Difference Scheme

For the construction of the two step difference schemes of an arbitrary high order of accuracy for the approximate solutions of the differential equation (1) we consider the uniform grid space

$$\mathbb{R}_\tau = (-\infty, \infty)_\tau = \{t_k = k\tau, k = 0, \pm 1, \ldots\}.$$

The construction of two-step difference schemes of an arbitrary high order of accuracy for the approximate solutions of the differential equation (1) is based on the following theorem.

Theorem 3.1. [14] *Let* $u(t_k)$ *be a solution of the differential equation (1) at the grid points* $t = t_k \in \mathbb{R}_\tau$. *Then* $\{u(t_k)\}_{-\infty}^{\infty}$ *is the solution of the following second order difference equations:*

$$-\tau^{-2}(u(t_{k+1}) - 2u(t_k) + u(t_{k-1})) + \tau^{-2}(I - \exp\{-\tau B\})(u(t_{k+1}) + u(t_{k-1})) \quad (12)$$

$$+\tau^{-2}(\exp\{-2\tau B\} - I)u(t_k) = \psi_k,$$

$$\psi_k = (2\tau B)^{-1}(\psi_{1,k} + \psi_{2,k+1}) - (2\tau B)^{-1}\exp\{-\tau B\}(\psi_{1,k+1} + \psi_{2,k}),$$

$$\psi_{1,k} = \tau^{-1} \int_{t_{k-1}}^{t_k} \exp\{-(t_k - s)B\}f(s)ds,$$

Applying the exact difference scheme (12), we obtain $(l+j)$−order of accuracy two-step difference schemes

$$-\tau^{-2}(u_{k+1} - 2u_k + u_{k-1}) + \tau^{-2}(I - R_{j,l}(\tau B))(u_{k+1} + u_{k-1}) \quad (13)$$

$$+\tau^{-2}(R_{j,l}^2(\tau B) - I)u_k = f_k^{l,j},$$

$$f_k^{l,j} = (2\tau B)^{-1}(f_{1,k}^{l,j} + f_{2,k+1}^{l,j}) - (2\tau B)^{-1}R_{j,l}(\tau B)(f_{1,k+1}^{l,j} + f_{2,k}^{l,j}),$$

$$f_{1,k}^{l,j} = \sum_{m=1}^{l+j}\sum_{\lambda=0}^{m}\binom{m}{\lambda}B^{m-\lambda}f^{(\lambda)}(t_k)\frac{(-1)^m\tau^m}{(m+1)!},$$

$$f_{2,k}^{l,j} = \sum_{m=1}^{l+j}\sum_{\lambda=0}^{m}\binom{m}{\lambda}(-B)^{m-\lambda}f^{(\lambda)}(t_{k-1})\frac{\tau^m}{(m+1)!}, k \in (-\infty, \infty).$$

Here, the function $R_{j,l}(z)$ is constructed on the base of Padé's fractions

$$R_{j,l}(z) = \frac{P_{j,l}(z)}{Q_{j,l}(z)},$$

respectively

$$P_{j,l} = 1 + a_1 z + \ldots + a_j z^j, \quad Q_{j,l}(z) = 1 + b_1 z + \ldots + b_j z^l,$$

where the coefficients $a_i, a_i = 1, \ldots, j$, and $b_i, i = 1, \ldots, l$, are uniquely defined from the conditio

$$| R_{j,l}(z) - e^{-z} | = o(| z |^{j+l+1})$$

for $| z | \to 0$. Note that the difference scheme (13) for $j = l$ and $j = l - 1$ include difference schemes of arbitrary high order of approximation. Moreover, the corresponding functions $R_{j,l+1}(z)$ tend to 0 as $z \to \infty$ for $j = l - 1, l$. Such difference schemes are simplest, in the sense that the degrees of the denominators of the corresponding Pade approximants of the function $\exp\{-z\}$ are minimal for a fixed order of approximation of the difference schemes. Let us investigate the well posedness of the exact two-step difference scheme (13) for $j = l-1, l-2$. Let us denote by $F_\tau(E) = F(\mathbb{R}_\tau, E)$ the space of grid functions $\varphi^\tau = \{\varphi_k\}_{k=-\infty}^{\infty}$ for fixed τ. Thus, $F_\tau(E)$ is the vector space whose elements are ordered elements of E. The space $F_\tau(E)$ can be equipped with various norms and thus become a normed space. Thus, for instance, the vector space $F_\tau(E)$ generates the normed space $C_\tau(E) = C(\mathbb{R}_\tau, E)$ with the norm

$$\| \varphi^\tau \|_{C_\tau(E)} = \sup_{-\infty < k < \infty} \| \varphi_k \|_E,$$

the normed space $C_\tau^\beta(E) = C^\beta(\mathbb{R}_\tau, E)$ $0 < \beta < 1$, with the norm

$$\| \varphi^\tau \|_{C_\tau^\beta(E)} = \| \varphi^\tau \|_{C_\tau(E)} + \sup_{-\infty < k < k+r < \infty} \| \varphi_{k+r} - \varphi_k \|_E \frac{1}{(r\tau)^\beta}.$$

The difference scheme (13) is uniquely solvable, and the following formula holds

$$u_k = \tau(I - R_{j,l}^2(\tau B))^{-1} \sum_{i=-\infty}^{\infty} R_{j,l}^{|k-i|}(\tau B)f_i^{l,j}, k \in (-\infty, \infty). \qquad (14)$$

Since the boundary value problem (1) in the space $C(E)$ of bounded continuous functions defined on the real line with values in E is not well-posed in the case of general positive operator A, then the well-posedness of the difference schemes (13) in $C(\tau, E)$ norm does not take place uniformly with respect to $\tau > 0$. This means that the coercive norm

$$\| u^\tau \|_{k(\tau, E)} = \| \{\tau^{-2}(u_{k+1} - 2u_k + u_{k-1})\}_{-\infty}^{\infty} \|_{C(\tau, E)}$$

$$+ \| \{\tau^{-2}(I - R_{j,l}(\tau B))(u_{k+1} + u_{k-1}) + \tau^{-2}(R_{j,l}^2(\tau B) - I)u_k\}_{-\infty}^{\infty} \|_{C(\tau, E)}$$

tends to ∞ as $\tau \to +0$. The investigation of difference schemes (13) permits us to establish the order of growth of this norm to ∞.

Lemma 3.1. [14] *Let B is a strongly positive operator in a Banach space E with spectral angle $\phi(B, E) < \frac{\pi}{T}$. Then the operator $I + R_{j,l}(\tau B)$ is invertible and one has the estimate*

$$\|(I + R_{j,l}(\tau B))^{-1}\|_{E \to E} \leq M.$$

We have that

Theorem 3.2. *Let A is a strongly positive operator in a Banach space E with spectral angle $\phi(A, E) < \frac{\pi}{T}$. Then the solutions of the difference problem (13) in $C_\tau(E)$ obey the almost coercive inequality*

$$\| u^\tau \|_{K_\tau(E)} \leq M_1 \min \left\{ \ln \frac{1}{\tau}, 1 + |\ln \| B \|_{E \to E}| \right\} \| f_{j,l}^\tau \|_{C_\tau(E)}.$$

Theorem 3.3. *Let A is a strongly positive operator in a Banach space E with spectral angle $\phi(A, E) < \frac{\pi}{T}$. Then for any $\alpha, \beta \in [0, 1], \alpha + \beta \neq 0, \alpha + \beta \neq 2$, the solutions of the difference problem (13) in $C_\tau^\beta(E_\alpha)$ obey the coercivity inequality*

$$\| \{\tau^{-2}(u_{k+1} - 2u_k + u_{k-1})\}_{-\infty}^{\infty} \|_{C_\tau^\beta(E_\alpha)}$$

$$+ \| \{\tau^{-2}(I - R_{j,l}(\tau B))(u_{k+1} + u_{k-1}) + \tau^{-2}(R_{j,l}^2(\tau B) - I)u_k\}_{-\infty}^{\infty} \|_{C_\tau^\beta(E_\alpha)}$$

$$\leq M \min \left\{ \frac{1}{\beta(1 - \beta)}, \frac{1}{\alpha(1 - \alpha)} \right\} \| f_{j,l}^\tau \|_{C_\tau^\beta(E_\alpha)}.$$

Now, the abstract theorems given from above are applied in the investigation of difference schemes of higher order of accuracy with respect to the set all variables for approximate solution of the boundary value problem (9). The discretization of problem (9) is carried out in two steps. In the first step let us define the grid space $\mathbb{R}_h^n$ $(0 < h \leq h_0)$ as the set of all points of the Euclidean space $\mathbb{R}^n$ whose coordinates are given by

$$x_k = s_k h, \qquad s_k = 0, \pm 1, \pm 2, \cdots, k = 1, \cdots, n.$$

The number h is called the step of the grid space. A function φ^h defined on $\mathbb{R}_h^n$ will be called a grid function. We introduce the space $C\left(\mathbb{R}_h^n\right)$ of all mesh functions $\varphi^h(x)$ defined on $\mathbb{R}_h^n$ with the norm

$$\left\|\varphi^h\right\|_{C_h} = \sup_{x \in \mathbb{R}_h^n} \left|\varphi^h(x)\right|.$$

Let $C_h^\beta = C^\beta\left(\mathbb{R}_h^n\right)$ be the Hölder space of all mesh functions $\varphi^h(x)$ defined on $\mathbb{R}_h^n$ satisfying a Hölder condition with the indicator $\beta \in (0,1)$ with the norm

$$\left\|\varphi^h\right\|_{C_h^\beta} = \left\|\varphi^h\right\|_{C_h} + \sup_{\substack{x,y \in \mathbb{R}_h^n \\ x \neq y}} \frac{\left|\varphi^h(x) - \varphi^h(y)\right|}{|x-y|^\beta}.$$

Then, let us give the difference operator A_h^x by the formula

$$A_h^x u^h(x) = \sum_{2m \leq |r| \leq S} b_r^x D_h^r u^h(x) + \delta u^h(x), x \in \mathbb{R}_h^n. \tag{15}$$

The coefficients are chosen in such a way that the operator A_h^x approximates in a specified way the operator

$$\sum_{|r|=2m} a_r(x) \frac{\partial^{|r|}}{\partial x_1^{r_1} ... \partial x_n^{r_n}} + \delta.$$

We shall assume that for $|\xi_k h| \leq \pi$ the symbol $A(\xi h, h)$ of the operator $A_h^x - \delta$ satisfies the inequalities

$$(-1)^m A^x(\xi h, h) \geq M_1 |\xi|^{2m}, |\arg A^x(\xi h, h)| \leq \phi < \phi_0 \leq \frac{\pi}{l}. \tag{16}$$

With the help of A_h^x we arrive at the boundary value problem

$$-\frac{d^2 v^h(y,x)}{dy^2} + A_h^x v^h(y,x) = f^h(y,x), y \in \mathbb{R}, x \in \mathbb{R}_h^n \tag{17}$$

for an infinite system of ordinary differential equations. In the second step we replace problem (17) by the difference scheme

$$-\frac{1}{\tau^2}(u_{k+1}^h(x) - 2u_k^h(x) + u_{k-1}^h(x)) + \frac{1}{\tau^2}(I - R_{j,l+1}(\tau B_h^x))(u_{k-1}^h(x) + u_{k+1}^h(x))$$

$$+\frac{1}{\tau^2}(R_{j,l+1}^2(\tau B_h^x) - I)u_k^h(x) = f_k^{j,l}(x), k \in (-\infty, \infty), \ x \in R_h^n, (B_h^x)^2 = A_h^x. \tag{18}$$

Let us give a number of corollaries of the abstract theorems given in the above

Theorem 3.4. *The solutions of the difference scheme (18) satisfy the following almost coercive stability estimates:*

$$\left\| \{\tau^{-2}(u_{k+1}^h - 2u_k^h + u_{k-1}^h)\}_{-\infty}^\infty \right\|_{C_\tau(C_h)} \leq M \ln \frac{1}{\tau + h} \left\| f_{j,l}^{\tau,h} \right\|_{C_\tau(C_h)}.$$

The proof of Theorem 3.4 is based on the abstract Theorem 3.2, the positivity of the operator A_h^x in C_h and on the estimate

$$\min\left\{\ln\frac{1}{\tau}, 1 + \left|\ln\| B_h^x \|_{C_h\to C_h}\right|\right\} \leq M\ln\frac{1}{\tau+h}. \tag{19}$$

Theorem 3.5. *The solutions of the difference scheme (18) satisfy the following coercivity estimate*

$$\| \{\tau^{-2}(u_{k+1}^h - 2u_k^h + u_{k-1}^h)\}_{-\infty}^\infty \|_{C^\beta(C_h^{2m\alpha})} \leq M(\alpha,\beta) \| f_{j,l}^{\tau,h} \|_{C^\beta_\tau(C_h^{2m\alpha})}.$$

The proof of Theorem 3.5 is based on the abstract Theorem 3.3, the positivity of the operator A_h^x in C_h^α and on the theorem about the well-posedness of elliptic difference equations [18] and on the fact that for any $0 < \alpha < \frac{1}{2m}$ the norms in the spaces $E_\beta'(A_h^x, C_h)$ and $C_h^{2m\alpha}$ are equivalent uniformly in h and on the following theorem on the structure of the fractional spaces $E_\alpha'((A_h^x)^{\frac{1}{2}}, C_h)$.

Theorem 3.6. *Let A is a strongly positive operator in a Banach space E with spectral angle $\phi(A, E) < \frac{\pi}{2}$. Then for $0 < \alpha < \frac{1}{2}$ the norms of the spaces $E_\alpha'(A^{\frac{1}{2}}, E)$ and $E_{\frac{\alpha}{2}}'(A, E)$ are equivalent.*

3 Difference Schemes Generated by Taylor's Decomposition on Three Points

We consider again the problem (1). Let the function $u(t)(t \in \mathbb{R})$ has a $(2l+2)$-th continuous derivative and $t_{k-1}, t_k, t_{k+1} \in \mathbb{R}_\tau$. We consider two-step difference schemes generated by Taylor's decomposition on three points

$$\begin{cases} -\tau^{-2}(u_{k+1} - 2u_k + u_{k-1}) + A_l u_k = f_k, -\infty < k < \infty, \\ A_l = \sum_{m=0}^{l-1} \frac{2\tau^{2m}}{(2m+2)!} A^{m+1}, f_k = \sum_{m=0}^{l-1} \frac{2\tau^{2m}}{(2m+2)!} \sum_{l=0}^{m} A^{m-i} f^{(2i)}(t_k), \\ t_k = k\tau, -\infty < k < \infty \end{cases} \tag{20}$$

of $2l$ -order of approximation of approximate solution of Eq. (1).

The step operator of difference scheme (20) is a operator $R(\tau B) = (I + \tau B)^{-1}$ and the B operator is denoted from the following formula

$$B = B(\tau, A) = \frac{1}{2}\tau^2 A_l + \sqrt{\frac{1}{4}(\tau^2 A_l)^2 + A_l}$$

However, for the investigation of (20) it is necessary to construct an operator $B = B(\tau, A)$ and to give estimates

$$\| R^k(\tau B) \|_{E\to E} \leq M(1 + \delta\tau)^{-k}, \quad \| k\tau B R^k(\tau B) \|_{E\to E} \leq M, \quad k \geq 1. \tag{21}$$

The difference scheme (20) is uniquely solvable, and the following formula holds

$$u_k = (I + \tau B)(2I + \tau B)^{-1}B^{-1} \sum_{i=-\infty}^{\infty} R^{|k-i|}(\tau B)f_i\tau, \quad -\infty < k < \infty, \qquad (22)$$

where

$$B = B(\tau, A) = \frac{\tau^2 A_l}{2} + \sqrt{\left(\frac{\tau^2 A_l}{2}\right)^2 + A_l}, \; A_l = \sum_{i=0}^{l-1} A^{i+1}\frac{2\tau^{2i}}{(2i+2)!}.$$

From the formula (22) it follows that the investigation of the stability and well-posedness of difference scheme (20) relies in an essential manner on a number of properties of the powers of the operator $(I + \tau B)^{-1}$. We were not able to obtain the estimates for powers of the operator $(I + \tau B)^{-1}$ in the general cases of operator A. We begin by deriving some estimates for powers of the operator $(I + \tau B)^{-1}$ and a strongly positive operator A in a Banach space E with spectral angle $\phi(A, E) < \frac{\pi}{2l}$. The proof of estimates (21) is based upon three lemmas.

Lemma 4.1. [11] *A necessary and sufficient condition for B to be strongly positive is that the estimates (21) are satisfied.*

Lemma 4.2. [11] *If A is a strongly positive operator, then the operator B denoted by*

$$B = \frac{1}{2}\tau A + \sqrt{\frac{1}{4}(\tau^2 A)^2 + A}$$

is a strongly positive operator.

Lemma 4.3. [20] *If A is a strongly positive operator with spectrum angle $\phi(A, E) \leq \frac{\pi}{2l}$ then the operator A_l denoted by formula (20) is also a strongly positive operator.*

Since the boundary value problem (1) in the space $C(E)$ of bounded continuous functions defined on the real line with values in E is not well-posed in the case of general positive operator A, then the coercive norm

$$\| u^\tau \|_{k(\tau, E)} = \| \{\tau^{-2}(u_{k+1} - 2u_k + u_{k-1})\}_{-\infty}^{\infty} \|_{C(\tau, E)} + \| \{A_l u_k\}_{-\infty}^{\infty} \|_{C(\tau, E)}$$

tends to ∞ as $\tau \to +0$. The investigation of difference schemes (20) permits us to establish the order of growth of this norm to ∞.

Theorem 4.1. *Let A is a strongly positive operator in a Banach space E with spectral angle $\phi(A, E) < \frac{\pi}{2l}$. Then the solutions of the difference scheme (20) in $C_\tau(E)$ obey the almost coercive inequality*

$$\| u^\tau \|_{K_\tau(E)} \leq M_1 \min\left\{\ln\frac{1}{\tau}, 1 + |\ln\| B \|_{E \to E}|\right\} \| f^\tau \|_{C_\tau(E)}.$$

Now let us study of the well-posedness of the difference problem (13) in $C_\tau^\alpha(E_\beta)$.

Theorem 4.2. *Suppose that the assumption of Theorem 4.1 holds. Then for any α, $\beta \in [0,1]$, $\alpha + \beta \neq 0, \alpha + \beta \neq 2$, the solutions of the difference problem (20) in $C_\tau^\beta(E_\alpha)$ obey the coercivity inequality*

$$\| \{\tau^{-2}(u_{k+1} - 2u_k + u_{k-1})\}_{-\infty}^\infty \|_{C_\tau^\beta(E_\alpha)} + \| \{A_l u_k\}_{-\infty}^\infty \|_{C_\tau^\beta(E_\alpha)}$$

$$\leq M \min \left\{ \frac{1}{\beta(1-\beta)}, \frac{1}{\alpha(1-\alpha)} \right\} \| f^\tau \|_{C_\tau^\beta(E_\alpha)}.$$

Now, the abstract theorems given from above are applied in the investigation of difference schemes of higher order of accuracy with respect to the set all variables for approximate solution of the boundary value problem (9). The discretization of problem (9) is carried out in two steps, too. In the first step we give discretization in space variable. It is done in the last section. In the second step we replace problem (17) by the difference scheme

$$\begin{cases} -\tau^{-2} \left(u_{k+1}^h(x) - 2u_k^h(x) + u_{k-1}^h(x)\right) + (A_h^x)_l u_k^h(x) = f_k^h(x), -\infty < k < \infty, \\ (A_h^x)_l = \sum\limits_{m=0}^{l-1} \frac{2\tau^{2m}}{(2m+2)!} (A_h^x)^{m+1}, f_k = \sum\limits_{m=0}^{l-1} \frac{2\tau^{2m}}{(2m+2)!} \sum\limits_{l=0}^m (A_h^x)^{m-i} \left(f_k^h\right)^{(2i)} (t_k, x), \\ t_k = k\tau, -\infty < k < \infty, \ x \in R_h^n. \end{cases}$$

$$(23)$$

Let us give a number of corollaries of the abstract theorems given in the above

Theorem 4.3. *The solutions of the difference scheme (23) satisfy the following almost coercive stability estimates:*

$$\| \{\tau^{-2}(u_{k+1}^h - 2u_k^h + u_{k-1}^h)\}_{-\infty}^\infty \|_{C_\tau(C_h)} \leq M \ln \frac{1}{\tau + h} \| f^{\tau,h} \|_{C_\tau(C_h)}.$$

The proof of Theorem 4.3 is based on the abstract Theorem 4.1, the positivity of the operator A_h^x in C_h and on the estimate (19).

Theorem 4.4. *The solutions of the difference scheme (23) satisfy the following coercivity estimate*

$$\| \{\tau^{-2}(u_{k+1}^h - 2u_k^h + u_{k-1}^h)\}_{-\infty}^\infty \|_{C_\tau^\beta(C_h^{2m\alpha})} \leq M(\alpha, \beta) \| f^{\tau,h} \|_{C_\tau^\beta(C_h^{2m\alpha})}.$$

The proof of Theorem 4.4 is based on the abstract Theorem 4.2, the positivity of the operator A_h^x in C_h^α and on the theorem about the well-posedness of elliptic equations [18] and on the fact that for any $0 < \alpha < \frac{1}{2m}$ the norms in the spaces $E_\beta'(A_h^x, C_h)$ and $C_h^{2m\alpha}$ are equivalent uniformly in h and on the Theorem 3.13 on the structure of the fractional spaces $E_\alpha'((A_h^x)^{\frac{1}{2}}, C_h)$

4 Conclusion

Applying method of present paper and papers [20–22], we can construct and investigate a high order of accuracy uniform two-step difference schemes for the approximate solution of the problem for the elliptic differential equation

$$-\varepsilon^2 u''(t) + Au(t) = f(t)(-\infty < t < \infty)$$

in an arbitrary Banach space E with the positive operator A with a small ε^2 parameter in derivative.

References

1. Ladyzhenskaya, O.A., Ural'tseva, N.N.: Linear and Quasilinear Equations of Elliptic Type. Nauka, Moscow (1973) (Russian)
2. Vishik, M.L., Myshkis, A.D., Oleinik, O.A.: Partial Differential Equations. Fizmatgiz, Moscow (1959) (Russian)
3. Grisvard, P.: Elliptic Problems in Nonsmooth Domains. Pitman Advanced Publishing Program, London (1986)
4. Agmon, S.: Lectures on Elliptic Boundary Value Problems. D Van Nostrand, Princeton (1965)
5. Krein, S.G.: Linear Differential Equations in a Banach Space. Nauka, Moscow (1966) (Russian)
6. Skubachevskii, A.L.: Elliptic Functional Differential Equations and Applications. Birkhauser Verlag, Boston (1997)
7. Gorbachuk, V.L., Gorbachuk, M.L.: Boundary Value Problems for Differential-Operator Equations. Naukova Dumka, Kiev (1984) (Russian)
8. Sobolevskii, P.E.: On elliptic equations in a Banach space. Differential'nye Uravneninya **4**(7), 1346–1348 (1969) (Russian)
9. Ashyralyev, A.: Method of positive operators of investigations of the high order of accuracy difference schemes for parabolic and elliptic equations. Doctor sciences thesis, Kiev (1992) (Russian)
10. Sobolevskii, P.E.: Well-posedness of difference elliptic equation. Discrete Dyn. Nat. Soc. **1**(3), 219–231 (1997)
11. Sobolevskii, P.E.: The theory of semigroups and the stability of difference schemes in operator theory in function spaces. Proc. School, Novosibirsk (1975). (pp. 304–307, "Nauka" Sibirsk. Otdel, Novosibirsk (1977)) (Russian)
12. Ashyralyev, A.: Well-posedness of the difference schemes for elliptic equations in $C_\tau^{\beta,\gamma}(E)$ spaces. Appl. Math. Lett. **22**, 390–395 (2009)
13. Ashyralyev, A.: A note on the Bitsadze-Samarskii type nonlocal boundary value problem in a Banach space. J. Math. Anal. Appl. **344**(1), 557–573 (2008)
14. Ashyralyev, A., Sobolevskii, P.E.: Well-posedness of the difference schemes of the high order of accuracy for elliptic equations. Discrete Dyn. Nat. Soc. **2006**, 1–12 (2006)
15. Agarwal, R., Bohner, M., Shakhmurov, V.B.: Maximal regular boundary value problems in Banach-valued weighted spaces. Bound. Value Prob. **1**, 9–42 (2005)
16. Ashyralyev, A., Cuevas, C., Piskarev, S.: On well-posedness of difference schemes for abstract elliptic problems in $L_p([0,1],E)$ spaces. Numer. Funct. Anal. Optim. **29**(1-2), 43–65 (2008)

17. Ashyralyev, A.: On well-posedness of the nonlocal boundary value problem for elliptic equations. Numer. Funct. Anal. Optim. **24**(1–2), 1–15 (2003)
18. Ashyralyev, A., Sobolevskii, P.E.: Well-Posedness of Parabolic Difference Equations, vol. 69. Birkhäuser Verlag, Basel-Boston-Berlin (1994)
19. Sobolevskii, P.E.: Imbedding theorems for elliptic and parabolic operators in C. Sov. Math., Dokl. **38**(2), 262–265 (1989). Translation from Dokl. Akad. Nauk SSSR **302**(1), 34–37 (1988)
20. Ashyralyev, A., Sobolevskii, P.E.: New Difference Schemes for Partial Differential Equations, vol. 148. Birkhäuser Verlag, Basel (2004)
21. Ashyralyev, A.: On the uniform difference schemes of a higher order of the approximation for elliptic equations with a small parameter. Appl. Anal. **36**(3–4), 211–220 (1990)
22. Ashyralyev, A., Fattorini, H.O.: On uniform difference schemes for second order singular perturbation problems in Banach spaces. SIAM J. Math. Anal. **23**(1), 29–54 (1992)

Operator Semigroups for Convergence Analysis

Petra Csomós[1], István Faragó[1,2]($\boxtimes$), and Imre Fekete[2,3]

[1] MTA-ELTE Numerical Analysis and Large Networks Research Group,
Hungarian Academy of Sciences, Budapest, Hungary
{csomos,faragois}@cs.elte.hu
[2] Institute of Mathematics, Eötvös Loránd University, Budapest, Hungary
[3] Department of Mathematics and Computer Science, Széchenyi István University,
Győr, Hungary
imrefekete1989@gmail.com

Abstract. The paper serves as a review on the basic results showing how functional analytic tools have been applied in numerical analysis. It deals with abstract Cauchy problems and present how their solutions are approximated by using space and time discretisations. To this end we introduce and apply the basic notions of operator semigroup theory. The convergence is analysed through the famous theorems of Trotter and Kato, Lax, and Chernoff. We also list some of their most important applications.

Keywords: Numerical analysis · Operator semigroups · Convergence analysis · Trotter–kato approximation theorem · Lax equivalence theorem · Chernoff's theorem

1 Introduction

In the present paper we will give an overview on how functional analytic tools have been applied in numerical analysis. In particular, we will consider well-posed partial differential equations and analyse how to ensure the convergence of their numerical solution to the exact one. To this end, we will treat the problem in a functional analytic framework and apply results from operator semigroup theory, for which our main reference is the monograph by Engel-Nagel [5].

We start with an example to motivate what kind of problems are to solved when seeking a numerical solution. In Sect. 3 the corresponding abstract problem and its solution, the operator semigroup, will be introduced. The convergence of the space and time discretisation methods are analysed in Sect. 4.1 and 4.2, respectively, based on the results of Trotter [15], Kato [9], Ito and Kappel [8], Lax and Richtmeyr [11], and Chernoff [4]. In Sect. 5 we show how the previous results can be combined and present the convergence result of Bátkai et al. [1] based on the work of Pazy [14]. Section 6 deals as an outlook on other topics in numerical analysis where operator semigroups play an important role.

© Springer International Publishing Switzerland 2015
I. Dimov et al. (Eds.): FDM 2014, LNCS 9045, pp. 38–49, 2015.
DOI: 10.1007/978-3-319-20239-6_4

2   Motivation

As a motivating example we consider the one-dimensional heat equation on the interval $[0, \pi]$ with homogeneous Dirichlet boundary condition

$$\begin{cases} \frac{\partial}{\partial t} w(t, x) = \frac{\partial^2}{\partial x^2} w(t, x), & t > 0, \ x \in (0, \pi), \\ w(0, x) = w_0(x), & x \in (0, \pi), \\ w(t, 0) = w(t, \pi) = 0 \end{cases} \tag{1}$$

with the given initial function $w_0 \in L^2(0, \pi)$. Its solution is obtained by separating the variables and has the form

$$w(t, x) = \sum_{j=1}^{\infty} c_j e^{-j^2 t} \sin(jx) \ \text{ with } \ c_j = \frac{2}{\pi} \int_0^\pi w_0(x) \sin(jx) \mathrm{d}x, \ j \in \mathbb{N}. \tag{2}$$

2.1   Numerical Solution

We show now two ways how to obtain an approximation to w, that is, the numerical solution to problem (1).

Example 1 (Finite differences). We approximate the partial derivatives in problem (1) by the usual finite difference schemes on equidistant spatial and temporal meshes with grid size $h = \frac{\pi}{m-1} > 0$, for some fixed $m \in \mathbb{N} \setminus \{1\}$, and time step $\tau > 0$:

$$\frac{\partial}{\partial t} w(t, x) \approx \frac{1}{\tau} \big(w(t + \tau, x) - w(t, x) \big),$$
$$\frac{\partial^2}{\partial x^2} w(t, x) \approx \frac{1}{h^2} \big(w(t, x - h) - 2w(t, x) + w(t, x + h) \big).$$

This leads to the following discrete problem for $w_j^{(\ell)} \approx w(\ell\tau, (j-1)h)$ for $j = 1, ..., m$ and $\ell \in \mathbb{N}$:

$$\frac{1}{\tau} \big(w_j^{(\ell)} - w_j^{(\ell-1)} \big) = \frac{1}{h^2} \big(w_{j+1}^{(\ell-1)} - 2w_j^{(\ell-1)} + w_{j-1}^{(\ell-1)} \big). \tag{3}$$

Due to the initial and boundary conditions $w_j^{(0)} = w_0((j-1)h), \ j = 1, ..., m$, and $w_1^{(\ell)} = w_m^{(\ell)} = 0, \ \ell \in \mathbb{N}$, the solution

$$w_j^{(\ell)} = w_j^{(\ell-1)} + \frac{\tau}{h^2} \big(w_{j+1}^{(\ell-1)} - 2w_j^{(\ell-1)} + w_{j-1}^{(\ell-1)} \big) \tag{4}$$

can be computed step by step for all indices $\ell \in \mathbb{N}$ and $j = 2, ..., m-1$. Then the approximation of w is obtained by certain interpolation schemes in space and time.

Example 2 (Spectral method). Let $\widehat{w}_j \in \mathbb{R}$ denote the jth Fourier coefficient of $w(\cdot, x)$ and $\langle \cdot, \cdot \rangle$ the inner product in $L^2(0, \pi)$. We define further the function

$\varphi_j(x) = \sqrt{\frac{2}{\pi}} \sin(jx)$, $j = 1, ..., m$, satisfying the boundary condition in problem (1). By taking the discrete Fourier transform of both sides of problem (1), one obtains the following initial value problem for the first m Fourier coefficients of w:

$$\begin{cases} \frac{d}{dt}\widehat{w}_j(t) = -j^2\widehat{w}_j(t), & t \in \mathbb{R}, \ j = 1, ..., m, \\ \widehat{w}_j(0) = \langle w_0, \varphi_j \rangle, & j = 1, ..., m \end{cases} \tag{5}$$

with the solution $\widehat{w}_j(t) = e^{-j^2 t}\widehat{w}_j(0) = e^{-j^2 t}\langle w_0, \varphi_j \rangle$. Then the approximation of w is obtained with the help of the inverse discrete Fourier transform, that is,

$$w(t, x) \approx \sum_{j=1}^{m} \widehat{w}_j(t)\varphi_j(x) = \sum_{j=1}^{m} e^{-j^2 t}\langle w_0, \varphi_j \rangle \sqrt{\frac{2}{\pi}} \sin(jx). \tag{6}$$

We remark that the formula above really seems to approximate w, since it corresponds to $c_j = \sqrt{\frac{2}{\pi}}\langle w_0, \varphi_j \rangle$ for $j = 1, ..., m$ and $c_j = 0$ for $j > m$ in (2). This means that in this case the inifinite sum is approximated by a finite one.

2.2   Abstract Setting

Problem (1) can also be handled in an abstract way. To this end we define the Banach space $X = L^2(0, \pi)$, the linear operator $A : X \to X$ as $Af = f''$ for all $f \in \{\eta \in L^2(0, \pi) : \eta(0) = \eta(\pi) = 0\}$, and the function $u : [0, \infty) \to X$ as $(u(t))(x) = w(t, x)$ for all $t \geq 0$ and $x \in [0, \pi]$. Then problem (1) corresponds to the following initial value problem on X:

$$\begin{cases} \frac{d}{dt}u(t) = Au(t), & t > 0, \\ u(0) = u_0 \end{cases} \tag{7}$$

with $u_0 = w_0$. In order to solve problem (7) numerically, for $m \in \mathbb{N}$ one defines Banach spaces X_m, some suitable (for the sake of simplicity linear) operators $P_m : X \to X_m$, $J_m : X_m \to X$, and a linear operator $A_m : X_m \to X_m$. Then the numerical solution $u_m : [0, \infty) \to X_m$ is obtained from the following initial value problem in X_m for all $m \in \mathbb{N}$:

$$\begin{cases} \frac{d}{dt}u_m(t) = A_m u_m(t), & t > 0, \\ u_m(0) = P_m u_0 \end{cases} \tag{8}$$

with $u_0 = w_0$. Problem (8) corresponds to the spatially discretised version of problem (7). The solution of the original problem (1) is obtained as $w(t, x) = (u(t))(x)$ where it is to analysed whether $u(t) = \lim_{m \to \infty} J_m u_m(t)$ holds uniformly for t in compact intervals. In some cases $u_m(t)$ is further approximated by $u_{m,k}$ by using certain time discretisation methods (see the examples below). Then

$$u(t) = \lim_{m \to \infty} J_m \lim_{k \to \infty} u_{m,k} \tag{9}$$

should hold uniformly for t in compact intervals. In Sect. 5 we will study under which conditions the limit (9) holds. The corresponding choices of the spaces and operators in Examples 1 and 2 are the following.

(a) *Example* 1. We choose $X_m = \mathbb{R}^m$, $(P_m w_0)_j = w_0((j-1)h)$ for $j = 1, ..., m$, and

$$A_m = \frac{1}{h^2} \begin{pmatrix} 1 & 0 & ... & 0 & 0 \\ 0 & D_{m-2} & & & \vdots \\ \vdots & & & & 0 \\ 0 & 0 & ... & 0 & 1 \end{pmatrix} \in \mathbb{R}^{m \times m}$$

with $D_{m-2} = \operatorname{tridiag}(1, -2, 1) \in \mathbb{R}^{(m-2) \times (m-2)}$. Then for all $t \geq 0$ and $j = 1, ..., m$, $(u_m(t))_j \in \mathbb{R}$ correspond to the approximate values at the grid points $(j-1)h$, and $u_m(t) \in \mathbb{R}^m$ is their vector. The solution to problem (8) in this case reads as $u_m(t) = e^{tA_m} P_m w_0$. Since

$$e^{tA_m} = \lim_{k \to \infty} \left(I_m + \tfrac{t}{k} A_m \right)^k,$$

where $I_m \in \mathbb{R}^{m \times m}$ denotes the identity matrix, one approximates the exponential matrix and obtaines the numerical solution

$$u_{m,k} = \left(I_m + \tfrac{t}{k} A_m \right)^k P_m w_0 \quad \text{for some } k \in \mathbb{N}.$$

If $k \in \mathbb{N}$ and $\tau = \tfrac{t}{k} > 0$ are fixed, we have

$$u_{m,k}^{(\ell)} = (I_m + \tau A_m)^\ell P_m w_0 = (I_m + \tau A_m) u_{m,k}^{(\ell-1)} \quad \text{for all } \ell \in \mathbb{N},$$

and this corresponds to formula (4), that is, $\left(u_{m,k}^{(\ell)} \right)_j = w_j^{(\ell)}$. The operator J_m describes an interpolation, such as the Lagrangian polynom, etc.

(b) *Example* 2. We choose $X_m = \mathbb{R}^m$, $(P_m w_0)_j = \langle w_0, \varphi_j \rangle$ for $j = 1, ..., m$, and $A_m \in \mathbb{R}^{m \times m}$ with diagonal elements $(A_m)_{jj} = -j^2$, $j = 1, ..., m$, and zero otherwise. Then for all $t \geq 0$, $(u_m(t))_j = c_j$ is the jth Fourier coefficient of w, and $u_m(t) \in \mathbb{R}^m$ is their vector. The solution to problem (8) reads then as

$$u_m(t) = e^{tA_m} P_m w_0 = \sum_{j=1}^m e^{-j^2 t} \langle w_0, \varphi_j \rangle.$$

The operator J_m corresponds now to the inverse discrete Fourier transform, that is,

$$\left(J_m u_m(t) \right)(x) = \sum_{j=1}^m \left(u_m(t) \right)_j \varphi_j(x) \quad \text{for all } t \geq 0, \ x \in [0, \pi],$$

which really gives back formula (6).

In the examples above, problem (8) could be easily solved because the spaces X_m were finite dimensional in both cases. However, problems like (7) and (8) can be treated even if X and X_m are infinite dimensional. Then the corresponding solutions are studied in an abstract way presented in the next section.

3 The Continuous Problem

This section is devoted to introduce the basic notions of operator semigroup theory needed later on. In order to study the convergence of space and time discretisations, the given partial differential equation should be formulated as an abstract Cauchy problem of the form (7) on an appropriate Banach space X with the linear operator $A : D(A) \to X$, where the connection to the unknown function w of a partial differential equation is given by $(u(t))(x) = w(t, x)$ for all $t \geq 0$ and x from the corresponding interval/domain (e.g. for all $x \in [0, \pi]$ for problem (1)). If A were a matrix or any bounded operator on X ($A \in \mathscr{L}(X, X)$ in notation), the solution to problem (7) would be simply the exponential e^{tA} applied to the initial value u_0. Since A is unbounded in general, its exponential cannot be defined as the infinite power series. One suspects, however, that the solutions properties should somehow reflect the properties of the exponential function.

Definition 1 (Definition I.5.1 in [5]). *Let $S : [0, \infty) \to \mathscr{L}(X, X)$ be a mapping with the following properties.*

(i) *The identity $S(t + s) = S(t)S(s)$ holds for all $t, s \geq 0$, and one has $S(0) = I$, the identity operator on X (semigroup property).*
(ii) *The mapping $t \to S(t)f \in X$ is continuous for all $f \in X$ (strong continuity).*

Then S is called a strongly continuous one-parameter semigroup of bounded linear operators on the Banach space X.

We note that there always exist constants $M \geq 1$ and $\omega \in \mathbb{R}$ such that the estimate $\|S(t)\| \leq Me^{\omega t}$ holds for all $t \geq 0$ (cf. Proposition I.5.5 in [5]). Consider the map $u(t) = S(t)f$ for $f \in X$ and note that if u is differentiable, then one has $\frac{d}{dt}u(t) = S(t)(\frac{d}{dt}u(t))|_{t=0}$ (cf. Lemma II.1.1 in [5]). Hence, the derivative of the map u at $t = 0$ determines the derivative at each point $t \geq 0$. This suggests us to give this object a name.

Definition 2 (Definition II.1.2 and Lemma II.1.1 in [5]). *The generator $A : D(A) \to X$ of a strongly continuous semigroup S on the Banach space X is the operator*

$$Af := \lim_{\tau \searrow 0} \tfrac{1}{\tau}\big(S(\tau)f - f\big)$$

defined for every f in its domain

$$D(A) := \Big\{f \in X : \lim_{\tau \searrow 0} \tfrac{1}{\tau}\big(S(\tau)f - f\big) \text{ exists}\Big\}.$$

The next result shows that the semigroup indeed yields the solution to the corresponding abstract Cauchy problem.

Theorem 1 (Theorem II.1.4 and Proposition II.6.2 in [5]). *The generator $A : D(A) \to X$ of a strongly continuous semigroup S has the following properties.*

(a) Operator A is linear, closed, and densely defined, and it determines the semi-group uniquely.

(b) For every $u_0 \in D(A)$, the solution to the abstract Cauchy problem (7) has the form $u(t) = S(t)u_0$.

This means that the solution of a partial differential equation, reformulated as an abstract Cauchy problem (7), is determined through the semigroup S generated by the operator A appearing in (7).

Example 3. Let $X = L^2(0, \pi)$ and $(Af)(x) = f''(x)$ for all

$$f \in D(A) = \{f \in L^2(0, \pi) : f(0) = f(\pi) = 0\}$$

as for the heat equation (1). Furthermore, let $\varphi_j(x) = \sqrt{\frac{2}{\pi}} \sin(jx)$ for $j \in \mathbb{N}$. One can show that then A generates the semigroup S of the form

$$S(t)f = \sum_{j=1}^{\infty} e^{-j^2 t} \langle f, \varphi_j \rangle \varphi_j,$$

where $\langle \cdot, \cdot \rangle$ denotes the inner product in $L^2(0, \pi)$. We note that the spectral method introduced in Example 2 follows this idea to approximate the solution to the heat Eq. (1).

4   Space and Time Discretisations

In Sect. 3 we saw that well-posed partial differential equations can be formulated as abstract Cauchy problems, and their solution is given by a strongly continuous semigroup. In Examples 1 and 2 we introduced two usual ways how the numerical solution to partial differential equations are usually obtained, that is, by using certain spatial and temporal discretisation schemes. We saw then that spatial discretisations mean the approximation of the generator A in problem (7). Discretisation in time is the approximation of the resulting semigroup.

4.1   Generator Approximations as Space Discretisations

Let X_m, $m \in \mathbb{N}$, be Banach spaces, and define some kind of projection and embedding operators as follows, see e.g. in Sect. 4.1 in [8].

Property 1. Let X and $X_m, m \in \mathbb{N}$ be Banach spaces. Consider the bounded linear operators $P_m \in \mathcal{L}(X, X_m)$ and $J_m \in \mathcal{L}(X_m, X)$ for $m \in \mathbb{N}$ with the properties

(i) $P_m J_m = I_m$, the identity on X_m, and
(ii) $\lim_{m \to \infty} \|J_m P_m f - f\| = 0$ for all $f \in X$.

One can show that operators P_m, J_m in Examples 1 and 2 possess Property 1.

The famous result of Trotter [15] and Kato [9] states that, under suitable conditions, if the generator A is approximated by a sequence of another generators A_m, then the corresponding semigroups S_m will approximate the semigroup S generated by A, as well.

Theorem 2 (First Trotter–Kato Approximation Theorem, Theorem 4.2 and Proposition 4.3 in [8], cf. Theorem III.4.8 in [5]). *For all $m \in \mathbb{N}$ let X and X_m be Banach spaces and let the operators P_m, J_m possess Property 1. Suppose that for all $m \in \mathbb{N}$, A and A_m generate the semigroups S and S_m in X and X_m, respectively. Suppose further that there exists constants $M \geq 1$ and $\omega \in \mathbb{R}$ such that $\|S(t)\|, \|S_m(t)\| \leq Me^{\omega t}$ holds for all $m \in \mathbb{N}$, $t \geq 0$. Then the following assertions are equivalent.*

(i) *There is a dense subspace $Y \subset D(A)$ such that there is $\lambda > 0$ with $(\lambda - A)Y$ being dense in X. Furthermore, for all $f \in Y$ there is a sequence with elements $f_m \in D(A_m)$ which satisfies $\lim\limits_{m \to \infty} \|f_m - P_m f\|_{X_m} = 0$ and*

$$\lim_{m \to \infty} \|A_m f_m - P_m A f\|_{X_m} = 0 .$$

(ii) *It holds that $\lim\limits_{m \to \infty} \|J_m S_m(t) P_m f - S(t)f\| = 0$ for all $f \in X$ uniformly for t in compact intervals.*

Since in both Examples 1 and 2 the sequence A_m converge to A in the sense of Theorem 2(a) and all the other conditions are satisfied as well, Theorem 2(b) implies that u_m converge to u.

4.2   Semigroup Approximations as Time Discretisations

We consider the abstract Cauchy problem (7) where A generates the strongly continuous semigroup S. Since multistep time discretisation schemes can also be treated as one-step methods (see [12]), we only consider one-step time discretisation methods. After some definitions, we will state the convergence results.

Property 2. Let Z be a Banach space, and let the map $V : [0, \infty) \to \mathscr{L}(Z, Z)$ possess the following properties.

(i) The map V is strongly continuous, that is, the function $[0, \infty) \ni \tau \mapsto V(\tau)f \in Z$ is continuous for all $f \in Z$.
(ii) $V(0) = I$, the identity on Z.

Definition 3. *Let S be the semigroup with generator A and consider the abstract Cauchy problem (7) on the Banach space X. Consider further a map $F : [0, \infty) \to \mathscr{L}(X, X)$ with Property 2, which is called then time discretisation.*

(a) *The time discretisation F is called consistent with S if*

$$\lim_{\tau \to 0} \tfrac{1}{\tau}\big(F(\tau)S(t)f - S(t+\tau)f\big) = 0$$

holds for all $f \in X$ and uniformly for t in compact intervals.
(b) *A time discretisation F is called stable, if there are constants $T > 0$ and $M \geq 1$ such that $\|F(\tau)^k\| \leq M$ holds for all $\tau \geq 0$ and $k \in \mathbb{N}$ with $k\tau \leq T$.*

(c) *A time discretisation F is called convergent, if for all $t \geq 0$, $\tau_n \to 0$, $k_n \to \infty$ with $k_n\tau_n \to t$ we have*

$$\lim_{n \to \infty} \|S(t)f - F(\tau_n)^{k_n}f\| = 0 \text{ for all } f \in X.$$

Theorem 1(b) states that the semigroup S corresponds to the solution operator of the abstract Cauchy problem (7). To get a reliable approximation to S (i.e., a numerical solution), one has to ensure the convergence of the time discretisation scheme F. The next celebrated result is the basic of the numerical convergence analysis.

Theorem 3 (Lax Equivalence Theorem, [11]). *A consistent time discretisation is convergent if and only if it is stable.*

One can also say something about the order of the convergence, however, maybe only on a smaller set of initial values.

Definition 4. *Let S be the semigroup with generator A and consider the abstract Cauchy problem (7) on the Banach space X. Consider further a map $F : [0, \infty) \to \mathscr{L}(X, X)$ with Property 2. Suppose that there is a densely and continuously embedded subspace $Y \subset X$, which is invariant under the semigroup, and let $p > 0$.*

(a) *The time discretisation F is called consistent with S of order p on Y if there is a constant $C > 0$ such that for all $f \in Y$ we have*

$$\|F(\tau)f - S(\tau)f\| \leq C\tau^{p+1}\|f\|_Y.$$

(b) *The time discretisation F is called convergent of order p on Y if for all $t \geq 0$ there is a constant $\widetilde{C} > 0$ such that for all $f \in Y$ we have*

$$\|F(\tau)^k f - S(k\tau)f\| \leq \widetilde{C}t\tau^p\|f\|_Y \tag{10}$$

for all $k \in \mathbb{N}$, $\tau \geq 0$ with $k\tau \leq t$.

We note that p may depend on the subspace Y. Essentially by the same way as proving Theorem 3, the next result can be shown.

Proposition 1. *Suppose that there is a densely and continuously embedded subspace $Y \subset X$ which is invariant under the semigroup operators $S(t)$ satisfying $\|S(t)\|_Y \leq Me^{\omega t}$ for some $M \geq 1$ and $\omega \in \mathbb{R}$ and for all $t \geq 0$. If there is $p > 0$ such that F is a stable time discretisation scheme which is consistent of order p on Y, then it is convergent of order p on Y.*

Example 4. Let $(A, D(A))$ generate the semigroup S with $\|S(t)\| \leq Me^{\omega t}$ for some $M \geq 1$ and $\omega \in \mathbb{R}$ and for all $t \geq 0$. For all $\tau \in (0, \frac{1}{\omega}]$, we define the implicit Euler time discretisation as $F(\tau) = (I - \tau A)^{-1}$ being consistent. Moreover, if $\omega = 0$ and $Y = D(A^2)$, one has $p = 1$ in (10).

Since the generator property of the operator A is equivalent to the well-posedness of the problem (7) (see Theorem II.6.7 in [5]), Theorem 3 and Proposition 1 concern only well-posed problems. There exist results, however, which prove the generator property of an operator through approximations. They are extremely important in numerical analysis. We present now one of the most famous ones by Chernoff [4].

Theorem 4 (Chernoff Product Formula, Cor. III.5.3 in [5]). *Let X be a Banach space and consider a map $F : [0, \infty) \to \mathscr{L}(X, X)$ with the following properties.*

(a) The map F has Property 2.
(b) There exist constants $M \geq 1$ and $\omega \in \mathbb{R}$ such that $\|F(t)^k\| \leq M e^{\omega k t}$ for all $t \geq 0$ and $k \in \mathbb{N}$.
(c) There is a subset $Y \subset X$ such that $(\lambda - A)Y$ is dense for some $\lambda > \omega$ and the limit

$$Af := \lim_{\tau \searrow 0} \tfrac{1}{\tau} \big(F(\tau)f - f\big) \tag{11}$$

exists for all $f \in Y$.

Then the closure $\overline{A}$ of A generates a strongly continuous semigroup S which is given by

$$S(t)f = \lim_{k \to \infty} F(\tfrac{t}{k})^k f \tag{12}$$

for all $f \in X$ and uniformly for t in compact intervals.

5   The Discrete Problem

In Sect. 3 we saw that the solution to the abstract Cauchy problem (7) is given by the semigroup generated by the operator A appearing in (7). Thus, if one aims to approximate the solution to problem (7), one has to approximate the corresponding semigroup S by the product of appropriate operators F_m depending on $m \in \mathbb{N}$. As already seen in Examples 1 and 2, one chooses a space discretisation scheme which corresponds to the approximation of the generator A by a sequence of generators A_m (cf. Section 4.1), then a time discretisation when the semigroup S_m is approximated by the product of F_m (cf. Section 4.2). The solution of a well-posed problem (7) is given by $u(t) = S(t)u_0$ for all $t \geq 0$. Application of a space discretisation means that $u(t)$ is approximated by $u_m(t) = S_m(t)P_m u_0$, $m \in \mathbb{N}$ (cf. Theorem 2). This is further approximated by using a time discretisation, that is, by $u_{m,k} = F_m(\tfrac{t}{k})^k P_m u_0$, $m, k \in \mathbb{N}$ (cf. Theorem 3 for the semigroup S_m on X_m).

Definition 5. *Let X be a Banach space and $A : D(A) \to X$ be the generator of the strongly continuous semigroup S on X. Furthermore, let $F_m : [0, \infty) \to \mathscr{L}(X_m, X_m)$ has Property 2 for all $m \in \mathbb{N}$. Then $u_{m,k} = J_m F_m(\tfrac{t}{k})^k P_m u_0 \in X_m$ is called the numerical solution at time $t \geq 0$ to the corresponding abstract*

Cauchy problem (7) *with initial value* u_0. *The numerical method is called convergent at time level* $t \geq 0$ *if for all* $u_0 \in X$ *one has*

$$\lim_{m,k\to\infty} \|u_{m,k} - u(t)\| = 0, \ that \ is, \ \lim_{m,k\to\infty} \left\|J_m F_m(\tfrac{t}{k})^k P_m u_0 - S(t)u_0\right\| = 0 \quad (13)$$

uniformly for t *in compact intervals.*

we note that the notaion $\lim\limits_{m,k\to\infty}$ stands for the usual limit for the double indexed sequences.

Remark 1. The following conditions are sufficient for the convergence (13).

(i) There exists $\bar{u}_m(t) \in X$ such that $\lim\limits_{k\to\infty} \|u_{m,k} - \bar{u}_m(t)\| = 0$ uniformly for $m \in \mathbb{N}$.

(ii) It holds that $\lim\limits_{m\to\infty} \|\bar{u}_m(t) - u(t)\|$ uniformly for t in compact intervals.

These conditions refer to the convergence of discretisations in time and space, respectively, studied in Sects. 4.2 and 4.1.

When considering well-posed problems (7), the Lax Equivalence Theorem 3 and the First Trotter–Kato Approximation Theorem 2 already imply the convergence.

Proposition 2. *Suppose that* A, A_m *generates the semigroups* S, S_m *on the Banach spaces* X, X_m, *respectively, for all* $m \in \mathbb{N}$, *and that the operators* P_m, J_m, $m \in \mathbb{N}$, *possess Property 1 such that* $P_m X \subset D(A_m)$. *Suppose further that*

$$\lim_{m\to\infty} \|A_m P_m f - P_m A f\|_{X_m} = 0 \quad (14)$$

holds for all $f \in Y$, *where* $Y \subset D(A)$ *and* $(\lambda - A)Y$ *are dense in* X *for some* $\lambda > 0$. *Moreover, let* $F_m : [0, \infty) \to X_m$ *be a stable time discretisation which is consistent with* S_m *for all* $m \in \mathbb{N}$. *Then* F_m *is convergent, more precisely, for all* $f \in X$ *one has*

$$\lim_{m,k\to\infty} \left\|J_m F_m(\tfrac{t}{k})^k P_m f - S(t)f\right\| = 0$$

uniformly for t *in compact intervals.*

Proof. Due to Remark 1, it suffices to study the limits separately. The Lax Equivalence Theorem 3 imply that F_m is convergent in X_m, that is,

$$\lim_{k\to\infty} \left\|F_m(\tfrac{t}{k})^k f_m - S_m(t)f_m\right\|_{X_m} = 0$$

holds for all $f_m \in X_m$. Since operators $J_m : X_m \to X$ are bounded and with the choice $f_m = P_m f$ for $f \in X$, we have that

$$\lim_{k\to\infty} \left\|J_m F_m(\tfrac{t}{k})^k P_m f - J_m S_m(t)P_m f\right\| = 0 \quad (15)$$

uniformly for t in compact intervals. From (14), the First Trotter–Kato Approximation Theorem 2 implies that

$$\lim_{m \to \infty} \| J_m S_m(t) P_m f - S(t) f \| = 0 \tag{16}$$

holds for all $f \in X$ uniformly for t in compact intervals. Hence, limits (15) and (16), and Remark 1 with $\bar{u}_m(t) := J_m S_m(t) P_m f$ yield the convergence.   □

The results above all concern well-posed problems. In case when the operator A is not known to be a generator of a semigroup, a modified version of Chernoff Product Formula 4 can be applied. The original theorem, presented in Pazy [14], states the result in the space X_m, however, we formulate it here as a result in the space X.

Theorem 5 (Modified Chernoff Product Formula, [1]). *Let X_m, $m \in \mathbb{N}$ be Banach spaces and consider a sequence of maps $F_m : [0, \infty) \to \mathscr{L}(X_m, X_m)$ with the following properties.*

(a) *The maps F_m have Property 2 for all $m \in \mathbb{N}$.*
(b) *There exist constants $M \geq 1$ and $\omega \in \mathbb{R}$ such that $\| F_m(t)^k \| \leq M e^{\omega k t}$ for all $t \geq 0$ and $m, k \in \mathbb{N}$.*
(c) *There is a subset $Y \subset X$ such that $(\lambda - A)Y$ is dense for some $\lambda > \omega$ and the limit*

$$\lim_{m \to \infty} \tfrac{1}{\tau} \big(J_m F_m(\tau) P_m f - J_m P_m f \big)$$

exists uniformly for τ in compact intervals, and

$$A f := \lim_{\tau \searrow 0} \lim_{m \to \infty} \tfrac{1}{\tau} \big(J_m F_m(\tau) P_m f - J_m P_m f \big)$$

exists for all $f \in Y$.

Then the closure $\overline{A}$ of A generates a strongly continuous semigroup S which is given by

$$S(t) f = \lim_{m, k \to \infty} J_m F_m(\tfrac{t}{k})^k P_m f \tag{17}$$

for all $f \in X$ and uniformly for t in compact intervals.

6   Outlook/Applications

With the help of similar techniques presented in Sect. 5, several numerical treatments can be proved to be convergent. We just mention here a few examples which are of great importance in practice. The convergence of the standard time discretisation methods, such as Runge–Kutta methods, were analysed by using Lax Equivalence Theorem 3. Even more general rational approximations are studied in Brenner and Thomée [3]. The convergence of various operator splitting methods were proved e.g. in Trotter [16], Kato [10], Faragó and Havasi [6], and Bátkai et al. [1] and [2] by using Chernoff Product Formula, Theorem 4 and its modified version, Theorem 5. Exponential integrators were also studied by using operator semigroup approach in Hochbruck and Ostermann [7]. Nonlinear problems are treated in Palencia and Sanz-Serna [13] contaning the Lax Equivalence Theorem 3 as a special case of well-posed linear initial value problems.

Acknowledgments. P. Csomós and I. Faragó kindly acknowledge the support of the bilateral Hungarian-Austrian Science and Technology program TET_10-1-2011-0728. I. Fekete was supported by the European Union and the State of Hungary, co-financed by the European Social Fund witihin the framework of TÁMOP-4.2.4.A/2-11/1-2012-0001 'National Program of Excellence'–convergence program.

References

1. Bátkai, A., Csomós, P., Nickel, G.: Operator splittings and spatial approximations for evolution equations. J. Evol. Equ. **9**, 613–636 (2009)
2. Bátkai, A., Csomós, P., Farkas, B.: Operator splitting with spatial-temporal discretization. In: Arendt, W., Ball, J.A., Behrndt, J., Förster, K.-H., Mehrmann, V., Trunk, C. (eds.) Spectral Theory, Mathematical System Theory, Evolution Equations, Differential and Difference Equations, 161–172. Springer, Basel (2012)
3. Brenner, P., Thomée, V.: On rational approximations of semigroups. SIAM J. Numer. Anal. **16**, 683–694 (1979)
4. Chernoff, P.R.: Note on product formulas for operator semigroups. J. Funct. Anal. **2**, 238–242 (1968)
5. Engel, K.-J.: One-Parameter Semigroups for Linear Evolution Equations. Graduate Texts in Mathematics. Springer-Verlag, New York (2000)
6. Faragó, I., Havasi, Á.: On the convergence and local splitting error of different splitting schemes. Prog. Comput. Fluid Dyn. **5**, 495–504 (2005)
7. Hochbruck, M., Ostermann, A.: Exponential integrators. Acta Numerica **19**, 209–286 (2010)
8. Ito, K., Kappel, F.: Evolution Equations and Approximations. World Scientific, Singapore (2002)
9. Kato, T.: On the trotter-lie product formula. Proc. Japan Acad. **50**, 694–698 (1974)
10. Kato, T.: Perturbation Theory for Linear Operators. Springer, New York (1976)
11. Lax, P.D., Richtmyer, R.D.: Survey of the stability of linear finite difference equations. Commun. Pure Appl. Math. **9**, 267–293 (1956)
12. Palencia, C.: Stability of rational multistep approximations of holomorphic semigroups. Math. Comput. **64**, 591–599 (1995)
13. Palencia, C., Sanz-Serna, J.M.: An extension of the lax-richtmyer theory. Numer. Math. **44**, 279–283 (1984)
14. Pazy, A.: Semigroups of Linear Operator and Applications to Partial Differential Equations. Springer, New York (1983)
15. Trotter, H.F.: Approximation of semi-groups of operators. Pac. J. Math. **8**, 887–919 (1958)
16. Trotter, H.F.: On the product of semi-groups of operators. Proc. Am. Math. Soc. **10**, 545–551 (1959)

The Power of Trefftz Approximations: Finite Difference, Boundary Difference and Discontinuous Galerkin Methods; Nonreflecting Conditions and Non-Asymptotic Homogenization

Fritz Kretzschmar[1,2], Sascha M. Schnepp[3], Herbert Egger[6], Farzad Ahmadi[4], Nabil Nowak[4], Vadim A. Markel[5], and Igor Tsukerman[4(✉)]

[1] Graduate School of Computational Engineering, Technische Universität Darmstadt, Dolivostrasse 15, Darmstadt, Germany
kretzschmar@gsc.tu-darmstadt.de

[2] Institut für Theorie Elektromagnetischer Felder, Technische Universitaet Darmstadt, Schlossgartenstrasse 8, Darmstadt, Germany

[3] Institut Für Geophysik, ETH Zürich, Sonneggstrasse 5, 8092 Zürich, Switzerland
schnepps@ethz.ch

[4] Department of Electrical and Computer Engineering, The University of Akron, Akron, OH 44325-3904, USA
igor@uakron.edu

[5] Graduate Group in Applied Mathematics and Computational Science, and Department of Radiology, Department of Bioengineering, University of Pennsylvania, Philadelphia, PA 19104, USA
vmarkel@mail.med.upenn.edu

[6] Department of Mathematics, TU Darmstadt, Dolivostrasse 15, Darmstadt, Germany
egger@mathematik.tu-darmstadt.de

Abstract. In problems of mathematical physics, Trefftz approximations by definition involve functions that satisfy the differential equation of the problem. The power and versatility of such approximations is illustrated with an overview of a number of application areas: (i) finite difference Trefftz schemes of arbitrarily high order; (ii) boundary difference Trefftz methods analogous to boundary integral equations but completely singularity-free; (iii) Discontinuous Galerkin (DG) Trefftz methods for Maxwell's electrodynamics; (iv) numerical and analytical nonreflecting Trefftz boundary conditions; (v) non-asymptotic homogenization of electromagnetic and photonic metamaterials.

Keywords: Trefftz functions · Finite difference schemes · Boundary difference schemes · Maxwell equations · Wave propagation · Effective medium theory · Discontinuous galerkin methods · Nonreflecting boundary conditions · Metamaterials

© Springer International Publishing Switzerland 2015
I. Dimov et al. (Eds.): FDM 2014, LNCS 9045, pp. 50–61, 2015.
DOI: 10.1007/978-3-319-20239-6_5

1   Introduction

In practical problems of mathematical physics and engineering, solution of partial differential equations relies on approximations of different types. This paper calls attention to local Trefftz approximations in subdomains covering the computational domain. Trefftz functions, by definition, satisfy the differential equations of the problem. Examples include harmonic polynomials for the Laplace equation; plane waves, cylindrical or spherical harmonics for wave problems, and so on. Trefftz approximations are well established in the context of pseudo-spectral and domain decomposition methods, but this overview is devoted to applications that are known less well or are entirely new:

1. Finite difference Trefftz schemes of arbitrarily high order, obtained by replacing classical Taylor expansions with local Trefftz approximations.
2. Boundary difference Trefftz methods analogous to boundary integral equations but completely singularity-free.
3. Discontinuous Galerkin – Trefftz methods for Maxwell's electrodynamics.
4. Numerical and analytical asymptotic boundary conditions based on Trefftz approximations.
5. Homogenization of electromagnetic and photonic metamaterials: a two-scale theory involving Trefftz approximations on both coarse and fine levels. This explains, in particular, artificial magnetism at high frequencies.

This discussion of the versatility and power of Trefftz methods is intended to stimulate their application in other areas of applied science and engineering.

2   Finite Difference Trefftz Schemes

Classical finite difference (FD) schemes are based upon, and derived from, Taylor expansions. These expansions are general but have limited approximation accuracy in cases where the solution is not sufficiently smooth – e.g. at material interfaces. This is the root cause of the notorious "staircase" effect at off-grid interface boundaries.

High-order schemes can be generated by replacing the Taylor expansions with Trefftz approximations which typically have much higher accuracy [1–6]. By definition, Trefftz functions satisfy the underlying differential equation of the problem and conditions at material interfaces. For example, waves in a homogeneous region can be represented as a superposition of plane waves; fields around spherical/cylindrical particles can be approximated by spherical/cylindrical harmonics; and so on.

One trivial example that helps to fix ideas is the Laplace equation in 1D. An obvious Trefftz basis in this case is $\psi_1(x) = 1$, $\psi_2(x) = x$, and the solution can be approximated as $u(x) = c_1\psi_1(x) + c_2\psi_2(x) = c_1 + c_2 x$. (This representation happens to be exact, but only because the problem is one-dimensional.) Consider a three-point stencil with nodes at, say, $x_1 = -h$, $x_2 = 0$, $x_3 = h$, where h is the grid size. Since there are three nodal values of u and only two free parameters

$c_{1,2}$, it is clear that these nodal values *must* be linearly dependent. This linear dependence yields the difference scheme we seek. It is easy to show that the coefficient vector $\underline{s} \in \mathbb{R}^3$ for the three-point stencil can be found as [1–3]

$$\underline{s} \in \text{Null } N_{1D,\text{Laplace}}^T, \quad N_{1D,\text{Laplace}}^T = \begin{pmatrix} \psi_1(x_1) & \psi_1(x_2) & \psi_1(x_3) \\ \psi_2(x_1) & \psi_2(x_2) & \psi_2(x_3) \end{pmatrix} \quad (1)$$

Substituting the nodal values of the ψs, one has

$$\text{Null } N_{1D,\text{Laplace}}^T = \text{Null } \begin{pmatrix} 1 & 1 & 1 \\ -h & 0 & h \end{pmatrix} = (1, -2, 1)^T$$

Thus one arrives at the standard scheme $\underline{s} = (1, -2, 1)^T$ (times an arbitrary factor) for the Laplace equation. This result is trivial but is a particular case of a more general nullspace formula of the Flexible Local Approximation MEthod ("FLAME")

$$\underline{s} \in \text{Null } N^T, \quad N_{\alpha\beta}^T = \psi_\alpha(\mathbf{r}_\beta) \quad (2)$$

For illustration, we briefly describe two representative examples from [1–3].

The first example is a nine-point (3×3) order-six scheme for the 2D Helmholtz equation. The basis set consists of eight plane waves traveling at angles $\phi_0 + m\pi/4$ ($m = 0, 1, \ldots, 7$), where ϕ_0 is a given angle; practical choices are $\phi_0 = 0$ or $\phi_0 = \pi/8$. Evaluating these plane waves at the stencil nodes, one obtains an 8×9 matrix N^T whose null vector – the FLAME scheme – can easily be found with symbolic algebra. The result for $\phi_0 = 0$ is given in [3]. For $\phi_0 = \pi/8$, one arrives at a scheme derived by Babuška *et al.* in 1995 [7] from very different considerations. Here it is an automatic particular case of FLAME.

The second example involves wave propagation and Bloch modes in periodic dielectric structures (photonic crystals). The most common physical arrangement in 2D is a square lattice of dielectric cylinders. In this case, an effective choice of the Trefftz-FLAME basis is cylindrical harmonics (Bessel/Hankel functions) centered at any given cylinder and matched at its boundary. On a 3×3 stencil, one then obtains a FLAME scheme of order six – that is, exactly of the same order as the respective scheme in a homogeneous region. The "staircase effect" which plagues classical FD schemes is thereby eliminated, and FLAME is routinely able to produce the level of accuracy unattainable with any traditional methods. As an example, the Bloch wavenumber in cylindrical photonic crystals can easily be obtained with 6 – 8 digits of accuracy even on coarse grids whose grid size is comparable with the radius of the cylinder. In contrast, classical FD schemes must accurately resolve the cylindrical boundary on a fine grid.

A clear and intuitive distinction between classical Taylor-based FD schemes and FLAME can be expressed as follows. Classical schemes approximate *the underlying differential equation*, while FLAME approximates the *solution* of the equation. Clearly, there is substantial redundancy in trying to approximate the operator on a whole class of sufficiently smooth functions, as only the solution is of real interest. FLAME greatly reduces this redundancy.

3   Boundary Difference Trefftz Schemes

Boundary difference methods (BDM) are discrete analogs of integral equation methods. Loosely speaking, BDM are to finite difference schemes what integral formulations are to partial differential equations. BDM date back to the work of Saltzer in the 1950s [8] but have been explored much less extensively than integral formulations.

There are two different paths for converting equations (either continuous or discrete) from the whole domain to its boundary. The first path (known as "direct" in the mathematical literature) is Calderón projections which on the continuous level have been extensively studied (e.g. [9]) and on the discrete level have been developed by Ryaben'kii and his coworkers [10, 11].

An alternative path ("indirect"), arguably less fundamental but more appealing to engineers and practitioners, is to introduce auxiliary boundary sources [12–15]. Fields in homogeneous subdomains can be expressed as convolutions of these sources with the appropriate Green functions and then coupled via interface boundary conditions. More specifically, our implementation of BDM, with 2D wave scattering as an example [12, 15], consists in the following steps:

1. Precompute discrete Green's functions (corresponding to the chosen FD scheme) for all homogeneous subdomains – the scatterer(s) and air. The relevant techniques are described in [12–15].
2. Introduce a uniform grid and determine the discrete boundary – two discrete layers (or more, for bigger grid stencils) immediately adjacent to the physical boundary of the scatterer. (See [10, 11] for precise mathematical definitions.) The grid is conceptually infinite, but the computational algorithm operates on the boundary nodes only.
3. Introduce auxiliary discrete sources on the discrete boundary. In each subdomain (scatterer(s), air) express the field as a convolution of these sources with the respective Green function.
4. Use FLAME to generate high-order matching conditions for fields across interfaces [12]. This can be done even for non-smooth and/or perfectly conducting boundaries [15].
5. Solve the resulting system of equations, with discrete sources as unknowns.
6. Post-process the results; in particular, find fields on the grid via discrete convolutions.

Advantages and disadvanatges of BDM stem from their dual nature as discrete and boundary methods; see Table 1.

4   Discontinuous Galerkin Trefftz Methods

For problems with discontinuous coefficients, it is usually impossible to construct *global* Trefftz functions that satisfy the governing equations everywhere in the domain. For illustration, consider propagation of an electromagnetic wave in an inhomogeneous time-independent dielectric linear medium Ω:

$$\epsilon\, \partial_t \mathbf{E} - \operatorname{curl} \mathbf{H} = 0, \quad \mu\, \partial_t \mathbf{H} + \operatorname{curl} \mathbf{E} = 0$$

Table 1. Advantages and disadvantages of boundary difference methods.

	Advantages	Disadvantages
As a discrete method	Singularity-free. (Discrete Green's functions, in contrast with their continuous counterparts, are nonsingular.)	Subject to discretization errors, in particular dispersion errors in wave problems
As a boundary method	Dimensionality of the problem is reduced. No artificial boundaries with absorbing conditions needed.	Full matrices. However, fast multipole acceleration for matrix-vector products is available

Frequency dispersion of the permittivity is disregarded, which is a good approximation for dielectrics in their spectral transparency window. Ways of handling dispersive media are well established in the finite difference time domain and DG literature.

The fields satisfy initial and boundary conditions $\mathbf{E}(0) = \mathbf{E}_0$, $\mathbf{H}(0) = \mathbf{H}_0$, and $\mathbf{n} \times \mathbf{E} = 0$ on $\partial\Omega$, where $\mathbf{n}$ is the outward unit normal vector on the boundary. We assume that the domain can be partitioned into subdomains (cells, elements) K_α ($\alpha = 1, 2, \ldots, n$) such that the coefficients ϵ, μ are constants within each subdomain. Construction of *local* Trefftz bases is straightforward, and on the time interval $I = [0, T]$ only the interface and boundary conditions remain to be enforced: the continuity of $\mathbf{n} \times \mathbf{E}$ and $\mathbf{n} \times \mathbf{H}$ over the interfaces f between subdomains and the condition $\mathbf{n} \times \mathbf{E} = 0$ on boundary faces f'. This is standard in the framework of Discontinuous Galerkin methods [16–18]:

$$\sum_f \int_{f \times I} [\mathbf{n} \times \mathbf{H}] \cdot \{\mathbf{v}\} - [\mathbf{n} \times \mathbf{E}] \cdot \{\mathbf{w}\} - \sum_{f'} \int_{f \times I} \mathbf{n} \times \mathbf{E} \cdot \mathbf{w}$$
$$+ \sum_K \int_K \epsilon \mathbf{E}(0) \cdot \mathbf{v}(0) + \mu \mathbf{H}(0) \cdot \mathbf{w}(0) = \sum_K \int_K \epsilon \mathbf{E}_0 \cdot \mathbf{v}(0) + \mu \mathbf{H}_0 \cdot \mathbf{w}(0).$$

Expansion of all fields in a local Trefftz basis then yields a linear implicit time stepping scheme that describes the evolution of the fields from $t_0 = 0$ to $t_1 = T$. This method can also be derived by starting from the standard space-time discontinuous Galerkin approximation [19–21] and utilizing Trefftz functions for the local approximations [18]. Frequency domain applications can be treated similarly [22,23]. As local Trefftz approximations one can use a complete set of *polynomial plane waves* of the form

$$(\mathbf{E}_{ip}, \mathbf{H}_{ip}) = (\mathbf{e}_{ip}, \mathbf{h}_{ip})(\mathbf{l}_{ip} \cdot \mathbf{r} - t/\sqrt{\epsilon\mu})^p,$$

where $\mathbf{h}_{ip} = \sqrt{\mu/\epsilon}\, \mathbf{l}_{ip} \times \mathbf{e}_{ip}$ and $\mathbf{l}_{ip}, \mathbf{e}_{ip}$ are orthonormal vectors. For an appropriate choice of directions $\mathbf{l}_{ip}, \mathbf{e}_{ip}$, these functions form a basis of the space of *Trefftz polynomials* which satisfy Maxwell's equations exactly in the respective subdomain. Clearly, even though there are much fewer Trefftz polynomials than generic polynomials of the same order, the same order of approximation is achieved, resulting in spectral convergence. Like the usual discontinuous Galerkin method, the Trefftz version can also be shown to be energy stable and slightly dissipative, with the numerical dissipation decreasing as the order of approximation

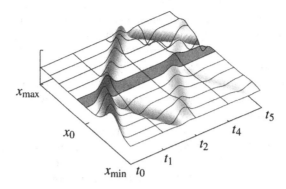

Fig. 1. Amplitude of the electric field in a 1D wave propagation problem. A plane wave travels (in the direction from x_{min} to x_{max}) through an inhomogeneous medium with a permittivity that jumps by a factor of four in the gray shaded area. The difference in phase velocities in the two parts of the domain can be clearly seen. Although the material interface is not resolved by the mesh, there is no loss of accuracy.

increases. Efficiency and accuracy of the discontinuous Galerkin Trefftz scheme outlined above is demonstrated in [17,18,22,23].

The Trefftz DG method is quite flexible with regard to geometric features and local refinement. Even non-conforming meshes with elements of very different size and shape can be handled without much difficulty. Moreover, due to the variational character of the method, various boundary conditions and special features can be modeled in a systematic way. For instance, the plane wave basis described above can be used to define transparent boundary conditions of arbitrary order [18]. If Trefftz functions for special configurations are available, these can also be incorporated into the basis.

Figure 1 shows the space-time solution corresponding to an incident plane wave in one spatial dimension, propagating through an off-grid material interface. In the middle of the gray shaded area, the dielectric permittivity jumps by a factor of four. The appropriate behavior of the field across the interface is incorporated into the local Trefftz basis for the corresponding element.

As another example, the weak coupling between subdomains allows one to interface waveguide modes with a polynomial plane wave approximation of interconnects; see [24] for a related application in the frequency domain.

5   Nonreflecting Trefftz Boundary Conditions

It is well known that finite difference or finite element solution of problems in unbounded domains requires artificial domain truncation and special absorbing/nonreflecting/transparent boundary conditions/Perfectly Matched Layers (PML) to be imposed [25–38]); see also excellent reviews by Givoli, Tsynkov and Hagstrom [39–41],

Let us consider the scalar wave equation in the frequency domain

$$\nabla^2 u + k_0^2 u = f \quad \text{in } R^n, \quad n = 1, 2, 3; \quad \text{supp} f \subset \Omega \subset R^n \tag{3}$$

We shall deal primarily with 2D problems, although all ideas can be extended to 3D, and some of them also carry over to time-dependent problems. For simplicity of exposition, we assume that $\partial\Omega$ is a rectangle.

Problem (3) requires radiation boundary conditions (e.g. Sommerfeld) at infinity, but our task is to replace these theoretical conditions with approximate but accurate and practical ones on the exterior surface $\partial\Omega$ away from the sources. This is to be done in such a way that the solution subject to these artificial conditions be by some measure close to the true solution in Ω.

The ideas of Trefftz approximation can be extended to analytical and numerical absorbing conditions. This section focuses on nonreflecting boundary conditions for finite difference methods. Similar ideas in the context of Trefftz DG are presented in [18]. Namely, let the exact solution of the unbounded problem be approximated in the vicinity of a given point at the artificial exterior boundary as a linear combination

$$u_a = \sum_\alpha c_\alpha \psi_\alpha = \underline{c}^T \underline{\psi} \tag{4}$$

where $\underline{c}$ is a vector of complex coefficients and $\underline{\psi}$ is a vector of basis functions, also in general complex. Coefficients $\underline{c}$ may be different at different boundary points, but for simplicity of notation this is not explicitly indicated. We seek a suitable boundary condition of the form

$$\sum_\beta s_\beta l_\beta(u_a) = 0 \tag{5}$$

where $\{l_\beta\}$ is a suitable set of m linear functionals, and $\underline{s} \in \mathbb{C}^m$ is a set of coefficients ("scheme") to be determined. We require that the scheme be exact for any u_a, i.e. for any linear combination of basis functions:

$$\sum_\beta s_\beta l_\beta \left(\sum_\alpha c_\alpha \psi_\alpha \right) = 0$$

or in matrix form

$$\underline{c}^T N^T \underline{s} = 0$$

where N^T is an $n \times m$ matrix with entries $N_{\alpha\beta}^T = l_\beta(\psi_\alpha)$. Since the above equality is required to hold for all $\underline{c}$, one must have

$$\underline{s} \in \text{Null } N^T \tag{6}$$

This derivation is completely parallel to that of FLAME (Sect. 2). (Gratkowski [42] uses similar ideas to derive analytical boundary conditions, albeit for static problems only and without the nullspace formula). However, it is beneficial to consider other degrees of freedom $\{l_\beta\}$, not limited to the nodal values. A few interesting combinations of basis functions and dof are summarized in Table 2. Details are given in [43,44] and partly in [2,3]. Here we highlight only a few points, to make the items in the table understandable.

Table 2. Absorbing conditions resulting from particular choices of bases and dof.

Basis	dof	Absorbing condition	Properties
Cylindrical harmonics	Solution and its radial derivatives	Bayliss-Turkel	Varying order
Plane wave and its parametric derivatives at normal incidence	Solution and its derivatives	Engquist-Majda	Varying order
A set of outgoing plane waves	Nodal values of the solution on a grid stencil	Trefftz-FLAME absorbing condition	Accuracy reasonable but convergence slow
A set of outgoing plane waves	Nodal values of the solution *and its radial derivatives* on a grid stencil	Generalized Trefftz-FLAME absorbing condition	Order 6 convergence in 2D for (6+3)-point stencils
A set of outgoing Hankel waves	Nodal values of the solution *and its radial derivatives* on a grid stencil	Generalized Trefftz-FLAME absorbing condition	Order 6 convergence as above, but *much* lower errors

By "parametric derivatives" of a plane wave we mean the following. Let

$$\psi(x, y, k, \theta) = \exp\left(-\mathrm{i}k(x\cos\theta + y\sin\theta)\right) \qquad (7)$$

be an outgoing plane wave relative to the half-space $x > 0$. Differentiating this wave successively with respect to k at $k = k_0$ or, alternatively, with respect to θ at $\theta = 0$, one obtains a Trefftz basis set, tailored toward accurate approximation of the solution near normal incidence. (In practice, however, this approximation tends to be good in a broad range of angles.) As shown in [43] and indicated in the table, this basis set, with appropriate derivatives as dof, leads to the well known Engquist-Majda condition.

Furthermore, as also indicated in the table, radial derivatives of the solution at the boundary nodes of the grid can be added to the set of dof. In practical terms, one could consider "double nodes" carrying two dof at the boundary: the solution and its radial derivative. These additional degrees of freedom improve the convergence and accuracy of the method dramatically. In numerical tests of 2D scattering from a PEC cylinder as an example, convergence of order six is easily attained, and the relative error of the numerical solution is on the order of 10^{-8} with 10 – 20 grid points per vacuum wavelength λ and the exterior boundary placed at $\sim 1.5\lambda$ from the scatterer.

In summary, the proposed Trefftz generator of high-order nonreflecting boundary conditions for wave problems has two main ingredients: a set of local Trefftz basis functions (outgoing waves) and a commensurate set of linear functionals (degrees of freedom). Particular cases of this generator include classical conditions due to Engquist-Majda and Bayliss-Turkel, possibly also their extensions [38], and novel highly accurate conditions involving additional dof on finite difference grids.

6   Non-Asymptotic Trefftz Homogenization of Periodic Structures

Trefftz approximations have also proved to be indispensable in a very different arena – non-asymptotic effective medium theory (homogenization) of periodic electromagnetic structures, over the last decade frequently referred to as "metamaterials". Such structures can be engineered to exhibit strong resonances at certain frequencies and, consequently, unusual behavior of effective material parameters. Most notably, appreciable magnetic response, not available in natural materials, can be engineered. Spectacular manifestations of that include negative index of refraction and cloaking (see e.g. [45] and references therein).

Many well-known mathematical and physical homogenization theories are *asymptotic* in their nature – that is, derived in the limit of the lattice cell size vanishingly small relative to the vacuum wavelength [46–48]. However, it has become clear that in this asymptotic limit the nontrivial magnetic effects disappear [49, 50]. Thus a non-asymptotic theory is called for.

We developed such a theory in a series of publications [51–53]. The main mathematical idea consists in approximating the fields on the fine (subcell) and coarse (multi-cell) scales with Trefftz functions [51], i.e. the respective eigenmodes that satisfy Maxwell's equations and boundary conditions as accurately as possible. These functions include, in particular, Bloch modes on the fine scale and plane waves (propagating in an equivalent medium, possibly anisotropic and magneto-electric) on the coarse scale. The effective material parameters are determined by minimizing approximation errors in Maxwell boundary conditions and dispersion relations.

A notable feature of this approach is that the effective parameters can be position-dependent and can be optimized for a defined range of illuminating conditions [51, 54]. Our method leads to a linear optimization problem, in contrast with the nonlinear inverse problem of S-parameter retrieval, and has a built-in error indicator characterizing the accuracy of homogenization. Finally, the proposed theory leads to analytical expressions of bulk material parameters in terms of Bloch fields on the boundary of a lattice cell.

The non-asymptotic homogenization procedure is valid for periodic dielectric structures with any composition and any size of cells (subwavelength but not necessarily small). Anisotropy and magnetoelectric effects are fully accounted for.

7   Conclusion

Trefftz approximations are beneficial in a variety of areas and can be used to generate high-order difference schemes, absorbing boundary conditions, boundary difference schemes, Discontinuous Galerkin methods, and homogenization procedures. In all of these cases, approximation accuracy can be significantly improved in comparison with more traditional techniques. The authors believe that Trefftz methods will find even broader applications in the near future.

Acknowledgment. IT thanks Prof. Ralf Hiptmair (ETH Zürich) for very helpful discussions and, in particular, for suggesting additional *discrete* dof on the absorbing boundary.

The work was supported in part by the following grants: German Research Foundation (DFG) GSC 233 (FK and HE); US National Science Foundation DMS-1216970 (IT and VAM); U.S. Army Research Office W911NF1110384 (IT). SMS acknowledges support by the Alexander von Humboldt-Foundation through a Feodor-Lynen research fellowship.

References

1. Tsukerman, I.: Computational Methods for Nanoscale Applications. Particles Plasmons and Waves. Springer, New York (2007)
2. Tsukerman, I.: Electromagnetic applications of a new finite-difference calculus. IEEE Trans. Magn. **41**(7), 2206–2225 (2005)
3. Tsukerman, I.: A class of difference schemes with flexible local approximation. J. Comput. Phys. **211**(2), 659–699 (2006)
4. Tsukerman, I.: Trefftz difference schemes on irregular stencils. J. Comput. Phys. **229**(8), 2948–2963 (2010)
5. Pinheiro, H., Webb, J., Tsukerman, I.: Flexible local approximation models for wave scattering in photonic crystal devices. IEEE Trans. Magn. **43**(4), 1321–1324 (2007)
6. Tsukerman, I., Čajko, F.: Photonic band structure computation using FLAME. IEEE Trans. Magn. **44**(6), 1382–1385 (2008)
7. Babuška, I., Ihlenburg, F., Paik, E.T., Sauter, S.A.: A generalized finite element method for solving the Helmholtz equation in two dimensions with minimal pollution. Comput. Meth. Appl. Mech. Eng. **128**, 325–359 (1995)
8. Saltzer, C.: Discrete potential theory for two-dimensional Laplace and Poisson difference equations. Technical report 4086, National Advisory Committee on Aeronautics (1958)
9. Hsiao, G., Wendland, W.L.: Boundary Integral Equations. Springer, Heidelberg (2008)
10. Ryaben'kii, V.S.: Method of Difference Potentials and Its Applications. Springer Series in Computational Mathematics, vol. 30. Springer-Verlag, Berlin (2002)
11. Tsynkov, S.V.: On the definition of surface potentials for finite-difference operators. J. Sci. Comput. **18**, 155–189 (2003)
12. Tsukerman, I.: A singularity-free boundary equation method for wave scattering. IEEE Trans. Antennas Propag. **59**(2), 555–562 (2011)
13. Martinsson, P.: Fast multiscale methods for lattice equations. Ph.D. thesis, The University of Texas at Austin (2002)
14. Martinsson, P., Rodin, G.: Boundary algebraic equations for lattice problems. Proc. R. Soc. A - Math. Phys. Eng. Sci. **465**(2108), 2489–2503 (2009)
15. AlKhateeb, O., Tsukerman, I.: A boundary difference method for electromagnetic scattering problems with perfect conductors and corners. IEEE Trans. Antennas Propag. **61**(10), 5117–5126 (2013)

16. Petersen, S., Farhat, C., Tezaur, R.: A space-time discontinuous Galerkin method for the solution of the wave equation in the time domain. Int. J. Numer. Meth. Eng. **78**(3), 275–295 (2009)

17. Kretzschmar, F., Schnepp, S., Tsukerman, I., Weiland, T.: Discontinuous Galerkin methods with Trefftz approximations. J. Comput. Appl. Math. **270**, 211–222 (2014)

18. Egger, H., Kretzschmar, F., Schnepp, S., Tsukerman, I., Weiland, T.: Transparent boundary conditions in a Discontinuous Galerkin Trefftz method. Appl. Math. Comput. **270** (submitted, 2014). http://arxiv.org/abs/1410.1899

19. Fezoui, L., Lanteri, S., Lohrengel, S., Piperno, S.: Convergence and stability of a discontinuous Galerkin time-domain method for the 3D heterogeneous Maxwell equations on unstructured meshes. ESAIM-Math. Model Numer. **39**(6), 1149–1176 (2005)

20. Griesmair, T., Monk, P.: Discretization of the wave equation using continuous elements in time and a hybridizable discontinuous Galerkin method in space. J. Sci. Comput. **58**, 472–498 (2014)

21. Lilienthal, M., Schnepp, S., Weiland, T.: Non-dissipative space-time hp - discontinuous Galerkin method for the time-dependent maxwell equations. J. Comput. Phys. **275**, 589–607 (2014)

22. Moiola, A., Hiptmair, R., Perugia, I.: Plane wave approximation of homogeneous Helmholtz solutions. Z. Angew. Math. Phys. **62**(5), 809–837 (2011)

23. Hiptmair, R., Moiola, A., Perugia, I.: Plane wave discontinuous Galerkin methods for the 2D Helmholtz equation: analysis of the p-version. SIAM J. Numer. Anal. **49**(1), 264–284 (2011)

24. Huber, M., Schöberl, J., Sinwel, A., Zaglmayr, S.: Simulation of diffraction in periodic media with a coupled finite element and plane wave approach. SIAM J. Sci. Comput. **31**, 1500–1517 (2009)

25. Berenger, J.P.: A perfectly matched layer for the absorption of electromagnetic waves. J. Comput. Phys. **127**, 363–379 (1996)

26. Teixeira, F.L., Chew, W.C.: General closed-form PML constitutive tensors to match arbitrary bianisotropic and dispersive linear media. IEEE Microwave Guided Wave Lett. **8**, 223–225 (1998)

27. Sacks, Z., Kingsland, D., Lee, R., Lee, J.F.: A perfectly matched anisotropic absorber for use as an absorbing boundary condition. IEEE Trans. Antennas Propag. **43**(12), 1460–1463 (1995)

28. Gedney, S.D.: An anisotropic perfectly matched layer-absorbing medium for the truncation of FDTD lattices. IEEE Trans. Antennas Propag. **44**(12), 1630–1639 (1996)

29. Collino, F., Monk, P.B.: Optimizing the perfectly matched layer. Comput. Meth. Appl. Mech. Eng. **164**, 157–171 (1998)

30. Higdon, R.L.: Absorbing boundary conditions for difference approximations to the multidimensional wave equation. Math. Comput. **47**(176), 437–459 (1986)

31. Higdon, R.L.: Numerical absorbing boundary conditions for the wave equation. Math. Comput. **49**(179), 65–90 (1987)

32. Bayliss, A., Turkel, E.: Radiation boundary-conditions for wave-like equations. Commun Pure Appl. Math. **33**(6), 707–725 (1980)

33. Bayliss, A., Gunzburger, M., Turkel, E.: Boundary conditions for the numerical solution of elliptic equations in exterior regions. SIAM J. Appl. Math. **42**(2), 430–451 (1982)

34. Hagstrom, T., Hariharan, S.I.: A formulation of asymptotic and exact boundary conditions using local operators. Appl. Numer. Math. **27**(4), 403–416 (1998)

35. Givoli, D.: High-order nonreflecting boundary conditions without high-order derivatives. J. Comput. Phys. **170**(2), 849–870 (2001)
36. Givoli, D., Neta, B.: High-order nonreflecting boundary conditions for the dispersive shallow water equations. J. Comput. Appl. Math. **158**(1), 49–60 (2003)
37. Hagstrom, T., Warburton, T.: A new auxiliary variable formulation of high-order local radiation boundary conditions: corner compatibility conditions and extensions to first-order systems. Wave Motion **39**(4), 327–338 (2004)
38. Zarmi, A., Turkel, E.: A general approach for high order absorbing boundary conditions for the Helmholtz equation. J. Comput. Phys. **242**, 387–404 (2013)
39. Givoli, D.: High-order local non-reflecting boundary conditions: a review. Wave Motion **39**(4), 319–326 (2004)
40. Tsynkov, S.V.: Numerical solution of problems on unbounded domains. Rev. Appl. Numer. Math. **27**, 465–532 (1998)
41. Hagstrom, T.: Radiation boundary conditions for the numerical simulation of waves. In: Iserlis, A. (ed.) Acta Numerica, vol. 8, pp. 47–106. Cambridge University Press, Cambridge (1999)
42. Gratkowski, S.: Asymptotyczne warunki brzegowe dla stacjonarnych zagadnień elektromagnetycznych w obszarach nieograniczonych - algorytmy metody elementów skończonych. Wydawnictwo Uczelniane Zachodniopomorskiego Uniwersytetu Technologicznego (2009)
43. Tsukerman, I.: A "Trefftz machine" for absorbing boundary conditions. Ann. Stat. **42**(3), 1070–1101 (2014). http://arxiv.org/abs/1406.0224
44. Paganini, A., Scarabosio, L., Hiptmair, R., Tsukerman, I.: tz approximations: a new framework for nonreflecting boundary conditions (in preparation, 2015)
45. Soukoulis, C.M., Wegener, M.: Past achievements and future challenges in the development of three-dimensional photonic metamaterials. Nat. Photonics **5**, 523–530 (2011)
46. Bensoussan, A., Lions, J., Papanicolaou, G.: Asymptotic Methods in Periodic Media. Elsevier, North Holland (1978)
47. Bakhvalov, N.S., Panasenko, G.: Homogenisation: Averaging Processes in Periodic Media. Mathematical Problems in the Mechanics of Composite Materials. Springer, The Netherlands (1989)
48. Milton, G.: The Theory of Composites. Cambridge University Press, Cambridge; New York (2002)
49. Bossavit, A., Griso, G., Miara, B.: Modelling of periodic electromagnetic structures bianisotropic materials with memory effects. J. Math. Pures Appl. **84**(7), 819–850 (2005)
50. Tsukerman, I.: Negative refraction and the minimum lattice cell size. J. Opt. Soc. Am. B **25**, 927–936 (2008)
51. Tsukerman, I., Markel, V.A.: A nonasymptotic homogenization theory for periodic electromagnetic structures. Proc. Royal Soc. A **470** 2014.0245 (2014)
52. Tsukerman, I.: Effective parameters of metamaterials: a rigorous homogenization theory via Whitney interpolation. J. Opt. Soc. Am. B **28**(3), 577–586 (2011)
53. Pors, A., Tsukerman, I., Bozhevolnyi, S.I.: Effective constitutive parameters of plasmonic metamaterials: homogenization by dual field interpolation. Phys. Rev. E **84**, 016609 (2011)
54. Xiong, X.Y., Jiang, L.J., Markel, V.A., Tsukerman, I.: Surface waves in three-dimensional electromagnetic composites and their effect on homogenization. Opt. Express **21**(9), 10412–10421 (2013)

On Extension of Asymptotic Comparison Principle for Time Periodic Reaction-Diffusion-Advection Systems with Boundary and Internal Layers

Nikolay Nefedov$^{(\boxtimes)}$ and Aleksei Yagremtsev

Department of Mathematics, Faculty of Physics,
Lomonosov Moscow State University, 119991 Moscow, Russia
nefedov@phys.msu.ru

Abstract. In this paper we present a further development of our asymptotic comparison principle, applying it for some new important classes of initial boundary value problem for the nonlinear singularly perturbed time periodic parabolic equations, which are called in applications as reaction-diffusion-advection equations. We illustrate our approach for the new problem with balanced nonlinearity. The theorems, which states the existence of the periodic solution with internal layer, gives it's asymptotic approximation and state their Lyapunov stability are proved.

Keywords: Singularly perturbed problems · Moving fronts · Time periodic reaction-diffusion-advection equations

1 Introduction

This work is devoted to nonlinear singularly perturbed periodic parabolic equations. In applications, these equations are often used as models reaction-diffusion and reaction-diffusion-advection processes in chemical kinetics, synergetic, astrophysics, biology, et al. It is well-known that such problems are extremely complicated for numerical treatment as well for asymptotic investigations and it needs to develop new asymptotic methods to investigate them formally as well as rigorously.

We present our recent extension of the asymptotic comparison principal to the new classes of singularly perturbed problems and present the application of our approach to the nonlinear singularly perturbed periodic reaction-diffusion-advection problem with internal layers. These results can be considered as a further development of our investigations which were published in the review paper [1] and papers [2] and [3].

Our rigorous investigation is based on the works on the comparison principal. These works are essentially using so called Krein-Rutman theorem and the results on the comparison principal (see [4]) and the results of M.A. Krasnoselskij on positive operators theory (see, for example, [5] and references therein).

© Springer International Publishing Switzerland 2015
I. Dimov et al. (Eds.): FDM 2014, LNCS 9045, pp. 62–71, 2015.
DOI: 10.1007/978-3-319-20239-6_6

In the present paper we discuss father development of the general scheme of asymptotic method of differential inequalities, the basic ideas of which were proposed in [6], for the periodic parabolic problems and illustrate it applying for new important cases of periodic boundary value problem for the equation

$$\varepsilon^2(\Delta u - \frac{\partial u}{\partial t}) = f(u, \nabla u, x, t, \varepsilon), \quad x \in \mathcal{D} \subset R^N, t \in R, \tag{1}$$

which is used in many applications and is called reaction-diffusion-advection equation. For this problem we state the conditions which imply the existence of periodic contrast structures - solutions with periodic internal layers. We use our approach to investigate a case when equation (1) is quasylinear. We investigate the following problems.

1. Existence and Lyapunov stability of the periodic contrast structures.
2. The analysis of local domain of stability of the stable contrast structures.

The basic idea of this approach is to construct lower and upper solutions to the problem by using formal asymptotics. By using these we state the existence of the solutions, estimate the accuracy of the asymptotics and investigate asymptotic stability of the periodic solutions in the sense of Lyapunov.

Another aspects of this work is to emphasize the possibility of use this analytical treatment for numerical approaches. The possibility of this is described in the our paper [7]. Moreover, analytic algorithms for asymptotic approximations of layered solutions were intensively used for creating numerical algorithms, see, e.g. the papers of H.-G. Roos, S. Franz and N. Kopteva (see [8–10] and references there for this intensively developing field). All these results mainly concern layered stationary solutions. We sagest the new classes of problems.

2 General Scheme of Asymptotic Method of Differential Inequalities for Periodic Reaction-Advection-Diffusion Equations

We consider some cases of initial boundary value problem

$$\varepsilon^2(\Delta u - \frac{\partial u}{\partial t}) = f(u, \nabla u, x, t, \varepsilon), \quad x \in \mathcal{D} \subset R^N, t \in R,$$
$$Bu = h(x, t), \quad x \in \partial\mathcal{D}, t \in R, \tag{2}$$

with periodicity condition $u(x, t, \varepsilon) = u(x, t, \varepsilon)$ where ε is a small parameter, f, h, and $\partial\mathcal{D}$ are sufficiently smooth, B is a boundary operator for Dirichlet, Neumann or third order boundary conditions. We also assume that f, h are T-periodic in t functions.

Denote by N the nonlinear operator in (2)

$$Nu \equiv \varepsilon^2(\Delta u - \frac{\partial u}{\partial t}) - f(u, \nabla u, x, t, \varepsilon).$$

We introduce the following definition for an upper and a lower solutions, which is more strong than classical definition.

Definition. The T-periodic functions $\beta(x, t, \varepsilon)$ and $\alpha(x, t, \varepsilon)$ we call an asymptotic upper and a lower solutions of order $q > 0$ of problem (2) if for sufficiently small ε they satisfy the inequalities

$$N\beta \leq -c\varepsilon^q, \ N\alpha \geq c\varepsilon^q, \quad x \in \mathcal{D}, t \in R, \tag{3}$$

$$B\alpha \leq h(x, t) \leq B\beta, \quad x \in \partial\mathcal{D}, t \in R, \tag{4}$$

where c is a positive constant.

Denote by L the linear operator which we get from N by linearizing f on the stationary solution, by L_f the linearization f on the stationary solution and by H the following characteristic of the nonlinearity

$$H \equiv f(\beta, \nabla\beta, x, t, \varepsilon) - f(\alpha, \nabla\alpha, x, t, \varepsilon) - L_f(\beta - \alpha)$$

The property of H depend on the asymptotic lower and upper solutions. Note that one of the most important our achievements is the method to construct them by using the formal asymptotic expansion. In what follows we describe this approach. Our assumption is

(A_1). *There exist asymptotic of order q an upper solution β and a lower solution α such that $\beta > \alpha$ and $|\beta - \alpha| \leq c\varepsilon^r$ for $x \in \mathcal{D}, t \in R$ and sufficiently small ε.*

We also assume

(A_2). $|H| \leq c\varepsilon^p$ for $x \in \mathcal{D}, t \in R$ and sufficiently small ε.

(A_3). $p \geq q$

It is clear that the estimates of the assumptions (A_2) (A_2) depend on the properties of the nonliniarity f and the lower and Under the assumptions above the following theorem take place.

Theorem 1. *Suppose the assumptions $(A_1) - (A_3)$ to be valid. Then, for sufficiently small ε there exists a solution $u(x, t, \varepsilon)$ of (2) which differ from the upper or lower solution on the value of order $O(\varepsilon^r)$ and is asymptotically stable in Lyapunov sense with the local domain of stability $[\alpha, \beta]$*

The proof of Theorem 1 is based on the revised maximum principal, which used Krein-Ruthman theorem.

From assumption (A_1) it follows the existence of the periodic solution $u(x, t, \varepsilon)$ of problem (2) satisfying the inequalities inequalities $\alpha \leq u(x, \varepsilon) \leq \beta$ (see [4]) and therefore we also have the asymptotic estimate for the solution. It differ from the upper or lower solution on the value of order $O(\varepsilon^r)$.

From (A_2), (A_3) it is easily follows that $L(\beta - \alpha) < 0$. From (8) we have $B(\beta - \alpha) > 0$ and therefore the pricipal eigenvalue (which exists and real) satisfy the estimate $\lambda_p < 0$, that imply the asymtotic stability of the periodic solution solution in the sense of Lyapunov (see [4]).

We note that the analogues of Theorem 1 are valid for stationary and nonlocal reaction-advection-diffusion equations (see [2]).

In order to get the upper and lower solutions satisfying the assumptions of Theorem 1 we use the formal asymptotics, which can be constructed in a lot of cases by our method described, for example, in the review papers [1, 11]. Under quite natural assumptions the formal asymptotics of internal layer solution is produced by the boundary layer operators L_B, regular expansion operators L_R and by the operators describing the location of transition layer A^Γ. To construct the formal asymptotic we assume that the operators are invertible.

For the construction of asymptotic lower and upper solutions we require that *these operators are monotone (order preserving) when they act in the same classes of functions in which we construct the asymptotic expansions by means of these operators.*

Finally we get $\alpha \equiv \alpha_n$, $\beta \equiv \beta_n$ – modified n-th order formal asymptotic. We illustrate our approach by the following example.

3   Periodic Solutions with Internal Layer in the Case of Balanced Advection

We consider the boundary value problem

$$N_\varepsilon(u) := \varepsilon \left(\frac{\partial^2 u}{\partial x^2} - \frac{\partial u}{\partial t} \right)$$

$$- A(u, x, t) \frac{\partial u}{\partial x} - B(u, x, t) = 0 \quad \text{for} \quad x \in (0, 1),\ t \in R, \tag{5}$$

$$u(0, t, \varepsilon) = u^{(-)}(t), \quad u(1, t, \varepsilon) = u^{(+)}(t) \quad \text{for} \quad t \in R,$$
$$u(x, t, \varepsilon) = u(x, t + T, \varepsilon) \quad \text{for} \quad t \in R,$$

where ε is a small parameter, A, B, $u^{(-)}$ and $u^{(+)}$ are sufficiently smooth and T-periodic in t. As we mentioned in the introduction the case of the non-balanced advection was considered in [3] where you can find more details on the asymptotic construction.

If we put $\varepsilon = 0$ in equation (9) we get the so-called degenerate equation

$$A(u, x, t) \frac{\partial u}{\partial x} + B(u, x, t) = 0, \tag{6}$$

where t has to be considered as a parameter. Equation (6) is a first order ordinary differential equation and can be considered with one of the following initial conditions from problem (9)

$$u(0, t) = u^{(-)}(t), \tag{7}$$

$$u(1, t) = u^{(+)}(t). \tag{8}$$

(H_1). The problems (6),(7) and (6),(8) have the solutions $u = \varphi^{(-)}(x, t)$ and $u = \varphi^{(+)}(x, t)$, respectively, which are defined for $0 \le x \le 1, t \in R$ and which are T-periodic in t. Additionally we assume

$$\varphi^{(-)}(x, t) < \varphi^{(+)}(x, t) \text{ for } x \in [0, 1],\ t \in R,$$

$$A(\varphi^{(+)}(x,t), x, t) < 0, \ A(\varphi^{(-)}(x,t), x, t) > 0,$$

$$x \in [0,1], \ t \in R.$$

To formulate the next assumptions we introduce the function $I(x,t)$ by

$$I(x,t) := \int_{\varphi^{(-)}(x,t)}^{\varphi^{(+)}(x,t)} A(u,x,t)du.$$

We assume

(H_2). The function

$$I(x,t) \equiv 0 \ for \ x \in [0,1], \ t \in R, \tag{9}$$

and

$$\int_{\varphi^{(-)}(x,t)}^{s} A(u,x,t) \, du > 0$$

for any $s \in \left(\varphi^{(-)}(x,t), \varphi^{(+)}(x,t) \right), \quad x \in [0,1], \quad t \in R.$

Assumption (H_2) differ our statement from the non-balanced case, where the equation $I(x,t) = 0$ a simple solution $x = x_0(t)$ $(\frac{\partial I}{\partial x}(x_0(t), t) < 0)$.

Construction of the Formal Asymptotics. The construction of the asymptotic follows the scheme proposed in [3], where it is possible to find some additional details.

To characterize the location of the interior layer we introduce the curve $x = x_*(t, \varepsilon)$ as locus of the intersection of the solution $u(x, t, \varepsilon)$ of (2) with the surface

$$u = \frac{1}{2} \left(\varphi^{(-)}(x,t) + \varphi^{(+)}(x,t) \right) =: \varphi(x,t).$$

In what follows we construct the asymptotic expansion of $x_*(t, \varepsilon)$ in the form

$$x_*(t, \varepsilon) = x_0(t) + \varepsilon \, x_1(t) + ..., \tag{10}$$

where $x_0(t)$ is the solution of equation (9) and $x_k(t), k = 1, 2, ...,$ are T-periodic functions to be determined. For the following we use the notations

$$\xi := \frac{x \quad x_*(t, \varepsilon)}{\varepsilon},$$

$$\overline{\mathcal{D}}^{(-)} := \{(x,t) \in R^2 : 0 \le x \le x_*(t, \varepsilon), t \in R\},$$

$$\overline{\mathcal{D}}^{(+)} := \{(x,t) \in R^2 : x_*(t, \varepsilon) \le x \le 1, t \in R\}.$$

First we consider in $\overline{\mathcal{D}}^{(-)}$ the boundary value problem

$$\varepsilon \left(\frac{\partial^2 u}{\partial x^2} - \frac{\partial u}{\partial t} \right) - A(u,x,t) \frac{\partial u}{\partial x} - B(u,x,t) = 0 \quad \text{for} \quad (x,t) \in \overline{\mathcal{D}}^{(-)},$$

$$u(0,t,\varepsilon) = u^0(t), \quad u(x(t,\varepsilon), t, \varepsilon) = \varphi(t, \varepsilon) \quad \text{for} \quad t \in R, \tag{11}$$

$$u(x,t,\varepsilon) = u(x, t+T, \varepsilon) \quad \text{for} \quad t \in R.$$

We look for the formal asymptotic expansion of the solution $U^{(-)}(x, t, \varepsilon)$ of this problem in the form

$$U^{(-)}(x, t, \varepsilon) = \bar{U}^{(-)}(x, t, \varepsilon) + Q^{(-)}(\xi, x_*, t, \varepsilon) =$$
$$\sum_{i=0}^{\infty} \varepsilon^i \left(\bar{U}_i^{(-)}(x, t) + Q_i^{(-)}(\xi, x_*, t) \right), \tag{12}$$

where $\bar{U}^{(-)}$ and $\bar{Q}^{(-)}$ denote the regular and the interior layer parts. Next we study in $\overline{\mathcal{D}}^{(+)}$ similar problem to construct $U^{(+)}(x, t, \varepsilon)$.

By using the standard procedure of boundary layer function method we can construct these expansions and to show that operators $L_R^{\pm}$, produsing regular part of the asymptotics have the form

$$L_R^{\pm} \equiv -A(\varphi^{(\pm)}(x, t), x, t) \frac{\partial}{\partial x} - \left(A_u(\varphi^{(\pm)}(x, t), x, t) \frac{\partial \bar{U}_0^{(\pm)}}{\partial x} \right.$$
$$+ B_u(\varphi^{(\pm)}(x, t), x, t) \Big), \tag{13}$$

and therefore is the first order differential operator and from assumption (H_1) it follows that it is positively invertible for a negative right hand part - inequality $L_R^{\pm}\bar{u} < 0$ has a positive solution.

In order to construct the boundary layer functions describing the internal layer near $x_*(t, \varepsilon)$ we represent the differential operators

$$\varepsilon \frac{\partial^2}{\partial x^2}, \quad \frac{\partial}{\partial x} \quad \text{and} \quad \varepsilon \frac{\partial}{\partial t}$$

in the form

$$\frac{1}{\varepsilon} \frac{\partial^2}{\partial \xi^2}, \quad \frac{1}{\varepsilon} \frac{\partial}{\partial \xi} \quad \text{and} \quad -\frac{dx_*}{dt} \frac{\partial}{\partial \xi} + \varepsilon \frac{\partial}{\partial t}.$$

For the transition layer functions we have equations

$$\frac{1}{\varepsilon} \frac{\partial^2 Q^{(\pm)}}{\partial \xi^2} + \frac{\partial x_*(t, \varepsilon)}{\partial t} \frac{\partial Q^{(\pm)}}{\partial \xi} - \varepsilon \frac{\partial Q^{(\pm)}}{\partial t} =$$

$$= \frac{1}{\varepsilon} \left[A \left(\bar{U}^{(\pm)}(x_*(t, \varepsilon) + \varepsilon\xi, t, \varepsilon) + Q^{(\pm)}, x_*(t, \varepsilon) + \varepsilon\xi, t \right) \frac{\partial}{\partial \xi} Q^{(\pm)} \right]$$

$$+ A \left(\bar{U}^{(\pm)}(x_*(t, \varepsilon) + \varepsilon\xi, t, \varepsilon) + Q^{(\pm)}, x_*(t, \varepsilon) + \varepsilon\xi, t \right) \frac{\partial \bar{U}^{(\pm)}}{\partial x}$$

$$- A \left(\bar{U}^{(\pm)}(x_*(t, \varepsilon) + \varepsilon\xi, t, \varepsilon), x_*(t, \varepsilon) + \varepsilon\xi, t \right) \frac{\partial}{\partial x} \bar{U}^{(\pm)}(x_*(t, \varepsilon) + \varepsilon\xi, t, \varepsilon)$$

$$+ [B \left(\bar{U}^{(\pm)}(x_*(t, \varepsilon) + \varepsilon\xi, t, \varepsilon) + Q^{(\pm)}, x_*(t, \varepsilon) + \varepsilon\xi, t \right)$$

$$- B \left(\bar{U}^{(\pm)}(x_*(t, \varepsilon) + \varepsilon\xi, t, \varepsilon), x_*(t, \varepsilon) + \varepsilon\xi, t \right) \Big]$$

with boundary conditions

$$Q^{(\pm)}(0, x_*, t, \varepsilon) + \bar{U}^{(\pm)}(x_*(t, \varepsilon), t, \varepsilon) = \varphi(x_*(t, \varepsilon), t).$$

For $Q_k^{(\pm)}(\xi, x_*, t)$ we use additional conditions at $\pm\infty$:

$$Q_k^{(\pm)}(\pm\infty, x_*, t) = 0, \ t \in R.$$

From this representation we have the problems for $Q_k^{(\pm)}(\xi, x_*, t)$ which were investigated in details in [3].

In order to find the terms $x_i(t)$ of the expansion $U^\pm$ we use C^1–matching condition of the i–th order in ε of the expression

$$\varepsilon \frac{\partial \bar{U}^{(+)}}{\partial x}(x_*(t, \varepsilon), t, \varepsilon) + \frac{\partial Q^{(+)}}{\partial \xi}(0, x_*, t, \varepsilon) =$$

$$\varepsilon \frac{\partial \bar{U}^{(-)}}{\partial x}(x_*(t, \varepsilon), t, \varepsilon) + \frac{\partial Q^{(-)}}{\partial \xi}(0, x_*, t, \varepsilon).$$

Different from the non-balanced case zero-th order C^1–matching condition is satisfied identically because the assumption (A_1), and $x_0(t)$ is determined from the first order C^1–matching condition. We get

$$\begin{cases} \frac{dx_0}{dt}\left(\varphi^{(-)}(x_0, t) - \varphi^{(+)}(x_0, t)\right) = H^{(+)}(t, x_0) - H^{(-)}(t, x_0), \\ x_0(t + T) = x_0(t), \end{cases} \tag{14}$$

where

$$H^{(\pm)}(t, x_*) = \frac{\partial \varphi^{(\pm)}(x_*, t)}{\partial x} +$$

$$\int\limits_{\pm\infty}^{0}\left(\Phi(\xi, t, x_*)\left(\frac{\partial \tilde{A}}{\partial u}\frac{\partial \varphi^{(\pm)}}{\partial x}\xi + \frac{\partial \tilde{A}}{\partial x}\xi\right) + \tilde{A}(\xi, t)\frac{\partial \varphi^{(\pm)}}{\partial x} + \tilde{B}(\xi, t)\right)d\xi$$

$$- \bar{U}_1^{(\pm)}\bar{A}^{(\pm)}(x_*, t).$$

Here we use the notations

$$\Phi(\xi, t, x_*) = \begin{cases} \frac{\partial Q_0^{(-)}}{\partial \xi}(0, x_*, t, \varepsilon) & \text{for } \xi \leq 0, \\ \frac{\partial Q_0^{(+)}}{\partial \xi}(0, x_*, t, \varepsilon) & \text{for } \xi \geq 0, \end{cases}$$

and functions $\tilde{A}(\xi, t)$ and $\tilde{B}(\xi, t)$ are functions A and B evaluated at the point $(Q_0^{(\pm)}(\xi, x_*, t) + \bar{U}_0^{(\pm)}(x_*, t), x_*, t, 0)$.

We assume

(H_3). Problem (14) has the solution $x_0(t) \in (0, 1)$.

We define also $H(t, x_*) = H^{(+)}(t, x_*) - H^{(-)}(t, x_*)$.

Using i-th order C^1-matching conditions, we get the problem to determine $x_i(t)$

$$\begin{cases} -\frac{dx_i}{dt} - Dx_i = G_i(t), \\ x_1(t + T) = x_i(t), \end{cases} \tag{15}$$

where

$$D(t) = \left(\frac{\partial H}{\partial x_*}(t, x_*) \Big|_{x_*=x_0} + v_0 \frac{\partial}{\partial x_*} \left(\varphi^{(+)}(x_*, t) - \varphi^{(-)}(x_*, t) \right) \Big|_{x_*=x_0} \right) \times$$

$$\left(\varphi^{(+)}(x_0, t) - \varphi^{(-)}(x_0, t) \right)^{-1},$$

and $G_i(t)$ is known on the each step function. If we assume

(H_4). The function $D(t)$ satisfies the inequality $\int_0^T D(t)dt > 0$,

we get that producing $x_i(t)$ first order periodic operator A^Γ is the order preserving operator ($A^\Gamma \delta(t) < 0$ has a positive solution).

Existence Results. We denote by $\mathcal{D}_n^{(-)}$ and $\mathcal{D}_n^{(-)}$ the domains

$$\mathcal{D}_n^{(-)} := \{(x, t) \in R^2 : 0 \le x \le \sum_{i=0}^{n+1} x_i(t)\varepsilon^i, \ t \in R\},$$

$$\mathcal{D}_n^{(+)} := \{(x, t) \in R^2 : \sum_{i=0}^{n+1} x_i(t)\varepsilon^i \le x \le 1, \ t \in R\}$$

and denote by $U_n^{(\pm)}$ the partial sums of order n of the expansions , where ξ is replaced by $\left(x - \sum_{i=0}^{n+1} x_i(t)\varepsilon^i \right)/\varepsilon$.
We introduce the notation

$$U_n(x, t, \varepsilon) = \begin{cases} U_n^{(-)}(x, t, \varepsilon) \text{ for } (x, t) \in \mathcal{D}_n^{(-)}, \\ U_n^{(+)}(x, t, \varepsilon) \text{ for } (x, t) \in \mathcal{D}_n^{(+)}. \end{cases}$$

Then we have the following existence theorem

Theorem 2. *Suppose the assumptions $(H_1) - (H_4)$ to be valid. Then, for sufficiently small ε there exists a solution $u(x, t, \varepsilon)$ of (9) which T-periodic in t, has an interior layer and satisfies*

$$|u(x, t, \varepsilon) - U_n(x, t, \varepsilon)| \le c\varepsilon^{n+1} \ (x, t) \in \overline{D}$$

where the positive constant c does not depend on ε.

Construction of the Upper and Lower Solutions. The proof of the theorem presented in the previous section is based on the technique of lower and upper solutions.

The upper and lower solutions satisfying the definition above are constructed by means of the modification of the formal asymptotics. In order to describe them we introduce the periodic curves $x = x_\beta(t, \varepsilon)$ and $x = x_\alpha(t, \varepsilon)$ as the n-th partial sums of the asymptotics of $x^*(t, \varepsilon)$ with a small shifts at the last term

$$x_\beta(t, \varepsilon) = x_0(t) + \varepsilon x_1(t) + \dots + \varepsilon^n(x_n(t) - \delta(t))$$

and
$$x_\alpha(t,\varepsilon) = x_0(t) + \varepsilon x_1(t) + ... + \varepsilon^n(x_n(t) + \delta(t))$$
where $\delta(t) > 0$ is a positive solution of the inequality $A^\Gamma \delta(t) < 0$. These curves divide our domain $\overline{D}$ into two subdomains $\overline{D}_\beta^{(-)}$, $\overline{D}_\beta^{(+)}$ and $\overline{D}_\alpha^{(-)}$, $\overline{D}_\alpha^{(+)}$ where

$$\overline{D}_\beta^{(-)} := \{(x,t) \in R^2 : 0 \le x \le x_\beta(t,\varepsilon),\ t \in R\},$$

$$\overline{D}_\beta^{(+)} := \{(x,t) \in R^2 : x_\beta(t,\varepsilon) \le x \le 1,\ t \in R\}.$$

The domains $\overline{D}_\alpha^{(\pm)}$ are defined similarly.

Now we can define the upper solution $\beta(x,t,\varepsilon) = \beta_n(x,t,\varepsilon)$ and the lower solution $\alpha(x,t,\varepsilon) = \alpha_n(x,t,\varepsilon)$ by the expressions

$$\beta_n(x,t,\varepsilon) = \beta_n^{(\pm)}(x,t,\varepsilon) = \bar{U}_0^{(\pm)}(x,t) + \varepsilon \bar{U}_1^{(\pm)}(x,t)$$
$$+ ... + \varepsilon^n \bar{U}_n^{(\pm)}(x,t) + \varepsilon^{n+1}(\bar{U}_{n+1}^{(\pm)}(x,t) + v(x))$$
$$+ Q_0^{(\pm)}(\xi_\beta,t) + \varepsilon Q_1^{(\pm)}(\xi_\beta,t) + ...$$
$$+ \varepsilon^{n+1} Q_{(n+1)\beta}^{(\pm)}(\xi_\beta,t) + \varepsilon^{n+2} Q_{(n+2)\beta}^{(\pm)}(\xi_\beta,t,\varepsilon)$$

and

$$\alpha_n(x,t,\varepsilon) = \alpha_n^{(\pm)}(x,t,\varepsilon) = \bar{U}_0^{(\pm)}(x,t) + \varepsilon \bar{U}_1^{(\pm)}(x,t)$$
$$+ ... + \varepsilon^n \bar{U}_n^{(\pm)}(x,t) + \varepsilon^{n+1}(\bar{U}_{n+1}^{(\pm)}(x,t) - v(x))$$
$$+ Q_0^{(\pm)}(\xi_\alpha,t) + \varepsilon Q_1^{(\pm)}(\xi_\alpha,t) + ...$$
$$+ \varepsilon^{n+1} Q_{(n+1)\alpha}^{(\pm)}(\xi_\alpha,t) + \varepsilon^{n+2} Q_{(n+2)\alpha}^{(\pm)}(\xi_\alpha,t,\varepsilon),$$

where $v(x) > 0$ is is a positive solution of the inequality $L_R^\pm v(x) < 0$, $\xi_\beta = (x - x_\beta)/\varepsilon$, $\xi_\alpha = (x - x_\alpha)/\varepsilon,$, the function $\beta_n(x,t,\varepsilon) = \beta_n^{(\pm)}(x,t,\varepsilon)$ in $\overline{D}_\beta^{(\pm)}$ and similarly we define $\alpha_n(x,t,\varepsilon)$. We can check it directly, that $\alpha_n(x,t,\varepsilon$ and $\beta_n(x,t,\varepsilon)$ satisfy the definition of the lower and upper solutions.

The existence theorem and its estimate for the solution follows from the differential inequalities theorem and from the structure of the upper and lower solutions (see, also, [3]), which we summarise in the following lemma.

Lemma 1. *The functions $\beta_n(x,t,\varepsilon)$ and $\alpha_n(x,t,\varepsilon)$ satisfies the following uniform in $\overline{D}$ estimates:*

$$\beta_n(x,t,\varepsilon) - \alpha_n(x,t,\varepsilon) = O(\varepsilon^n),$$
$$|\alpha_n(x,t,\varepsilon) - u_p(x,t,\varepsilon)| = O(\varepsilon^n),$$
$$|\beta_n(x,t,\varepsilon) - u_p(x,t,\varepsilon)| = O(\varepsilon^n) \tag{16}$$
$$\frac{\partial \alpha_n}{\partial x} = \frac{\partial u_p}{\partial x} + O(\varepsilon^{n-1}),\quad \frac{\partial \beta_n}{\partial x} = \frac{\partial u_p}{\partial x} + O(\varepsilon^{n-1}).$$

where $u_p(x,t,\varepsilon)$ is the periodic internal layer solution of problem (9), stated in the Theorem 2.

Stability Results. In this section we investigate the stability (in the sense of Lyapunov) of the periodic solution $u_p(x, t, \varepsilon)$ established by Theorem 2. We use the Theorem 1. From the construction of lower and upper solutions it follows that they are asymptotic order of $q = n + 1$. Direct calculations shows that the constant p in our case is $2n - 2$ (analogous calculations one can found in [2]). Therefore inequality $p \geq q$ is satisfied for $n \geq 3$. We formulate our result as the theorem.

Theorem 3. *Suppose the assumptions $(H_1) - (H_4)$ to be satisfied. Then for sufficiently small ε the periodic solution of problem (9) with interior layer is asymptotically stable with a local region of attraction $[\alpha_3(x, t, \varepsilon), \beta_3(x, t, \varepsilon)]$.*

Acknowledgements. This work is supported by RFBR, pr. N 13-01-91333.

References

1. Vasilieva, A.B., Butuzov, V.F., Nefedov, N.N.: Contrast structures in singularly perturbed problems. Fundamentalnaja i prikladnala matematika **4**(3), 799–851 (1998)
2. Nefedov, N.: Comparison principle for reaction-diffusion-advection problems with boundary and internal layers. In: Dimov, I., Faragó, I., Vulkov, L. (eds.) NAA 2012. LNCS, vol. 8236, pp. 62–72. Springer, Heidelberg (2013)
3. Nefedov, N.N., Recke, L., Schnieder, K.R.: Existence and asymptotic stability of periodic solutions with an interior layer of reaction-advection-diffusion equations. J. Math. Anal. Appl. **405**, 90–103 (2013)
4. Hess, P.: Periodic-Parabolic Boundary Value Problems and Positivity, Pitman Research Notes in Math. Series 247. Longman Scientific and Technical, Harlow (1991)
5. Zabrejko, P.P., Koshelev, A.I., et al.: Integral Equations. M.Nauka, Moscow (1968)
6. Nefedov, N.N.: The method of differential inequalities for some classes of nonlinear singularly perturbed problems with internal layers. Differ. Uravn. **31**(7), 1142–1149 (1995)
7. Volkov, V., Nefedov, N.: Asymptotic-numerical investigation of generation and motion of fronts in phase transition models. In: Dimov, I., Faragó, I., Vulkov, L. (eds.) NAA 2012. LNCS, vol. 8236, pp. 524–531. Springer, Heidelberg (2013)
8. Franz, S., Roos, H.-G.: The capriciousness of numerical methods for singular perturbations. SIAM rev. **53**, 157–173 (2011)
9. Franz, S., Kopteva, N.: Green's function estimates for a singularly perturbed convection-diffusion problem. J. Diff. Eq. **252**, 1521–1545 (2012)
10. Kopteva, N.: Numerical analysis of a 2d singularly perturbed semilinear reaction-diffusion problem. In: Margenov, S., Vulkov, L.G., Waśniewski, J. (eds.) NAA 2008. LNCS, vol. 5434, pp. 80–91. Springer, Heidelberg (2009)
11. Vasileva, A.B., Butuzov, V.F., Nefedov, N.N.: Singularly perturbed problems with boundary and internal layers. Proc. Steklov Inst. Math. **268**, 258–273 (2010)

The Finite Difference Method for Boundary Value Problem with Singularity

Viktor A. Rukavishnikov$^{(\boxtimes)}$ and Elena I. Rukavishnikova

Computing Center, Far-Eastern Branch, Russian Academy of Sciences,
Kim-Yu-Chena St. 65, 680000 Khabarovsk, Russian Federation
vark0102@mail.ru

Abstract. For boundary value problems with singularity, we developed the theory of finite difference schemes based on concept of an R_ν-generalized solution. The difference scheme is constructed, the rate of convergence of the approximate solution to the R_ν-generalized solution in the norm of the Sobolev weighted space is established.

Keywords: The R_ν-generalized solution for BVP · Finite difference method

1 Introduction

For boundary value problems with a singular solution, we developed the theory of numerical methods on the concept of an R_ν-generalized solution (see, e.g., [1–18]). We can define a weighted space or set containing the unique R_ν-generalized solution depending on the singularities of the input data (the coefficients and right-hand sides of the equation and the boundary conditions) and on the geometry of the domain boundary. For boundary value problems with singularity, it was constructed of the numerical methods with the convergence rate independent of the size singularity, for example, of the reentrant angle on the boundary of the domain.

In this paper we construct a finite difference scheme for boundary value problem in the rectangle. We study the rate with which an approximate solution by the proposed difference scheme converges toward an exact R_ν-generalized solution in the difference norm of the Sobolev weighted space.

2 Notation and Auxiliary Statements

Let $\Omega = \{x : x = (x_1, x_2), -l_1 < x_1 < l_2, 0 < x_2 < l_2\}$ be a rectangle with boundary $\partial\Omega$; and let $\overline{\Omega}$ be the closure of Ω, i.e. $\overline{\Omega} = \Omega \cup \partial\Omega$. The boundary $\partial\Omega = \bigcup_{i=1}^{4} \partial\Omega^{(i)}$, where sides of the rectangle $\partial\Omega^{(i)}$, $i = 1, 2, 3, 4$ are defined by the equations $x_1 = -l_1$, $x_2 = 0$, $x_1 = l_1$, $x_2 = l_2$ respectively. In addition, we denote by $\partial\Omega^{(5)} = \{x : x = (x_1, 0), -l_1 \leq x_1 \leq 0\}$, $\Gamma = \partial\Omega^{(2)} \setminus \partial\Omega^{(5)}$. Let $\partial\Omega^0$ be a set

© Springer International Publishing Switzerland 2015
I. Dimov et al. (Eds.): FDM 2014, LNCS 9045, pp. 72–83, 2015.
DOI: 10.1007/978-3-319-20239-6_7

of points $\{\tau_i\}_{i=1}^5 = \left\{ \bigcup_{i=1}^{4} \partial\Omega^{(i-1,i)} \cup \partial\Omega^{(4,5)} \right\}$. Here $\partial\Omega^{(i-1,i)} = \partial\Omega^{(i-1)} \cap \partial\Omega^{(i)}$
$(i = 1,\ldots,4)$ and $\partial\Omega^{(0,1)} = \partial\Omega^{(4,1)}$, $\partial\Omega^{(0)} = \partial\Omega^{(4)}$, $\partial\Omega^{(4,5)} = \Gamma \cap \partial\Omega^{(5)}$.

Let O_i^δ be a disk of radius $\delta > 0$ with its center in τ_i $(i = 1,\ldots,5)$, i.e. $O_i^\delta = \{x : \|x - \tau_i\| \le \delta\}$, and supopse that $O_i^\delta \cap O_j^\delta = \emptyset$, $i \ne j$. Let $\Omega_1 = \Omega \cap \bigcup_{i-1}^{n} O_i^\delta$.

We introduce a weight function $\rho(x)$ which coincides with the distance to the point τ_i in O_i^δ and it is equal to δ outside Ω_1.

We introduce the weighted spaces with norms:

$$\|v\|^2_{W_{2,\alpha}^k(\Omega)} = \sum_{|\lambda|\le k} \int_\Omega \rho^{2\alpha} |D^\lambda v|^2 dx, \quad \|v\|_{W_{2,0}^k(\Omega)} = \|v\|_{W_2^k(\Omega)},$$

$$\|v\|_{W_{2,\alpha}^0(\Omega)} = \|v\|_{L_{2,\alpha}(\Omega)}, \quad |v|^2_{W_{2,\alpha}^s(\Omega)} = \int_\Omega \rho^{2\alpha} |D^s v|^2 dx, \ s \le k,$$

$$\|v\|^2_{H_{2,\alpha}^k(\Omega)} = \sum_{|\lambda|\le k} \int_\Omega \rho^{2\alpha+2|\lambda|-2k} |D^\lambda v|^2 dx, \tag{1}$$

where $D^\lambda = \dfrac{\partial^{|\lambda|}}{\partial x_1^{\lambda_1} \partial x_2^{\lambda_2}}$, $\lambda = (\lambda_1, \lambda_2)$ and $|\lambda| = \lambda_1 + \lambda_2$, k is a nonnegative integer, and α is a real number.

The space $\overline{H}_{2,\alpha}^1(\Omega)$ is defined as the closures of the set of infinitely differentiable in $\mathbb{R}^2$ functions with support Ω and vanish in a neighbourhood $\partial\Omega \setminus \Gamma$.

Let γ is subset $\partial\Omega$. We say that $\varphi \in H_{2,\alpha}^{k-1/2}(\gamma)$ if there exists a function $\Phi(x)$ from $H_{2,\alpha}^k(\Omega)$ such that $\Phi(x)|_\gamma = \varphi(x)$ and

$$\|\varphi\|_{H_{2,\alpha}^{k-1/2}(\gamma)} = \inf_{\Phi|_\gamma = \varphi} \|\Phi\|_{H_{2,\alpha}^k(\Omega)}.$$

Lemma 1 [2]. *Let k be a nonnegative integer:*

(A) If $v \in H_{2,\alpha}^k(\Omega)$, then $\rho^{\alpha-(k-s)} v \in W_{2,0}^s(\Omega)$ $(s = 0,\ldots,k)$ and

$$|\rho^\alpha v|_{W_{2,0}^k(\Omega)} + |\rho^{\alpha-1} v|_{W_{2,0}^{k-1}(\Omega)} + \ldots + |\rho^{\alpha-k} v|_{W_{2,0}^{-1}(\Omega)} \le C_1 \|v\|_{H_{2,\alpha}^k(\Omega)},$$

where C_1 is a positive constant independent of v.
(B) If $\rho^{\alpha-(k-s)} v \in W_{2,0}^s(\Omega)$ $(s = 0,\ldots,k)$, then $v \in H_{2,\alpha}^k(\Omega)$ and there exist positive constants $C_0^*,\ldots,C_k^*$ independent of v such that

$$C_k^* |\rho^\alpha v|_{W_{2,0}^k(\Omega)} + C_{k-1}^* |\rho^{\alpha-1} v|_{W_{2,0}^{k-1}(\Omega)} + \ldots + C_0^* |\rho^{\alpha-k} v|_{L_{2,0}(\Omega)} \le \|v\|_{H_{2,\alpha}^k(\Omega)}.$$

Lemma 2. *Let $Q \subset \mathbb{R}^2$ be an open subset with the Lipschitz continuous boundary and let $F(v)$ is a bounded continuous linear functional on $W_2^{k+1}(Q)$ $(k \ge 0)$. Suppose that*

$$\forall p \in P_k = \left\{ \sum_{i=0}^{k} \sum_{j=0}^{k-i} a_{ij} x_1^i x_2^j \right\} \quad |F(p)| \le \delta_2, \ \delta_2 = const.$$

Then there exists constant $\delta_3(Q)$ such that

$$\forall v \in W_2^{k+1}(Q) \quad |F(v)| \le \delta_3(Q)|v|_{W_2^{k+1}(Q)} + \delta_2.$$

The proof of this statement follows from the Bramble-Hilbert lemma directly [19].

3   Problem Formulation

Consider the differential equation

$$Av \equiv -\sum_{l=1}^{2} \frac{\partial}{\partial x_l}\left(a_l(x)\frac{\partial v}{\partial x_l}\right) + a(x)v(x) = f(x), \quad x \in \Omega \tag{2}$$

with boundary conditions

$$\frac{\partial v}{\partial N} = \varphi_1(x), \quad x \in \Gamma, \qquad v = \varphi_2(x), \quad x \in \partial\Omega \setminus \Gamma. \tag{3}$$

Here $\dfrac{\partial v}{\partial N}$ is a conormal derivative.

Assume that the coefficients of the equation and the right-hand sides (2) and (3) satisfy

$$a_l \in C_3(\Omega), \quad a \in C_2(\Omega), \quad a(x) \ge a_0 > 0, \tag{4}$$

$$\sum_{l=1}^{2} a_l(x)\xi_l^2 \ge \lambda \sum_{l=1}^{2} \xi_l^2, \quad x \in \overline{\Omega}, \tag{5}$$

$$f \in H_{2,\mu}^2(\Omega), \quad \varphi_1 \in H_{2,\mu}^{5/2}(\Gamma), \quad \varphi_2 \in H_{2,\mu}^{7/2}(\partial\Omega \setminus \Gamma), \tag{6}$$

where λ is positive constant independent of x, ξ_1 and ξ_2 are any real parameters, and μ is some nonnegative real number.

Definition 1. *A function v_ν form the space $H_{2,\nu}^1(\Omega)$ is called an R_ν-generalized solution of boundary value problem (2), (3) if $v_\nu = \varphi_2$ almost everywhere on $\partial\Omega \setminus \Gamma$ and for all g from $\overline{H}_{2,\nu}^1(\Omega)$ the following integral identity*

$$\int_{\Omega}\left[\sum_{l=1}^{2}\rho^{2\nu}a_l\frac{\partial v_\nu}{\partial x_l}\frac{\partial g}{\partial x_l} + a_l\frac{\partial v_\nu}{\partial x_l}\frac{\partial \rho^{2\nu}}{\partial x_l}g + a\rho^{2\nu}v_\nu g\right]dx$$

$$= \int_{\Omega}\rho^{2\nu}fgdx + \int_{\Gamma}\rho^{2\nu}\varphi_1 gdx_1, \tag{7}$$

holds, where ν is arbitrary but fixed and satisfies the inequality $\nu \ge \mu - 3$.

The existence and uniqueness, the coercitive and differential properties of the R_ν-generalized solution for the differential equation (2) with different boundary conditions were established in [6,8,9,18]. In this paper we do not propose to study the differentiability properties of the R_ν-generalized solution. Therefore we assume that v_ν belongs to the space $H_{2,\mu}^4(\Omega)$.

4   Difference Scheme

Let us construct a difference mesh in $\overline{\Omega}$ by analogy with [1]. We briefly explain the construction: $\overline{\Omega}_h = \{x_h : x_h = ((i_1 - 0.5\,\mathrm{sign}(i_1))h_1, i_2 h_2), h_1 = 2l_1/2N_1 - 1, h_2 = l_2/N_2, i_1 = -N_1, \ldots, -1, 1, \ldots, N_1, i_2 = 0, \ldots, N_2\}$, $\Omega_h = \{x_h : x_h = ((i_1 - 0.5\,\mathrm{sign}(i_1))h_1, i_2 h_2), -N_1 < i_1 < N_1, i_1 \neq 0, 0 < i_2 < N_2\}$, $\partial\Omega_h = \overline{\Omega}_h \backslash \Omega_h$, $\Gamma_h = \{x_h : x_h = ((i_1 - 0.5\,\mathrm{sign}(i_1))h_1, 0), 1 \leq i_1 < N_1\}$, $\partial\Omega_h^{(1)} = \{x_h : x_h = (-l_1, i_2 h_2), 0 \leq i_2 \leq N_2\}$, $\partial\Omega_h^{(1+2)} = \{x_h : x_h = (l_1, i_2 h_2), 0 < i_2 < N_2\}$, $\partial\Omega_h^{(5)} = \partial\Omega_h^{(2)} \backslash \Gamma_h$, $\partial\Omega_h^{(1,2)} = \partial\overline{\Omega}_h^{(1)} \cap \partial\overline{\Omega}_h^{(2)}$, $\partial\Omega_h^{(5)} \cap \Gamma_h = \{x_h : x_h = (-h_1, 0)\}$.

Let u and w be mesh functions given on $\overline{\Omega}_h$. The difference quotients, inner products, and norms on $\overline{\Omega}_h$ are defined as follows:

$$u = u(x_h) = u(x_{1h}, x_{2h}), \quad I_{\pm 1}u(x_h) = u(x_{1h} \pm h_1, x_{2h}), \quad I_{\pm 2}u(x_h) = u(x_{1h}, x_{2h} \pm h_2),$$

$$\partial_1^+ u = (I_{+1}u - u)/h_1, \quad \partial_1^- u = (u - I_{-1}u)/h_1, \quad \partial_1^+ \partial_1^- u = (I_{+1}u - 2u + I_{-1}u)/h_1^2,$$

$$(u, w)_{L_{2,\alpha}(\overline{\Omega}_h)} = \sum_{x_h \in \overline{\Omega}_h} \rho_h^{2\alpha} u w \hbar, \quad \|u\|_{L_{2,\alpha}(\overline{\Omega}_h)}^2 = (u, u)_{L_{2,\alpha}(\overline{\Omega}_h)},$$

$$(\partial_l^+ u, \partial_l^+ w)_{L_{2,\alpha}(\overline{\Omega}_h)} = \sum_{x_h \in \overline{\Omega}_h\binom{+}{l}} \rho_h^{2\alpha} \partial_l^+ u \partial_l^+ w \hbar,$$

$$|u|_{W_{2,\alpha}^1(\overline{\Omega}_h)}^2 = \frac{1}{2} \sum_{l=1}^2 \left[(\partial_l^+ u, \partial_l^+ u)_{L_{2,\alpha}(\overline{\Omega}_h)} + (\partial_l^- u, \partial_l^- u)_{L_{2,\alpha}(\overline{\Omega}_h)} \right],$$

$$\|u\|_{W_{2,\alpha}^1(\overline{\Omega}_h)}^2 = |u|_{W_{2,\alpha}^1(\overline{\Omega}_h)}^2 + \|u\|_{L_{2,\alpha}(\overline{\Omega}_h)}^2,$$

$$/u, w/_{L_{2,\alpha}(\partial\overline{\Omega}_h)} = \sum_{l=1}^2 \sum_{x_h \in \partial\overline{\Omega}_h\binom{l}{l+2}} \rho_h^{2\alpha} u w \hbar^l, \quad \|u\|'_{L_{2,\alpha}(\partial\overline{\Omega}_h)} = /u, u/_{L_{2,\alpha}(\partial\overline{\Omega}_h)}^{1/2}.$$

Here $\Omega_h\binom{+}{l}$, $\Omega_h\binom{-}{l}$ are the sets of nodes of the mesh $\overline{\Omega}_h$ at which the difference quotients ∂_l^+ and ∂_l^- are defined, $\partial\Omega_h\binom{l}{l+2} = \partial\Omega_h^{(l)} \cup \partial\Omega_h^{(l+2)}$, and

$$\hbar = \hbar_1 \hbar_2, \quad \hbar^l = \prod_{\substack{i=1 \\ i \neq l}}^2 \hbar_i, \quad \hbar_l = \begin{cases} \hbar_l, & x_h \in \overline{\Omega}_h \backslash \partial\overline{\Omega}_h\binom{l}{l+2}, \\ \hbar_l/2, & x_h \in \partial\overline{\Omega}_h\binom{l}{l+2}. \end{cases}$$

In this paper we will assume $h_1 = h_2$.
The mesh function $\rho_h(x_h)$ is defined of the following rule:

$$\rho_h(x_h) = \min\left\{ h^{1-\beta/2}, \min_{i=1,\ldots,5} \mathrm{dist}(x_h, \tau_i) \right\}, \quad 0 \leq \beta \leq 1.$$

Let Ω_h' is the set of nodes of the mesh at which function $\rho_h(x_h)$ is $\mathrm{dist}(x_h, \tau_i)$. We choose β such that $\rho_h(x_h) \leq \rho(x_h)$ for $x_h \in \overline{\Omega}_h$ and $\Omega_h' \subset \overline{\Omega}_1$.

To determine an R_ν-generalized solution of problem (2), (3) we use the difference scheme

$$\overline{A}_{h\rho}u = F_{h\rho}, \quad x_h \in \overline{\Omega}_h, \tag{8}$$

where

$$\overline{A}_{h\rho}u = \begin{cases} A_{h\rho}u, \, x_h \in \Omega_h, \\ a_{h\rho}u, \, x_h \in \partial\Omega_h; \end{cases} \quad F_{h\rho}u = \begin{cases} \rho_h^\nu f_h, & x_h \in \Omega_h, \\ (h_2/2)\rho_h^\mu f_h + \rho_h^\mu \varphi_{1h}, & x_h \in \Gamma_h, \\ \rho_h^\nu \varphi_{2h}, \, x_h \in \partial\Omega_h \setminus \Gamma_h; \end{cases}$$

$$A_{h\rho}u \equiv -\frac{1}{2}\sum_{l=1}^{2}\left[\partial_l^+\left(a_l\rho_h^\nu\partial_l^- u\right) + \partial_l^-\left(a_l\rho_h^\nu\partial_l^+ u\right)\right] + a\rho_h^\nu u,$$

$$a_{h\rho} \equiv -\frac{\rho_h^\nu a_2 + I_{+2}(\rho_h^\nu a_2)}{2}\partial_2^+ u - \frac{h_2}{2}\frac{1}{2}\left(\partial_1^+\left(a_1\rho_h^\nu\partial_1^- u\right) + \partial_1^-\left(a_1\rho_h^\nu\partial_1^+ u\right)\right)$$

$$+ \frac{h_2}{2}a\rho_h^\nu u, \, x_h \in \Gamma_h,$$

$$a_{h\rho}u \equiv \rho_h^\nu u, \, x_h \in \partial\Omega_h \setminus \Gamma_h.$$

5   Approximation Error

For the research of the approximation error of the difference problem (5), for $z = [v_\nu]_h - u$ we pose the problem

$$\overline{A}_{h\rho}z = \psi, \quad x_h \in \overline{\Omega}_h, \tag{9}$$

where $\psi(x_h) = \overline{A}_{h\rho}[v_\nu]_h - F_{h\rho}(x_h)$ is the approximation error of the difference scheme at an R_ν-generalized solution of the differential problem.
We represent $\psi(x_h)$ in the form of a sum

$$\psi = \psi_1 + \psi_2, \tag{10}$$

here for $x_h \in \Omega_h$

$$\psi_1(x_h) = \frac{1}{2}\sum_{l=1}^{2}\rho_h^\nu\left[\partial_l^+\left(a_l\partial_l^-[v_\nu]_h\right) + \partial_l^-\left(a_l\partial_l^+[v_\nu]_h\right)\right] + a\rho_h^\nu[v_\nu]_h - \rho_h^\nu f_h,$$

$$\psi_2(x_h) = -\frac{1}{2}\sum_{l=1}^{2}\left[\partial_l^-\rho_h^\nu a_l\partial_l^-[v_\nu]_h + \partial_l^+\rho_h^\nu a_l\partial_l^+[v_\nu]_h\right];$$

for $x_h \in \Gamma_2$

$$\psi_1(x_h) = -\rho_h^\nu\frac{a_2 + I_{+2}a_2}{2}\partial_2^+[v_\nu]_h - \frac{h_2}{2}\frac{1}{2}\left(\rho_h^\nu\partial_1^+\left(a_1\partial_1^-[v_\nu]_h\right) + \partial_1^-\left(a_1\partial_1^+[v_\nu]_h\right)\right)$$

$$+ \frac{h_2}{2}a\rho_h^\nu[v_\nu]_h - \frac{h_2}{2}\rho_h^\nu f_h - \rho_h^\nu\varphi_{1h},$$

$$\psi_2(x_h) = -\frac{h}{2}I_{+2}a_2\partial_2^+\rho_h^\nu\partial_2^+[v_\nu]_h - \frac{h_2}{4}\left(a_1\partial_1^+\rho_h^\nu\partial_1^+[v_\nu]_h + a_1\partial_1^-\rho_h^\nu\partial_1^-[v_\nu]_h\right);$$

for $x_h \in \partial\Omega_h \setminus \Gamma_h$ error is $\psi(x_h) = 0$.

Lemma 3. *Let the R_ν-generalized solution v_ν of the boundary value problem* (2), (3) *belongs to the space $H_{2,\nu}^4(\Omega)$ ($\nu \geq \mu$) and conditions* (4)–(6) *be satisfied. Then the estimate*

$$|\psi_1(\overline{x}_h)| \leq c_2|h|^2(h_1h_2)^{-1/2}\|v_\nu\|_{H_{2,\nu}^4(\omega_{i_1i_2})} + O(|h|^2), \quad \overline{x}_h \in \Omega_h \cap \omega_{i_1i_2}, \quad (11)$$

where c_2 is independent of h and v_ν, $\omega_{i_1i_2} = \{x : (x_1,x_2), (i_k - 1)h_k < x_k < (i_k + 1)h_k, k = 1,2\}$.

Lemma 4. *Suppose $v_\nu \in H_{2,\nu}^4(\Omega)$ and the conditions* (3) *be realized. Then for $\psi_2(x_h)$, $x_h \in \Omega_h'$ the inequality*

$$|\psi_2(x_h)| \leq c_3\rho^2(x_h)(h_1h_2)^{-1/2}\left(|v_\nu|_{W_{2,\nu-3}^1(\omega_{i_1i_2})} + |v_\nu|_{W_{2,\nu-2}^2(\omega_{i_1i_2})}\right) \quad (12)$$

holds, where c_3 is independent of h and v_ν.

The Lemma 2 was proved in [3,4]. The proof of Lemma 3 for $a_l = 1$ can be found in [4]. The condition (4) ensures the validity of the Lemma 3 and Lemma 4 for problem (9). Moreover for $a_l \neq 1$ $\psi_1(p) = O(|h|^2)$ for any $p(x)$ from P_3. This explains the presence of the term $O(|h|^2)$ on the right-hand side (11).

Lemma 5. *Suppose that the conditions of Lemma 3 are satisfied. Then for $\psi_1(\overline{x}_h)$, $\overline{x}_h \in \omega_{i_10} \cap \Gamma_h$ the following estimate is valid*

$$|\psi_1(\overline{x}_h)| \leq c_4|h|^3(h_1h_2)^{-1/2}\|v_\nu\|_{H_{2,\nu}^4(\omega_{i_10})} + O(|h|^2), \quad (13)$$

where c_4 is a positive constant not dependent the functions f and v_ν, $\omega_{i_10} = \{x : x = (x_1,x_2), (i_1 - 1)h_1 < x_1 < (i_1 + 1)h_1, 0 \leq x_2 < h\}$.

Proof. We map ω_{i_10} onto the rectangle $\Pi = \{y : y = (y_1,y_2), -1 < y_1 < 1, 0 < y_2 < 1\}$ by means of the linear transformation $y_1 = (x_1 - i_1h_1)/h_1$, $y_2 = x_2/h_2$. If the function $\varphi(x)$ is defined on ω_{i_10}, then we denote by $\overline{\varphi}(y)$ the function which is defined on Π by $\overline{\varphi}(y) = \varphi(y_1h_1 + i_1h_1, y_2h_2)$. We estimate $|\psi_1(\overline{x}_h)|$, $x_h \in \omega_{i_10} \cap \Gamma_h$, by the norm of the function $\rho^\nu(y)v_\nu(y)$ in the space $W_2^4(\Pi)$. We have

$$\left|\rho_h^\nu(\overline{x}_h)\frac{a_2(\overline{x}_h) + I_{+2}a_2(\overline{x}_h)}{2}\partial_2^+[v_\nu]_h\right| \leq \frac{c_5}{h_2}\max_{x \in \overline{\omega}_{i_10}}|\rho^\nu(x)v_\nu(x)|$$

$$\leq \frac{c_5}{h_2}\max_{y \in \overline{\Pi}}|\overline{\rho}^\nu(y)\overline{v}_\nu(y)| \leq c_5|h|^{-1}\|\overline{\rho}^\nu\overline{v}_\nu\|_{W_2^2(\Pi)}, \quad (14)$$

$$\left|\frac{h_2}{4}\rho^\nu(\overline{x}_h)\left(\partial_1^+(a_1\partial_1^-[v_\nu]_h) + \partial_1^-(a_1\partial_1^+[v_\nu]_h)\right)\right| \leq c_6|h|^{-1}\|\overline{\rho}^\nu\overline{v}_\nu\|_{W_2^2(\Pi)}, \quad (15)$$

$$\left|\frac{h_2}{2}a(\overline{x}_h)\rho_h^\nu(\overline{x}_h)[v_\nu]_h\right| \leq c_7|h|^{-1}\|\overline{\rho}^\nu\overline{v}\|_{W_2^2(\Pi)}. \quad (16)$$

Due to conditons (4) and the Sobolev embedding theorems the inequalities (14)–(16) are valid; the norm of the function $\overline{\rho}^\nu \overline{v}$ in the space $W_2^2(\Pi)$ is defined because $v \in H_{2,\nu}^4(\Omega)$.

Note that, if the function $\varphi_1 \in H_{2,\mu}^{5/2}(\Gamma)$, then there exists its extension $\Phi_1(x)$ to Ω in the space $H_{2,\mu}^3(\Omega)$. It is obvious that $\Phi_1 \in H_{2,\nu}^2(\Omega)$ and by Lemma 1 $\rho^\nu \Phi_1$ belongs to $H_2^2(\Omega)$. Besides the inequalities (14)–(16) we have

$$\left| \frac{h_2}{2} \rho_h^\nu(\overline{x}_h) f_h(\overline{x}_h) + \rho_h^\nu(\overline{x}_h)\varphi_{1h}(\overline{x}_h) \right| \le \max_{x \in \overline{\omega}_{i_1 0\varepsilon}} \left| \frac{h_2}{2} \rho^\nu(x) f(x) + \rho^\nu(x)\Phi_1(x) \right|$$

$$\le \max_{y \in \overline{\Pi}_\varepsilon} \left| \frac{h_2}{2} \overline{\rho}^\nu(y)\overline{f}(y) + \overline{\rho}^\nu(y)\overline{\Phi}_1(y) \right| \le c_8 \left\| \frac{h_2}{2}\overline{\rho}^\nu \overline{f} + \overline{\rho}^\nu \overline{\Phi}_1 \right\|_{W_2^2(\Pi_\varepsilon)}. \quad (17)$$

Here $\omega_{i_1 0\varepsilon}$ is a set of points in Ω belonging to a semi-circle with radius $h/2 - \varepsilon$ and with center at the point $\overline{x}_h$ (ε is an arbitrary positive number such that $\varepsilon < h/2$); Π_ε is the image of the domain $\omega_{i_1 0\varepsilon}$ which is obtained by means of the transformation $y_1 = (x_1 - i_1 h_1)/h_1$, $y_2 = x_2/h_2$.

We estimate the semi-norms of the function $\frac{h_2}{2}\overline{\rho}^\nu(y)\overline{f}(y) + \overline{\rho}^\nu(y)\overline{\Phi}_1(y)$ in the spaces $L_2(\Pi_\varepsilon)$, $W_2^1(\Pi_\varepsilon)$, $W_2^2(\Pi_\varepsilon)$ by the norm $\|\overline{\rho}^\nu \overline{v}_\nu\|_{W_2^4(\Pi)}$.

We denote by $Q_{i_1 0\varepsilon}$ a set which is defined by the equality $Q_{i_1 0\varepsilon} = \omega_{i_1 0} \setminus \omega_{i_1 0\varepsilon/2}$. Let $\delta_\varepsilon(x)$ be a cut-off function that is infinitely differentiable and for a sufficiently small ε satisfies the following conditions: $\delta_\varepsilon(x) = 1$ for $x \in \omega_{i_1 0\varepsilon}$; $\delta_\varepsilon(x) = 0$ for $x \in Q_{i_1 0\varepsilon}$; $0 \le \delta_\varepsilon(x) \le 1$ for $x \in \omega_{i_1 0}$.

Then, we substitute $g(x)$ by the function $\delta(x)g_0(x)$ in (7) (where $\delta(x)$ is the function $\delta_\varepsilon(x)$, which is extended by zero on Ω, $g_0(x)$ is an arbitrary function in the space $H_{2,\nu}^1$) and integrate the obtained equality over x_2 on $(0, h_2/2)$. As a result we obtain

$$\frac{h_2}{2} \int_{\omega_{i_1 0}} \left[\sum_{l=1}^{2} \rho^{2\nu} a_l \frac{\partial v_\nu}{\partial x_l} \frac{\partial(\delta g_0)}{\partial x_l} + a_l \frac{\partial \rho^{2\nu}}{\partial x_l} \frac{\partial v_\nu}{\partial x_l} \delta g_0 + \rho^{2\nu} a v_\nu \delta g_0 \right] dx$$

$$= \frac{h_2}{2} \int_{\omega_{i_1 0}} \rho^{2\nu} f \delta g_0 dx + \int_{\omega_{i_1 0}} \rho^{2\nu} \Phi_1 \delta g_0 dx. \quad (18)$$

Successively, we substitute in equation (18) instead of g_0 the functions

$$\left(\frac{h_2}{2} f + \Phi_1 \right)\delta, \quad \sum_{s=1}^{2} \frac{1}{\rho^\nu} \Delta_s^- \Delta_s^+ \left(\delta \rho^\nu \left(\frac{h_2}{2} f + \Phi_1 \right) \right),$$

$$\sum_{a,l=1}^{2} \frac{1}{\rho^\nu} \Delta_l^- \Delta_s^- \Delta_l^+ \Delta_s^+ \left(\delta \rho^\nu \left(\frac{h_2}{2} f + \Phi_1 \right) \right),$$

where $\Delta_1^- \varphi(x) = \dfrac{\varphi(x_1 - H, x_2) - \varphi(x)}{-H}$, $\Delta_1^+ \varphi(x) = \dfrac{\varphi(x_1 + H, x_2) - \varphi(x)}{H}$, $0 < H < \dfrac{\varepsilon}{2}$.

As a result, by analogy with Lemma 1 in [3], we establish the estimate

$$\left| \frac{h_2}{2} \rho_h^\nu(\overline{x}_h) f(\overline{x}_h) + \rho_h^\nu(\overline{x}_h) \varphi_{1h}(\overline{x}_h) \right| \leq c_9 |h|^{-1} \|\overline{\rho}^\nu \overline{v}_\nu\|_{W_2^4(\Pi)}. \tag{19}$$

Now, we denote that, if the R_ν-generalized solution $v_\nu \in P_3(\overline{\Pi})$, then it is a classical solution of the problem (2), (3), moreover

$$\frac{-a_2(x_h) + I_{+2} a_2(x_h)}{2} \partial_2^+[v_\nu]_h = -a_2(x_h) \frac{\partial v_\nu(x_h)}{\partial x_2} - \frac{h_2}{2} \frac{\partial}{\partial x_2} \left(a_2(x_h) \frac{\partial v_\nu(x_h)}{\partial x_2} \right) + O(h^2),$$

$$-\frac{1}{2} \left(\partial_1^+(a_1(x_h) \partial_1^-[v_\nu]_h) + \partial_1^-(a_1(x_h) \partial_1^+[v_\nu]_h) \right) = -\frac{\partial}{\partial x_1} \left(a_1(x_h) \frac{\partial v_\nu(x_h)}{\partial x_1} \right) + O(h^2),$$

$$x_h \in \Gamma_h,$$

and $\varphi_1(x_h) = O(h^2)$ for $x_h \in \Gamma_h$.
By using this remark and inequalities (14)–(16), (19), we obtain the estimate

$$|\psi_1(\overline{x}_h)| \leq c_{10} |h|^{-1} |\overline{\rho}^\nu \overline{v}_\nu|_{W_2^4(\Pi)} + O(h^2) \leq c_{11} |h|^3 (h_1 h_2)^{-1/2} |\rho^\nu v_\nu|_{W_2^4(\omega_{i_1 0})} + O(h^2)$$

$$\leq c_4 |h|^3 (h_1 h_2)^{-1/2} \|v_\nu\|_{H_{2,\nu}^4(\omega_{i_1 0})} + O(h^2).$$

$\square$

Lemma 6. *Let $v_\nu \in H_{2,\nu}^4(\Omega)$. Then for $\psi_2(x_h)$, $x_h \in \Gamma_h \cap \overline{\Omega}_h'$ the following estimate is valid*

$$|\psi_2(x_h)| \leq c_{12} \rho_h^2(x_h) \left(|v_\nu|_{H_{2,\nu-3}^1(\omega_{i_1 0})} + |v_\nu|_{H_{2,\nu-2}^2(\omega_{i_1 0})} \right), \tag{20}$$

where c_{12} is a positive constant independent of h, ρ_h and v_ν.

By analogy with Lemma 2 from [4], we estimate each addend of $\psi_2(x_h)$ for $x_h \in \Gamma_h \cap \overline{\Omega}_h'$ in absolute value, except of the term $\left| \frac{h_2 a_1}{4} \partial_1^- \rho_h^\nu \partial_1^-[v_\nu]_h \right|$. But this term is equal zero, since $\partial_1^- \rho_h^\nu \left(\frac{h_1}{2}, 0 \right) = 0$.

6   Convergence and Stability

In this section we obtain the weak stability of the difference scheme (8) and we get the estimate of the rate of convergence of the solution u this scheme to an R_ν-generalized solution of the original problem (2), (3).

Lemma 7. *Let $\overline{A}_{h\rho}$ be the operator defined in (8); assume that conditions (4), (5) are satisfied. Then for any function w given on $\overline{\Omega}_h$ the inequality*

$$|h|^{-\frac{\nu\beta}{4}} \left(\|A_{h\rho} w\|_{L_2(\Omega_h)} + \|a_{h\rho} w\|'_{L_2(\Gamma_h)} + |h|^{-1} \|a_{h\rho} w\|'_{L_2(\partial \Omega_h \backslash \Gamma_h)} \right)$$

$$\geq c_{13} \|w\|_{W_{2,\nu}^1(\overline{\Omega}_h)} \tag{21}$$

holds, where c_{13} is positive constant independent of h and w.

Proof. By transforming the first term in $(A_{h\rho}w, w)_{\Omega_h}$ with the use of summation by parts and taking into account of boundary conditions, we get

$$(A_{h\rho}w, w)_{\Omega_h} = \left(-\frac{1}{2}\sum_{l=1}^{2}[\partial_l^+(a_l\rho_h^\nu\partial_l^-w) + \partial_l^-(a_l\rho_h^\nu\partial_l^+w)] + a\rho_h^\nu w, w\right)_{\Omega_h}$$

$$=\frac{1}{2}\sum_{l=1}^{2}\left[(a_l\rho_h^\nu\partial_l^-w, \partial_l^-w)_{\Omega_h\cup\partial\Omega_h^{(l+2)}} + (a_l\rho_h^\nu\partial_l^+w, \partial_l^+w)_{\Omega_h\cup\partial\Omega_h^{(l)}}\right] + (a\rho^\nu w, w)_{\Omega_h}$$

$$+\sum_{\substack{k,l=1\\k\neq l}}^{2}\left[\frac{a_l\rho_h^\nu + I_{+l}(a_l\rho_h^\nu)}{2}h_k\partial_l^+ww\bigg|_{\partial\Omega_h^{(l)}} - \frac{a_l\rho_h^\nu + I_{-l}(a_l\rho_h^\nu)}{2}h_k\partial_l^-ww\bigg|_{\partial\Omega_h^{(l+2)}}\right].$$

$$(22)$$

By virtue of boundary condition on Γ_h we have the equality

$$\bigg/\frac{a_2\rho_h^\nu + I_{+2}(a_2\rho_h^\nu)}{2}\partial_2^+w, w\bigg/_{\Gamma_h} = -\bigg/\frac{h_2}{2}\frac{1}{2}\left(\partial_1^+(a_1\rho_h^\nu\partial_1^-w) + \partial_1^-(a_1\rho_h^\nu\partial_1^+w)\right), w\bigg/_{\Gamma_h}$$

$$+\bigg/\frac{h_2}{2}a\rho_h^\nu w, w\bigg/_{\Gamma_h} - \bigg/\frac{h_2}{2}\rho_h^\nu f_h + \rho_h^\nu\varphi_{1h}, w\bigg/_{\Gamma_h}. \qquad (23)$$

By transforming the first term on the right-hand side (23) with the use of summation by parts, we have

$$-\bigg/\frac{h_2}{2}\frac{1}{2}\left(\partial_1^+(a_1\rho_h^\nu\partial_1^-w) + \partial_1^-(a_1\rho_h^\nu\partial_1^+w)\right), w\bigg/_{\Gamma_h}$$

$$= \frac{1}{2}\left[(a_1\rho_h^\nu\partial_1^-w, \partial_1^-w)_{\Gamma_h\cup(\partial\overline{\Omega}_h^{(3)}\cap\overline{\Gamma}_h)} + (a_1\rho_h^\nu\partial_1^+w, \partial_1^+w)_{\Gamma_h\cup(\partial\Omega_h^{(5)}\cap\overline{\Gamma}_h)}\right]$$

$$+\frac{h_2}{2}\frac{a_1\rho_h^\nu + I_{+1}(a_1\rho_h^\nu)}{2}\partial_1^+ww\bigg|_{\partial\Omega_h^{(5)}\cap\overline{\Gamma}_h} - \frac{h_2}{2}\frac{a_1\rho_h^\nu + I_{-1}(a_1\rho_h^\nu)}{2}\partial_1^-ww\bigg|_{\partial\overline{\Omega}_h^{(3)}\cap\overline{\Gamma}_h}.$$

$$(24)$$

Using (23), (24), we rewrite the equality (22) in the form

$$(A_{h\rho}w, w)_{\Omega_h} + /a_{h\rho}w, w/_{\Gamma_h}$$

$$= \frac{1}{2}\sum_{l=1}^{2}\left[(a_l\rho_h^\nu\partial_l^-w, \partial_l^-w)_{\Omega_h\cup\partial\Omega_h^{(l+2)}} + (a_l\rho_h^\nu\partial_l^+w, \partial_l^+w)_{\Omega_h\cup\partial\Omega_h^{(l)}}\right]$$

$$+\frac{1}{2}\left[(a_1\rho_h^\nu\partial_1^-w, \partial_1^-w)_{\Gamma_h\cup(\partial\overline{\Omega}_h^{(3)}\cap\overline{\Gamma}_h)} + (a_1\rho_h^\nu\partial_1^+w, \partial_1^+w)_{\Gamma_h\cup(\partial\Omega_h^{(5)}\cap\overline{\Gamma}_h)}\right]$$

$$+(a\rho^\nu w, w)_{\Omega_h\cap\Gamma_h} + \frac{a_1\rho_h^\nu + I_{+1}(a_1\rho_h^\nu)}{2}h_2\partial_1^+ww\bigg|_{\partial\Omega_h^{(1)}} - \frac{a_1\rho_h^\nu + I_{-1}(a_1\rho_h^\nu)}{2}h_2\partial_1^-ww\bigg|_{\partial\Omega_h^{(3)}}$$

$$+\frac{h_2}{2}\frac{a_1\rho_h^\nu + I_{+1}(a_1\rho_h^\nu)}{2}\partial_1^+ww\bigg|_{\partial\Omega_h^{(5)}\cap\overline{\Gamma}_h} - \frac{h_2}{2}\frac{a_1\rho_h^\nu + I_{-1}(a_1\rho_h^\nu)}{2}\partial_1^-ww\bigg|_{\partial\overline{\Omega}_h^{(3)}\cap\overline{\Gamma}_h}$$

$$+\frac{a_2\rho_h^\nu + I_{+2}(a_2\rho_h^\nu)}{2}h_1\partial_2^+ww\bigg|_{\partial\Omega_h^{(5)}} - \frac{a_2\rho_h^\nu + I_{-2}(a_2\rho_h^\nu)}{2}h_1\partial_2^-ww\bigg|_{\partial\Omega_h^{(4)}}. \qquad (25)$$

We multiply and divide by function $\rho_h^{\nu/2}$ the inner products of the left-hand side of the equality (25) and we estimate them with the use of the Cauchy-Schwartz inequality, the ε-inequality and the inequality

$$\|w\|'^2_{L_{2,\nu/2}(\Gamma_h)} \le c_{14}\left(|w|^2_{W^1_{2,\nu/2}(\Omega_h \cup \Gamma_h)} + \|w\|'^2_{L_{2,\nu/2}(\partial\Omega_h^{(4)})}\right).$$

As a result we obtain

$$(A_{h\rho}w, w)_{\Omega_h} + /a_{h\rho}w, w/_{\Gamma_h} \le \frac{1}{4\varepsilon_1}\left(\|A_{h\rho}w\|^2_{L_{2,-\nu/2}(\Omega_h)} + \|a_{h\rho}w\|'^2_{L_{2,-\nu/2}(\Gamma_h)}\right)$$
$$+ \varepsilon_1\left(\|w\|^2_{L_{2,\nu/2}(\Omega_h)} + C_{14}\left(|w|^2_{W^1_{2,\nu/2}(\Omega_h\cup\Gamma_h)} + \|w\|'^2_{L_{2,\nu/2}(\partial\Omega_h^{(4)})}\right)\right). \quad (26)$$

Note that

$$\|\partial_1^+ w\|^2_{L_{2,\nu/2}(\partial\overline{\Omega}_h^{(4)}\cup\partial\Omega_h^{(5)})} + \|\partial_2^+ w\|^2_{L_{2,\nu/2}(\partial\Omega_h^{(1)}\cup\partial\Omega_h^{(3)})}$$
$$\le \varepsilon_2\left(\|\partial_1^+ w\|^2_{L_{2,\nu/2}(\partial\overline{\Omega}_h^{(4)}\cup\partial\Omega_h^{(5)})} + \|\partial_2^+ w\|^2_{L_{2,\nu/2}(\partial\Omega_h^{(1)}\cup\partial\Omega_h^{(3)})}\right)$$
$$+ \frac{1}{4\varepsilon_2}|h|^{-1}\|w\|'^2_{L_{2,\nu/2}(\partial\Omega_h\backslash\Gamma_h)}, \quad (27)$$

$$\left|h_2\frac{a_1\rho_h^\nu + I_{+1}(a_1\rho_h^\nu)}{2}\partial_1^+ ww\,\Big|_{\partial\Omega_h^{(1)}}\right|$$
$$\le c_{15}\left|(\rho_h^{\nu/2}h_1^{1/2}h_2^{1/2}\partial_1^+ w)(\rho_h^{\nu/2}h_2^{1/2}h_1^{-1/2}w)\,\Big|_{\partial\Omega_h^{(1)}}\right|$$
$$\le c_{16}\left(\varepsilon_3\|\partial_1^+ w\|'^2_{L_{2,\nu/2}(\partial\Omega^{(1)})} + \frac{1}{4\varepsilon_3}h_1^{-1}\|w\|'^2_{L_{2,\nu/2}(\partial\Omega_h^{(1)})}\right). \quad (28)$$

We carry the last six terms from the right-hand side of the equality (25) to its left-hand side and we estimate the modulus of each of them by analogy with (28). Then, using estimations (26), (27) and conditions (4), (5), we get

$$\frac{1}{4\varepsilon}\left(\|A_{h\rho}w\|^2_{L_{2,-\nu/2}(\Omega_h)} + \|a_{h\rho}w\|'^2_{L_{2,-\nu/2}(\Gamma_h)}\right) + c_{17}\varepsilon\|w\|^2_{W^1_{2,\nu/2}(\overline{\Omega}_h)}$$
$$+ \frac{c_{18}|h|^{-1}}{\varepsilon}\|w\|'^2_{L_{2,\nu/2}(\partial\Omega_h\backslash\Gamma_h)} \ge \min(\lambda, a_0)\|w\|^2_{W^1_{2,\nu/2}(\overline{\Omega}_h)}. \quad (29)$$

By choosing ε sufficiently small and by extracting the square root of both sides of inequality (29), we obtain

$$\|A_{h\rho}w\|^2_{L_{2,-\nu/2}(\Omega_h)} + \|a_{h\rho}w\|'_{L_{2,-\nu/2}(\Gamma_h)} + |h|^{-1}\|w\|'_{L_{2,\nu/2}(\partial\Omega_h\backslash\Gamma_h)}$$
$$\ge C_{19}\|w\|^2_{W^1_{2,\nu/2}(\overline{\Omega}_h)}.$$

From this inequality we get the estimate (21).

On the basis of the lemms 3–7 we establish the estimate of the convergence rate.

Theorem 1. *Let $\overline{A}_{h\rho}$ be the operator defined in (8); assume that conditions (4), (5) are satisfied and $v_\nu \in H^4_{2,\nu}(\Omega)$. Then the approximate R_ν-generalized solution u obtained by the difference scheme (8) converges to the R_ν-generalized solution v_ν to the original problem (2), (3) in the norm $W^1_{2,\nu}(\overline{\Omega}_h)$ and the following estimate holds:*

$$\left\| u - [v_\nu]_h \right\|_{W^1_{2,\nu}(\overline{\Omega}_h)} \leq c_{20} |h|^{2-\beta-\frac{\nu\beta}{4}} \left\| v_\nu \right\|_{H^4_{2,\nu}(\Omega)},$$

where C_{20} is a positive constant not depending on u, v_ν and h.

References

1. Rukavishnikov, V.A.: On a weighted estimate of the rate of convergence of difference schemes. Sov. Math. Dokl. **22**, 826–829 (1986)
2. Rukavishnikov, V.A.: On differentiability properties of an R_ν-generalized solution of Dirichlet problem. Sov. Math. Dokl. **40**, 653–655 (1990)
3. Rukavishnikov, V.A.: On a weighted estimates of error of difference schemes for Helmholtz's equation. Numer. Anal. Math. Model. **24**, 397–408 (1990). Banach Center Publications, Warsaw
4. Rukavishnikov, V.A.: Study of difference schemes for Dirichlet problem Sobolev's weight spaces. Sibiring J. Comput. Math. **1**, 191–204 (1992)
5. Rukavishnikov, V.A.: On the Dirichlet problem for the second order elliptic equation with noncoordinated degeneration of input data. Differ. Equ. **32**, 406–412 (1996)
6. Rukavishnikov, V.A.: On the uniqueness of R_ν-generalized solution for boundary value problem with non-coordinated degeneration of the input data. Dokl. Math. **63**, 68–70 (2001)
7. Rukavishnikov, V.A., Ereklintsev, A.G.: On the coercivity of the R_ν-generalized solution of the first boundary value problem with coordinated degeneration of the input data. Differ. Equ. **41**, 1757–1767 (2005)
8. Rukavishnikov, V.A., Kuznetsova, E.V.: Coercive estimate for a boundary value problem with noncoordinated degeneration of the data. Differ. Equ. **43**, 550–560 (2007)
9. Rukavishnikov, V.A., Kuznetsova, E.V.: The R_ν-generalized solution of a boundary value problem with a singularity belongs to the space $W^{k+2}_{2,\nu+\beta/2+k+1}(\Omega,\delta)$. Differ. Equ. **45**, 913–917 (2009)
10. Rukavishnikov, V.A., Rukavishnikova, E.I.: The finite element method for the first boundary value problem with compatible degeneracy of the input data. Russ. Acad. Sci. Dokl. Math. **50**, 335–339 (1995)
11. Rukavishnikov, V.A., Kuznetsova, E.V.: A finite element method scheme for boundary value problems with noncoordinated degeneration of input data. Numer. Anal. Appl. **2**, 250–259 (2009)
12. Rukavishnikov, V.A., Rukavishnikova, H.I.: The finite element method for a boundary value problem with strong singularity. J. Comp. Appl. Math. **234**, 2870–2882 (2010)

13. Rukavishnikov, V.A.: On differential properties R_ν-generalized solution of the Dirichlet problem with coordinated degeneration of the input data. ISRN Math. Anal. **243724**, 18 (2011). doi:10.5402/2011/243724

14. Rukavishnikov, V.A., Mosolapov, A.O.: New numerical method for solving time-harmonic Maxwell equations with strong singularity. J. Comput. Phys. **231**, 2438–2448 (2012)

15. Rukavishnikov, V.A., Mosolapov, A.O.: Weighted edge finite element method for Maxwell's equations with strong singularity. Dokl. Math. **87**, 156–159 (2013)

16. Rukavishnikov, V.A., Nikolaev, S.G.: Weighted finite element method for an elasticity problem with singularity. Dokl. Math. **88**, 705–709 (2013)

17. Rukavishnikov, V.A., Rukavishnikova, H.I.: On the error estimation of the finite element method for the boundary value problems with singularity in the lebesgue weighted space. Numer. Funct. Anal. Optim. **34**, 1328–1347 (2013)

18. Rukavishnikov, V.A.: On the existence and uniqueness of R_ν-generalized solution for boundary value problem with the non-coordinated degeneration of the initial data. Dokl. Akad. Nauk. **458**(3), 261–263 (2014)

19. Bramble, J.H., Hilbert, S.R.: Estimation of linear functional on Sobolev spaces with applications to Fourier transforms and spline interpolation. SIAM J. Numer. Anal. **7**, 112–124 (1970)

Superconvergence of Some Linear and Quadratic Functionals for Higher-Order Finite Elements

Vladimir Shaydurov[1,2]([✉]) and Tianshi Xu[2]

[1] Institute of Computational Modeling of Siberian Branch of Russian Academy of Sciences, Krasnoyarsk, Russia
[2] Beihang University, Beijing, China
shaidurov04@mail.ru

Abstract. This paper deals with the calculation of linear and quadratic functionals of approximate solutions obtained by the finite element method. It is shown that under certain conditions the output functionals of an approximate solution are computed with higher order of accuracy than that of the solution itself. These abstract results are illustrated by two numerical examples for the Poisson equation.

Keywords: Finite element method · Output functionals · Dual problems · Hermite finite elements · Bogner-Fox-Schmit element · Convergence order

1 Introduction

Traditional methods for solving equations of mathematical physics, such as the finite element method, are to find a solution in the entire domain. Meanwhile, in a number of applications, researchers are interested not in the solution as a whole, but only in its goal-oriented output functionals. For example, in air flow around the body, engineers are interested in lift and drag rather than in the solution at every point in the space [1,2]. In such cases, one would be interested in the precision of these output functionals rather than of the entire solution. Moreover, with appropriate triangulation in the finite element method one can achieve a significant increase in the accuracy of the required functionals without increasing the computational time for the problem as a whole [2,3].

It has long been noted that the finite elements of higher degrees provide a higher order of convergence for an approximate solutions (under sufficient smoothness of the exact solution) [4–6]. And the weaker the norm in which the error between the exact and approximate solutions $u - u^h$ is estimated, the higher the rate of convergence. For example, in the norms of the Sobolev spaces $H^m(\Omega)$, the less m, the higher the attainable convergence rate

$$\left\| u - u^h \right\|_{H^m(\Omega)} \le ch^{k+1-m} \|u\|_{H^{k+1}(\Omega)}, \quad 0 \le m \le k+1,$$

where k is the full degree of the polynomials involved in the approximation of the solution. When solving second-order elliptic equations, the use of linear

© Springer International Publishing Switzerland 2015
I. Dimov et al. (Eds.): FDM 2014, LNCS 9045, pp. 84–95, 2015.
DOI: 10.1007/978-3-319-20239-6_8

or polylinear polynomials corresponding to $k = 1$ quite simply leads to this estimate for $m = 1$. The Aubin-Nitsche approach (independently discovered and described in [7–9]) leads to this estimate for $m = 0$.

In this paper first we prove some abstract results and then consider a two-dimensional model problem solved by the finite element method with cubic Hermite elements. Unexpectedly the ultrahigh order of convergence was achieved for some output linear and quadratic functionals of an approximate solution which does not directly followed from the accuracy order of an approximate solution. Thus, the paper is devoted to the theoretical justification of this beneficial effect with the help of dual problems.

2   An Abstract Results

Let V and W be the Banach spaces with norms $\|\cdot\|_V$ and $\|\cdot\|_W$ respectively and V_h and W_h be the families of their finite-dimensional subspaces (trial and test subspaces of the finite element method) with a discrete set of h approaching 0.

Let $a(v, w) : V \times W \to R$ be a bounded bilinear form

$$|a(v, w)| \le c_1 \|v\|_V \|w\|_W \quad \forall v \in V, \ w \in W \tag{1}$$

with a constant c_1 independent of v and u. And let $f(w) : W \to R$ be a linear functional.

Suppose that we solve the problem:
find $u \in V$ such that

$$a(u, \varphi) = f(\varphi) \ \forall \varphi \in W. \tag{2}$$

But as it mentioned above, let the main purpose consist in the computation of the value $J(u)$ of an (output) linear functional $J(v) : V \to R$.

Instead of problem (2) we solve the following one (for example, by the finite element method):
find $u^h \in V_h$ such that

$$a(u^h, \varphi) = f(\varphi) \ \forall \varphi \in W^h. \tag{3}$$

Thereafter we compute the approximate value $J(u^h)$.

For a moment, suppose that the functional $J(v)$ is bounded:

$$|J(v)| \le c_2 \|v\|_V \ \forall v \in V \tag{4}$$

with a constant c_2 independent of v. Then we get the estimate

$$\left| J(u) - J(u^h) \right| \le c_2 \left\| u - u^h \right\|_V. \tag{5}$$

We see that this estimate gives a rather modest result.

To improve this situation, consider the auxiliary dual problem:
find $w \in W$ such that

$$a(\psi, w) = J(\psi) \ \forall \psi \in V. \tag{6}$$

This problem is indeed auxiliary: we need not solve it either analytically or numerically.

Theorem 1. *Let the problems (2), (3), and (6) with the condition (1) have unique solutions u, u^h, and w, respectively. Besides, the approximation properties of the subspaces V_h and W_h provide the following estimate:*

$$\left\| u - u^h \right\|_V \leq c_3 h^r \tag{7}$$

and there exists an element $w^h \in W_h$ such that

$$\left\| w - w^h \right\|_W \leq c_4 h^s \tag{8}$$

with constants $c_3, c_4, r > 0, s > 0$ independent of h. Then

$$\left| J(u) - J(u^h) \right| \leq c_1 c_3 c_4 h^{r+s}. \tag{9}$$

Proof. Due to linearity

$$\left| J(u) - J(u^h) \right| = \left| J(u - u^h) \right|. \tag{10}$$

From (6) we have

$$J(u - u^h) = a(u - u^h, w). \tag{11}$$

From the problems (2) and (3) it follows that $a(u - u^h, w^h) = 0$. Subtract this from (11):

$$J(u - u^h) = a(u - u^h, w - w^h). \tag{12}$$

Then due to (1), (7), (8) we get

$$\left| J(u - u^h) \right| \leq c_1 \left\| u - u^h \right\|_V \left\| w - w^h \right\|_W \leq c_1 c_3 c_4 h^{r+s}. \qquad \square \tag{13}$$

Thus, the estimate (9) demonstrates a higher order of accuracy which is improved by the order of approximation in the dual problem.

Now consider the case where we need to find a *quadratic* functional of an approximate solution. For this purpose we introduce a *symmetric* bilinear form $b(v, w) : V \times V \rightarrow R$ and try to find the value $I(u) = b(u, u)$. Solving the problem (3) we get an the approximate solution $u^h \in V_h$ for which we can compute $I(u^h) = b(u^h, u^h)$. Show that under some simple conditions we again get a higher order of accuracy like for the linear functional.

For this purpose consider the auxiliary dual problem:
find $w \in W$ such that

$$a(\psi, w) = b(u + u^h, \psi) \ \forall \psi \in V. \tag{14}$$

Again this problem is indeed auxiliary and we need not solve it either analytically or numerically.

Theorem 2. *Let the problems (2), (3), and (14) with the condition (1) have unique solutions u, u^h, and w, respectively. And let the approximation properties of subspaces V_h and W_h provide the estimates (7) and (8). Then*

$$\left| I(u) - I(u^h) \right| \leq c_1 c_3 c_4 h^{r+s}. \tag{15}$$

Proof. Due to linearity in each arguments and symmetry between them we get

$$\left|I(u) - I(u^h)\right| = \left|b(u,u) - b(u^h,u^h)\right| = \left|b(u+u^h, u-u^h)\right|. \qquad (16)$$

From (14) we have

$$b(u+u^h, u-u^h) = a(u-u^h, w). \qquad (17)$$

From the problems (2) and (3) it follows that $a(u-u^h, w^h) = 0$. Subtract this from (17):

$$b(u+u^h, u-u^h) = a(u-u^h, w-w^h).$$

Then due to (1), (7), (8) we get

$$\left|b(u+u^h, u-u^h)\right| \le c_1 \left\|u-u^h\right\|_V \left\|w-w^h\right\|_W \le c_1 c_3 c_4 h^{r+s}. \qquad \square$$

Note that we did not use in the direct way the boundedness of the bilinear form

$$|b(v,w)| \le c_5 \|v\|_V \|w\|_V \quad \forall v, w \in V.$$

From here on, constants c_i are independent of functions in the right-hand side and of h. The usage of the above inequality gives a much weaker order of accuracy:

$$\left|I(u) - I(u^h)\right| = \left|b(u+u^h, u-u^h)\right| \le c_5 \left\|u+u^h\right\|_V \left\|u-u^h\right\|_V \le$$
$$\le c_3 c_5 h^r \left\|u+u^h\right\|_V.$$

3   Formulations of Test Problems

Let Ω be the square $[0,1] \times [0,1]$ with the boundary Γ. For our further consideration we use the usual notations for Sobolev spaces [10]. Let $H^0(\Omega) = L^2(\Omega)$ be the Hilbert space of functions Lebesgue measurable on Ω and equipped with the inner product

$$(u,v)_\Omega = \int_\Omega u\,v\,d\Omega, \quad u, v \in H^0(\Omega),$$

and the finite norm

$$\|u\|_{0,\Omega} = (u,u)_\Omega^{1/2}, \quad u \in H^0(\Omega).$$

For integer *positive* k, $H^k(\Omega)$ is the Hilbert space of functions $u \in H^0(\Omega)$ whose weak derivatives up to order k inclusive belong to $H^0(\Omega)$. The norm in this space is defined by the formula

$$\|u\|_{k,\Omega} = \left(\sum_{0 \le s+r \le k} \left| \frac{\partial^{s+r} u}{\partial x^s \, \partial y^r} \right|^2_{0,\Omega} \right)^{1/2}.$$

Introduce also the functional space $H_0^1(\Omega)$ as the closure in the norm $\|\cdot\|_{1,\Omega}$ of all infinitely differentiable functions with support in Ω.

Consider the following model problem: *find $u(x,y) \in H^2(\Omega)$ such that*

$$- \Delta u = f(x,y) \text{ in } \Omega, \tag{18}$$

$$u = 0 \text{ on } \Gamma. \tag{19}$$

Let the solution u be smooth enough: $u \in H^4(\Omega)$. Then $f \in H^2(\Omega)$.

First take the output functional

$$J(u) := \int_\Omega ug \, d\Omega \tag{20}$$

with some function $g \in H^0(\Omega)$ and show that this functional is computed by bicubic finite elements with higher order of accuracy than a solution as a whole.

To get the weak form of this problem, multiply the equation (18) by an arbitrary function $\varphi \in H_0^1(\Omega)$ and integrate by parts with the help of the boundary conditions (19). As a result we get the equality

$$\int_\Omega \left(\frac{\partial u}{\partial x} \frac{\partial \varphi}{\partial x} + \frac{\partial u}{\partial y} \frac{\partial \varphi}{\partial y} \right) d\Omega = \int_\Omega f\varphi \, d\Omega. \tag{21}$$

In the weak form [5, 6, 11] this problem is reformulated as follows: *find $u \in H_0^1(\Omega)$ such that*

$$a(u, \varphi) = (f, \varphi)_\Omega \ \forall \varphi \in H_0^1(\Omega) \tag{22}$$

with the bilinear form

$$a(u, \varphi) = \int_\Omega \left(\frac{\partial u}{\partial x} \frac{\partial \varphi}{\partial x} + \frac{\partial u}{\partial y} \frac{\partial \varphi}{\partial y} \right) d\Omega.$$

We construct a uniform triangulation $\Im_h$ by subdividing Ω into N^2 closed "rectangles" by the lines

$$x_i = ih, \ i = 0, ..., N; \ y_j = jh, \ j = 0, ..., N; \text{ where } h = 1/N.$$

Here we shall describe a finite element by the triple (e, P_e, Σ_e) [6] where e is a "reference" cell (in this paper we put $e = [0,1]^2$); P_e is a space of polynomials on e; and Σ_e is the set of linear functionals called degrees of freedom (DoF).

Denote by P_k with positive *integer* k the space of all polynomials in two variables of full degree k:

$$\sum_{0 \le i+j \le k} a_{i,j} x^i y^j.$$

And denote by Q_k the space of all polynomials of degree k for each variable:

$$\sum_{0 \le i,j \le k} a_{i,j} x^i y^j.$$

First consider the possible implementation of the *bilinear* finite elements for solving the problem (22) by the Bubnov-Galerkin finite element method. Due to the approximation properties of these elements we can get only

$$\left\| u - u^h \right\|_{1,\Omega} \le c_3 h \quad \text{and} \quad \left\| w - w^h \right\|_{1,\Omega} \le c_4 h$$

with an interpolant w^h. Then from Theorem 1 we obtain

$$\left| J(u) - J(u^h) \right| \le c_1 c_3 c_4 h^2.$$

But this estimate does not provide us an improvement in comparison with the direct analysis by Aubin-Nitsche trick.

The situation is different for finite elements of higher order. First consider Lagrange elements on square $e = [0,1]^2$ (Fig. 1). Introduce the corresponding grid of nodes

$$S^{(2)} = S^{(1)} \times S^{(1)} \quad \text{where} \quad S^{(1)} = \{a_i : \ a_i = i/3, \ i = 0, ..., 3\}.$$

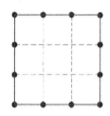

a) full element   b) serendipity element

Fig. 1. Nodes of the full and incomplete "serendipity" Lagrange cubic elements.

Then the bicubic element is described by the triple (e, Q_3, Σ_3) where

$$\Sigma_3 = \{\psi_{i,j} : \ \psi_{i,j}(p) = p(a_i, a_j) \ \forall \ i,j = 0, ..., 3 \ \forall p \in Q_3\} . \tag{23}$$

It has 16 degrees of freedom on one elementary cell.

The usual mapping of the two-dimensional "reference" element into an elementary cell $[x_i, x_{i+1}] \times [y_j, y_{j+1}]$ of the triangulation $\Im_h$ has the form

$$\begin{cases} x' = x_i + hx, \\ y' = y_j + hy. \end{cases} \tag{24}$$

Generally speaking, the numbers of DoF are excessive to obtain the corresponding approximation order. Indeed, to achieve the same order of approximation it is sufficient to take polynomials P_3 on e [6] with the number of DoF equal to 10. Therefore the incomplete Lagrange "serendipity" element is often used. In this case, the DoF are omitted which lie strictly inside the cell e and have no influence on interelement continuity (Fig. 1b) [5,6]. The number of DoF for the serendipity element decreases and becomes equal 12. Since the polynomial spaces satisfy the condition $Q_{3'} \supset P_3$, the serendipity element provides the same order of approximation as the full Lagrange element and is more effective because of less number of DoF.

Now consider a simple Hermite bicubic element [12] (Fig. 2a). The number of DoF and the space of polynomials of this element coincide with those of the cubic serendipity element. Therefore, it may seem that they have identical properties. In fact, this is not the case! The Hermite element appears to be more efficient. To show this, we compare the global number of DoF for the interpolation of a smooth function u by these elements on the triangulation $\Im_h$ of the rectangle Ω.

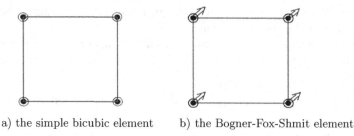

a) the simple bicubic element   b) the Bogner-Fox-Shmit element

Fig. 2. The cubic Hermite elements. A circle means that DoF involve both first-order derivatives; a double arrow means that DoF involve the second-order mixed derivative.

The global number of DoF of the interpolant u_I^h on the triangulation $\Im_h$ is not proportional to the number of DoF on an element. A part of DoF for different elements coincides along interelement boundaries. Therefore, as the local characteristics of the global number of DoF we take the number M of DoF for the element on the half-closed set $[0,1)^2$ (Fig. 3). When mapping the element on the cells of $\Im_h$, these DoF are not repeated and exhaust all nodes inside Ω. Their total number is MN^2.

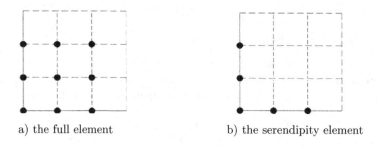

a) the full element   b) the serendipity element

Fig. 3. Nodes of the Lagrange elements on the half-closed set $[0,1)^2$.

An open question remains on the number of nodes on the boundary Γ. In the Dirichlet problem, these DoF are excessive, but they are necessary for the Neumann problem. In both cases their number is of $O(N)$. Thus, MN^2 is the principal term of the asymptotic number of unknowns and equations in the finite element method, e.g., for a second-order elliptic equation.

The index M for the full Lagrange, serendipity, and Hermite bicubic elements is 9, 5, and 3 (Fig. 4a), respectively. Thus, the number of unknowns and equations in the finite element method for the Hermite element is approximately 3 times less than that for the full Lagrange element and 5/3 times less than that for the serendipity element. Such is the case despite the fact that they have the same order of approximation.

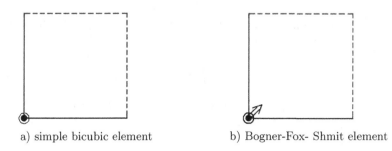

a) simple bicubic element   b) Bogner-Fox- Shmit element

Fig. 4. Nodes of the Hermite elements on the half-closed set $[0, 1)^2$.

The Bogner-Fox-Schmit element [12–14] is a more complicated Hermite bicubic finite element. It is defined by the triple (Fig. 2b)

$$(e,\ Q_3,\ \Sigma_{3'})\ \text{where}\ \Sigma_{3'} = \{\psi_{s,i,j}\ (s = 0, 1, 2, 3):\ \psi_{0,i,j}(p) = p(a_i, a_j),$$

$$\psi_{1,i,j}(p) = \partial p/\partial x(a_i, a_j),\ \ \psi_{2,i,j}(p) = \partial p/\partial y(a_i, a_j), \tag{25}$$

$$\psi_{3,i,j}(p) = \partial^2 p/\partial x \partial y(a_i, a_j)\ \ \forall i, j = 0, 3\}.$$

For the triangulation $\Im_h$, it provides continuity of an approximation u^h as well as of its first-order derivatives [12–14]. Thus, this element belongs to $H^2(\Omega)$ [11]. At the same time, it has the index $M = 4$ (Fig. 4b) which is less than the indices M of the Lagrange full and serendipity elements.

Now consider the implementation of any cubic finite elements for solving the problem (22) by the Bubnov-Galerkin finite element method. Due to the approximation properties of these elements (under sufficient smoothness) we can get

$$\left\|u - u^h\right\|_{1,\Omega} \le c_3 h^3 \quad \text{and} \quad \left\|w - w^h\right\|_{1,\Omega} \le c_4 h^3$$

for an interpolant w^h. Then from Theorem 1 we obtain

$$\left|J(u) - J(u^h)\right| \le c_1 c_3 c_4 h^6. \tag{26}$$

4   Numerical Results

Now consider the concrete problem (18)–(19) with the right-hand side

$$f(x, y) = 16(1 - x)(1 - y)(x^2 + y^2)\sin(4xy) + 8(x - x^2 + y - y^2)\cos(4xy). \tag{27}$$

This problem has the exact solution

$$u(x, y) = \sin(4xy)(1 - x)(1 - y).$$

And assume that we need to compute the output functional

$$J(u) := \int_\Omega u \, d\Omega. \tag{28}$$

We solve the problem (22) by the finite element method with the help of the Bogner-Fox-Schmit finite element. And then calculate (28) for an approximate solution u^h :

$$J(u^h) := \int_\Omega u^h \, d\Omega. \tag{29}$$

We perform these computations for $h = 1/8, 1/16, 1/32$ and determine the error $\varepsilon_1^h = \left| J(u) - J(u^h) \right|$. We demonstrate this error in Table 1 together with its decreasing exponent $d_1^h = \ln_2 \left| \varepsilon_1^{2h} / \varepsilon_1^h \right|$.

Table 1. The approximation errors and their decreasing exponent.

i	h	$\varepsilon_1^h = \left\| J(u) - J(u^h) \right\|$	d_1^h	$\varepsilon_2^h = \left\| I(u) - I(u^h) \right\|$	d_2^h
1	1/8	5.81×10^{-9}	–	3.92×10^{-9}	–
2	1/16	9.64×10^{-11}	5.91	6.94×10^{-11}	5.81
3	1/32	1.53×10^{-12}	5.97	1.15×10^{-12}	5.91

From this Table we can see that $\varepsilon_1^h = \left| J(u) - J(u^h) \right|$ tends to zero asymptotically as $O(h^6)$. But this does not follow from Theorem 1 in the direct way. Indeed, in this case the problem (6) has the form

$$-\Delta w = 1 \text{ in } \Omega,$$
$$w = 0 \text{ on } \Gamma. \tag{30}$$

Despite the smoothness of the right hand side the solution w does not belong to space $H^3(\Omega)$ because of singularities in four angles of the rectangle [15].

There are two ways to avoid these singularities. One of them is in special condensation of mesh in the vicinity of singularities. This is a really productive way in some cases [3]. But in our situation we got the sixth order without any condensation of mesh. This means that the justification must be finer. It may be transformed in different ways. One of them consists in the introduction of weighted norms in spaces $\|\cdot\|_V$ and $\|\cdot\|_W$. The reasoning is very tedious. We simplify it by some transformation of the theorem proof. Take the right-hand side of equality (12) and transform it in following way:

$$a(u - u^h, w - w^h) = \int_\Omega -\Delta(w - w^h)(u - u^h) \, d\Omega. \tag{31}$$

Introduce the weight function

$$\rho(x,y) = x(1-x)y(1-y)$$

and use it in the following way:

$$\left| \int_\Omega -\rho\Delta(w - w^h)\rho^{-1}(u - u^h)\, d\Omega \right| \le \left\| \rho\Delta(w - w^h) \right\|_{0,\Omega} \left\| \rho^{-1}(u - u^h) \right\|_{0,\Omega}.$$

The first norm becomes small enough because of weight degenerating in the vicinity of each angle:

$$\left\| \rho\Delta(w - w^h) \right\|_{0,\Omega} \le ch^2.$$

And the second norm becomes small enough because of the Aubin-Nitsche trick and degenerating both functions in the vicinity of the boundary:

$$\left\| \rho^{-1}(u - u^h) \right\|_{0,\Omega} \le ch^4.$$

Therefore

$$\left| J(u - u^h) \right| = \left| a(u - u^h, w - w^h) \right| \le ch^6.$$

This estimate indeed is consistent with numerical results.

For the quadratic functional the situation with Theorem 2 is simpler. Let we need to compute the output functional

$$I(u) := \int_\Omega u^2\, d\Omega = b(u, u) \quad \text{where} \quad b(v, w) = \int_\Omega vw\, d\Omega. \tag{32}$$

Let we solved the problem (22) by the finite element method with the help of the Bogner-Fox-Schmit finite element again. And then calculate the required value (32) for an approximate solution u^h :

$$I(u^h) := \int_\Omega (u^h)^2\, d\Omega. \tag{33}$$

We perform there computations for $h = 1/8,\ 1/16,\ 1/32$ and determine the error $\varepsilon_2^h = \left| I(u) - I(u^h) \right|$. This error is demonstrated in Table 1 together with its decreasing exponent $d_2^h = \ln_2 \left| \varepsilon_2^{2h}/\varepsilon_2^h \right|$. From this Table we can see that $\varepsilon_2^h = \left| I(u) - I(u^h) \right|$ tends to zero asymptotically as $O(h^6)$. Moreover, this follows directly from Theorem 2. Indeed, the function w is a solution of the problem

$$-\Delta w = u + u^h \quad \text{in} \quad \Omega,$$
$$w = 0 \quad \text{on} \quad \Gamma. \tag{34}$$

First, this time the right-hand side of the problem belongs to $H^2(\Omega)$ due to the application of the Bogner-Fox-Schmit element. Second, it equals zero in each angle of the rectangle Ω. These properties ensure that $w \in H^4(\Omega)$ [15] and

$$\left\| w - w^h \right\|_{1,\Omega} \le c_4 h^3.$$

Function u also belongs to $H^4(\Omega)$ and provides the estimate [6]

$$\left\| u - u^h \right\|_{1,\Omega} \le c_4 h^3.$$

Thus, Theorem 2 indeed guarantees the sixth order of accuracy.

5 Resume

By virtue of the dual problems, for some linear and quadratic functionals we prove convergence of higher order than follows from the standard theory of the finite element method. Note that this effect becomes possible for more complicated (for example, cubic) finite elements than linear ones. For linear elements on triangles and for bilinear ones on quadrangles this approach does not give higher order of accuracy than it follows from the usual implementation of the Aubin-Nitsche trick.

Moreover, once again we remind that Hermite finite elements are more effective in comparison with the Lagrange ones of the same degrees of polynomials due to a smaller number of unknowns and of discrete equations in the finite element method. Besides, the Bogner-Fox-Schmit finite element is more effective than the Lagrange cubic elements and belongs to $H^2(\Omega)$ which simplifies the justification of higher order convergence and gives some useful possibilities like direct computation of a residual for an approximate solution u^h. This provides necessary and visual information for condensation of a triangulation. Usually, their use is limited to domains consisting of rectangles. But the complementing of these elements by suitable triangular elements near the boundary [16] extends the possible range of their application.

This work is supported by Project 14-11-00147 of Russian Scientific Foundation.

References

1. Shaydurov, V., Liu, T., Zheng, Z.: Four-stage computational technology with adaptive numerical methods for computational aerodynamics. Am. Inst. Phys. Conf. Proc. **1484**, 42–48 (2012)
2. ADIGMA. A European Initiative on the Development of Adaptive Higher-Order Variational Methods for Aerospace Applications. Notes on Numerical Fluid Mechanics and Multidisciplinary Design. **113**, Berlin: Springer (2010)
3. Bangerth, W., Rannacher, R.: Adaptive Finite Element Methods for Differential Equations. Birkhäuser, Berlin (2003)
4. Yue, H., Shaydurov, V.: Superconvergence of some output functionals for Hermitian finite elements. Young Scientist **12**, 15–19 (2012)
5. Strang, G., Fix, G.J.: An Analysis of the Finite Element Method. Prentice-Hall, New York (1973)
6. Ciarlet, P.G.: The Finite Element Method for Elliptic Problems. North Holland, New York (1978)
7. Aubin, J.P.: Behavior of the error of the approximate solutions of boundary value problems for linear elliptic operators by Galerkin's and finite difference methods. Ann. Scuola Norm. Sp. Pisa. **21**, 599–637 (1967)
8. Nitsche, J.A.: Ein Kriterium für die Quasi-Optimalitat des Ritzchen Verfahrens. Numer. Math. **11**, 346–348 (1968)
9. Oganesyan, L.A.: Investigation of the convergence rate of variational-difference schemes for elliptic second order equations in a two-dimensional domain with a smooth boundary. Zh. Vychisl. Mat. i Mat. Fiz. **9**, 1102–1120 (1969)

10. Adams, R.A., Fournier, J.J.F.: Sobolev spaces. Academic Press, New York (2003)
11. Chen, Z., Wu, H.: Selected Topics in Finite Element Methods. Science Press, Beijing (2010)
12. Shaydurov, V., Shut, S., Gileva, L.: Some properties of Hermite finite elements on rectangles. Am. Inst. Phys. Cof. Proc. **1629**, 32–43 (2014)
13. Bogner, F.K., Fox, R.L., Schmit, L.A.: The generation of interelement compatible stiffness and mass matrices by the use of interpolation formulas. In: Proceedings of the Conference on Matrix Methods in Structural Mechanics, pp. 397–444. Wright-Patterson Air Force Base, Ohio (1965)
14. Zhang, S.: On the full C_1-Q_k finite element spaces on rectangles and cuboids. Adv. Appl. Math. Mech. **2**(6), 701–721 (2010). doi:10.4208/aamm.09-m0993
15. Kondrat'ev, V.A.: Boundary-value problems for elliptic equations in domain with conic and angular points. Trans. Moscow Math. Soc. **16**, 209–292 (1967)
16. Gileva, L., Shaydurov, V., Dobronets, B.: The triangular Hermite finite element complementing the Bogner-Fox-Schmit rectangle. Appl. Math. **5**(12A), 50–56 (2013)

Time Step for Numerically Solving Parabolic Problems

Petr N. Vabishchevich[1,2]([✉])

[1] Nuclear Safety Institute, 52, B. Tulskaya, 115191 Moscow, Russia
[2] North-Eastern Federal University, 58, Belinskogo, 677000 Yakutsk, Russia
vabishchevich@gmail.com

Abstract. This work deals with the problem of choosing a time step for the numerical solution of boundary value problems for parabolic equations. The problem solution is derived using the fully implicit scheme, whereas a time step is selected via explicit calculations. Using the explicit scheme, we calculate the solution at a new time level. We employ this solution in order to obtain the solution at the previous time level (the implicit scheme, explicit calculations). This solution should be close to the solution of our problem at this time level with a prescribed accuracy. Such an algorithm leads to explicit formulas for the calculation of the time step and takes into account both the dynamics of the problem solution and changes in coefficients of the equation and in its right-hand side.

1 Introduction

The problem of the control over a time step is relatively well resolved for the numerically solving Cauchy problem for systems of differential equations [1–3]. The basic approach involves the following stages. First, we perform additional calculations in order to estimate the error of the approximate solution at a new time level. Further, a time step is estimated using the theoretical asymptotic dependence of accuracy on a time step. After that we decide is it necessary to correct the time step and to repeat calculations.

Additional calculations for estimating the error of the approximate solution may be performed in a different way. In particular, it is possible to obtain an approximate solution using two different schemes that have the same theoretical order of accuracy. The most famous example of this strategy involves the solution of the problem on a separate time interval using a preliminary step (the first solution) and the step reduced by half (the second solution). In numerically solving the Cauchy problem for systems of ordinary differential equations, there are also applied nested methods, where two approximate solutions of different orders of accuracy are compared.

In this paper, we consider an a priori selection of a time step for the approximate solution of boundary value problems (BVPs) for parabolic equations. To obtain the solution at a new time level, the backward Euler scheme is employed. The time step at the new time level is explicitly calculated using two previous time levels and takes into account changes in the equation coefficients and in its right-hand side.

© Springer International Publishing Switzerland 2015
I. Dimov et al. (Eds.): FDM 2014, LNCS 9045, pp. 96–103, 2015.
DOI: 10.1007/978-3-319-20239-6_9

2   Differential Problem

Let us consider the Cauchy problem for the linear equation

$$B(t)\frac{du}{dt} + A(t)u = f(t), \quad 0 < t \leq T, \tag{1}$$

supplemented with the initial condition

$$u(0) = u_0. \tag{2}$$

The problem is investigated in a finite-dimensional Hilbert space H. Assume that in H, we have

$$A(t) \geq 0, \quad B(t) = B^*(t) \geq \delta E, \quad \delta > 0,$$

where E is the unit (identity) operator.

We present an a priori estimate for the solution of the problem (1), (2). Assume that

$$\frac{d}{dt}B(t) \leq \beta B(t). \tag{3}$$

Multiply equation (1) scalarly in H by u. Due to the non-negativity of the operator A, we have

$$(B(t)u, u) \leq (f, u).$$

In view of

$$\left(B(t)\frac{du}{dt}, u\right) = \frac{1}{2}\frac{d}{dt}(B(t)u, u) - \frac{1}{2}\left(\frac{d}{dt}B(t)u, u\right),$$

we get

$$\frac{1}{2}\frac{d}{dt}(B(t)u, u) \leq \frac{1}{2}\left(\frac{d}{dt}B(t)u, u\right) + (f, u).$$

Taking into account (1) and

$$(f, u) \leq \|u\|_B \|f\|_{B^{-1}},$$

we obtain the inequality

$$\frac{d}{dt}\|u\|_B \leq \frac{\beta}{2}\|u\|_B + \|f\|_{B^{-1}}.$$

From this inequality, we have (Gronwall's lemma) the following estimate for stability with respect for the initial data and the right-hand side:

$$\|u(t)\|_B \leq \exp\left(\frac{\beta}{2}t\right)\|u_0\|_B + \int_0^t \exp\left(\frac{\beta}{2}(t - \theta)\right)\|f(\theta)\|_{B^{-1}}d\theta.$$

For the right-hand side, we apply a simpler estimate

$$\|f(t)\|_{B^{-1}} \leq \frac{1}{\delta}\|f(t)\|,$$

which results in

$$\|u(t)\|_B \leq \exp\left(\frac{\beta}{2}t\right)\|u_0\|_B + \frac{1}{\delta}\int_0^t \exp\left(\frac{\beta}{2}(t-\theta)\right)\|f(\theta)\|d\theta. \qquad (4)$$

The estimate (4) serves us as a reference point for constructing discretization in time.

The problem (1), (2) can be associated with the BVP for the parabolic equation. In this problem, an unknown function $u(x,t)$ satisfies the equation

$$c(x,t)\frac{\partial u}{\partial t} - \sum_{\alpha=1}^m \frac{\partial}{\partial x_\alpha}\left(k(x,t)\frac{\partial u}{\partial x_\alpha}\right) + d(x,t)u = f(x,t), \quad x \in \Omega, \quad 0 < t \leq T,$$

where $\underline{k} \leq k(x) \leq \overline{k}$, $x \in \Omega$, $\underline{k} > 0$, $c(x,t) \geq c_0 > 0$, $d(x,t) \geq 0$. The equation is complemented by the Dirichlet boundary conditions

$$u(x,t) = g(x,t), \quad x \in \partial\Omega, \quad 0 < t \leq T,$$

and the initial condition

$$u(x,0) = u_0(x), \quad x \in \Omega.$$

The problem (1), (2) results from finite difference, finite volume or finite element approximations (lumped masses scheme [4–6]) for numerically solving BVPs for a parabolic equation of second order.

3   Time-Stepping Technique

To solve numerically this time-dependent problem, we introduce a non-uniform grid in time:

$$t_0 = 0, \quad t_{n+1} = t_n + \tau_{n+1}, \quad n = 0, 1, ..., N-1, \quad t_N = T.$$

We will employ notation $f_n = f(t_n)$. For the problem (1), (2), we apply the fully implicit scheme, where the transition from the current time level to the next one is performed as follows:

$$B_{n+1}\frac{y_{n+1} - y_n}{\tau_{n+1}} + A_{n+1}y_{n+1} = f_{n+1}, \quad n = 0, 1, ..., N-1, \qquad (5)$$

starting from the initial condition

$$y_0 = u_0. \qquad (6)$$

Let us obtain a discrete analogue of the estimate (4). At the discrete level, we associate the condition (3) with the following Lipschitz-continuity condition:

$$((B_{n+1} - B_n)y, y) \leq \tau_{n+1}\beta(B_n y, y). \qquad (7)$$

Multiplying (5) by $\tau_{n+1} y_{n+1}$, under the restriction $A_{n+1} \geq 0$, we get

$$(B_{n+1} y_{n+1}, y_{n+1}) \leq (B_{n+1} y_n, y_{n+1}) + \tau_{n+1}(f_{n+1}, y_{n+1}). \tag{8}$$

For the second term in the right-hand side, we use the estimates

$$(f_{n+1}, y_{n+1}) \leq \|f_{n+1}\|_{B_{n+1}^{-1}} \|y_{n+1}\|_{B_{n+1}}, \quad \|f_{n+1}\|_{B_{n+1}^{-1}} \leq \frac{1}{\delta} \|f_{n+1}\|.$$

For the first term in the right-hand side of (8), we have

$$(B_{n+1} y_n, y_{n+1}) \leq \|y_n\|_{B_{n+1}} \|y_{n+1}\|_{B_{n+1}}.$$

In view of (8), we arrive at

$$\|y_n\|_{B_{n+1}}^2 = (B_{n+1} y_n, y_n)$$
$$= (B_n y_n, y_n) + ((B_{n+1} - B_n) y_n, y_n) \leq (1 + \beta \tau_{n+1}) \|y_n\|_{B_n}^2.$$

Thus, from (8), we arrive at the level-wise estimate

$$\|y_{n+1}\|_{B_{n+1}} \leq \varrho \|y_n\|_{B_n} + \frac{\tau_{n+1}}{\delta} \|f_{n+1}\|, \quad \varrho = 1 + \frac{\beta}{2} \tau_{n+1}.$$

Thus, we obtain the discrete analog of the estimate (4):

$$\|y_{n+1}\|_{B_{n+1}} \leq \varrho^{n+1} \|u_0\|_{B_0} + \frac{1}{\delta} \sum_{k=0}^{n} \varrho^{n-k} \tau_{k+1} \|f_{k+1}\| \tag{9}$$

corresponding to the problem (5), (6).

The a priori estimate (9) is used to study the convergence of the approximate solution to the exact one. For the error $z_n = y_n - u_n$ of the approximate solution, we have the problem

$$B_{n+1} \frac{z_{n+1} - z_n}{\tau_{n+1}} + A_{n+1} z_{n+1} = \psi_{n+1}, \quad n = 0, 1, ..., N - 1,$$

$$z_0 = 0.$$

Here ψ_{n+1} stands for the truncation error:

$$\psi_{n+1} = f_{n+1} - B_{n+1} \frac{u_{n+1} - u_n}{\tau_{n+1}} + A_{n+1} u_{n+1}. \tag{10}$$

Similarly to (9), we get the estimate for error:

$$\|z_{n+1}\|_{B_{n+1}} \leq \frac{1}{\delta} \sum_{k=0}^{n} \varrho^{n-k} \tau_{k+1} \|\psi_{k+1}\|.$$

Therefore, to control the error, we can employ the summarized error $\tau_{n+1} \varepsilon$ over the interval $t_n \leq t \leq t_{n+1}$. In this case, a value ε defines the same level of the error over the entire interval of integration.

4   Algorithm for the Estimation of a Time Step

If we will be able to calculate the truncation error ψ_{n+1}, then it will be possible to get a posteriori estimate for the error. Comparing $\|\psi_{n+1}\|$ with the prescribed error level δ, this makes possible to evaluate the quality of the choice of the time step τ_{n+1}. Namely, if $\|\psi_{n+1}\|$ is much larger (smaller) than δ, then the time step is taken too large (small), and if $\|\psi_{n+1}\|$ is close to δ, then this time step is optimal. Thus, we have

$$\tau_{n+1} : \|\psi_{n+1}\| \approx \delta. \tag{11}$$

The problem is that we cannot evaluate the truncation error, since it is determined using the exact solution that is unknown. Because of this, we must focus on some estimates for the truncation error that guarantee the fulfilment of (11).

The general approach to the adaptive choice of a time step for solving unsteady problems consists of the following key elements: begin itemize item a selection of a predicted time step via an analysis of the solution at the previous time levels; item conduction of calculations with the predicted time step; item a study on accuracy for the obtained approximate solution and the possible recalculation with a smaller time step, if necessary. end itemize This general strategy is usually implemented (see, e.g., [1–3]) employing the asymptotic analysis for the error of the approximate solution in the assumption that the error does not vary essentially in time. The main features of our approach to selecting the time step proposed in [7] are presented below.

In our case, the predicted time step is constant. To estimate the time step at the new time level (in the transition from the time level t_n to the time level t_{n+1})), we focus on the previous time step $\tau_n = t_n - t_{n-1}$. First of all, we are interested in the possibility of using a larger time step at new time level. In view of this, we define the predicted time step as follows :

$$\widetilde{\tau}_{n+1} = \gamma \tau_n, \tag{12}$$

where γ is a numerical parameter. The factor γ for the maximum increasing the time step is defined, for example, to be equal to 1.25 or 1.5. The problem parameters (the coefficients of the equation and the right-hand side) are estimated over the interval $[t_n, t_n + \widetilde{\tau}_{n+1}]$. In estimating the time step, we should not miss the time moment, when changes in the parameters of the problem are observed.

The choice of the time step under the restriction $\tau_{n+1} \leq \widetilde{\tau}_{n+1}$ is performed using calculation formulas based on the implicit error estimate at the new time level. The approximate solution at the new time level is evaluated by the implicit scheme (5), whereas estimating the time step is carried out via the explicit scheme. Both the implicit and explicit schemes have the same order of approximation and they are considered with the same initial conditions (at $t = t_n$). We perform a small number (one or two) of time steps, and therefore, possible computational instability for the explicit scheme has no time to appear. Because of this, we can expect that such approximate solutions are close to each other. On the basis of this closeness, we evaluate the error of the approximate solution and obtain the calculation formula for the time step.

Among possible variants for the correction of the time step, we consider the following technique. The step τ_{n+1} is selected from the conditions:

Forward Step. Using the explicit scheme, we calculate the solution v_{n+1} at the time level t_{n+1};

Backward Step. From the obtained v_{n+1}, applying the implicit scheme, we determine v_n at the time level t_n (explicit calculations);

Step Selecting. The step τ_{n+1} is evaluated via closeness between v_n and y_n.

In fact, we carry out the back analysis of the error of the approximate solution over the interval $t_n \leq t \leq t_{n+1}$ using two schemes (explicit and implicit) of the same accuracy.

Let us present the formulas for selecting a time step. The solution v_{n+1} is determined from the equation

$$B_n \frac{v_{n+1} - y_n}{\tau_{n+1}} + A_n y_n = f_n. \tag{13}$$

For v_n, we have

$$B_{n+1} \frac{v_{n+1} - v_n}{\tau_{n+1}} + A_{n+1} v_{n+1} = f_{n+1}. \tag{14}$$

Note that for explicit schemes, we must calculate $B_n^{-1} v$. We believe that this problem is much easier than the evaluation of $A_n^{-1} v$.

The equations (13), (14) may be rewritten in the form

$$\frac{v_{n+1} - y_n}{\tau_{n+1}} + \tilde{A}_n y_n = \tilde{f}_n, \tag{15}$$

$$\frac{v_{n+1} - v_n}{\tau_{n+1}} + \tilde{A}_{n+1} v_{n+1} = \tilde{f}_{n+1}, \tag{16}$$

where

$$\tilde{A}_n = B_n^{-1} A_n, \quad \tilde{f}_n = B_n^{-1} f_n.$$

From (15), (16), we immediately get

$$v_n - y_n = \tau_{n+1}(\tilde{A}_{n+1} - \tilde{A}_n) y_n - \tau_{n+1}(\tilde{f}_{n+1} - \tilde{f}_n) \tag{17}$$
$$+ \tau_{n+1}^2 \tilde{A}_{n+1}(\tilde{f}_n - \tilde{A}_n y_n).$$

The first two terms are associated with the time derivative applied to the problem operator and to the right-hand side. To evaluate them approximately, it seems reasonable to use the time step from the previous time level. But this may be inconvenient to implement.

For instance, we have

$$\tau_{n+1}(\tilde{f}_{n+1} - \tilde{f}_n) = \tau_{n+1}^2 \frac{\tilde{f}_{n+1} - \tilde{f}_n}{\tau_{n+1}},$$

and therefore we have to evaluate the difference derivative of the right-hand side for $t_n \leq t \leq t_{n+1}$. The problem is that the derivation of such estimates involves the unknown value t_{n+1}. The simplest approach is to evaluate this derivative using the previous time step:

$$\frac{\widetilde{f}_{n+1} - \widetilde{f}_n}{\tau_{n+1}} \approx \frac{\widetilde{f}(t_n + \tau_n) - \widetilde{f}_n}{\tau_n}.$$

But in this case, if $\tau_{n+1} > \tau_n$, then we cannot detect significant changes in the right-hand side for $t_n + \tau_n < t \le t_n + \tau_{n+1}$.

It seems natural to evaluate changes in the problem operator and the right-hand side over the whole time interval $[t_n, t_n + \widetilde{\tau}_{n+1}]$. Under the assumption (12), we can estimate the time derivative of the right-hand side, putting

$$\frac{\widetilde{f}_{n+1} - \widetilde{f}_n}{\tau_{n+1}} \approx \frac{\widetilde{f}(t_n + \gamma\tau_n) - \widetilde{f}_n}{\gamma\tau_n}.$$

Therefore

$$\tau_{n+1}(\widetilde{f}_{n+1} - \widetilde{f}_n) \approx \frac{\tau_{n+1}^2}{\gamma\tau_n}(\widetilde{f}(t_n + \gamma\tau_n) - \widetilde{f}_n),$$

$$\tau_{n+1}(\widetilde{A}_{n+1} - \widetilde{A}_n)y_n \approx \frac{\tau_{n+1}^2}{\gamma\tau_n}(\widetilde{A}(t_n + \gamma\tau_n) - \widetilde{A}_n)y_n.$$

For the last term in the right-hand side of (17), in view of (5), we have

$$\tau_{n+1}^2 \widetilde{A}_{n+1}(\widetilde{f}_n - \widetilde{A}_n y_n) = \frac{\tau_{n+1}^2}{\tau_n}\widetilde{A}_{n+1}(y_n - y_{n-1}).$$

With accuracy up to $O(\tau_{n+1}^3)$, we put

$$\tau_{n+1}^2 \widetilde{A}_{n+1}(\widetilde{f}_n - \widetilde{A}_n y_n) \approx \frac{\tau_{n+1}^2}{\tau_n}\widetilde{A}(t_n + \gamma\tau_n)(y_n - y_{n-1}).$$

With this in mind, the equality (17) is replaced by the approximate equality:

$$v_n - y_n \approx \frac{\tau_{n+1}^2}{\tau_n}\left(\frac{1}{\gamma}(\widetilde{A}(t_n + \gamma\tau_n) - \widetilde{A}_n)y_n - \frac{1}{\gamma}(\widetilde{f}(t_n + \gamma\tau_n) - \widetilde{f}_n)\right.$$
$$\left. + \widetilde{A}(t_n + \gamma\tau_n)(y_n - y_{n-1})\right). \tag{18}$$

The value of $v_n - y_n$ we associate with the solution error over the interval $t_n \le t \le t_{n+1}$. Because of this, we set

$$\|v_n - y_n\| \le \tau_{n+1}\varepsilon. \tag{19}$$

From (18), we have

$$\|v_n - y_n\| \lesssim \frac{\tau_{n+1}^2}{\tau_n}\left(\frac{1}{\gamma}\|\widetilde{A}(t_n + \gamma\tau_n) - \widetilde{A}_n)y_n\|\right.$$
$$\left. + \frac{1}{\gamma}\|\widetilde{f}(t_n + \gamma\tau_n) - \widetilde{f}_n\| + \|\widetilde{A}(t_n + \gamma\tau_n)(y_n - y_{n-1})\|\right). \tag{20}$$

For a predicted time step, we also require that it will be not too small. Because of this, we put

$$\tau_{n+1} \geq \tau_0, \tag{21}$$

where τ_0 is a specified minimum time step.

In view of (21) and $\tau_{n+1} \leq \widetilde{\tau}_{n+1}$ from (18), we obtain the following formula for calculating the time step:

$$\tau_{n+1} = \max\{\tau_0, \min\{\gamma, \gamma_{n+1}\}\tau_n\},$$

$$\gamma_{n+1} = \varepsilon \left(\frac{1}{\gamma} \|(\widetilde{A}(t_n + \gamma\tau_n) - \widetilde{A}_n)y_n\| \right.$$

$$\left. + \frac{1}{\gamma} \|\widetilde{f}(t_n + \gamma\tau_n) - \widetilde{f}_n\| + \|\widetilde{A}(t_n + \gamma\tau_n)(y_n - y_{n-1})\| \right)^{-1}.$$

This formula for selecting a time step reflects clearly (see the denominator in the expression for γ_{n+1}) corrective actions, which are related to the time-dependence of the problem operator (the first part) and the right-hand side (the second part) as well as to the time-variation of the solution itself (the third part).

Acknowledgements. This work was supported by RFBR (project 14-01-00785).

References

1. Ascher, U.M., Petzold, L.R.: Computer Methods for Ordinary Differential Equations and Differential-Algebraic Equations. Society for Industrial and Applied Mathematics (1998)
2. Gear, C.W.: Numerical initial value problems in ordinary differential equations. Prentice Hall (1971)
3. Hairer, E., Norsett, S.P., Wanner, G.: Solving Ordinary Differential Equations I: Nonstiff Problems. Springer, Heidelberg (1987)
4. Samarskii, A.A.: The Theory of Difference Schemes. Marcel Dekker, New York (2001)
5. LeVeque, R.J.: Finite Difference Methods for Ordinary and Partial Differential Equations: Steady-state and Time-dependent Problems. Society for Industrial and Applied Mathematics, Philadelphia (2007)
6. Thomée, V.: Galerkin Finite Element Methods for Parabolic Problems. Springer-Verlag, Heidelberg (2010)
7. Vabishchevich, P.N.: A priori estimation of a time step for numerically solving parabolic problems. Appl. Math. Comput. **250**, 424–431 (2015)

Recent Advances in Numerical Solution of HJB Equations Arising in Option Pricing

Song Wang$^{(\boxtimes)}$ and Wen Li

Department of Mathematics and Statistics, Curtin University,
GPO Box U1987, 6845 Perth, Australia
{Song.Wang,Wen.Li}@curtin.edu.au

Abstract. This paper provides a brief survey on some of the recent numerical techniques and schemes for solving Hamilton-Jacobi-Bellman equations arising in pricing various options. These include optimization methods in both infinite and finite dimensions and discretization schemes for nonlinear parabolic PDEs.

1 Introduction

Financial derivative securities consist of three major parts: *Forwards and Future* (obligation to buy or sell), *Options* (right to buy or sell) and *Swaps* (simultaneous selling and purchasing). The first two form the basis of derivative securities. It is known that an option is a contract which gives to its owner the right, not obligation, to buy (*call*) or sell (*put*) a fixed quantity of assets of a specified stock at a fixed price called *exercise/strike price* on or before a given date (*expiry date*). There are two major types of options – European options which can be exercised only on the expiry date and American options that are exercisable on or before the expiry date.

An option has both intrinsic and time values, and can be traded on a secondary financial market even though it may not be exercisable at the time point. How to accurately price options has long been a hot topic for mathematicians and financial engineers. It was shown by Black and Scholes [7] that the value of a European option on a stock satisfies a second order parabolic partial differential equation with respect to the time t and the underlying asset price S in a complete market with constant volatility and interest rate and without transaction costs on trading the option and its underlying stock. This equation is now known as the Black-Scholes (BS) equation. A more comprehensive discussion of this model can be found in [32]. The BS equation can be solved exactly when the coefficients are constants. However, for problems of practical importance, numerical solutions to them are normally sought. Therefore, efficient and accurate numerical algorithms are essential for solving such a problem accurately.

The value of an American call option is usually the same as that of its European counterpart. However, the value $V(S,t)$ of an American put option on an asset/stock whose price S follows a geometric Brownian motion is governed by the following linear complementarity problem (LCP) (cf., e.g., [45,47])

© Springer International Publishing Switzerland 2015
I. Dimov et al. (Eds.): FDM 2014, LNCS 9045, pp. 104–116, 2015.
DOI: 10.1007/978-3-319-20239-6_10

$$LV := -\frac{\partial V}{\partial t} - \frac{1}{2}\sigma^2(t)S^2\frac{\partial^2 V}{\partial S^2} - r(t)S\frac{\partial V}{\partial S} + r(t)V \geq 0, \quad (1)$$

$$V - V^*(S) \geq 0, \quad LV \cdot (V - V^*(S)) = 0 \quad (2)$$

for $(S,t) \in \Omega := I \times [0,T)$ almost everywhere $(a.e.)$ with the payoff/terminal and boundary conditions

$$V(S,T) = V^*(S), \quad V(0,t) = V^*(0), \quad V(S_{max},t) = 0, \quad (3)$$

where $I = (0, S_{max}) \subset \mathbb{R}$ with S_{max} a positive constant usually much greater than the strike price K of the option, $\sigma(t)$ denotes the volatility of the asset, $r(t)$ the interest rate, and $V^*(S)$ is the final or payoff condition of the option. There are various payoff conditions depending types of options [47]. For example the payoff function for a vanilla American put is

$$V^*(S) = \max\{K - S, 0\}, \quad S \in I.$$

The LCP (1)–(2) can also be written as

$$\min\{LV(S,t), V(S,t) - V^*(S)\} = 0 \quad (S,t) \in \Omega \quad (4)$$

with (3). This equation is called a Hamilton-Jacobi-Bellman (HJB) equation which is usually unsolvable analytically.

When selling and buying a put whose underlying stock incurs transaction costs, the price of the put is no longer governed by the LCP or (4). Instead, a Nonlineaar Complementarity Problem (NCP) needs to be solved to determine the value of such an option. More specifically, the NCP is of the same form as (1)–(2) with $\sigma(t)$ replaced with $\sigma(S,t,V_S,V_{SS})$. Various models for the nonlinear volatility have been proposed, for example [4,8,19,23,24]. A notable one is the following nonlinear volatility model proposed in [4]:

$$\sigma^2(t,S,V_{SS}) = \sigma_0^2\left(1 + \Psi\left(e^{r(T-t)}a^2S^2V_{SS}\right)\right) \quad (5)$$

where σ_0 is a constant, $a = \kappa\sqrt{\nu N}$ with κ being the transaction cost parameter, ν a risk aversion factor and N the number of options to be sold. In the rest of this paper, we simply refer to a as the transaction parameter. The function Ψ in (5) is the solution to the following nonlinear initial value problem

$$\Psi'(z) = \frac{\Psi(z) + 1}{2\sqrt{z\Psi(z)} - z} \quad \text{for} \quad z \neq 0 \quad \text{and} \quad \Psi(0) = 0$$

to which an implicit exact solution is derived in [13].

HJB equations also arise in determination of the reservation price of a European or American option under proportional transaction costs [15–17] and valuation of American options under a Levy process [9,10], with uncertain valotility [31,48], or with stochastic valotility [18,53], just to name a few. All of these problems are of the form:

$$\min\{L_1(V) - f_1, L_2V - f_2, ..., L_mV - f_m\} = 0 \quad (6)$$

in a given solution domain with a set of boundary and terminal/payoff conditions, where m is a positive integer and, usually, L_1 is a 2nd-order nonlinear differential operator and L_i and f_i are respectively a linear 1st- or 0th-order differential operator and a given function for each of $i = 2, ..., m$.

Because of an optimization process involved and non-smooth payoff conditions, (6) in general does not have any classic (twice continuously differentiable) solutions. Instead, it has the so-called viscosity solutions [14]. Uniqueness of the solution to (6) can usually be proved. For some simple cases, it is also possible to prove the unique solvability of (6) using a conventional technique. For instance, if we introduce a weighted Sobolev space $H_{0,w}^1(I)$ and a convex set $\mathcal{K} = \{v \in H_{0,w}^1(I) : v \leq u^*\}$ with $u^* = e^{\beta t}(V_0 - V^*)$, where $V_0 = (1 - S/S_{\max})K$ and $\beta = \sup_{0 < t < T} \sigma^2(t)$, the LCP (1) and (2) can be cast into the following Variational Inequality (VI) (cf. [45] for details).

Problem 1. Find $u(t) \in \mathcal{K}$ such that, for all $v \in \mathcal{K}$,

$$\left(-\frac{\partial u(t)}{\partial t}, v - u(t)\right) + A(u(t), v - u(t); t) \geq (f(t), v - u(t)) \tag{7}$$

a.e. in $(0, T)$, where $A(\cdot, \cdot)$ is a bilinear form defined by

$$A(u, v; t) = \left(aS^2 u' + bSu, v'\right) + (cu, v), \quad u, v \in H_{0,w}^1(I) \tag{8}$$

with $(\cdot, \cdot)$ denoting the usual inner product, $u = -e^{\beta t}(V - V_0)$, $f(t) = e^{\beta t}LV_0$, $a = \sigma^2/2$, $b = r - \sigma^2$, and $c = 2r + \beta - \sigma^2$.

For this VI we can show that the bilinear form $A(\cdot, \cdot)$ is coercive and Lipschitz continuous, and thus Problem 1 has a unique solution by a standard argument.

2 The Penalty Method in Infinite Dimensions

The HJB Eq. (6), particularly (7), may be viewed as a constrained optimization problem and it can be solved using an optimization technique. A popular choice is a penalty approach in which a constrained optimization problem is approximated by an unconstrained one with a penalty term in the objective function. Since the resulting optimization problem is unconstrained, it is easier to solve than the original. The linear penalty method for HJB equations was discussed in detail in [6] and its extension to arbitrary power penalty has been proposed and analyzed in [42, 45, 52] for various HJB equations.

Let us demonstrate the power penalty method using (1) and (2) which can be rewritten in the following standard form

$$\mathcal{L}u(x, t) := \frac{\partial u}{\partial t} + \frac{\partial}{\partial S}\left[a(t)S^2\frac{\partial u}{\partial S} + b(t)Su\right] - c(t)u \leq f(S, t), \tag{9}$$

$$u(S, t) - u^*(S, t) \leq 0, \quad (\mathcal{L}u(S, t) - f(S, t)) \cdot (u(S, t) - u^*(S, t)) = 0, \tag{10}$$

in Ω with the boundary and terminal conditions

$$u_\lambda(0, t) = 0 = u_\lambda(X, t) \quad \text{and} \quad u_\lambda(S, T) = u^*(S, T), \tag{11}$$

where the coefficients functions and given data are defined in Problem 1. In fact, (7) is the variational form of this LCP, and (9)–(10) can be viewed as the optimality conditions of a constrained functional optimization problem with the constraint in (10). (We will show this in finite dimensions later.) This motivates us to devise the following penalty equation

$$\mathcal{L}u_\lambda(S,t) + \lambda[u_\lambda(S,t) - u^*(S,t)]_+^{1/k} = f(S,t), \quad (S,t) \in \Omega \tag{12}$$

satisfying (11), where $\lambda > 0$ and $k > 0$ are parameters and $[z]_+ = \max\{z, 0\}$ for any z. In (12) the power penalty term $\lambda[u_\lambda(S,t) - u^*(S,t)]_+^{1/k}$ penalizes the positive part of $u_\lambda - u^*$.

Equation (12) is a nonlinear parabolic PDE even when $k = 1$ and the variational problem corresponding to (12) is

Problem 2. Find $u_\lambda(t) \in H_{0,w}^1(I)$ such that, for all $v \in H_{0,w}^1(I)$,

$$\left(-\frac{\partial u_\lambda(t)}{\partial t}, v\right) + A(u_\lambda(t), v; t) + \lambda\left([u_\lambda(t) - u^*(t)]_+^{1/k}, v\right) = (f(t), v) \tag{13}$$

a.e. in $(0, T)$, where A is the bilinear form defined in (8).

The unique solvability of Problem 2 can be proved by showing that the mapping on the LHS of (13) is strongly monotone and continuous [45].

The solution to Problem 2 is in general not equal to that of Problem 1, but we expect that when $\lambda \to \infty$, $u_\lambda \to u$ at some rate depending on λ and k. A convergence theory for this penalty method is established in [6] for $k = 1$ and in [45] for any $k > 0$, which requires the introduction of some function spaces and norms.

For any Hilbert space $H(I)$, let $L^p(0, T; H(I))$ denote the space defined by

$$L^p(0, T; H(I)) = \{v(\cdot, t) : v(\cdot, t) \in H(I) \text{ a.e. in } (0, T); \|v(\cdot, t)\|_H \in L^p((0, T))\},$$

where $1 \leq p \leq \infty$ and $\|\cdot\|_H$ denotes the natural norm on $H(I)$. The space $L^p(0, T; H(I))$ is equipped with the norm $\|v\|_{L^p(0,T;H(I))} = \left(\int_0^T \|v(\cdot, t)\|_H^p \, dt\right)^{1/p}$. Clearly, $L^p(0, T; L^p(I)) = L^p(I \times (0, T)) = L^p(\Omega)$. Using this space, it is possible to establish the following theorem.

Theorem 1. *Let u and u_λ be the solutions to Problems 1 and 2, respectively. If $u_\lambda \in L^{1+1/k}(\Omega)$ and $\frac{\partial u}{\partial t} \in L^{k+1}(\Omega)$, then there exists a constant $C > 0$, independent of u, u_λ and λ, such that*

$$\|u - u_\lambda\|_{L^\infty(0,T;L^2(I))} + \|u - u_\lambda\|_{L^2(0,T;H_{0,w}^1(I))} \leq \frac{C}{\lambda^{k/2}}, \tag{14}$$

where k is the parameter used in (13).

Theorem 1 tells us that $u_\lambda \to u$ at the rate of $\mathcal{O}(\lambda^{-k/2})$ as λ or/and k goes to ∞. Similar results for Nonlinear Complementarity Problems (NCPs) and bounded

NCPs are given in [42,52]. The idea of the above penalty approach can also be used for solving (6). More specifically, (6) can be approximated by the following penalty equation

$$L_1(V_\lambda) - \sum_{i=1}^{m} \lambda_i [L_i V_\lambda - f_i]_-^{1/k} = f_1, \tag{15}$$

where $[z]_- = \max\{0, -z\}$ for any z, $\lambda = (\lambda_1, ..., \lambda_m)^\top$ is a set of penalty parameters and $k > 0$ is a power parameter. (Clearly, we may use different k's for different penalty terms in the above equation.) It would be thought that, as established above for Problem 2, (15) is uniquely solvable and its solution converges exponentially to that of (6). However, for some cases we are only able to prove that (15) has a unique viscosity solution and the solution converges to that of (6), but unable to establish the rates of convergence. For example, the penalty method for the HJB equations arising from determining the reservation prices of European and American options with transaction costs [28–30] in which the constraints contain derivatives of the solution. The main reason for this is that when the penalty terms contain differential operators, we may not be able to prove the strong monotonicity of the operator on the LHS of (15).

We also comment that though the solution to the penalty equation converges to that of the HJB equation, it does not mean that the constraints are strictly satisfied for any fixed (λ, k). Instead, they are satisfied up to an approximation error. Thus, the above method is sometimes called an exterior penalty method. It is possible to construct an interior penalty method such as that proposed in [35] and analyzed in [49] in which an approximation always satisfies the constraints.

3   Discretization Schemes

LCPs and HJB equations in infinite dimensions such as (6, 9, 10 and 13) can hardly be solved exactly unless for some trivial cases. Therefore, a numerical scheme is needed to for the discretization of such a system so that the discretized system can be solved by linear/nonlinear algebraic system solver in finite dimensions. Various discretization schemes can be used for the PDEs depending on the problem in question. Popular spatial discretization schemes for (12) are

    upwind finite difference schemes [11, 25, 26, 29, 30, 37],
 – fitted finite volume method [3, 12, 22, 36, 39, 46, 50], and
 – finite element methods [1, 2, 38].

In designing a descretization scheme for (12), the main requirements are as follows.

1. The scheme should be unconditionally or conditionally stable and the solution to the discretized system should converge to the viscosity solution to (13).
2. The solution to the discretized system should be non-negative irrespectively of choices of mesh or other parameters, as by nature prices are non-negative.
3. The finite dimensional linear/nonlinear system can be solved efficiently by an advanced, usually iterative, solver.

Item 2 is guaranteed if the discretiztion is monotone or the system matrix of the discrtized equation is an M-matrix in the linear case. In this case, a discrete maximum principle is satisfied by the scheme and so the solution to the discrtized system attains its extrema at the boundary of the solution domain.

Note that the BS operator $\mathcal{L}$ in (12) becomes degenerate as $S \to 0^+$. Mathematically, the weak solution to (12) cannot take a trace (boundary condition) at $S = 0$. This is why we needed the introduction of a weighted Sobolev space $H_{0,w}^1(I)$ in the previous section. Also, because of this difficulty, one usually needs to truncate the spatial domain I into $(S_{\min}, S_{\max})$ for a small positive number $S_{\min} < S_{\max}$ if a conventional scheme is used to discretize (12). Equivalently, a common practice is to use $x = \ln S$ to transform I into $-\infty < x < \ln S_{\max}$ and solve the transformed problem on a finite interval. A fitted finite volume is proposed in [39] for solving the BS equation governing European options without this domain transformation or truncation. The scheme has the merit that it is unconditionally stable, has the first-order convergence rate in mesh parameters and yields a system of which the coefficient matrix is an M-matrix. We now demonstrate this scheme using (12).

Let I be divided into N sub-intervals $I_i := (S_i, S_{i+1})$, $i = 0, 1, .., N-1$ with $0 = S_0 < S_1 < \cdots < S_N = S_{\max}$. For each $i = 0, 1, ..., N-1$, we put $h_i = S_{i+1} - S_i$ and $h = \max_{0 \le i \le N-1} h_i$. Dual to this mesh, we define another mesh with nodes $S_{i-1/2} = (S_{i-1} + S_i)/2$ for $i = 1, 2, ..., N$, $S_{-1/2} = 0$ and $S_{N+1/2} = S_{\max}$. Integrating both sides of (12) over $(S_{i-1/2}, S_{i+1/2})$ and applying the mid-point quadrature rule to the first, third, fourth and last terms, we obtain

$$-\frac{\partial u_i}{\partial t} l_i - \left[S_{i+1/2} \rho(u)|_{S_{i+1/2}} - S_{i-1/2} \rho(u)|_{S_{i-1/2}} \right] + \left[c_i u_i + \lambda [u_i - u_i^*]_+^{1/k} \right] l_i = f_i l_i$$

(16)

for $i = 1, 2, ..., N-1$, where $l_i = S_{i+1/2} - S_{i-1/2}$, $c_i = c(S_i, t)$, $f_i = f(S_i, t)$, $u_i^* = u^*(S_i)$, u_i is the nodal approximation to $u(S_i, t)$ to be determined and $\rho(u)$ is a flux associated with u defined by $\rho(u) := aSu_S + bu$.

To derive an approximation to the flux at the two end-points $S_{i+1/2}$ and $S_{i-1/2}$, let us consider the following two-point boundary value problem

$$(\rho_i)' := (aSv' + b_{i+1/2}v)' = 0, \quad S \in I_i, \quad v(S_i) = u_i, \quad v(S_{i+1}) = u_{i+1}, \qquad (17)$$

where $b_{i+1/2} = b(S_{i+1/2}, t)$. This is motivated by the technique used for singularly perturbed convection-diffusion equations (cf. [33,34]). When $i \ge 1$, (17) has the exact solution

$$\rho_i = b_{i+1/2} \frac{S_{i+1}^{\alpha_i} u_{i+1} - S_i^{\alpha_i} u_i}{S_{i+1}^{\alpha_i} - S_i^{\alpha_i}}, \quad v = \frac{\rho_i}{b_{i+1/2}} - \frac{u_{i+1} - u_i}{S_{i+1}^{\alpha_i} - S_i^{\alpha_i}} (S_i S_{i+1})^{\alpha_i} S^{-\alpha_i}, \quad (18)$$

where $\alpha_i = b_{i+1/2}/a$. Obviously, ρ_i i provides an approximation to the flux $\rho(u)$ at $S_{i+1/2}$ for $i = 1, ..., N-1$.

When $i = 0$, (17) becomes degenerate at $S = 0$, and we need to look into the asymptotic behaviour of ρ_0 as $S_0 \to 0^+$. This is given in the following two cases.

If $\alpha_0 < 0$, it is easy t see to verify that $\lim_{S_0 \to 0^+} \rho_0 = b_{1/2} u_0$. Similarly, if $\alpha_0 > 0$, we have from (18) $\lim_{x_0 \to 0^+} \rho_0 = b_{1/2} u_1$. Combining these two cases we have

$$\rho_0 = b_{1/2} \frac{1 - \text{sign}(b_{1/2})}{2} u_0 + b_{1/2} \frac{1 + \text{sign}(b_{1/2})}{2} u_1, \tag{19}$$

since $\alpha_0 = b_{1/2}/a$ and $b_{1/2}$ have the same sign pattern.

Using (18) and (19), we have from (16)

$$-\frac{\partial u_i}{\partial t} l_i + e_{i,i-1} u_{i-1} + e_{i,i} u_i + e_{i,i+1} u_{i+1} + d_i(u_i) = f_i l_i, \tag{20}$$

where, $d_i(u_i) = \lambda l_i [u_i - u_i^*]_+^{1/k}$,

$$e_{1,0} = -\frac{x_1}{2} b_{1/2} \frac{1 - \text{sign}(b_{1/2})}{2}, \quad e_{1,2} = -\frac{b_{1+1/2} x_{1+1/2} x_2^{\alpha_1}}{x_2^{\alpha_1} - x_1^{\alpha_1}},$$

$$e_{1,1} = \frac{x_1}{2} b_{1/2} \frac{1 + \text{sign}(b_{1/2})}{2} + \frac{b_{1+1/2} x_{1+1/2} x_1^{\alpha_1}}{x_2^{\alpha_1} - x_1^{\alpha_1}} + c_1 l_1,$$

$$e_{i,i-1} = -\frac{b_{i-1} x_{i-1/2} x_{i-1}^{\alpha_{i-1}}}{x_i^{\alpha_{i-1}} - x_{i-1}^{\alpha_{i-1}}}, \quad e_{i,i+1} = -\frac{b_{i+1/2} x_{i+1/2} x_{i+1}^{\alpha_i}}{x_{i+1}^{\alpha_i} - x_i^{\alpha_i}},$$

$$e_{i,i} = \frac{b_{i-1} x_{i-1/2} x_i^{\alpha_{i-1}}}{x_i^{\alpha_{i-1}} - x_{i-1}^{\alpha_{i-1}}} + \frac{b_{i+1/2} x_{i+1/2} x_i^{\alpha_i}}{x_{i+1}^{\alpha_i} - x_i^{\alpha_i}} + c_i l_i$$

for $i = 2, 3, ..., N - 1$. These form an $N - 1$ nonlinear ODE system for $U(t) := (u_1(t), ..., u_N(t))^\top$ with the homogeneous boundary condition $u_0(t) = 0 = u_N(t)$.

Let E_i be a row vector defined by $E_i = (0, .., 0, e_{i,i-1}(t), e_{i,i}(t), e_{i,i+1}(t), 0, ..., 0)$ fir $i = 1, ..., N - 1$, where $e_{i,i-1}, e_{i,i}, e_{i,i+1}$ are defined above and those not defined are zeros. Obviously, using E_i, (20) can be rewritten as

$$-\frac{\partial u_i(t)}{\partial t} l_i + E_i(t) U(t) + d_i(u_i(t)) = f_i(t) l_i, \tag{21}$$

for $i = 1, 2, ..., N - 1$. This is a first order ODE system.

To discretize (21), we choose t_i $(i = 0, 1, ..., M)$ satisfying $T = t_0 > t_1 > \cdots > t_M = 0$, and apply the two-level implicit time stepping method with a splitting parameter $\theta \in [1/2, 1]$ to (21) to yield

$$\frac{u_i^{m+1} - u_i^m}{-\Delta t_m} l_i + \theta \left[E_i^{m+1} U^{m+1} + d_i(u_i^{m+1}) \right] + (1 - \theta) \left[E_i^m U^m + d_i(u_i^m) \right]$$

$$= (\theta f_i^{m+1} + (1 - \theta) f_i^m) l_i$$

for $m = 0, 1, ..., M - 1$, where $\Delta t_m = t_{m+1} - t_m < 0$, $E_i^m = E_i(t_m)$, $f_i^m = f(x_i, t_m)$ and $U^m = (u_1^m, u_2^m, ..., u_{N-1}^m)^\top$. This nonlinear system can be rewritten as the following matrix form

$$(\theta E^{m+1} + G^m) U^{m+1} + \theta D(U^{m+1}) = F^m + [G^m - (1 - \theta) E^m] U^m - (1 - \theta) D(U^m)$$

for $m = 0, 1, ..., M - 1$, where the coefficient matrices are self-explanatory. The boundary and terminal conditions are $u_0^m = 0 = u_N^m$ and $U^0 = (u_1^*, u_2^*, ..., u_{N-1}^*)^\top$.

When $\theta = 1/2$, the time-stepping scheme becomes that of the Crank-Nicolson and when $\theta = 1$, it is the backward Euler scheme. Both of the two cases are unconditionally stable. It is easy to show that the linear part of the coefficient matrix of the above system is an M-matrix and the nonlinear part is strongly monotone. Thus, the solution to the system is non-negative. An upper error bound of order $\mathcal{O}(h + \Delta t)$ in a discrete analogue of the norm in (14) for the solution to the above system has been proved in [3] under certain conditions, where h and Δt denote the maximal mesh sizes in space and time. A superconvergent fitted finite volume method for (12) with the linear penalty, based on a judicious choice of the dual mesh in the above scheme, has been recently proposed in [46] which has the merit that the scheme yields a superconvergent derivative (Delta of an option) with almost no additional computational costs.

Upwind finite difference schemes in space have been used for solving HJB equations arising from pricing other types of options such as those in [26,29,30]. In these cases, we showed the convergence of the numerical schemes by proving they are consistent, stable and monotone [5]. However, convergence rates for these schemes have not been established. We comment that, the use of upwind finite difference methods for multi-dimensional HJB equations such as those arising from pricing options on multiple assets or with stochastic volatility [22,51] does not in general yield systems whose coefficient matrices are M-matrix. In this case, the finite volume method has to be used.

4 The Penalty Method in Finite Dimensions

The process for solving HJB equations in the previous sections is to use the penalty equation to approximate an HJB equation and then solve the penalty equation by a discretization scheme. This process is reversible, i.e., one may discretize an HJB equation first to yield a finite-dimensional one and then devise a penalty method for solving the HJB equation in finite-dimensions. Let us demonstrate this procedure using (9) and (10).

The application of the fitted finite volume method and time-stepping scheme in the previous section to the LCP (9) and (10) yields, at each time step, an LCP of the form

$$Ax \leq b, \quad x \leq 0 \quad \text{and} \quad x^\top (Ax - b) = 0, \tag{22}$$

where $x \in \mathbb{R}^n$ for a positive integer n, A is an $n \times n$ positive-definite matrix and $b \in \mathbb{R}^n$ is a known vector. In (22), x represents an approximation of the values of $u - u^*$ at the interior spatial mesh nodes. Let us consider the minimization problem

$$\min_{x \in \mathbb{R}^n} Q(x) \quad \text{subject to} \quad x \leq 0, \tag{23}$$

where $Q(x)$ is a quadratic function of x such that $\nabla Q(x) = Ax - b$. The Karush–Kuhn–Tucker (KKT) conditions for this problem are

$$Ax - b + \mu = 0, \quad \mu^\top x = 0, \quad \mu \geq 0, \quad x \leq 0,$$

where $\mu \in \mathbb{R}^n$ is the multiplier. From the first and third expressions in the above we have $Ax - b = -\mu \le 0$. Using this inequality and eliminating μ from the above yields (22). Therefore, (22) is an optimality condition for (23) and thus both have the same solutions. To find an approximation to the solution of (23), we consider the following unconstrained problem

$$\min_{x \in \mathbb{R}^n} \left(Q(x) + \frac{\lambda}{1+1/k}[x]_+^{1+1/k} \right),$$

where $\lambda > 1$ is the penalty constant and $k > 0$ is a parameter. The 1st-order necessary optimality condition for this problem is

$$Ax_\lambda - b + \lambda[x_\lambda]_+^{1/k} = 0 \quad \text{or} \quad Ax_\lambda + \lambda[x_\lambda]_+^{1/k} = b.$$

This is a (power) penalty equation approximating (22). Clearly, it is a finite-dimensional analogue of (12). The solution x_λ is an approximation to that of (22).

Discretized HJB equations are often of the form: $f(x) \le b$, $x \le 0$ and $x^\top f(x) = 0$ ([27]). Following the above discussion, the penalty equation approximating this NCP is

$$f(x_\lambda) + \lambda[x_\lambda]_+^{1/k} = 0, \tag{24}$$

where $f : \mathbb{R}^n \mapsto \mathbb{R}^n$. If, for some $\alpha, \beta > 0$, $\gamma \in (0,1]$ and $\xi \in (1,2]$, f satisfies

1. Holder Continuity: $||f(x_1) - f(x_2)||_2 \le \beta ||x_1 - x_2||_2^\gamma$, $\quad \forall x_1, x_2 \in \mathbb{R}^n$,
2. ξ-monotonicity: $(x_1 - x_2)^\top (f(x_1) - f(x_2)) \ge \alpha ||x_1 - x_2||_2^\xi$, $\quad \forall x_1, x_2 \in \mathbb{R}^n$,

then one can show that there exist a positive constant C, independent of λ, such that

$$||x_\lambda - x||_2 \le \frac{C}{\lambda^{k/(\xi-\gamma)}}, \tag{25}$$

where x and x_λ are respectively the solutions to the NCP and (24) and $|| \cdot ||_2$ denotes the l_2-norm on $\mathbb{R}^n$ (cf. [20,43]).

Note that the convergence rate of the power penalty method in finite-dimensions established above is higher than the one in (14) of its infinite-dimensional counterpart, particularly when $f(x)$ is strongly monotone and Lipschitz continuous (i.e., $\xi = 2$ and $\gamma = 1$). However, the arbitrary constant C in (25) is dependent on the dimensionality n, since in the proof of (25) we used the fact that all norms in $\mathbb{R}^n$ are equivalent which is not true in infinite dimensions.

Discretization of some HJB equations such as those arising from determining reservation price of an option under proportional transaction costs (cf. [29,30]) gives rise to optimization problems with bound constraints on Bx for a non-square matrix B, rather than on x. This kind of HJB equations also arises in optimization problems with bound constraints on derivatives. Power penalty methods have been extended to NCPs and mixed NCPs with either unbounded or bounded linear constraints (cf. [21,40,41,44]) and the upper error bounds in these cases are essentially the same as that in (25).

Note that the nonlinear penalty term in (24) becomes non-Lipschitz when $k > 1$. When solve (24) using a gradient-based method such Newton's method, the penalty term needs to be smoothed out locally in $[0, \varepsilon]$ with ε a small positive number, i.e., we replace $\lambda[x_\lambda]_+^{1/k}$ with $\lambda\phi(x_\lambda)$, where ϕ is given by

$$\phi(z) = \begin{cases} z^{\frac{1}{k}}, & z \geq \varepsilon, \\ \left[\varepsilon^{\frac{1}{k}-2}(3 - \frac{1}{k})z^2 + \varepsilon^{\frac{1}{k}-3}(\frac{1}{k} - 2)z^3\right], & z < \varepsilon, \end{cases}$$

The coefficient matrix of the linearized system of (24) is usually an M-matrix and thus a preconditioned conjugate gradient based iterative method can be used for solving it, particularly when (24) is large-scale.

5   Concluding Remarks

Pricing financial options often involves numerical solution of PDE-constrained nonlinear and non-smooth optimization problems. An efficient numerical technique for option pricing should contain three components - discretization of differential operators, techniques for constrained optimization and numerical solution of nonlinear and non-smooth algebraic systems. In this work we have presented some of our recent advances in the development of efficient and accurate numerical methods for pricing options. Extensive numerical experiments on these methods have been carried out and we refer the reader to the listed references for details.

References

1. Achdou, Y., Pironneau, O.: Computational Methods for Option Pricing. SIAM series in Applied Math, Philadelphia (2005)
2. Allegretto, W., Lin, Y., Yang, H.: Finite element error estimates for a nonlocal problem in American option valuation. SIAM J. Numer. Anal. **39**, 834–857 (2001)
3. Angermann, L., Wang, S.: Convergence of a fitted finite volume method for European and American Option Valuation. Numerische Mathematik **106**, 1–40 (2007)
4. Barles, G., Soner, H.M.: Option pricing with transaction costs and a nonlinear Black-Scholes equation. Finance Stochast. **2**(4), 369–397 (1998)
5. Barles, G., Souganidis, P.E.: Convergence of approximation schemes for fully nonlinear second order equations. Asymptotic Analysis **4**, 271–283 (1991)
6. Bensoussan, A., Lions, J.L.: Applications of Variational Inequalities in Stochastic Control. North-Holland, Amsterdam (1978)
7. Black, F., Scholes, M.: The pricing of options and corporate liabilities. J. Polit. Econ. **81**(3), 637–654 (1973)
8. Boyle, P.P., Vorst, T.: Option replication in discrete time with transaction costs. J. Finan. **XLVI**(1), 271–293 (1992)
9. Cartea, A., del-Castillo-Negrete, D.: Fractional diffusion models of option prices in markets with jumps. Physica A **374**(2), 749–763 (2007)
10. Chen, W., Wang, S.: A penalty method for a fractional order parabolic variational inequality governing American put option valuation. Comp. Math. Appl. **67**(1), 77–90 (2014)

11. Chen, W., Wang, S.: A finite difference method for pricing European and American options under a geometric Levy process. J. Ind. Manag. Optim. **11**, 241–264 (2015)
12. Chernogorova, T., Valkov, R.: Finite volume difference scheme for a degenerate parabolic equation in the zero-coupon bond pricing. Math. Comp. Model. **54**, 2659–2671 (2011)
13. Company, R., Navarro, E., Pintos, J.R., Ponsoda, E.: Numerical solution of linear and nonlinear Black-Scholes option pricing equation. Comp. Math. Appl. **56**, 813–821 (2008)
14. Crandall, M.G., Lions, P.L.: Viscosity Solution of Hamilton-Jacobi-Equations. Trans. Am. Math. Soc. **277**, 1–42 (1983)
15. Damgaard, A.: Computation of Reservation Prices of Options with Proportional Transaction Costs. J. of Econ. Dyn. Control **30**, 415–444 (2006)
16. Davis, M.H.A., Panas, V.G., Zariphopoulou, T.: European Option Pricing with Transaction Costs. SIAM J. Control Optim. **31**, 470–493 (1993)
17. Davis, M.H.A., Zariphopoulou, T.: In mathemtical finance. In: Davis, M.H.A., et al. (eds.) American Options and Transaction Fees, pp. 47–62. Springer, New York (1995)
18. Hanson, F.B., Yan, G.: American put option pricing for stochastic-volatility, jump-diffusion models. In: Proceedings of 2007 American Control Conference, pp. 384–389 (2007)
19. Hoggard, T., Whalley, A.W., Wilmott, P.: Hedging option portfolios in the presence of transaction costs. Adv. Futures Options Research **7**, 21–35 (1994)
20. Huang, C.C., Wang, S.: A Power Penalty approach to a nonlinear complementarity problem. Oper. Res. Lett. **38**, 72–76 (2010)
21. Huang, C.C., Wang, S.: A penalty method for a mixed nonlinear complementarity problem. Nonlinear Anal. TMA **75**, 588–597 (2012)
22. Huang, C.-S., Hung, C.-H., Wang, S.: A fitted finite volume method for the valuation of options on assets with stochastic volatilities. Comput. **77**, 297–320 (2006)
23. Jandačka, M., Ševčovič, D.: On the risk-adjusted pricing-methodology-based valuation of vanilla options and explanation of the volatility smile. J. Appl. Math. **3**, 235–258 (2005)
24. Leland, H.E.: Option pricing and replication with transaction costs. J. Finan. **40**, 1283–1301 (1985)
25. Koleva, M.N., Vulkov, L.G.: A positive flux limited difference scheme for the uncertain correlation 2D Black-Scholes problem, J. Comp. Appl. Math. (in press)
26. Lesmana, D.C., Wang, S.: An upwind finite difference method for a nonlinear Black-Scholes equation governing European option valuation. Appl. Math. Comp. **219**, 8818–8828 (2013)
27. Lesmana, D.C., Wang, S.: Penalty approach to a nonlinear obstacle problem governing American put option valuation under transaction costs. Appl. Math. Comp. **251**, 318–330 (2015)
28. Li, W., Wang, S.: Penalty approach to the HJB equation arising in European stock option pricing with proportional transaction Costs. J. Optim. Theory Appl. **143**, 279–293 (2009)
29. Li, W., Wang, S.: Pricing American options under proportional transaction costs using a penalty approach and a finite difference scheme. J. Ind. Manag. Optim. **9**, 365–398 (2013)
30. Li, W., Wang, S.: A numerical method for pricing European options with proportional transaction costs. J. Glob. Optim. **60**, 59–78 (2014)

31. Lyons, T.J.: Uncertain volatility and the risk free synthesis of derivatives. Appl. Math. Finance **2**, 117–133 (1995)
32. Merton, R.C.: Theory of rational option pricing. Bell J. Econ. Manage. Sci. **4**, 141–183 (1973)
33. Miller, J.J.H., Wang, S.: A new non-conforming Petrov-Galerkin method with triangular elements for a singularly perturbed advection-diffusion problem. IMA J. Numer. Anal. **14**, 257–276 (1994)
34. Miller, J.J.H., Wang, S.: An exponentially fitted finite element volume method for the numerical solution of 2D unsteady incompressible flow problems. J. Comput. Phys. **115**, 56–64 (1994)
35. Nielsen, B.F., Skavhaug, O., Tveito, A.: Penalty and front-fixing methods for the numerical solution of American option problems. J. Comp. Fin. **5**, 69–97 (2001)
36. Valkov, R.: Fitted finite volume method for a generalized Black Scholes equation transformed on finite interval. Numer. Algorithms **65**, 195–220 (2014)
37. Vazquez, C.: An upwind numerical approach for an American and European option pricing model. Appl. Math. Comp. **97**, 273–286 (1998)
38. Wang, G., Wang, S.: Convergence of a finite element approximation to a degenerate parabolic variational inequality with non-smooth data Arising from American option valuation. Optim. Method Softw. **25**, 699–723 (2010)
39. Wang, S.: A novel fitted finite volume method for the Black-Scholes equation governing option pricing. IMA J. Nuner. Anal. **24**, 669–720 (2004)
40. Wang, S.: A power penalty method for a finite-dimensional obstacle problem with derivative constraints. Optim. Lett. **8**, 1799–1811 (2014)
41. Wang, S.: A penalty approach to a discretized double obstacle problem with derivative constraints. J. Glob. Optim. (2014). doi:10.1007/s10898-014-0262-3
42. Wang, S., Huang, C.-S.: A power penalty method for a nonlinear parabolic complementarity problem. Nonlinear Anal. TMA **69**, 1125–1137 (2008)
43. Wang, S., Yang, X.Q.: A power penalty method for linear complementarity problems. Oper. Res. Lett. **36**, 211–214 (2008)
44. Wang, S., Yang, X.Q.: A power penalty method for a bounded nonlinear complementarity problem. Optimization (2014). doi:10.1080/02331934.2014.967236
45. Wang, S., Yang, X.Q., Teo, K.L.: Power penalty method for a linear complementarity problem arising from American option valuation. J. Optim. Theory Appl. **129**(2), 227–254 (2006)
46. Wang, S., Zhang, S., Fang, Z.: A superconvergent fitted finite volume method for black scholes equations governing European and American option valuation. Numer. Meth. Part. D.E. **31**, 1190 (2015)
47. Wilmott, P., Dewynne, J., Howison, S.: Option Pricing Math. Models Comput. Oxford Financial Press, Oxford (1993)
48. Zhang, K., Wang, S.: A computational scheme for uncertain volatility model in option pricing. Appl. Numer. Math. **59**, 1754–1767 (2009)
49. Zhang, K., Wang, S.: Convergence property of an interior penalty approach to pricing American option. J. Ind. Manag. Optim. **7**, 435–447 (2011)
50. Zhang, K., Wang, S.: Pricing American bond options using a penalty method. Automatica **48**, 472–479 (2012)
51. Zhang, K., Wang, S., Yang, X.Q., Teo, K.L.: A power penalty approach to numerical solutions of two-asset American options. Numer. Math. Theory Methods Appl. **2**, 127–140 (2009)

52. Zhou, Y.Y., Wang, S., Yang, X.Q.: A penalty approximation method for a semi-linear parabolic double obstacle problem. J. Glob. Optim. **60**, 531–550 (2014)
53. Zvan, R., Forsyth, P.A., Vetzal, K.R.: Penalty methods for American options with stochastic volatility. Comput. Appl. Math. **91**, 199–218 (1998)

Applications of Numerical Methods for Stochastic Controlled Switching Diffusions with a Hidden Markov Chain: Case Studies on Distributed Power Management and Communication Resource Allocation

Zhixin Yang[1], Le Yi Wang[2], George Yin[3](✉), Qing Zhang[4], and Hongwei Zhang[5]

[1] Department of Mathematics, University of Wisconsin-Eau Claire, Eau Claire, WI 54701, USA
yangzhix@uwec.edu
[2] Department of Electrical and Computer Engineering, Wayne State University, Detroit, MI 48202, USA
lywang@wayne.edu
[3] Department of Mathematics, Wayne State University, Detroit, MI 48202, USA
gyin@math.wayne.edu
[4] Department of Mathematics, The University of Georgia, Athens, GA 30602, USA
qingz@math.uga.edu
[5] Department of Computer Science, Wayne State University, Detroit, MI 48202, USA
hongwei@wayne.edu

1 Introduction

Recently, considerable attention has been drawn to stochastic controlled systems with hidden Markov chains. Much motivation stems from applications in distributed power management and platoon inter-vehicle distance maintenance, among others. The dynamic systems of interest are controlled diffusions with switching, known as switching diffusions [6]. Different from the extensive studies contained in the aforementioned reference, the switching process in this paper is assumed to be a continuous-time Markov chain that is hidden. We can only observe the state of the Markov chain with additive noise. Mean-variance control problems were first considered in the Nobel prize winning paper of Markowitz [2]. It was subsequently considered by a host of researchers. The recent advances in backward stochastic differential equations enable the treatment of the mean-variance controls in continuous time, which is otherwise impossible because of the so-called indefinite control weights; see Zhou and Li [7] for the first paper in this direction and further details. Further work in conjunction with regime-switching models can be found in Zhou and Yin [8], among others.

As a new twist of the mean-variance portfolio selections, our recent work focuses on using the mean-variance formulation to treat networked control systems. That is, we borrow the idea in financial engineering to treat problems

© Springer International Publishing Switzerland 2015
I. Dimov et al. (Eds.): FDM 2014, LNCS 9045, pp. 117–128, 2015.
DOI: 10.1007/978-3-319-20239-6_11

arising in networked control problems. Much of the motivation stems from applications arising in cyber-physical systems. It has been observed in [4] that a large class of problems arising from networked systems and platoon controls can be formulated as such systems, similar to the mean variance control problems that were originally pursued in financial engineering [8]. In [4], we outlined three potential applications in platoon controls based on mean-variance controls. The first problem concerns the longitudinal inter-vehicle distance control. To increase highway utility, it is desirable to reduce the total length of a platoon, resulting in smaller overall inter-vehicle distances. The drawback of this strategy, however, is the increase in the risk of collision due to traffic uncertainties. The task of minimizing the risk with desired inter-vehicle distance fits naturally to a mean-variance optimization framework. The second one is communication resource allocation of bandwidths for vehicle-to-vehicle (V2V) communications. For a given maximum throughput of a platoon communication system, the communication system operator must find a way to assign this resource to different V2V channels, which may also be formulated as a mean-variance control problem. The third one is the platoon fuel consumption. Due to variations in vehicle sizes and speeds, each vehicle's fuel consumption is a controlled random process. Tradeoff between a platoon's team acceleration/maneuver capability and fuel consumption can be summarized in a desired platoon fuel consumption rate. Assigning fuels to different vehicles results in coordination of vehicle operations modeled by subsystem fuel rate dynamics. This problem may also be casted into the framework of mean-variance control. Such problems are highly nonlinear, it is virtually impossible to find closed-form solutions. Our objective is thus devoted to finding feasible algorithms for the desired tasks. Recently, in our work [5], numerical approximation methods have been developed. The convergence of the algorithms is proved. The basic idea is first to convert the partially observable stochastic control problems to completely observed systems by means of the Wonham filtering methodologies. Then we use relaxed controls and Markov chain approximation techniques to build convergent numerical schemes. Based on that work, this paper aims to provide case studies of two typical problems in applications. Our main effort is to demonstrate using numerical methods solving the problems arising in the specific applications.

The rest of the paper is arranged as follows. Section 2 formulates the problem. Section 3 introduces the Markov chain approximation methods and provides the approximation of the optimal controls. Sections 4 and 5 present two case studies to illustrate the wide applications of the scheme developed in our work.

2   Problem Formulation

Consider a given probability space $(\Omega, \mathcal{F}, P)$ in which there is $w_1(t)$, a standard ρ-dimensional Brownian motion with $w_1(t) = (w_1^1(t), w_1^2(t), \ldots, w_1^\rho(t))'$, where z' denotes the transpose of z. Let $\alpha(t)$ be a continuous-time finite-state Markov chain, independent of $w_1(t)$, taking values in $\mathcal{M} = \{1, 2, \ldots, m\}$ with generator $Q = (q_{ij})_{m \times m}$. We consider a networked system that consists of $\rho + 1$ nodes (subsystems), which is modeled for $t \in [s, T]$ by

$$
\begin{aligned}
dx_0(t) &= \mu_0(t, \alpha(t))x_0(t)dt, \ x_0(s) = x_0, \\
dx_l(t) &= x_l(t)\mu_l(t, \alpha(t))dt + x_l(t)\bar{\sigma}_l(t, \alpha(t))dw_1(t), \ x_l(s) = x_l, l = 1, \ldots, \rho,
\end{aligned}
\tag{1.1}
$$

where for each i, $\mu_l(t, i)$ is the drift and $\bar{\sigma}_l(t, i) = (\bar{\sigma}_{l1}(t, i), \ldots, \bar{\sigma}_{l\rho}(t, i))$ is the volatility for the lth node. In our framework, instead of having full information of the Markov chain, we can only observe

$$
dy(t) = g(\alpha(t))dt + \beta dw_2(t), \ y(s) = 0,
\tag{1.2}
$$

where $\beta > 0$ and $w_2(\cdot)$ is a standard scalar Brownian motion, $w_1(\cdot)$, $w_2(\cdot)$, and $\alpha(\cdot)$ are independent. Moreover, the initial data $p(s) = p = (p^1, p^2, \ldots, p^m)$ in which $p^i = p^i(s) = P(\alpha(s) = i)$ is given for $1 \leq i \leq m$. By distributing the portion $N_l(t)$ of the lth node's flow $x_l(t)$ at time t and denoting the total flows for the whole networked system as $x(t)$, we have $x(t) = \sum_{l=0}^{\rho} N_l(t)x_l(t), t \geq s$. With $x(s) = \sum_{l=0}^{\rho} N_l(s)x_l(s) = x$, the dynamics of $x(t)$ are given as

$$
dx(t) = [x(t)\mu_0(t, \alpha(t)) + M(t, \alpha(t))\pi(t)]dt + \pi'(t)\bar{\sigma}(t, \alpha(t))dw_1(t),
\tag{1.3}
$$

in which $\pi(t) = (\pi_1(t), \ldots, \pi_\rho(t))'$ and $\pi_l(t) = N_l(t)x_l(t)$ for $l = 1, \ldots, \rho$ is the actual flow of the network system for the lth node and $\pi_0(t) = x(t) - \sum_{l=1}^{\rho} \pi_l(t)$ is the actual flow of the networked system for the first node, and $M(t, \alpha(t)) = (\mu_i(t, \alpha(t)) - \mu_0(t, \alpha(t)) : i = 1, \ldots, \rho)$ and $\bar{\sigma}(t, \alpha(t)) = (\bar{\sigma}_{lj}(t, \alpha(t)))_{\rho \times \rho}$. We define $\mathcal{F}_t = \sigma\{w_1(\tilde{s}), y(\tilde{s}), x(s) : s \leq \tilde{s} \leq t\}$. Our objective is to find an $\mathcal{F}_t$ admissible control $\pi(\cdot)$ in a compact set ϖ under the constraint that the expected terminal flow is $Ex(T) = \kappa$ for some given $\kappa \in \mathbb{R}$, so that the risk measured by the variance of the terminal flow is minimized. Specifically, we have the following goal

$$
\begin{aligned}
\min \ &J(s, x, p, \pi(\cdot)) := E[x(T) - \kappa]^2 \\
&\text{subject to } Ex(T) = \kappa.
\end{aligned}
\tag{1.4}
$$

We apply the Lagrange multiplier techniques (see, e.g.,[7]) to arrive at the unconstrained optimization problem

$$
\begin{aligned}
\min \ &J(s, x, p, \pi(\cdot), \lambda) := E[x(T) + \lambda - \kappa]^2 - \lambda^2 \\
&\text{subject to } (x(\cdot), \pi(\cdot)) \text{ admissible,}
\end{aligned}
\tag{1.5}
$$

where λ is the Lagrange multiplier. A pair $(\sqrt{\text{Var}\,(x(T))}, \kappa) \in \mathbb{R}^2$, corresponding to the optimal control if it exists, is called an *efficient point*.

Next, to treat the partially observed control problem, let $p^i(t) = P(\alpha(t) = i | \mathcal{F}^y(t))$ for $i = 1, 2, \ldots, m$, with $p(t) = (p^1(t), \ldots, p^m(t)) \in \mathbb{R}^{1 \times m}$ and $\mathcal{F}^y(t) = \sigma\{y(\tilde{s}) : s \leq \tilde{s} \leq t\}$. It was shown in [3] that this conditional probability satisfies the following system of stochastic differential equations

$$
dp^i(t) = \sum_{j=1}^m q^{ji}p^j(t)dt + \frac{1}{\beta}p^i(t)(g(i) - \bar{\alpha}(t))d\hat{w}_2(t), \ p^i(s) = p^i, i = 1, \ldots, m
\tag{1.6}
$$

where $\overline{\alpha}(t) = \sum_{i=1}^{m} g(i)p^i(t)$ and $\widehat{w}_2(t)$ is the innovation process. Now we have a completely observable system so that $x(s) = x$, $p^i(s) = p^i$, and

$$dx(t) = \mu(x(t), p(t), \pi(t))dt + \sigma(x(t), p(t), \pi(t))dw_1(t)$$

$$dp^i(t) = \sum_{j=1}^{m} q^{ji}p^j(t)dt + \frac{1}{\beta}p^i(t)(g(i) - \overline{\alpha}(t))d\widehat{w}_2(t), \text{ for } i \in \{1, \ldots, m\} \quad (1.7)$$

where

$$\mu(x(t), p(t), \pi(t)) = \sum_{i=1}^{m} \mu_0(t, i)p^i(t)x(t) + \sum_{l=1}^{\rho}\sum_{i=1}^{m}(\mu_l(t, i) - \mu_0(t, i))p^i(t)\pi_l(t)$$

$$\sigma(x(t), p(t), \pi(t))dw_1(t) = \sum_{l=1}^{\rho}\sum_{j=1}^{\rho}\sum_{i=1}^{m} \pi_l(t)\overline{\sigma}_{lj}(t, i)p^i(t)dw_1^j(t).$$

For an arbitrary π and $\phi(\cdot, \cdot, \cdot) \in C^{1,2,2}(\mathbb{R})$, consider the operator

$$\mathcal{L}^r\phi(s, x, p) = \frac{\partial\phi}{\partial s} + \frac{\partial\phi}{\partial x}\mu(x, p, r) + \frac{1}{2}\frac{\partial^2\phi}{\partial x^2}[\sigma(x, p, r)\sigma'(x, p, r)]$$

$$+ \sum_{i=1}^{m}\frac{\partial\phi}{\partial p^i}\sum_{j=1}^{m}q^{ji}p^j + \frac{1}{2}\sum_{i=1}^{m}\frac{\partial^2\phi}{\partial (p^i)^2}\frac{1}{\beta^2}[p^i(g(i) - \overline{\alpha})]^2. \quad (1.8)$$

Let $W(s, x, p, \pi)$ be the objective function with $E^\pi_{s,x,p}$ denoting the expectation of functionals on $[s, T]$ with $x(s) = x, p(s) = p$, the admissible control $\pi = \pi(\cdot)$, and the value function $V(s, x, p)$

$$V(s, x, p) = \inf_{\pi \in \pi} W(s, x, p, \pi) = \inf_{\pi \in \pi} E^\pi_{s,x,p}(x(T) + \lambda - k)^2 - \lambda^2. \quad (1.9)$$

The value function is a solution of the following equation

$$\inf_{r \in \pi} \mathcal{L}^r V(s, x, p) = 0, \quad (1.10)$$

with boundary condition $V(T, x, p) = (x(T) + \lambda - \kappa)^2 - \lambda^2$. Note that (1.10) is known as the Hamilton-Jacobi-Bellman (HJB) equation. To proceed, we use the relaxed control representation. For the σ-algebra $\mathcal{B}(\pi)$ and $\mathcal{B}(\pi \times [s, T])$ of Borel subsets of π and $\pi \times [s, T]$, an admissible relaxed control or simply a relaxed control $m(\cdot)$ is a measure on $\mathcal{B}(\pi \times [s, T])$ such that $m(\pi \times [s, t]) = t - s$ for all $t \in [s, T]$ for all $t \in [s, T]$. For notational simplicity, for any $B \in \mathcal{B}(\pi)$, we write $m(B \times [s, T])$ as $m(B, T - s)$. Since $m(\pi \times [s, t]) = t - s$ for all $t \in [s, T]$ and $m(B, \cdot)$ is nondecreasing, it is absolutely continuous. Hence the derivative $\dot{m}(B, t) = m_t(B)$ exists almost everywhere for each B. We can further define the relaxed control representation $m(\cdot)$ of $\pi(\cdot)$ by $m_t(B) = I_{\{\pi(t) \in B\}}$ for any $B \in \mathcal{B}(\pi)$. We say that $M(\cdot)$ is a measure-value $\mathcal{F}_t$ martingale with values $M(B, t)$ if $M(B, \cdot)$ is an $\mathcal{F}_t$ martingale for each $B \in \pi$, and for each t, the following holds: $\sup_{B \in \pi} EM^2(B, t) < \infty$,

$M(A \cup B, t) = M(A, t) + M(B, t)$ w.p.1. for all disjoint $A, B \in \mathcal{B}(\pi)$, and $EM^2(B_n, t) \to 0$ if $B_n \to \emptyset$. We say that $M(\cdot)$ is orthogonal if $M(A, \cdot)$ and $M(B, \cdot)$ are $\mathcal{F}_t$ martingales whenever $A \cap B = \emptyset$. If $M(\cdot)$, $\bar{M}(\cdot)$ are $\mathcal{F}_t$ martingale measures and $M(A, \cdot)$, $\bar{M}(B, \cdot)$ are $\mathcal{F}_t$ martingales for any Borel set A, B, then $M(\cdot)$ and $\bar{M}(\cdot)$ are said to be strongly orthogonal. Letting $M(\cdot) = (M_1(\cdot), \ldots, M_\rho(\cdot))'$, a vector valued martingale measure, we impose the following conditions.

(A1) $M(\cdot) = (M_1(\cdot), \ldots, M_\rho(\cdot))'$ is square integrable and continuous; each component is orthogonal; and the pairs are strongly orthogonal.

Under (A1), there are measure-valued random processes $m_i(\cdot)$ such that the quadratic variation processes satisfy, for each t and $A, B \in \mathcal{B}(\pi)$, $\langle M_i(A, \cdot), M_j(B, \cdot) \rangle(t) = \delta_{ij} m_i(A \cap B, t)$.

(A2) m_i does not depend on i, $m_i(\cdot) = m(\cdot)$, and $m(\pi, t) = t$ for all t.

With the help of the martingale measures and relaxed controls, we can represent our control system in the following way:

$$x(t) = x + \int_s^t \int_\pi \mu(x(z), p(z), c) m_z(dc) dz + \int_s^t \int_\pi \sigma(x(z), p(z), c) M(dc, dz)$$

$$p^i(t) = \int_s^t \sum_{j=1}^m q^{ji} p^j(z) dz + \int_s^t \frac{1}{\beta} [p^i(z)(g(i) - \alpha(z))] d\widehat{w}_2(z), \; i \in \{1, \ldots, m\}$$

(A3) $\mu(\cdot, \cdot, \cdot)$ and $\sigma(\cdot, \cdot, \cdot)$ are continuous; $\mu(\cdot, p, c)$ and $\sigma(\cdot, p, c)$ are Lipschitz continuous uniformly in p, c and bounded.

(A4) $\sigma(x, p, c) = (\sigma_1(x, p, c), \ldots, \sigma_\rho(x, p, c)) > 0$.

3 Approximation Algorithms

To facilitate subsequent numerical computations, let $v^i(t) = \log p^i(t)$. Itô's rule leads to the dynamics of $v^i(t)$. We can then obtain the following discrete-time approximation of the Wonham filter

$$v_{n+1}^{h_2, i} = v_n^{h_2, i} + h_2 [\sum_{j=1}^m q^{ji} \frac{p_n^{h_2, j}}{p_n^{h_2, i}} - \frac{1}{2\beta^2}(g(i) - \bar{\alpha}_n^{h_2})^2] + \sqrt{h_2} \frac{1}{\beta}(g(i) - \bar{\alpha}_n^{h_2}) \varepsilon_n, \quad (1.11)$$

$$v_0^{h_2, i} = \log(p^i), \; p_{n+1}^{h_2, i} = \exp(v_{n+1}^{h_2, i}),$$

where $\bar{\alpha}_n^{h_2} = \sum_{i=1}^m g(i) p_n^{h_2, i}$ and $\{\varepsilon_n\}$ is a sequence of i.i.d. random variables satisfying $E\varepsilon_n = 0$, $E\varepsilon_n^2 = 1$, and $E|\varepsilon_n|^{2+\gamma} < \infty$ for some $\gamma > 0$ with $\varepsilon_n = \frac{\widehat{w}_2((n+1)h_2) - \widehat{w}_2(nh_2)}{\sqrt{h_2}}$. Here $p_n^{h_2, i}$ appeared as a denominator in (1.11) and we have concentrated on the case that $p_n^{h_2, i}$ stays away from 0. Let $h_1 > 0$ be a discretization parameter for state variables, and recall that $h_2 > 0$ is the step size for the time variable. We construct a discrete-time finite-states Markov chain to approximate the controlled diffusion process, $x(t)$. Let $N_{h_2} = (T - s)/h_2$ be an integer and define $S_{h_1} = \{x : x = kh_1, k = 0, \pm 1, \pm 2, \ldots\}$. We use $\pi_n^{h_1,h_2}$ to denote the random variable that is the control action for the chain at discrete time n. Let $\pi^{h_1,h_2} = (\pi_0^{h_1,h_2}, \pi_1^{h_1,h_2}, \ldots)$ denote the sequence of $\mathbb{u}$-valued random variables which are the control actions at time $0, 1, \ldots$ and $p^{h_2} = (p_0^{h_2}, p_1^{h_2}, \ldots)$ be the corresponding posterior probabilities in which $p_n^{h_2} = (p_n^{h_2,1}, p_n^{h_2,2}, \ldots, p_n^{h_2,m})$. We define the difference $\Delta\xi_n^{h_1,h_2} = \xi_{n+1}^{h_1,h_2} - \xi_n^{h_1,h_2}$ and let $E_n^{h_1,h_2,r}$, $V_n^{h_1,h_2,r}$ denote the conditional expectation and variance given $\{\xi_k^{h_1,h_2}, \pi_k^{h_1,h_2}, p_k^{h_2}, k \leq n, \xi_n^{h_1,h_2} = x, p_n^{h_2} = p, \pi_n^{h_1,h_2} = r\}$. By stating that $\{\xi_n^{h_1,h_2}, n < \infty\}$ is a controlled discrete-time Markov chain on a discrete-time state space S_{h_1} with transition probabilities denoted by $p^{h_1,h_2}((x, y)|r, p)$, we mean that the transition probabilities are functions of a control variable r and posterior probability p. The sequence $\{\xi_n^{h_1,h_2}, n < \infty\}$ is said to be locally consistent with (1.7) if it satisfies

$$
\begin{aligned}
E_{x,p,n}^{h_1,h_2,r} \Delta\xi_n^{h_1,h_2} &= \mu(x, p, r)h_2 + o(h_2), \\
V_{x,p,n}^{h_1,h_2,r} \Delta\xi_n^{h_1,h_2} &= \sigma(x, p, r)\sigma'(x, p, r)h_2 + o(h_2), \\
\sup_n |\Delta\xi_n^{h_1,h_2}| &\to 0, \quad \text{as } h_1, h_2 \to 0.
\end{aligned}
\tag{1.12}
$$

With the approximating Markov chain given above, we can approximate the cost function $W^{h_1,h_2}(s, x, p, \pi^{h_1,h_2})$ in which $x(T)$ is replaced by $\xi_{N_{h_2}}^{h_1,h_2}$ and can find approximation of $V(s, x, p)$. Now we will proceed to find a reasonable Markov chain that is locally consistent. We first suppose that the control space has a unique admissible control $\pi^{h_1,h_2} \in \mathbb{u}^{h_1,h_2}$, so that we can drop inf in (1.10). We discretize (1.8) by a finite difference method using step-size $h_1 > 0$ for the state variable and $h_2 > 0$ for the time variable as mentioned above. For simplicity, we omit the details. We can show that the approximating Markov chain constructed above satisfies local consistency. Note that we have used local transitions here so that we can avoid the problem of "numerical noise" or "numerical viscosity", which appears in non-local transitions cases, and is even more serious in higher dimension scenarios, see [1] for more details. We omit most of the details and please refer to [5] for further demonstration.

It can be shown that the Markov chain $\{\xi_n^{h_1,h_2}, n < \infty\}$ with transition probabilities $p^{h_1,h_2}(\cdot)$ properly defined is locally consistent with (1.7). Next, we give the discrete-time approximation algorithm for the controlled Markov chain. Based on the local consistency, we can represent $\xi_{n+1}^{h_1,h_2}$ as

$$
\xi_{n+1}^{h_1,h_2} = \xi_n^{h_1,h_2} + \mu(\xi_n^{h_1,h_2}, p_n^{h_2}, \pi_n^{h_1,h_2})h_2 + \sigma(\xi_n^{h_1,h_2}, p_n^{h_2}, \pi_n^{h_1,h_2})\Delta w_n^{h_1,h_2} + o(1), \tag{1.13}
$$

where $o(1)$ can be written as $\varepsilon_n^{h_1,h_2}$ in which $\varepsilon_n^{h_1,h_2} \to 0$ as $h_1, h_2 \to 0$. To approximate the continuous-time process $(x(t), p(t), m(t), M(t))$, we use continuous-time interpolation. For $t \in [nh_2, (n + 1)h_2)$, we define the piecewise constant interpolations by

$$\xi^{h_1,h_2}(t) = \xi_n^{h_1,h_2}, \; p^{h_2}(t) = p_n^{h_2}, \; \bar{a}^{h_1,h_2}(t) = \sum_{i=1}^{m} g(i)p_n^{h_2}, \; \pi^{h_1,h_2}(t) = \pi_n^{h_1,h_2},$$

$$z^{h_2}(t) = n, \; w_l^{h_1,h_2}(t) = \sum_{k=0}^{z^{h_2}(t)-1} \Delta w_{l,k}^{h_1,h_2}, \; \varepsilon^{h_1,h_2}(t) = \varepsilon_n^{h_1,h_2}.$$

(1.14)

With most of the technical details omitted, which can be found in [5], we present the main approximation theorem below.

Theorem 1. *Assuming* (A1)-(A4), *let* $\{\xi_n^{h_1,h_2}, n < \infty\}$, *the approximating chain be constructed with transition probabilities properly defined. Let* $\{\pi_n^{h_1,h_2}, n < \infty\}$ *be a sequence of admissible controls,* $\xi^{h_1,h_2}(\cdot)$ *and* $p^{h_2}(\cdot)$ *be the continuous time interpolation defined in* (1.14), $m^{h_1,h_2}(\cdot)$ *be the relaxed control representation of* $\pi^{h_1,h_2}(\cdot)$ *(continuous time interpolation of* $\pi_n^{h_1,h_2}$*). Then*

$$(\xi^{h_1,h_2}(\cdot), p^{h_2}(\cdot), m^{h_1,h_2}(\cdot), M^{h_1,h_2}(\cdot)) \; is tight,$$

$(\xi^{h_1,h_2}(\cdot), p^{h_2}(\cdot), m^{h_1,h_2}(\cdot), M^{h_1,h_2}(\cdot))$ *converges weakly to* $(x(\cdot), p(\cdot), m(\cdot), M(\cdot))$, *and* $W(s, x, p, m^{h_1,h_2}) \to W(s, x, p, m)$. *Denoting the limit of a weakly convergent subsequence by* $(x(\cdot), p(\cdot), m(\cdot), M(\cdot))$, *the martingale measure* $M(\cdot)$ *has quadratic variation process given by* $m(\cdot)$ *and the desired limit dynamics hold. Moreover,* $V^{h_1,h_2}(s, x, p) \to V(s, x, p)$ *as* $h_1 \to 0$ *and* $h_2 \to 0$.

4   Case Study I: Distributed Power Management

Consider a distribution network of three renewable energy generators and energy storage devices. Typically, the distributed generators can be photovoltaic (PV) systems, wind turbines, bio-engines, fuel cells, etc. Energy storage devices can be batteries, super-capacitors, etc. To be concrete, let $x_i(t)$, $i = 1, 2, 3$ be the maximum power generating capacity of the ith generator at time t. In addition, $x_0(t)$ is the available maximum capacity that is allowed to be purchased from the main grid at t.

Let $N_i(t)$ be the portion of the power generated by the ith generator that is used to satisfy total power demand, Then, the total locally generated power at time t is $\sum_{i=1}^{3} N_i(t)x_i(t)$. Implicitly, the remaining power will be purchased from the main grid, i.e., $\pi_0(t) = N_0(t)x_0(t) = x(t) - \sum_{i=1}^{3} N_i(t)x_i(t)$. A renewable generator's maximum capacity is a stochastic process. For example, a wind turbine's maximum power is determined by the wind speed and direction. Similarly, a PV system's output is determined by how much solar radiation is available at a given time, weather condition, and the angle that the sunlight is shining on the solar panels. Here, $\{x_i(t) : i = 0, 1, \ldots, 3\}$ is given by (1.1) with $\alpha(t)$ being a 3-state switching process which takes values in $\{1, 2, 3\}$ with generator

$$Q = \begin{pmatrix} -0.5 & 0,2 & 0.3 \\ 0.3 & -0.6 & 0.3 \\ 0.4 & 0.4 & -0.8 \end{pmatrix}, \; \mu_1(\alpha) = 2\alpha, \; \mu_2(\alpha) = \alpha + 1, \; \mu_3(\alpha) = \alpha + 2,$$

$\sigma_1(\alpha) = (\alpha, 0, 0)$, $\sigma_2(\alpha) = (0, \frac{\alpha}{2}, 0)$ and $\sigma_3(\alpha) = (0, 0, \frac{\alpha}{3})$, for $\alpha = 1, 2, 3$, and $w_1(t) \in \mathbb{R}^3$. Here, the drift term represents average solar radiation values throughout a day; and diffusion term represents solar radiation fluctuations which are caused by many factors such as clouds, weather conditions, etc. The dynamics of the process depend on an event variable α which reflects system structural changes. This is exemplified by scheduled or emergency maintenance of solar modules, failure of a battery cell, addition of super-capacitor banks, tap changes in transformer actions, etc.

It is noted that sometimes such switching actions α cannot be observed directly, such as solar or battery cell failures. However, such switching actions will affect certain measured variables. For example, battery cell failures will cause a jump in terminal voltages. In this study, instead of direct access to α, we assume (1.2) is observable where $g(1) = 1, g(2) = 2$ and $g(3) = 3$, $\beta = 1 > 0$ is a constant, and $w_2(t)$ is a Brownian motion, independent of $w_1(t)$. $y(t)$ is a measured quantity. Distributed power management aims to decide dispatching parameters $N_i(t)$, $i = 1, \ldots, 3$. This can be formulated as a mean-variance control problem. To meet the total power consumption demand $z = 1\,\mathrm{MW}$ (mega watts), it is required that we have the constraint $Ex(T) = z$. On the other hand, to maintain grid stability, smooth operations, and reduced waste, it is desirable that generation-consumption disparity in transient be as small as possible. It is well understood in traditional power flow analysis that transient power fluctuations cause energy loss on lines, affect voltage and frequency stability. In view of (1.4), the Lagrange multiplier technique leads to (1.5). The value function and corresponding control are in Fig. 1 in which x axis is the possible consumption demand of all the generators in the system at $T = 2$ and y axis represent the feedback control π_1 for the first generator and value function V, respectively. The efficient frontier is demonstrated in Fig. 2 in which the x axis is the standard deviation of total generation-consumption of the system and y axis is the expected power consumption. We use the simplex method to find out the optimal λ.

5   Case Study II: Communication Resource Allocation

The second case study is concerned with communication resource allocation of bandwidths for vehicle-to-vehicle (V2V) communications. For a given maximum throughput of a platoon communication system, the communication system must find a way to assign this resource to different V2V channels. If the total bandwidth used is lower than the assigned bandwidth, there will be a waste of resource. Conversely, usage of bandwidths over the budget may incur high costs or interfere with other platoons' operations. In this case, each channel's bandwidth usage is the state of the subsystem. Their summation is a random process and is desired to approach the maximum throughput (the desired mean at the terminal time) with variations as small as possible. Consequently, it becomes a mean-variance control problem.

Consider a platoon of five vehicles. Let $B_i(t)$, $i = 0, 1, \ldots, 4$ be the maximum transmission data rate of vehicle i at time t. In practice, the maximum

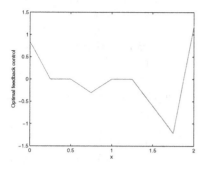

(a) Optimal feedback control $\pi_1(t) = N_1(t)x_1(t)$ for the first generator by using the step size $h_1 = 0.25$ for the state variable and step size $h_2 = 0.001$ for the time variable with the fixed expectation $z = 1$ MW

(b) Approximate value function V by using the step size $h_1 = 0.25$ for the state variable and step size $h_2 = 0.001$ for the time variable with fixed expectation $z = 1$ MW

Fig. 1. Optimal control for the first generator and value function V for the power management system

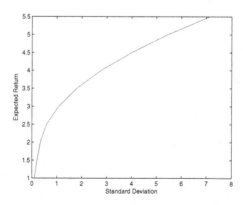

Fig. 2. Mean variance efficient frontier for power management system in which step sizes for the state variable and time variable are $h_1 = 0.25$ and $h_2 = 0.001$, respectively.

data rate is determined by the processing capability limits, the resources used by other tasks of the vehicle's communication system, and the bandwidth allocation scheme between vehicles (e.g., through wireless transmission scheduling). If the platoon is assigned with the total data rate $B(t)$ Mbps (mega bits per second), which must be shared by all the vehicles within the platoon. Let $N_i(t)$ be the portion of $B_i(t)$ that is used in the actual transmission by vehicle i. Then, $N_i(t)B_i(t)$ is the data rate of vehicle i and the total data rate of the entire platoon is desired to be $B(t) = \sum_{i=0}^{4} N_i(t)B_i(t)$. Due to dynamics of many tasks, $B_i(t)$ is a stochastic process. In addition, since vehicles move along roads, we have a communication network whose topology switches. Assume that $\{B_i(t) : i = 0, 1, \ldots, 4\}$ obeys the stochastic system (1.1) with the Markov chain $\alpha(t)$ having m states, representing m possible network topologies. To be concrete, suppose that $m = 1, 2, 3, 4$ and the switching process has the generator

$$Q = \begin{pmatrix} -0.7 & 0.5 & 0.1 & 0.1 \\ 0.4 & -0.8 & 0.2 & 0.2 \\ 0.2 & 0.1 & -0.5 & 0.2 \\ 0.1 & 0.2 & 0.3 & -0.6 \end{pmatrix} \text{ and } \mu_0(\alpha(t), t) = 0.5\alpha, \ \mu_1(\alpha(t), t) = \alpha + t,$$

$\mu_2(\alpha(t), t) = 2\alpha + 1.5t$, $\mu_3(\alpha, t) = \alpha - t$, $\sigma_1(\alpha(t), t) = (\alpha, 0, 0, 0)$, $\sigma_2(\alpha(t), t) = (0, \frac{\alpha}{2}, 0, 0)$, $\sigma_3(\alpha(t), t) = (0, 0, \frac{\alpha}{3}, 0)$ for $\alpha = 1, 2, 3, 4$, and $w_1(t) \in \mathbb{R}^4$. Here, the drift term represents average maximum data rates during an operating time interval of the communication system and $B_i(t)\sigma_i(\alpha(t), t)dw_1(t)$ represents fluctuations on B_i, which are determined by other communication tasks such as coding, data compression, packet formation, etc. The dynamics of the process depend on the event variable α which reflects communication network topology changes. Communication link changes typically contain both observable and unobservable elements. It is noted that a communication link can be terminated by the associated vehicles, which is an observable event. However, packet loss can cause a link to be broken which is not observable directly until the data transmission is completed and data were lost. In this sense, this unobservable event can be partially observed from data flows and receipt acknowledgement. Consequently, the event α can be modeled by (1.2) where $g(1) = 2$, $g(2) = 1.5$, $g(3) = 3$ and $g(4) = -1$, and $\beta = 1 > 0$ is a constant. Here $y(t)$ is a measured variable for the event.

Communication system management decides data rate allocation strategies by assigning $N_i(t)$ proportion of data rate to vehicle i, $i = 1, \ldots, 4$. This can be formulated as a mean-variance control problem. To use efficiently the total available data rate $z = 2$ Mbps, we require that at the end of the resource assignment period T, $EB(T) = z$. To ensure that the platoon does not overuse resources (causing interruptions to other platoons, incurring penalty, etc.) or waste resources, it is desirable that the platoon's actual total data rate is as close to 2 Mbps as possible. This is consistent to (1.4), or equivalently (1.5).

The value function and corresponding control are in Fig. 3 in which x axis is the possible value for the resource assignment at $T = 2$ in the platoon communication system and y axis represents π_1- the feedback control or in other words, the data rate of the first vehicle and value function V, respectively. The efficient

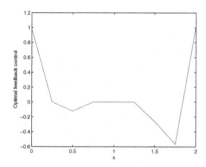

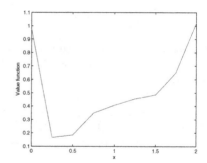

(a) Optimal feedback control (data rate) $\pi_1(t)$ for the first vehicle by using the step size $h_1 = 0.25$ for state variable and step size $h_2 = 0.001$ for the time variable with fixed expectation $B = 2$

(b) Approximate value function V by using the step size $h_1 = 0.25$ for state variable and step size $h_2 = 0.001$ for the time variable with fixed expectation $B = 2$

Fig. 3. Optimal control for the first vehicle and value function V for the entire platoon system

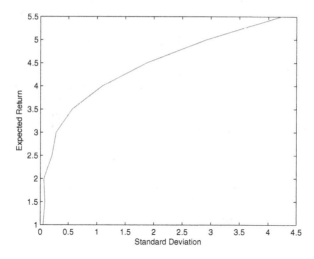

Fig. 4. Mean variance efficient frontier for communication system in which step sizes for the state variable and time variable are $h_1 = 0.25$ and $h_2 = 0.001$, respectively.

frontier is demonstrated in Fig. 4 in which the x axis is the standard deviation of the total data rate of the entire platoon and y axis is the standard deviation of the total data rate allocation for the V2V communications at the end of the resource assignment period.

6   Concluding Remarks

This paper presented case studies on two applications. The main characteristics of the problems are regime-switching diffusions with a hidden Markov chain. Our effort was devoted to the numerical solutions of the problems. After converting the problems into completely observed systems, based on Markov chain approximation techniques, controlled discrete-time Markov chains were constructed for the intended task. Although only two examples have been presented, the techniques used and the methods of approximation can be applied to a wide range of applications.

Acknowledgements. This research was supported in part by the National Science Foundation under CNS-1136007.

References

1. Kushner, H.J.: Consistency issues for numerical methods for variance control with applications to optimization in finance. IEEE. Trans. Automat. Control **44**, 2283–2296 (2000)
2. Markowitz, H.: Portfolio selection. J. Finance **7**, 77–91 (1952)
3. Wonham, W.M.: Some applications of stochastic differential equations to optimal nonlinear filtering. SIAM J. Control **2**, 347–369 (1965)
4. Yang, Z., Yin, G., Wang, L.Y., Zhang, H.: Near-optimal mean-variance controls under two-time-scale formulations and applications. Stochastics **85**, 723–741 (2013)
5. Yang, Z., Yin, G., Zhang, Q.: Mean-variance type controls involving a hidden Markov chain: Models and numerical approximation. IMA J. Control Inf. (to appear)
6. Yin, G., Zhu, C.: Hybrid Switching Diffusions: Properties and Applications. Springer, New York (2010)
7. Zhou, X.Y., Li, D.: Continuous time mean variance portfolio selection: a stochastic LQ framework. Appl. Math. Optim. **42**, 19–33 (2000)
8. Zhou, X.Y., Yin, G.: Markowitz mean-variance portfolio selection with regime switching: a continuous time model. SIAM J. Control Optim. **42**, 1466–1482 (2003)

Error Estimates of the Crank-Nicolson-Polylinear FEM with the Discrete TBC for the Generalized Schrödinger Equation in an Unbounded Parallelepiped

Alexander Zlotnik$^{(\boxtimes)}$

Department of Higher Mathematics at Faculty of Economics,
National Research University Higher School of Economics, Myasnitskaya 20,
101000 Moscow, Russia
azlotnik2008@gmail.com

Abstract. We deal with an initial-boundary value problem for the generalized time-dependent Schrödinger equation with variable coefficients in an unbounded n–dimensional parallelepiped ($n \geq 1$). To solve it, the Crank-Nicolson in time and the polylinear finite element in space method with the discrete transparent boundary conditions is considered. We present its stability properties and derive new error estimates $O(\tau^2 + |h|^2)$ uniformly in time in L^2 space norm, for $n \geq 1$, and mesh H^1 space norm, for $1 \leq n \leq 3$ (a superconvergence result), under the Sobolev-type assumptions on the initial function. Such estimates are proved for methods with the discrete TBCs for the first time.

Keywords: Time-dependent Schrödinger equation · Unbounded domain · Crank-Nicolson scheme · Finite element method · Discrete transparent boundary conditions · Stability · Error estimates · Superconvergence

1 Introduction

The linear time-dependent Schrödinger equation is the key one in many physical fields. It should be often solved in unbounded space domains. A number of approaches were developed to deal with such problems using approximate transparent boundary conditions (TBCs) at the artificial boundaries.

Among the best methods of such kind are those using the so-called discrete TBCs remarkable by the clear mathematical background and the corresponding rigorous stability results in theory as well as the complete absence of spurious reflections in practice. They first were constructed and studied for the standard Crank-Nicolson in time finite-difference schemes, see [1,5] and also [2,3], in the cases of the infinite or semi-infinite axis and strip. Later families of finite-difference schemes with general space averages were treated in [4,12,15]. In particular, they include the linear and bilinear FEMs in space.

© Springer International Publishing Switzerland 2015
I. Dimov et al. (Eds.): FDM 2014, LNCS 9045, pp. 129–141, 2015.
DOI: 10.1007/978-3-319-20239-6_12

In this paper, we consider the Crank-Nicolson-polylinear FEM in an unbounded n–dimensional parallelepiped ($n \geq 1$), present results on its stability with respect to the initial data and a free term as well as on exploiting the discrete TBCs and mainly derive the corresponding new error estimates $O(\tau^2 + |h|^2)$ uniform in time and in L^2 norm (for $n \geq 1$) and mesh H^1 norm (for $1 \leq n \leq 3$) in space under the Sobolev-type assumptions on the initial function. The latter estimate is a superconvergence result. Such estimates are proved for the methods with the discrete TBCs for the first time. Importantly, the error estimates contain no mesh steps in negative powers like for other approximate TBCs, see [6,7], that is one more advantage of using the discrete TBCs.

2 The IBVP and Numerical Methods to Solve it

We deal with the initial-boundary value problem (IBVP) for the time-dependent generalized Schrödinger equation with $n \geq 1$ space variables

$$i\hbar \rho D_t \psi = \mathcal{H}\psi := -\frac{\hbar^2}{2}\mathrm{div}(B\nabla\psi) + V\psi \quad \text{on} \quad \Pi \times (0,T), \tag{1}$$

$$\psi|_{\partial\Pi} = 0, \quad \psi|_{t=0} = \psi^0(x) \quad \text{on} \quad \Pi. \tag{2}$$

Hereafter $\psi = \psi(x,t)$ is the complex-valued unknown wave function, i is the imaginary unit and $\hbar > 0$ is a physical constant. The $x = (x_1, \ldots, x_n)$-depending coefficients $\rho, V \in L^\infty(\Pi)$ and the $n \times n$ matrix $B \in L^\infty(\Pi)$ are real-valued and satisfy $\rho(x) \geq \rho > 0$ and $B(x) \geq \underline{B}I > 0$ on Π, where I is the unit matrix (whereas V can have any sign in general). Here $\Pi := \mathbb{R}$ for $n = 1$ or, for $n \geq 2$, $\Pi := \mathbb{R} \times \Pi_{\hat{1}}$ is the infinite parallelepiped, with $\Pi_{\hat{1}} := (0, X_2) \times \cdots \times (0, X_n)$.

Also $D_t = \frac{\partial}{\partial t}$ and $D_i = \frac{\partial}{\partial x_i}$ are the partial derivatives, and the operators div and ∇ are taken with respect to space variables.

We also assume that, for some (sufficiently large) $X_0 > 0$,

$$\rho(x) = \rho_\infty, \quad B(x) = \mathrm{diag}(B_{1\infty}, \ldots, B_{n\infty}), \quad V(x) = V_\infty \quad \text{for} \quad |x_1| \geq X_0, \tag{3}$$

where $\mathrm{diag}(B_{1\infty}, \ldots, B_{n\infty})$ is the diagonal matrix with the listed positive diagonal entries. More generally, it could be easily assumed that ρ, B and V have different constant values for $x_1 \leq -X_0$ and for $x_1 \geq X_0$. Let $X_1 > X_0$, and $\Omega = \Omega_X = (-X_1, X_1)$ for $n = 1$ or $\Omega = \Omega_X = (-X_1, X_1) \times \Pi_{\hat{1}}$ for $n \geq 2$.

We consider the weak solution $\psi \in C([0,T]; H_0^1(\Pi))$ having $D_t\psi \in C([0,T]; L^2(\Pi))$ and satisfying the integral identity

$$i\hbar(D_t\psi(\cdot,t), \varphi)_{L^{2,\rho}(\Pi)} = \mathcal{L}_\Pi(\psi(\cdot,t), \varphi) \quad \text{for any} \quad \varphi \in H_0^1(\Pi), \quad \text{on} \quad [0,T], \tag{4}$$

and the initial condition $\psi|_{t=0} = \psi^0 \in H_0^1(\Pi)$. Hereafter we use the standard complex Lebesgue and Sobolev spaces (and subspaces), the weighted complex Lebesgue space $L^{2,\rho}(G)$ endowed by the inner product $(w, \varphi)_{L^{2,\rho}(G)} := (\rho w, \varphi)_{L^2(G)}$ and the $\mathcal{H}$-related Hermitian-symmetric sesquilinear form

$$\mathcal{L}_G(w, \varphi) := \frac{\hbar^2}{2}(B\nabla w, \nabla\varphi)_{L^2(G)} + (Vw, \varphi)_{L^2(G)}, \quad \text{with} \quad G = \Pi, \Omega, \text{etc.}$$

We define a non-uniform mesh in x_1 on $\mathbb{R}$ containing the points $\pm X_1$ and being uniform with a step $0 < h_1 < X_1$ outside $[-X_1 + h_1, X_1 - h_1] \supset [-X_0, X_0]$. We also define non-uniform meshes in $x_2, \ldots, x_n$ respectively on $[0, X_2], \ldots, [0, X_n]$ (containing the ends of the segments). They induce the partition of $\bar{\Pi}$ into finite elements that are rectangular parallelepipeds without common internal points. Let $|h|$ be their maximal diagonal length. Let $S_h(\bar{\Pi})$ be the (infinite-dimensional) subspace of functions in $H_0^1(\Pi)$ that are polylinear over each element. Clearly $S_h(\bar{\Pi}) \subset C(\bar{\Pi}) \cap L^2(\Pi)$. Let S_h be the restriction of $S_h(\bar{\Pi})$ to $\bar{\Omega}$.

Let $\bar{\omega}_M^\tau$ be the non-uniform mesh $0 = t_0 < \ldots < t_M = T$ with steps $\tau_m := t_m - t_{m-1}$. We put $\tau_{\max} := \max_{1 \le m \le M} \tau_m$ and $\hat{\tau}_m := \frac{\tau_m + \tau_{m+1}}{2}$ for $1 \le m \le M-1$ and $\hat{\tau}_0 := \frac{\tau_1}{2}$. We define the time mesh operators

$$\bar{\partial}_t Y^m := \frac{Y^m - Y^{m-1}}{\tau_m}, \quad \hat{\partial}_t Y^m := \frac{Y^{m+1} - Y^m}{\hat{\tau}_m}, \quad \bar{s}_t Y^m := \frac{Y^{m-1} + Y^m}{2}.$$

We introduce *the Crank-Nicolson-polylinear FEM* approximate solution Ψ: $\bar{\omega}_M^\tau \to S_h(\bar{\Pi})$ satisfying the integral identity

$$i\hbar(\bar{\partial}_t \Psi^m, \varphi)_{L^{2,\rho}(\Pi)} = \mathcal{L}_\Pi(\bar{s}_t \Psi^m, \varphi) \text{ for any } \varphi \in S_h(\bar{\Pi}) \text{ and } 1 \le m \le M, \quad (5)$$

compare with (4), and the initial condition $\Psi|_{t=0} = \Psi^0 \in S_h(\bar{\Pi})$, where Ψ^0 is an approximation for ψ^0.

Let $\ell_\infty^m(\varphi)$ be a conjugate linear functional on $S_h(\bar{\Pi})$ that we add to the right-hand side of (5) to study stability in more detail and to derive error estimates.

Proposition 1. *Let* $\ell_\infty^m(\varphi) = (F^m, \varphi)_{L^2(\Pi)}$ *with* $F^m \in L^2(\Pi)$ *for* $1 \le m \le M$. *Then there exists a unique approximate solution* Ψ *and the following first stability bound holds*

$$\max_{0 \le m \le M} \|\Psi^m\|_{L^{2,\rho}(\Pi)} \le \|\Psi^0\|_{L^{2,\rho}(\Pi)} + \frac{2}{\hbar} \sum_{m=1}^{M} \|F^m\|_{L^{2,1/\rho}(\Pi)} \tau_m. \quad (6)$$

We introduce also the "energy" norm such that

$$\|w\|_{\mathcal{H}+\hat{v}\rho;\,\Pi}^2 := \mathcal{L}_\Pi(w, w) + \hat{v}\|w\|_{L^{2,\rho}(\Pi)}^2 \ge \hat{\delta}\|w\|_{L^{2,\rho}(\Pi)}^2 \text{ for any } w \in H_0^1(\Pi), \quad (7)$$

with some real numbers $\hat{v}$ and $\hat{\delta} > 0$. Inequality (7) is knowingly valid for $\hat{v}$ so large that $\frac{\hbar^2}{2}\underline{B}\lambda_0 + V(x) + (\hat{v} - \hat{\delta})\rho(x) \ge 0$ on Ω with $\lambda_0 := \sum_{k=2}^{n} \left(\frac{\pi}{X_k}\right)^2$ (here $\lambda_0 = 0$ for $n = 1$). We define also the corresponding dual mesh depending norm

$$\|w\|_{H_h^{-1}(\Pi)} := \max_{\varphi \in S_h(\bar{\Pi}):\,\|\varphi\|_{\mathcal{H}+\hat{v}\rho;\,\Pi}=1} |\langle w, \varphi \rangle_\Pi| \le c\|w\|_{H^{-1}(\Pi)},$$

where $\langle w, \varphi \rangle_\Pi$ is the conjugate duality relation on $H^{-1}(\Pi) \times H_0^1(\Pi)$ and $H^{-1}(\Pi) = [H_0^1(\Pi)]^*$. Hereafter c and c_1 are generic constants independent of the meshes, any functions and T whereas c_0 denotes absolute constants (fixed numbers).

Proposition 2. *Let* $\ell_\infty^m(\varphi) = \langle F^m, \varphi \rangle_\Pi$ *with* $F^m \in H^{-1}(\Pi)$ *for* $1 \leq m \leq M$ *and* $F^0 \in H^{-1}(\Pi)$ *be arbitrary. Then there exists a unique approximate solution* Ψ *and the following second stability bound holds*

$$\max_{0 \leq m \leq M} \|\Psi^m\|_{\mathcal{H}+\hat{v}\rho;\, \Pi} \leq \|\Psi^0\|_{\mathcal{H}+\hat{v}\rho;\, \Pi}$$

$$+ 4 \sum_{m=1}^{M} \left(\frac{|\hat{v}|}{\hbar} \|F^m\|_{H_h^{-1}(\Pi)} + \|\overline{\partial}_t F^m\|_{H_h^{-1}(\Pi)} \right) \tau_m + 4 \|F^0\|_h^{(-1)}. \quad (8)$$

Method (5) cannot be directly used in practice because of the infinite number of unknowns at each time level. Nevertheless it is possible to restrict the method to $\bar{\Omega}$ provided that $\Psi^0 \in S_{0h} := \{\varphi \in S_h; \varphi(x) = 0 \text{ on } \Omega \setminus \Omega_0\}$, where $\Omega_0 := \Omega_{X_1-h_1, X_2,\ldots,X_n}$, and $\overline{\omega}_M^\tau$ is uniform with the step $\tau = \frac{T}{M}$. Let both assumptions be valid up to the end of the section.

By definition, *the discrete TBCs* are conditions at the artificial boundaries $x_1 = \pm X_1$ allowing to accomplish the restriction (they are non-local in $x_2, \ldots, x_n$ and t). To write down them explicitly, for clarity, we confine ourselves by the case of the uniform mesh in x_k with the step $h_k = \frac{X_k}{J_k}$, for $2 \leq k \leq n$, and define the related well-known direct and inverse discrete sine Fourier transforms

$$P^{(q)} = (\mathcal{F}_k P)^{(q)} := \frac{2}{J_k} \sum_{j=1}^{J_k-1} P_j \sin \frac{\pi q j}{J_k}, \quad 1 \leq q \leq J_k - 1,$$

$$P_j = (\mathcal{F}_k^{-1} P^{(\cdot)})_j := \sum_{q=1}^{J_k-1} P^{(q)} \sin \frac{\pi q j}{J_k}, \quad 1 \leq j \leq J_k - 1.$$

The related eigenvalues of the 1D linear in x_k FEM counterparts of the operators $-D_k^2$ and the unit one (for zero Dirichlet boundary values at $x_k = 0, X_k$) are

$$\lambda_q^{(k)} = \left(\frac{2}{h_k} \sin \frac{\pi q h_k}{2X_k} \right)^2, \quad \sigma_q^{(k)} = 1 - \frac{2}{3} \sin^2 \frac{\pi q h_k}{2X_k} \in \left(\frac{1}{3}, 1 \right).$$

Denote by $\omega_{h\hat{1}}$ the internal part of the introduced uniform mesh in $\bar{\Pi}_{\hat{1}}$ and define the related mesh inner product

$$(U, W)_{\omega_{h\hat{1}}} := \sum_{j_2=1}^{J_2-1} \cdots \sum_{j_n=1}^{J_n-1} U_{j_2,\ldots,j_n} W_{j_2,\ldots,j_n}^* h_2 \ldots h_n \quad \text{for } n \geq 2$$

or set $(U, W)_{\omega_{h\hat{1}}} := UW^*$ for $n = 1$, where W^* is the complex conjugate for W.

Recall that the discrete convolution of mesh functions $R, Q: \overline{\omega}_M^\tau \to \mathbb{C}$ is given by $(R * Q)^m := \sum_{p=0}^{m} R^p Q^{m-p}$ for $0 \leq m \leq M$.

Proposition 3. *The restriction* $\Psi|_{\bar{\Omega}}$ *of the above approximate solution obeys the integral identity on* Ω

$$i\hbar \, (\overline{\partial}_t \Psi^m, \varphi)_{L^{2,\rho}(\Omega)} = \mathcal{L}_\Omega(\overline{s}_t \Psi^m, \varphi)$$

$$- \frac{\hbar^2}{2} B_{1\infty}(S_{\text{ref}}^m \Psi_{X_1}^m, \varphi|_{x_1=X_1})_{\omega_{h\hat{1}}} + \frac{\hbar^2}{2} B_{1\infty}(S_{\text{ref}}^m \Psi_{-X_1}^m, \varphi|_{x_1=-X_1})_{\omega_{h\hat{1}}} \quad (9)$$

for any $\varphi \in S_h$ and $1 \leq m \leq M$, and the initial condition

$$\Psi|_{t=0} = \Psi^0|_{\bar{\Omega}} \in S_h. \tag{10}$$

Here $\Psi^m_{\pm X_1} = \left\{ \Psi^0\big|_{x_1=\pm X_1}, \ldots, \Psi^m\big|_{x_1=\pm X_1} \right\}$ is a vector-function.

The operator $\mathcal{S}_{\mathrm{ref}}$ in the discrete TBC has the form

$$\mathcal{S}^m_{\mathrm{ref}}\Phi^m := \mathcal{F}_2^{-1} \ldots \mathcal{F}_n^{-1} \left[\sigma^{(2)}_{q_2} \ldots \sigma^{(n)}_{q_n} R_{\mathbf{q}} * \Phi^{\mathbf{q}} \right]^m \quad \text{on } \overline{\omega}^{\tau}_M \tag{11}$$

for any $\Phi \colon \omega_{h\hat{1}} \times \overline{\omega}^{\tau}_M \to \mathbb{C}$ such that $\Phi^0 = 0$, where $\Phi^m := \{\Phi^0, \ldots, \Phi^m\}$, $\Phi^{\mathbf{q}} := (\mathcal{F}_n \ldots (\mathcal{F}_2\Phi)^{(q_2)} \ldots)^{(q_n)}$ and $\mathbf{q} = (q_2, \ldots, q_n)$. Here the kernel $R_{\mathbf{q}}$ can be computed by the recurrent formulas

$$R^0_{\mathbf{q}} = c_{1\mathbf{q}}, \quad R^1_{\mathbf{q}} = -c_{1\mathbf{q}}\kappa_{\mathbf{q}}\mu_{\mathbf{q}}, \quad R^m_{\mathbf{q}} = \frac{2m-3}{m}\kappa_{\mathbf{q}}\mu_{\mathbf{q}}R^{m-1}_{\mathbf{q}} - \frac{m-3}{m}\kappa^2_{\mathbf{q}}R^{m-2}_{\mathbf{q}}, \quad m \geq 2,$$

with the coefficients defined by

$$c_{1\mathbf{q}} = -\frac{|\alpha_{\mathbf{q}}|^{1/2}}{2}e^{-i(\arg\alpha_{\mathbf{q}})/2}, \quad \kappa_{\mathbf{q}} = -e^{i\arg\alpha_{\mathbf{q}}}, \quad \mu_{\mathbf{q}} = \frac{\beta_{\mathbf{q}}}{|\alpha_{\mathbf{q}}|} \in (-1,1), \tag{12}$$

$$\alpha_{\mathbf{q}} = 2a_{\mathbf{q}} + \frac{1}{3}h_1^2 a_{\mathbf{q}}^2 \neq 0, \quad \arg\alpha_{\mathbf{q}} \in (0, 2\pi), \quad \beta_{\mathbf{q}} = 2\,\mathrm{Re}\,a_{\mathbf{q}} + \frac{1}{3}h_1^2|a_{\mathbf{q}}|^2,$$

$$a_{\mathbf{q}} = \frac{V_\infty}{B_{1\infty}\hbar^2} + \frac{1}{2B_{1\infty}}\left(B_{2\infty}\frac{\lambda^{(2)}_{q_2}}{\sigma^{(2)}_{q_2}} + \cdots + B_{n\infty}\frac{\lambda^{(n)}_{q_n}}{\sigma^{(n)}_{q_n}} \right) + i\frac{2\rho_\infty}{\tau\hbar B_{1\infty}}.$$

The next lemma is important to prove stability results for method (9), (10).

Lemma 1. *The operator $\mathcal{S}^m_{\mathrm{ref}}$ satisfies the inequalities [2, 3]*

$$\mathrm{Im}\sum_{l=1}^{m}\left(\mathcal{S}^l_{\mathrm{ref}}\Phi^l, \bar{s}_t\Phi^l\right)_{\omega_{h\hat{1}}}\tau \geq 0, \quad \mathrm{Im}\sum_{l=1}^{m}\left(\mathcal{S}^l_{\mathrm{ref}}\Phi^l, (i\hbar\overline{\partial}_t + \hat{v}\overline{s}_t)\Phi^l\right)_{\omega_{h\hat{1}}}\tau \geq 0$$

on $\overline{\omega}^{\tau}_M$, for any $\Phi \colon \omega_{h\hat{1}} \times \overline{\omega}^{\tau}_M \to \mathbb{C}$ such that $\Phi^0 = 0$ and $\hat{v} \geq -\frac{V_\infty}{\rho_\infty}$ (see (3)).

Let $\ell^m(\varphi)$ be a conjugate linear functional on S_h that we add to the right-hand side of (9) to study stability in more detail.

Proposition 4. *Let $\ell^m(\varphi) = (F^m, \varphi)_{L^2(\Omega)}$ with $F^m \in L^2(\Omega)$ for $1 \leq m \leq M$. Then the solution to (9), (10) is unique and satisfies the first stability bound*

$$\max_{0\leq m\leq M}\|\Psi^m\|_{L^{2,\rho}(\Omega)} \leq \|\Psi^0\|_{L^{2,\rho}(\Omega)} + \frac{2}{\hbar}\sum_{m=1}^{M}\|F^m\|_{L^{2,1/\rho}(\Omega)}\tau. \tag{13}$$

We introduce the "energy" norm on Ω such that

$$\|w\|^2_{\mathcal{H}+\hat{v}\rho;\,\Omega} := \mathcal{L}_\Omega(w,w) + \hat{v}\|\sqrt{\rho}\,w\|^2_{L^2(\Omega)} > 0, \tag{14}$$

for any $w \in \tilde{H}^1(\Omega) := \{H^1(\Omega); \ w|_{(-X_1,X_1) \times \partial \Pi_{\tilde{1}}} = 0\}$ except for $w = 0$, and some real number $\hat{v} \geq -\frac{V_\infty}{\rho_\infty}$. In particular, for $\hat{v}$ so large that $\frac{\hbar^2}{2} \underline{B} \lambda_0 + V(x) + \hat{v}\rho(x) > 0$ on Ω, (14) is valid. Define also the respective dual mesh depending norm

$$\|w\|_{H_h^{-1}(\Omega)} := \max_{\varphi \in S_h: \|\varphi\|_{\mathcal{H}+\hat{v}\rho; \ \Omega}=1} |\langle w, \varphi \rangle_\Omega| \leq c\|w\|_{H^{-1}(\Omega)}, \quad H^{-1}(\Omega) = [\tilde{H}^1(\Omega)]^*,$$

where $\langle w, \varphi \rangle_\Omega$ is the conjugate duality relation on $H^{-1}(\Omega) \times \tilde{H}^1(\Omega)$.

Proposition 5. *Let $\ell^m(\varphi) = \langle F^m, \varphi \rangle_\Omega$ with $F^m \in H^{-1}(\Omega)$ for $1 \leq m \leq M$ and $F^0 \in H^{-1}(\Omega)$ be arbitrary. Then the solution to (9), (10) is unique and satisfies the second stability bound*

$$\max_{0 \leq m \leq M} \|\Psi^m\|_{\mathcal{H}+\hat{v}\rho; \ \Omega} \lesssim \|\Psi^0\|_{\mathcal{H}+\hat{v}\rho; \ \Omega}$$

$$+4 \sum_{m=1}^M \left(\frac{|\hat{v}|}{\hbar} \|F^m\|_{H_h^{-1}(\Omega)} + \|\bar{\partial}_t F^m\|_{H_h^{-1}(\Omega)} \right) \tau + 4 \|F^0\|_{H_h^{-1}(\Omega)}. \quad (15)$$

Propositions 1–5 and Lemma 1 in the quite similar cases of the semi-infinite $\Pi = (0, \infty)$ $(n = 1)$ and $\Pi = (0, \infty) \times (0, X_2)$ $(n = 2)$ were proved respectively in [4] and [12,15] (see also [16]), where families of finite-difference schemes with space averages depending on a parameter θ were treated covering, in particular, the linear and bilinear FEMs (for $\theta = \frac{1}{6}$). For the presented improvement in formulas (12), see also [14]. For $n = 1$, the results are as well particular cases of those from [13] (given specifically for general FEM). The case $n \geq 3$ can be treated in the same manner as $n = 2$ (for such an example, see [11]).

The numerical results for the method can be found in [4,15,16].

3   Error Estimates

Let condition (7) be valid and σw be *the elliptic projection* of $w \in H_0^1(\Pi)$ onto $S_h(\bar{\Pi})$ such that

$$\mathcal{L}_\Pi(\sigma w, \varphi) + \hat{v}(\sigma w, \varphi)_{L^{2,\rho}(\Pi)} = \mathcal{L}_\Pi(w, \varphi) + \hat{v}(w, \varphi)_{L^{2,\rho}(\Pi)} \quad (16)$$

for any $\varphi \in S_h(\bar{\Pi})$. Note that σw exists and is unique. We also assume below that $B \in W^{1,\infty}(\Pi)$ and then the following error estimate holds

$$\|w - \sigma w\|_{L^{2,\rho}(\Pi)} \leq c|h|^2 \|(\mathcal{H}_\rho + \hat{v})w\|_{L^{2,\rho}(\Pi)} \text{ for any } w \in H^2(\Pi) \cap H_0^1(\Pi). (17)$$

We consider $\mathcal{H}_\rho := \frac{1}{\rho}\mathcal{H}$ as an unbounded operator in $L^{2,\rho}(\Pi)$ with $\mathcal{D}(\mathcal{H}_\rho) = H^2(\Pi) \cap H_0^1(\Pi)$. Assume below that $\psi^0 \in \mathcal{D}(\mathcal{H}_\rho^3)$.

Proposition 6. *The following first error estimate holds*

$$\max_{0 \leq m \leq M} \|(\psi - \Psi)^m\|_{L^{2,\rho}(\Pi)} \leq \|\Psi^0 - \sigma\psi^0\|_{L^{2,\rho}(\Pi)}$$

$$+c(1+T) \left\{ \tau_{\max}^2 \|\mathcal{H}_\rho^3 \psi^0\|_{L^{2,\rho}(\Pi)} + |h|^2 \left(\|\mathcal{H}_\rho^2 \psi^0\|_{L^{2,\rho}(\Pi)} + \|\psi^0\|_{L^{2,\rho}(\Pi)} \right) \right\}. (18)$$

Here $\sigma\psi^0$ can be replaced by ψ^0.

Proof. 1. For $y \in L^1(0,T)$, define the average (the projection on the time mesh) $[y]^m := \frac{1}{\tau_m} \int_{t_{m-1}}^{t_m} y(t)dt$, $1 \le m \le M$, and notice that

$$\sum_{m=1}^{M} \|[u]^m\|_{\mathbf{B}} \tau_m \le \int_0^T \|u(\cdot,t)\|_{\mathbf{B}} \, dt, \tag{19}$$

where $\|\cdot\|_{\mathbf{B}} = \|\cdot\|_{L^{2,\rho}(\Pi)}$ or $\|\cdot\|_{\mathcal{H}+\hat{v}\rho;\, \Pi}$ and $u \in L^1(0,T;\mathbf{B})$, and

$$|[y]^m - \overline{s}_t y^m| \le c_0 \tau_m^2 [|D_t^2 y|]^m, \quad 1 \le m \le M, \text{ for } y \in W^{2,1}(0,T). \tag{20}$$

2. Applying $[\cdot]$ to identity (4), we get

$$i\hbar([D_t\psi]^m, \varphi)_{L^{2,\rho}(\Pi)} = \mathcal{L}_\Pi([\psi]^m, \varphi) \text{ for any } \varphi \in H_0^1(\Pi) \text{ and } 1 \le m \le M. \tag{21}$$

Then for any $\eta \colon \overline{\omega}_M^\tau \to S_h(\bar{\Pi})$ from identities (5) and (21) it follows that

$$i\hbar(\overline{\partial}_t(\Psi - \eta)^m, \varphi)_{L^{2,\rho}(\Pi)} - \mathcal{L}_\Pi(\overline{s}_t(\Psi - \eta)^m, \varphi)$$
$$= i\hbar(([D_t\psi] - \overline{\partial}_t\eta)^m, \varphi)_{L^{2,\rho}(\Pi)} - \mathcal{L}_\Pi(([\psi] - \overline{s}_t\eta)^m, \varphi) \text{ for any } \varphi \in S_h(\bar{\Pi}).$$

Let $\eta^m := \sigma\psi^m$, $0 \le m \le M$. By identity (16) and $[D_t y] = \overline{\partial}_t y$ we get

$$i\hbar(\rho\overline{\partial}_t(\Psi - \sigma\psi)^m, \varphi)_{L^2(\Pi)} - \mathcal{L}_\Pi(\overline{s}_t(\Psi - \sigma\psi)^m, \varphi)$$
$$= i\hbar(\rho([D_t\psi]^m - [D_t\sigma\psi]^m), \varphi)_{L^2(\Pi)} - \mathcal{L}_\Pi(([\psi] - \overline{s}_t\psi)^m, \varphi)$$
$$+\hat{v}(\rho\overline{s}_t(\psi - \sigma\psi)^m, \varphi)_{L^2(\Pi)} =: (F, \varphi)_{L^2(\Pi)} \text{ for any } 1 \le m \le M, \tag{22}$$

where (after rearranging the summands)

$$F = -([\mathcal{H}\psi] - \overline{s}_t\mathcal{H}\psi) + i\hbar\rho[D_t(\psi - \sigma\psi)] + \hat{v}\rho\overline{s}_t(\psi - \sigma\psi). \tag{23}$$

Let now on $\hbar = 1$. Proposition 1 together with (19) lead to the bound

$$\max_{0 \le m \le M} \|(\Psi - \sigma\psi)^m\|_{L^{2,\rho}(\Pi)} \le \|\Psi^0 - \sigma\psi^0\|_{L^{2,\rho}(\Pi)} + 2\sum_{m=1}^{M} \|F^m\|_{L^{2,1/\rho}(\Pi)} \tau_m$$

$$\le \|\Psi^0 - \sigma\psi^0\|_{L^{2,\rho}(\Pi)} + 2\sum_{m=1}^{M} \|([\mathcal{H}_\rho\psi] - \overline{s}_t\mathcal{H}_\rho\psi)^m\|_{L^{2,\rho}(\Pi)} \tau_m$$

$$+2\int_0^T \|D_t(\psi - \sigma\psi)\|_{L^{2,\rho}(\Pi)} \, dt + 2|\hat{v}|T \max_{0 \le m \le M} \|(\psi - \sigma\psi)^m\|_{L^{2,\rho}(\Pi)}.$$

The formula $\psi - \Psi = \psi - \sigma\psi - (\Psi - \sigma\psi)$ and estimates (17) and (20) imply

$$\max_{0 \le m \le M} \|(\psi - \Psi)^m\|_{L^{2,\rho}(\Pi)} \le \|\Psi^0 - \sigma\psi^0\|_{L^{2,\rho}(\Pi)}$$

$$+c\left\{\tau_{\max}^2 \int_0^T \|D_t^2\mathcal{H}_\rho\psi\|_{L^{2,\rho}(\Pi)} \, dt + |h|^2 \int_0^T \|D_t(\mathcal{H}_\rho + \hat{v})\psi\|_{L^{2,\rho}(\Pi)} \, dt \right.$$

$$\left. +(1 + |\hat{v}|T)|h|^2 \max_{0 \le m \le M} \|(\mathcal{H}_\rho + \hat{v})\psi^m\|_{L^{2,\rho}(\Pi)}\right\}. \tag{24}$$

Under the above assumptions, the solution to problem (1), (2) satisfies the bound

$$\max_{0\le t\le T} \|D_t^k\psi\|_{L^{2,\rho}(\Pi)} \le \|(D_t^k\psi)|_{t=0}\|_{L^{2,\rho}(\Pi)} = \|\mathcal{H}_\rho^k\psi^0\|_{L^{2,\rho}(\Pi)}, \quad 0\le k\le 3, (25)$$

and the property $D_t^k\psi = D_t^{k-l}(-i\mathcal{H}_\rho)^l\psi$ for $1\le l\le k$. Therefore

$$\max_{0\le m\le M}\|(\psi-\Psi)^m\|_{L^{2,\rho}(\Pi)} \le \|\Psi^0 - \sigma\psi^0\|_{L^{2,\rho}(\Pi)} + c\left\{T\tau_{\max}^2\|\mathcal{H}_\rho^3\psi^0\|_{L^{2,\rho}(\Pi)}\right.$$

$$\left. + T|h|^2\|\mathcal{H}_\rho^2\psi^0\|_{L^{2,\rho}(\Pi)} + (1+|\hat{v}|T)|h|^2\left(\|\mathcal{H}_\rho\psi^0\|_{L^{2,\rho}(\Pi)} + |\hat{v}|\|\psi^0\|_{L^{2,\rho}(\Pi)}\right)\right\}.$$

Note that for $\hat{v} = 0$ the estimate is simplified.

The following multiplicative inequality holds

$$\|\mathcal{H}_\rho^l\psi^0\|_{L^{2,\rho}(\Pi)}^2 \le \|\mathcal{H}_\rho^{l+1}\psi^0\|_{L^{2,\rho}(\Pi)}\|\mathcal{H}_\rho^{l-1}\psi^0\|_{L^{2,\rho}(\Pi)} \quad \text{for } l=1,2. \qquad (26)$$

Using (17) for $w = \psi^0$ together with (26) for $l = 1$, we complete the proof. $\quad\square$

Corollary 1. *Let $\psi^0(x) = 0$ for $|x_1| \ge X_0$, $\Psi^0 \in S_{0h}$ and $\overline{\omega}_M^\tau$ be uniform. Then for the solution to (9), (10) the following first error estimate holds*

$$\max_{0\le m\le M}\|(\psi-\Psi)^m\|_{L^{2,\rho}(\Omega)} \le \|\Psi^0 - \psi^0\|_{L^{2,\rho}(\Omega)}$$

$$+ c(1+T)\left\{\tau^2\|\mathcal{H}_\rho^3\psi^0\|_{L^{2,\rho}(\Omega)} + |h|^2\left(\|\mathcal{H}_\rho^2\psi^0\|_{L^{2,\rho}(\Omega)} + \|\psi^0\|_{L^{2,\rho}(\Omega)}\right)\right\}.$$

The result immediately follows from Proposition 6. Notice also that, for $1 \le n \le 3$, for the interpolant $s\psi^0$ in S_{0h} for ψ^0, the following error estimate holds

$$\|s\psi^0 - \psi^0\|_{L^{2,\rho}(\Omega)} \le c|h|^2\left(\|\mathcal{H}_\rho\psi^0\|_{L^{2,\rho}(\Omega)} + \|\psi^0\|_{L^{2,\rho}(\Omega)}\right),$$

thus one can set simply $\Psi^0 := s\psi^0$. Other possible choices of Ψ^0 with the same error estimate, for any $n \ge 1$, are the $L^2(\Omega_0)$ (possibly with a weight like ρ) projection of ψ^0 onto S_{0h} or its elliptic projection onto S_{0h} like (16) (with Π replaced by Ω_0 and any $\varphi \in S_{0h}$).

Let $B \in W^{2,\infty}(\Pi)$ and $\rho, V \in W^{1,\infty}(\Pi)$.

Proposition 7. *Let $\mathcal{H}_\rho^3\psi^0 \in H_0^1(\Pi)$, $1 \le n \le 3$ and $\Psi^0 := s\psi^0$. Then the following second error estimate holds*

$$\max_{0\le m\le M}\|s\psi^m - \Psi^m\|_{\mathcal{H}+\hat{v}\rho;\,\Pi} \le c(1+T)\left\{\tau_{\max}^2\left(\|\mathcal{H}_\rho^3\psi^0\|_{\mathcal{H}+\hat{v}\rho;\,\Pi} + \|\mathcal{H}_\rho^2\psi^0\|_{\mathcal{H}+\hat{v}\rho;\,\Pi}\right)\right.$$

$$\left. + |h|^2\left(\|\mathcal{H}_\rho^3\psi^0\|_{L^{2,\rho}(\Pi)} + \|\psi^0\|_{L^{2,\rho}(\Pi)}\right)\right\}. \qquad (27)$$

Here $s\psi^m$ is the interpolant in $S_h(\bar{\Pi})$ for ψ^m, $0 \le m \le M$.

Proof. 1. Let first $n \ge 1$. Inequality (7) implies

$$\|\rho w\|_{H_h^{-1}(\Pi)} \le \hat{\delta}^{-1/2}\|w\|_{L^{2,\rho}(\Pi)}, \quad \|w\|_{L^{2,\rho}(\Pi)} \le \hat{\delta}^{-1/2}\|w\|_{\mathcal{H}+\hat{v}\rho;\,\Pi}. \qquad (28)$$

Then setting $\hat{c} := 1 + \frac{|\hat{v}|}{\delta}$, we also get

$$\|\mathcal{H}w\|_{H_h^{-1}(\Pi)} \leq \|(\mathcal{H} + \hat{v}\rho)w\|_{H_h^{-1}(\Pi)} + |\hat{v}|\|\rho w\|_{H_h^{-1}(\Pi)} \leq \hat{c}\|w\|_{\mathcal{H}+\hat{v}\rho;\,\Pi}. \quad (29)$$

For $y \in L^1(0,T)$, define two more averages (projections on the time mesh)

$$[y]_2^m := \frac{1}{2\hat{\tau}_m} \int_{t_{m-1}}^{t_{m+1}} y(t)dt, \quad \langle y \rangle^m := \frac{1}{\hat{\tau}_m} \int_{t_{m-1}}^{t_{m+1}} y(t)e_m(t)dt, \quad 1 \leq m \leq M - 1,$$

$$[y]_2^0 := [y]^1, \quad \langle y \rangle^0 := \frac{2}{\tau_1} \int_0^{t_1} y(t)e_0(t)dt,$$

where $e_m(t)$ is the "hat" function linear on all segments $[t_{l-1}, t_l]$ and such that $e_m(t_m) = 1$ and $e_m(t_l) = 0$ for all $l \neq m$.

Notice that the following relations hold

$$\hat{\partial}_t[y]^m = \langle D_t y \rangle^m, \quad (\hat{\partial}_t \overline{s}_t y)^m = [D_t y]_2^m, \quad 1 \leq m \leq M - 1, \quad (30)$$

$$[y]^1 - y(0) = \frac{\tau_1}{2}\langle D_t y \rangle^0, \quad (\overline{s}_t y)^1 - y(0) = \frac{\tau_1}{2}[D_t y]^1, \quad \text{for } y \in W^{1,1}(0,T), (31)$$

$$\sum_{m=0}^{M-1} \|\langle u \rangle^m\|\hat{\tau}_m \leq \int_0^T \|u(\cdot,t)\| \, dt, \quad \sum_{m=0}^{M-1} \|[u]_2^m\|\hat{\tau}_m \leq \int_0^T \|u(\cdot,t)\| \, dt, \quad (32)$$

where $\|\cdot\| = \|\cdot\|_{L^{2,\rho}(\Pi)}$ and $u \in L^{2,1}(\Pi \times (0,T))$, and

$$|\langle y \rangle^m - [y]_2^m| \leq c_0 \tau_m^2 [|D_t^2 y|]_2^m, \quad 1 \leq m \leq M - 1, \quad \text{for } y \in W^{2,1}(0,T). \quad (33)$$

2. Let first $\Psi^0 \in S_h(\bar{\Pi})$ be arbitrary. We go back to the error identity (22). Applying Proposition 2, now we get

$$\max_{0 \leq m \leq M} \|(\Psi - \sigma\psi)^m\|_{\mathcal{H}+\hat{v}\rho;\,\Pi} \leq \|\Psi^0 - \sigma\psi^0\|_{\mathcal{H}+\hat{v}\rho;\,\Pi}$$

$$+4\sum_{m=0}^{M-1} \|\hat{\partial}_t F^m\|_{H_h^{-1}(\Pi)}\hat{\tau}_m + 4|\hat{v}|\sum_{m=1}^{M} \|F^m\|_{H_h^{-1}(\Pi)}\tau_m + 4\|F^0\|_{H_h^{-1}(\Pi)}, \quad (34)$$

where the right-hand side is slightly transformed and F is given by (23). We introduce the decomposition

$$F = F_\tau + \rho F_h, \quad F_\tau := -([\mathcal{H}\psi] - \overline{s}_t\mathcal{H}\psi), \quad F_h := i[D_t(\psi - \sigma\psi)] + \hat{v}\overline{s}_t(\psi - \sigma\psi)$$

as well as set $F_\tau^0 := 0$ and $F_h^0 = iD_t(\psi - \sigma\psi)|_{t=0} + \hat{v}(\psi^0 - \sigma\psi^0)$.

Applying sequentially relations (29), (30), (20), (33), (19) and (32), we obtain

$$S_\tau := |\hat{v}| \sum_{m=1}^{M} \|F_\tau^m\|_{H_h^{-1}(\Pi)} \tau_m + \sum_{m=0}^{M-1} \|\hat{\partial}_t F_\tau^m\|_{H_h^{-1}(\Pi)} \hat{\tau}_m$$

$$\leq \hat{c}\Big(|\hat{v}| \sum_{m=1}^{M} \|([\psi] - \overline{s}_t\psi)^m\|_{\mathcal{H}+\hat{v}\rho;\, \Pi} \tau_m + \sum_{m=1}^{M-1} \|(\langle D_t\psi\rangle - [D_t\psi]_2)^m\|_{\mathcal{H}+\hat{v}\rho;\, \Pi} \hat{\tau}_m$$

$$+ \|([\psi] - \overline{s}_t\psi)^1\|_{\mathcal{H}+\hat{v}\rho;\, \Pi}\Big)$$

$$\leq \hat{c}c_0\tau_{\max}^2\Big(|\hat{v}| \sum_{m=1}^{M} \|[D_t^2\psi]^m\|_{\mathcal{H}+\hat{v}\rho;\, \Pi} \tau_m + \sum_{m=1}^{M-1} \|[D_t^3\psi]_2^m\|_{\mathcal{H}+\hat{v}\rho;\, \Pi} \hat{\tau}_m$$

$$+ \|[D_t^2\psi]^1\|_{\mathcal{H}+\hat{v}\rho;\, \Pi}\Big)$$

$$\leq \hat{c}c_0\tau_{\max}^2\Big\{\|(D_t^2\psi)|_{t=0}\|_{\mathcal{H}+\hat{v}\rho;\, \Pi} + \int_0^T \big(|\hat{v}|\|D_t^2\psi\|_{\mathcal{H}+\hat{v}\rho;\, \Pi} + 2\|D_t^3\psi\|_{\mathcal{H}+\hat{v}\rho;\, \Pi}\big)\, dt\Big\}. \quad (35)$$

The left inequality (28) implies

$$S_h := |\hat{v}| \sum_{m=1}^{M} \|\rho F_h^m\|_{H_h^{-1}(\Pi)} \tau_m + \|\rho F_h^0\|_{H_h^{-1}(\Pi)} + \sum_{m=0}^{M-1} \|\rho\hat{\partial}_t F_h^m\|_{H_h^{-1}(\Pi)} \hat{\tau}_m$$

$$\leq \hat{\delta}^{-1/2}\Big(|\hat{v}| \sum_{m=1}^{M} \|F_h^m\|_{L^{2,\rho}(\Pi)} \tau_m + \|F_h^0\|_{L^{2,\rho}(\Pi)} + \sum_{m=0}^{M-1} \|\hat{\partial}_t F_h^m\|_{L^{2,\rho}(\Pi)} \hat{\tau}_m\Big), \quad (36)$$

and further the error estimate (17) leads to

$$\|F_h^0\|_{L^{2,\rho}(\Pi)} \leq \|D_t(\psi - \sigma\psi)|_{t=0}\|_{L^{2,\rho}(\Pi)} + |\hat{v}|\|\psi^0 - \sigma\psi^0\|_{L^{2,\rho}(\Pi)}$$

$$\leq c|h|^2\big(\|D_t(\mathcal{H}_\rho + \hat{v})\psi|_{t=0}\|_{L^{2,\rho}(\Pi)} + |\hat{v}|\|(\mathcal{H}_\rho + \hat{v})\psi^0\|_{L^{2,\rho}(\Pi)}\big). \quad (37)$$

Applying sequentially relations (30), (31), (17) and (32), we also obtain

$$\sum_{m=0}^{M-1} \|\hat{\partial}_t F_h^m\|_{L^{2,\rho}(\Pi)} \hat{\tau}_m$$

$$\leq \sum_{m=0}^{M-1} \big(\|\langle D_t^2(\psi - \sigma\psi)\rangle^m\|_{L^{2,\rho}(\Pi)} + |\hat{v}|\|[D_t(\psi - \sigma\psi)]_2^m\|_{L^{2,\rho}(\Pi)}\big)\hat{\tau}_m$$

$$\leq c|h|^2 \sum_{m=0}^{M-1} \big(\|\langle D_t^2(\mathcal{H}_\rho + \hat{v})\psi\rangle^m\|_{L^{2,\rho}(\Pi)} + |\hat{v}|\|[D_t(\mathcal{H}_\rho + \hat{v})\psi]_2^m\|_{L^{2,\rho}(\Pi)}\big)\hat{\tau}_m$$

$$\leq c|h|^2 \int_0^T \big(\|D_t^2(\mathcal{H}_\rho + \hat{v})\psi\|_{L^{2,\rho}(\Pi)} + |\hat{v}|\|D_t(\mathcal{H}_\rho + \hat{v})\psi\|_{L^{2,\rho}(\Pi)}\big)\, dt. \quad (38)$$

Inserting into (34) all the estimates (35)-(38) together with the estimate for $\sum_{m=1}^{M} \|F_h^m\|_{L^{2,\rho}(\Pi)} \tau_m$ used in the preceding proof in (24), we derive

$$\max_{0 \le m \le M} \|(\Psi - \sigma\psi)^m\|_{\mathcal{H}+\hat{v}\rho;\,\Pi} \le \|\Psi^0 - \sigma\psi^0\|_{\mathcal{H}+\hat{v}\rho;\,\Pi} + 4S_\tau + 4S_h$$

$$\le c\tau_{max}^2 \left\{ \|(D_t^2\psi)|_{t=0}\|_{\mathcal{H}+\hat{v}\rho;\,\Pi} + \int_0^T \left(|\hat{v}| \|D_t^2\psi\|_{\mathcal{H}+\hat{v}\rho;\,\Pi} + \|D_t^3\psi\|_{\mathcal{H}+\hat{v}\rho;\,\Pi} \right) dt \right\}$$

$$+ c|h|^2 \left\{ |\hat{v}| \int_0^T \|D_t(\mathcal{H}_\rho + \hat{v})\psi\|_{L^{2,\rho}(\Pi)} \, dt + \hat{v}^2 T \max_{0 \le t \le T} \|(\mathcal{H}_\rho + \hat{v})\psi^m\|_{L^{2,\rho}(\Pi)} \right.$$

$$+ \|D_t(\mathcal{H}_\rho + \hat{v})\psi|_{t=0}\|_{L^{2,\rho}(\Pi)} + |\hat{v}| \|(\mathcal{H}_\rho + \hat{v})\psi^0\|_{L^{2,\rho}(\Pi)}$$

$$\left. + \int_0^T \left(\|D_t^2(\mathcal{H}_\rho + \hat{v})\psi\|_{L^{2,\rho}(\Pi)} + |\hat{v}| \|D_t(\mathcal{H}_\rho + \hat{v})\psi\|_{L^{2,\rho}(\Pi)} \right) dt \right\}.$$

Once again for $\hat{v} = 0$ the estimate is essentially simplified.

Under all the above assumptions, the solution to problem (1), (2) satisfies the bound

$$\max_{0 \le t \le T} \|D_t^k\psi\|_{\mathcal{H}+\hat{v}\rho;\,\Pi} \le \|(D_t^k\psi)|_{t=0}\|_{\mathcal{H}+\hat{v}\rho;\,\Pi} = \|\mathcal{H}_\rho^k\psi^0\|_{\mathcal{H}+\hat{v}\rho;\,\Pi}, \quad 0 \le k \le 3.$$

This bound and (25) and the property $D_t^k\psi = D_t^{k-l}(-i\mathcal{H}_\rho)^l\psi$, $1 \le l \le k$, imply

$$\max_{0 \le m \le M} \|(\Psi - \sigma\psi)^m\|_{\mathcal{H}+\hat{v}\rho;\,\Pi} \le \|\Psi^0 - \sigma\psi^0\|_{\mathcal{H}+\hat{v}\rho;\,\Pi}$$

$$\le c(1+T) \left\{ \tau_{max}^2 (\|\mathcal{H}_\rho^2\psi^0\|_{\mathcal{H}+\hat{v}\rho;\,\Pi} + \|\mathcal{H}_\rho^3\psi^0\|_{\mathcal{H}+\hat{v}\rho;\,\Pi}) + |h|^2 \sum_{k=0}^{3} \|\mathcal{H}_\rho^k\psi^0\|_{L^{2,\rho}(\Pi)} \right\}. \tag{39}$$

3. Let now $1 \le n \le 3$ and $\Psi^0 = s\psi^0$. Similarly to [9], Lemma 5.1 for $n = 1$ and [10], Theorem 2.1 for $n = 2$ and 3 (see also [8]), the following elliptic FEM error estimate holds

$$\|sw - \sigma w\|_{\mathcal{H}+\hat{v}\rho;\,\Pi} \le c|h|^2 \left(\sum_{p \ne q} \|D_p^2 D_q w\|_{L^2(\Pi)} + \sum_{p=1}^{n} \|D_p^2 w\|_{L^2(\Pi)} \right)$$

$$\le c_1 |h|^2 \left(\|\mathcal{H}_\rho w\|_{\mathcal{H}+\hat{v}\rho;\,\Pi} + \|w\|_{\mathcal{H}+\hat{v}\rho;\,\Pi} \right)$$

for any $w \in \mathcal{D}(\mathcal{H}_\rho)$ such that $\mathcal{H}_\rho w \in H_0^1(\Pi)$, taking into account the above regularity assumptions on B, ρ and V, where the first sum is taken over all p and q from 1 to n excluding $p = q$ and disappears for $n = 1$. This estimate allows to pass from (39) to the final estimate (27) by the triangle inequality together with inequalities (26) and

$$\|\mathcal{H}_\rho^l\psi^0\|_{\mathcal{H}+\hat{v}\rho;\,\Pi}^2 \le \|(\mathcal{H}_\rho + \hat{v})\mathcal{H}_\rho^l\psi^0\|_{L^{2,\rho}(\Pi)} \|\mathcal{H}_\rho^l\psi^0\|_{L^{2,\rho}(\Pi)}, \quad l = 0, 1.$$

$$\square$$

Corollary 2. *Let $\psi^0(x) = 0$ for $|x_1| \ge X_0$, $1 \le n \le 3$ and $\Psi^0 = s\psi^0$ on $\bar{\Omega}$ and $\bar{\omega}_M^\tau$ be uniform. Then for the solution to (9), (10) the following second error estimate holds*

$$\max_{0\le m\le M} \|s\psi^m - \Psi^m\|_{\mathcal{H}+\hat{v}\rho;\,\Omega} \le c(1+T)\Big\{\tau^2\big(\|\mathcal{H}_\rho^3\psi^0\|_{\mathcal{H}+\hat{v}\rho;\,\Omega} + \|\mathcal{H}_\rho^2\psi^0\|_{\mathcal{H}+\hat{v}\rho;\,\Omega}\big)$$

$$+ |h|^2\big(\|\mathcal{H}_\rho^3\psi^0\|_{L^{2,\rho}(\Omega)} + \|\psi^0\|_{L^{2,\rho}(\Omega)}\big)\Big\}.$$

The result immediately follows from Proposition 7.

Note that the norm $\|\cdot\|_{\mathcal{H}+\hat{v}\rho;\,\Omega}$ is equivalent to $\|\cdot\|_{H^1(\Omega)}$ and $\|s\cdot\|_{H^1(\Omega)}$ is actually the mesh counterpart of the latter norm.

Acknowledgments. The study is supported by The National Research University – Higher School of Economics' Academic Fund Program in 2014–2015, research grant No. 14-01-0014.

References

1. Arnold, A., Ehrhardt, M., Sofronov, I.: Discrete transparent boundary conditions for the Schrödinger equation: fast calculations, approximation and stability. Comm. Math. Sci. **1**, 501–556 (2003)
2. Ducomet, B., Zlotnik, A.: On stability of the Crank-Nicolson scheme with approximate transparent boundary conditions for the Schrödinger equation. Part I. Commun. Math. Sci. **4**(4), 741–766 (2006)
3. Ducomet, B., Zlotnik, A.: On stability of the Crank-Nicolson scheme with approximate transparent boundary conditions for the Schrödinger equation. Part II. Commun. Math. Sci. **5**(2), 267–298 (2007)
4. Ducomet, B., Zlotnik, A., Zlotnik, I.: On a family of finite-difference schemes with discrete transparent boundary conditions for a generalized 1D Schrödinger equation. Kinetic Relat. Models **2**(1), 151–179 (2009)
5. Ehrhardt, M., Arnold, A.: Discrete transparent boundary conditions for the Schrödinger equation. Riv. Mat. Univ. Parma. **6**, 57–108 (2001)
6. Jin, J., Wu, X.: Analysis of finite element method for one-dimensional time-dependent Schrödinger equation on unbounded domain. J. Comput. Appl. Math. **220**, 240–256 (2008)
7. Jin, J., Wu, X.: Convergence of a finite element scheme for the two-dimensional time-dependent Schrödinger equation in a long strip. J. Comput. Appl. Math. **234**, 777–793 (2010)
8. Zlotnik, A.A.: On the rate of convergence in $W_{2,h}^1$ of the variational-difference method for elliptic equations. Soviet Math. Dokl. **28**(1), 143–148 (1983)
9. Zlotnik, A.A.: Convergence rate estimates of finite-element methods for second-order hyperbolic equations. In: Marchuk, G.I. (ed.) Numerical Methods and Applications, pp. 155–220. CRC Press, Boca Raton (1994)
10. Zlotnik, A.A.: On superconvergence of a gradient for finite element methods for an elliptic equation with the nonsmooth right-hand side. Comput. Meth. Appl. Math. **2**(3), 295–321 (2002)
11. Zlotnik, A., Ducomet, B., Zlotnik, I., Romanova, A.: Splitting in potential finite-difference schemes with discrete transparent boundary conditions for the time-dependent Schrödinger equation. In: Abdulle, A., Deparis, S., Kressner, D., Nobile, F., Picasso, M. (eds.) Numerical Mathematics and Advanced Applications - ENUMATH 2013. Lecture Notes in Computational Science and Engineering, vol. 103, pp. 203–211. Springer, Switzerland (2015)

12. Zlotnik, A.A., Zlotnik, I.A.: Family of finite-difference schemes with transparent boundary conditions for the nonstationary Schrödinger equation in a semi-infinite strip. Dokl. Math. **83**(1), 12–18 (2011)
13. Zlotnik, A., Zlotnik, I.: Finite element method with discrete transparent boundary conditions for the time-dependent 1D Schrödinger equation. Kinetic Relat. Models **5**(3), 639–667 (2012)
14. Zlotnik, A., Zlotnik, I.: Remarks on discrete and semi-discrete transparent boundary conditions for solving the time-dependent Schrödinger equation on the half-axis. Russ. J. Numer. Anal. Math. Model. (2016, to appear)
15. Zlotnik, I.A.: Family of finite-difference schemes with approximate transparent boundary conditions for the generalized nonstationary Schrödinger equation in a semi-infinite strip. Comput. Math. Math. Phys. **51**(3), 355–376 (2011)
16. Zlotnik, I.A.: Numerical methods for solving the generalized time-dependent Schrödinger equation in unbounded domains. PhD thesis, Moscow Power Engineering Institute (2013) (in Russian)

Contributed Papers

Difference Schemes
for Delay Parabolic Equations
with Periodic Boundary Conditions

Allaberen Ashyralyev[1,2] and Deniz Agirseven[3(⊠)]

[1] Department of Mathematics, Fatih University, Istanbul, Turkey
aashyr@fatih.edu.tr
[2] ITTU, Ashgabat, Turkmenistan
[3] Department of Mathematics, Trakya University, Edirne, Turkey
denizagirseven@trakya.edu.tr

Abstract. The initial-boundary value problem for the delay parabolic partial differential equation with nonlocal conditions is studied. The convergence estimates for solutions of first and second order of accuracy difference schemes in Hölder norms are obtained. The theoretical statements are supported by a numerical example.

Keywords: Difference schemes · Delay parabolic equation · Hölder spaces · Convergence

1 Introduction

Delay differential equations provide a mathematical model for a physical or biological system in which the rate of change of the system depends on the past history. The theory of these equations is well developed and has numerous applications in natural and engineering sciences. Typical examples that delay differential equations appear are diffusive population models with temporal averages over the past,tumor growth, neural networks, control theory, climate models etc. Numerical solutions of delay ordinary differential equations have been studied mostly for ordinary differential equations (cf., e.g., [1–9] and the references therein). Generally, delay partial differential equations get less attention than delay ordinary differential equations.

In recent years, A. Ashyralyev and P. E. Sobolevskii obtained the stability estimates in Hölder norms for solutions of the initial-value problem of delay differential and difference equations of the parabolic type [10,11]. D.B. Gabriella used extrapolation spaces to solve delay differential equations with unbounded delay operators [12]. Different kinds of problems for delay partial differential equations are solved by using operator approach (see, e.g., [13–20]).

I. Dimov et al. (Eds.): FDM 2014, LNCS 9045, pp. 145–152, 2015.
DOI: 10.1007/978-3-319-20239-6_13

In this paper, the initial-boundary value problem for the delay differential equation of the parabolic type

$$\begin{cases} u_t(t,x) - a(x)u_{xx}(t,x) + c(x)u(t,x) \\ = d(t)(-a(x)u_{xx}(t-\omega,x) + c(x)u(t-\omega,x)), \\ 0 < t < \infty, \ x \in (0,L), \\ u(t,x) = g(t,x), \ -\omega \le t \le 0, \ x \in [0,L], \\ u(t,0) = u(t,L), u_x(t,0) = u_x(t,L), \ t \ge 0 \end{cases} \tag{1}$$

is studied. Here $g(t,x)$ $(t \in (-\infty, 0), x \in [0,L])$, $a(x), c(x)$ $(x \in (0,L))$ are given smooth bounded functions and $a(x) \ge a > 0$, $c(x) \ge c > 0$. Difference schemes of first and second order of accuracy for the numerical solutions of Problem (1) are presented. The convergence of these difference schemes are studied. The numerical solutions are found by using MATLAB programs.

2   Difference Schemes, Convergence Estimate

The discretization of Problem (1) is carried out in two steps. In the first step, we define the grid space

$$[0,L]_h = \{x = x_n : \ x_n = nh, \ 0 \le n \le M, \ Mh = L\}.$$

To formulate our results, we introduce the Banach space $\overset{\circ}{C}_h^\alpha = \overset{\circ}{C}{}^\alpha [0,L]_h$, $\alpha \in [0,1)$, of all grid functions $\varphi^h = \{\varphi_n\}_{n=0}^M$ defined on $[0,L]_h$ with $\varphi_0 = \varphi_M$, $\varphi_1 - \varphi_0 = \varphi_M - \varphi_{M-1}$ or $3\varphi_0 - 4\varphi_1 + \varphi_2 = -3\varphi_M + 4\varphi_{M-1} - \varphi_{M-2}$ equipped with the norm

$$\left\| \varphi^h \right\|_{\overset{\circ}{C}_h^\alpha} = \left\| \varphi^h \right\|_{C_h} + \sup_{1 \le n < n+r \le M-1} |\varphi_{n+r} - \varphi_n| (rh)^{-\alpha},$$

$$\left\| \varphi^h \right\|_{C_h} = \max_{1 \le n \le M-1} |\varphi_n|.$$

Moreover, $C_\tau(E) = C([0,\infty)_\tau, E)$ is the Banach space of all grid functions $f^\tau = \{f_k\}_{k=1}^\infty$ defined on

$$[0,\infty)_\tau = \{t_k = k\tau, \ k = 0, 1, \cdots\}$$

with values in E equipped with the norm

$$\left\| f^\tau \right\|_{C_\tau(E)} = \sup_{1 \le k < \infty} \left\| f_k \right\|_E.$$

To the differential operator A generated by Problem (1), we assign the difference operators A_h^x, B_h^x by the formulas

$$A_h^x \varphi^h(x) = \left\{ -a(x_n) \frac{\varphi_{n+1} - 2\varphi_n + \varphi_{n-1}}{h^2} + c(x_n)\varphi_n \right\}_1^{M-1},$$

$$B_h^x(t)\,\varphi^h(x) = d(t)\,A_h^x\varphi^h(x),$$

acting in the space of grid functions $\varphi^h(x) = \{\varphi_n\}_0^M$ satisfying the conditions $\varphi_0 = \varphi_M, \varphi_1 - \varphi_0 = \varphi_M - \varphi_{M-1}$ for the first order of approximation of difference operator A_h^x and the conditions $\varphi_0 = \varphi_M, 3\varphi_0 - 4\varphi_1 + \varphi_2 = -3\varphi_M + 4\varphi_{M-1} - \varphi_{M-2}$ for the second order of approximation of difference operator A_h^x. It is well known that A_h^x is a strongly positive operator in C_h. With the help of A_h^x and $d(t)\,A_h^x$, we arrive at the initial value problem

$$\begin{cases} \frac{du^h(t,x)}{dt} + A_h^x u^h(t,x) = d(t)\,A_h^x u^h(t-w,x), \ 0 < t < \infty, \\ u^h(t,x) = g^h(t,x), \ -\omega \leq t \leq 0. \end{cases} \tag{2}$$

In the second step, we consider difference schemes of first and second order of accuracy

$$\begin{cases} \frac{1}{\tau}\left(u_k^h(x) - u_{k-1}^h(x)\right) + A_h^x u_k^h(x) = d(t_k)\,A_h^x u_{k-N}^h(x), \\ t_k = k\tau, \ 1 \leq k, \ N\tau = w, \\ u_k^h(x) = g^h(t_k,x), \ t_k = k\tau, \ -N \leq k \leq 0, \end{cases} \tag{3}$$

$$\begin{cases} \frac{1}{\tau}(u_k^h(x) - u_{k-1}^h(x)) + (A_h^x + \frac{1}{2}\tau\,(A_h^x)^2)u_k^h(x) \\ = \frac{1}{2}(I + \frac{\tau}{2}A_h^x)d\left(t_k - \frac{\tau}{2}\right)A_h^x(u_{k-N}^h(x) + u_{k-N-1}^h(x)), \\ t_k = k\tau, 1 \leq k, u_k^h = g^h(t_k,x), \ t_k = k\tau, \ -N \leq k \leq 0. \end{cases} \tag{4}$$

Note that in (4), we assign the difference operator $(A_h^x)^2 = A_h^x \cdot A_h^x$ acting in the space of grid functions $\varphi^h(x) = \{\varphi_n\}_0^M$ satisfying the conditions

$$\varphi_0 = \varphi_M, 3\varphi_0 - 4\varphi_1 + \varphi_2 = -3\varphi_M + 4\varphi_{M-1} - \varphi_{M-2},$$

$$2\varphi_0 - 5\varphi_1 + 4\varphi_2 - \varphi_3 = 2\varphi_M - 5\varphi_{M-1} + 4\varphi_{M-2} - \varphi_{M-3},$$

$$10\varphi_0 - 15\varphi_1 + 6\varphi_2 - \varphi_3 = -10\varphi_M + 15\varphi_{M-1} - 6\varphi_{M-2} + \varphi_{M-3}$$

generated by second order approximations of conditions $\varphi_0 = \varphi_L, \varphi_0' = \varphi_L', \varphi_0'' = \varphi_L''$ and $\varphi_0''' = \varphi_L'''$.

Theorem 1. Assume that

$$\sup_{0 \leq t < \infty} |d(t)| \leq \frac{1-\alpha}{M2^{2-\alpha}}. \tag{5}$$

Suppose that Problem (1) has a smooth solution $u(t,x)$ and

$$\int_0^\infty \left[\max_{0 \leq x \leq L} |u_{ss}(s,x)| + \sup_{0 < x < x+y < L} \frac{|u_{ss}(s,x+y) - u_{ss}(s,x)|}{y^{2\alpha}} \right] ds < \infty,$$

$$\int_0^\infty \left[\max_{0 \leq x \leq L} |u_{xxxx}(s,x)| + \sup_{0 < x < x+y < L} \frac{|u_{xxxx}(s,x+y) - u_{xxxx}(s,x)|}{y^{2\alpha}} \right] ds < \infty.$$

Then, for the solution of difference scheme (3), the following convergence estimate holds

$$\sup_k \left\| u_k^h - u^h(t_k,\cdot) \right\|_{\overset{\circ}{C}_h^{2\alpha}} \leq M_1\left(\tau + h^2\right)$$

with M_1 is a real number independent of τ, α and h.

Proof. Using notations of A_h^x and $B_h^x(t_k)$, we can obtain the following formula for the solution

$$u_k^h(x) = R^k g^h(0,x) + \sum_{j=1}^{k} R^{k-j+1} B_h^x(t_j) g^h(t_{j-N},x) \tau, \ 1 \le k \le N, \quad (6)$$

and

$$u_k^h(x) = R^{k-nN} u_{nN}^h(x)$$

$$+ \sum_{j=nN+1}^{k} R^{k-j+1} B_h^x(t_j) u_{j-N}^h(x) \tau, \ nN \le k \le (n+1)N, \quad (7)$$

where $R = (I + \tau A_x^h)^{-1}$. The proof of the Theorem 1 is based on the Formulas (6) and (7) on the convergence theorem for difference schemes in $C_\tau(E_\alpha^h)$, on the estimate

$$\| \exp\{-t_k A_h^x\} \|_{C_h \to C_h} \le M, \ k \ge 0, \quad (8)$$

and on the fact that in the $E_\alpha^h = E_\alpha(A_h^x, C_h)$- norms are equivalent to the norms $\overset{\circ}{C}_h^{2\alpha}$ uniformly in h for $0 < \alpha < \frac{1}{2}$ (see, [13]).

Theorem 2. Assume that assumption (5) of the Theorem 1 and the following conditions hold.

$$\int_0^\infty \left[\max_{0 \le x \le L} |u_{sss}(s,x)| + \sup_{0<x<x+y<L} \frac{|u_{sss}(s,x+y) - u_{sss}(s,x)|}{y^{2\alpha}} \right] ds < \infty,$$

$$\int_0^\infty \left[\max_{0 \le x \le L} |u_{xxss}(s,x)| + \sup_{0<x<x+y<L} \frac{|u_{xxss}(s,x+y) - u_{xxss}(s,x)|}{y^{2\alpha}} \right] ds < \infty,$$

$$\int_0^\infty \left[\max_{0 \le x \le L} |u_{xxxxs}(s,x)| + \sup_{0<x<x+y<L} \frac{|u_{xxxxs}(s,x+y) - u_{xxxxs}(s,x)|}{y^{2\alpha}} \right] ds < \infty.$$

Then for the solution of difference scheme (4), the following convergence estimate is satisfied

$$\sup_k \left\| u_k^h - u^h(t_k, \cdot) \right\|_{\overset{\circ}{C}_h^{2\alpha}} \le M_2 (\tau^2 + h^2)$$

with M_2 is a real number independent of τ, α and h.

Proof. Using notations of A_h^x and $B_h^x(t_k)$ again, we can obtain the following formula for the solution

$$u_k^h(x) = R^k g^h(0,x)$$

$$+ \sum_{j=1}^{k} R^{k-j+1} \left(I + \frac{\tau A_h^x}{2} \right) \left(g^h(t_{j-N},x) + g^h(t_{j-N-1},x) \right) \tau, \ 1 \le k \le N, \quad (9)$$

and

$$u_k^h(x) = R^{k-nN} u_{nN}^h(x)$$

$$+ \sum_{j=nN+1}^{k} R^{k-j+1}\left(I + \frac{\tau A_h^x}{2}\right) B_h^x(t_j)\frac{1}{2}\left(u_{j-N}^h(x) + u_{j-N-1}^h(x)\right)\tau,$$

$$nN \le k \le (n+1)N. \tag{10}$$

where $R = \left(I + \tau A_h^x + \frac{(\tau A_x^h)^2}{2}\right)^{-1}$. The proof of the Theorem 2 is based on the Formulas (9) and (10) on the convergence theorem for difference schemes in $C_\tau(E_\alpha^h)$, on the estimate (8) and on the equivalence of the norms as in Theorem 1.

Finally, the numerical methods are given in the following section for the solution of delay parabolic differential equation with the nonlocal condition. The method is illustrated by numerical examples.

3   Numerical Applications

We consider the initial-boundary value problem

$$\begin{cases} \frac{\partial u(t,x)}{\partial t} - \frac{\partial^2 u(t,x)}{\partial x^2} = -(0.1)\frac{\partial^2 u(t-1,x)}{\partial x^2}, \ t > 0, \ 0 < x < \pi, \\[2mm] u(t,x) = e^{-4t}\sin 2x, \ -1 \le t \le 0, \ 0 \le x \le \pi, \\[2mm] u(t,0) = u(t,\pi), u_x(t,0) = u_x(t,\pi), \ t \ge 0 \end{cases} \tag{11}$$

for the delay parabolic differential equation. The exact solution of this problem for $t \in [n-1,n]$, $n = 0,1,2,\cdots$, $x \in [0,\pi]$ is

$$u(t,x) = \begin{cases} e^{-4t}\sin 2x, & -1 \le t \le 0, \\[2mm] e^{-4t}\left\{1 + 4(0.1)e^4 t\right\}\sin 2x, & 0 \le t \le 1, \\[2mm] e^{-4t}\left\{1 + 4(0.1)e^4 t + \frac{[4(0.1)e^4(t-1)]^2}{2!}\right\}\sin 2x, & 1 \le t \le 2, \\[2mm] \cdots\cdots\cdots\cdots\cdots \\[2mm] e^{-4t}\left\{1 + 4(0.1)e^4 t + \cdots + \frac{[4(0.1)e^4(t-n)]^{n+1}}{(n+1)!}\right\}\sin 2x, & n \le t \le n+1, \\[2mm] \cdots\cdots\cdots\cdots\cdots \end{cases}$$

We get the following first order of accuracy difference scheme for the approximate solution of the initial-boundary value problem for the delay parabolic Eq. (11)

$$\begin{cases} \dfrac{u_n^k - u_n^{k-1}}{\tau} - \dfrac{u_{n+1}^k - 2u_n^k + u_{n-1}^k}{h^2} = -0.1 \dfrac{u_{n+1}^{k-N} - 2u_n^{k-N} + u_{n-1}^{k-N}}{h^2}, \\[2mm] mN + 1 \le k \le (m+1)\, N, \ m = 0, 1, \cdots, \ 1 \le n \le M - 1, \\[2mm] u_n^k = e^{-4t_k} \sin 2x_n, \ -N \le k \le 0, \ 0 \le n \le M, \\[2mm] u_0^k = u_M^k, \ u_1^k - u_0^k = u_M^k - u_{M-1}^k, \ k \ge 0. \end{cases} \tag{12}$$

We can rewrite (12) in matrix form

$$AU^k + BU^{k-1} = R\varphi^k, \ 1 \le k \le N, \ U^0 = \varphi. \tag{13}$$

From (13) it follows that

$$U^k = -A^{-1}BU^{k-1} + A^{-1}R\varphi^k, \ 1 \le k \le N. \tag{14}$$

Second, using the second order of accuracy difference scheme for the approximate solution of Problem (11) we obtain the following system of equations

$$\begin{cases} \dfrac{u_n^k - u_n^{k-1}}{\tau} - \dfrac{u_{n+1}^k - 2u_n^k + u_{n-1}^k}{h^2} + \dfrac{\tau}{2}\left(\dfrac{u_{n+2}^k - 4u_{n+1}^k + 6u_n^k - 4u_{n-1}^k + u_{n-2}^k}{h^4} \right) \\[3mm] = -(0.1)\left\{ \dfrac{u_{n+1}^{k-N} - 2u_n^{k-N} + u_{n-1}^{k-N}}{2h^2} + \dfrac{u_{n+1}^{k-1-N} - 2u_n^{k-1-N} + u_{n-1}^{k-1-N}}{2h^2} \right. \\[3mm] \left. - \dfrac{\tau}{2}\left[\dfrac{u_{n+2}^{k-N} - 4u_{n+1}^{k-N} + 6u_n^{k-N} - 4u_{n-1}^{k-N} + u_{n-2}^{k-N}}{2h^4} \right. \right. \\[3mm] \left. \left. + \dfrac{u_{n+2}^{k-1-N} - 4u_{n+1}^{k-1-N} + 6u_n^{k-1-N} - 4u_{n-1}^{k-1-N} + u_{n-2}^{k-1-N}}{2h^4} \right] \right\} = 0, \\[3mm] mN + 1 \le k \le (m+1)N, \ m = 0, 1, \cdots, \ 2 \le n \le M - 2, \\[2mm] u_n^k = e^{-4t_k} \sin 2x_n, \ -N \le k \le 0, \ 0 \le n \le M, \\[2mm] u_0^k = u_M^k, \ k \ge 0, \\[2mm] 3u_0^k - 4u_1^k + u_2^k = -3u_M^k + 4u_{M-1}^k - u_{M-2}^k, \ k \ge 0, \\[2mm] 2u_0^k - 5u_1^k + 4u_2^k - u_3^k = 2u_M^k - 5u_{M-1}^k + 4u_{M-2}^k - u_{M-3}^k, \ k \ge 0, \\[2mm] 10u_0^k - 15u_1^k + 6u_2^k - u_3^k = -10u_M^k + 15u_{M-1}^k - 6u_{M-2}^k + u_{M-3}^k, \ k \ge 0. \end{cases} \tag{15}$$

We have again $(M+1) \times (M+1)$ system of linear equations and we rewrite them in the matrix form (13). Now, we give numerical results for different values of N and M and u_n^k represent the numerical solutions of these difference schemes at (t_k, x_n). Tables 1, 2, 3, and 4 are constructed for $N = M = 50, 100, 200$ in

Table 1. Comparison of the errors of different difference schemes in $t \in [0, 1]$

Method	N=M=50	N=M=100	N=M=200
Difference scheme (3)	0.1073487831	0.0547768623	0.0276391767
Difference scheme (4)	0.0022364343	0.0005798409	0.0001463692

Table 2. Comparison of the errors of different difference schemes in $t \in [1, 2]$

Method	N=M=50	N=M=100	N=M=200
Difference scheme (3)	0.0044792100	0.0023497584	0.0012013854
Difference scheme (4)	0.0014763957	0.0003821482	0.0000965218

Table 3. Comparison of the errors of different difference schemes in $t \in [2, 3]$

Method	N=M=50	N=M=100	N=M=200
Difference scheme (3)	0.0030208607	0.0015155583	0.0007582183
Difference scheme (4)	0.0003769232	0.0000973214	0.0000245676

Table 4. Comparison of the errors of different difference schemes in $t \in [3, 4]$

Method	N=M=50	N=M=100	N=M=200
Difference scheme (3)	0.0008494067	0.0004217773	0.0002100074
Difference scheme (4)	0.0000762753	0.0000196839	0.0000049727

$t \in [0, 1]$, $t \in [1, 2]$, $t \in [2, 3]$, $t \in [3, 4]$ respectively and the error is computed by the following formula.

$$E_M^N = \max_{\substack{-N \leq k \leq N \\ 1 \leq n \leq M}} \left| u\left(t_k, x_n\right) - u_n^k \right|.$$

Thus the numerical results of this section support the theoretical arguments in Theorems 1 and 2.

References

1. Al-Mutib, A.N.: Stability properties of numerical methods for solving delay differential equations. J. Comput. and Appl. Math. **10**(1), 71–79 (1984)
2. Bellen, A., Jackiewicz, Z., Zennaro, M.: Stability analysis of one-step methods for neutral delay-differential equations. Numer. Math. **52**(6), 605–619 (1988)
3. Cooke, K.L., Györi, I.: Numerical approximation of the solutions of delay differential equations on an infinite interval using piecewise constant arguments. Comput. Math. Appl. **28**, 81–92 (1994)
4. Torelli, L.: Stability of numerical methods for delay differential equations. J. Comput. and Appl. Math. **25**, 15–26 (1989)

5. Yenicerioglu, A.F.: The behavior of solutions of second order delay differential equations. J. Math. Anal. Appl. **332**(2), 1278–1290 (2007)
6. Ashyralyev, A., Akca, H.: Stability estimates of difference schemes for neutral delay differential equations. Nonlinear Anal.: Theory Methods Appl. **44**(4), 443–452 (2001)
7. Ashyralyev, A., Akca, H., Yenicerioglu, A.F.: Stability properties of difference schemes for neutral differential equations. Differ. Equ. Appl. **3**, 57–66 (2003)
8. Liu J., Dong P., Shang G.: Sufficient conditions for inverse anticipating synchronization of unidirectional coupled chaotic systems with multiple time delays. In: Proceedings of the Chinese Control and Decision Conference, pp. 751–756. IEEE (2010)
9. Mohamad, S., Akca, H., Covachev, V.: Discrete-time cohen-grossberg neural networks with transmission delays and impulses. Differ. Differ. Equ. Appl. Book Ser.: Tatra Mountains Math. Publ. **43**, 145–161 (2009)
10. Ashyralyev, A., Sobolevskii, P.E.: On the stability of the delay differential and difference equations. Abstr. Appl. Anal. **6**(5), 267–297 (2001)
11. Ashyralyev, A., Sobolevskii, P.E.: New Difference Schemes for Partial Differential Equations. Birkhäuser Verlag, Boston (2004)
12. Gabriella, D.B.: Delay differential equations with unbounded operators acting on delay terms. Nonlinear Anal.: Theory Methods Appl. **52**(1), 1–18 (2003)
13. Ashyralyev, A.: Fractional spaces generated by the positive differential and difference operator in a banach space. In: Tas, K., Tenreiro Machado, J.A., Baleanu, D. (eds.) Mathematical Methods in Engineering, pp. 13–22. Springer, Netherlands (2007)
14. Ashyralyev, A., Agirseven, D.: On convergence of difference schemes for delay parabolic equations. Comput. Math. Appl. **66**(7), 1232–1244 (2013)
15. Akca, H., Shakhmurov, V.B., Arslan, G.: Differential-operator equations with bounded delay. Nonlinear Times Dig. **2**, 179–190 (1995)
16. Akca, H., Covachev, V.: Spatial discretization of an impulsive Cohen-Grossberg neural network with time-varying and distributed delays and reaction-diffusion terms. Analele Stiintifice Ale Universitatii Ovidius Constanta-Seria Mat. **17**(3), 15–16 (2009)
17. Ashyralyev A., Agirseven D.: Finite difference method for delay parabolic equations. In: AIP Conference Proceedings of 1389 Numerical Analysis And Applied Mathematics ICNAAM 2011: International Conference on Numerical Analysis and Applied Mathematics, pp. 573–576 (2011)
18. Agirseven, D.: Approximate solutions of delay parabolic equations with the Dirichlet condition. Abstr. Appl. Anal. **2012**, 1–31 (2012). doi:10.1155/2012/682752. Article ID 682752
19. Ashyralyev, A., Agirseven, D.: Well-posedness of delay parabolic difference equations. Adv. Differ. Equ. **2014**, 1–18 (2014)
20. Ashyralyev, A., Agirseven, D.: Approximate solutions of delay parabolic equations with the Neumann condition, In: AIP Conference Proceedings, ICNAAM 2012, vol. 1479, pp. 555–558 (2012)

Some Results of FEM Schemes Analysis by Finite Difference Method

Dmitry T. Chekmarev$^{(\boxtimes)}$

Lobachevsky State University of Nizhni Novgorod, Nizhny Novgorod, Russia
`4ekm@mm.unn.ru`

Abstract. A method of investigation of numerical schemes deriving from the variational formulation of the problem (variational- difference method and FEM) is discusses. The method is based on the reduction of the numerical schemes to the canonical finite difference form. The resulting numerical scheme standard notation in the form of a grid operator equality is used for analyzing its approximation, stability and other properties. The application of this approach to a wider classes of finite elements (from the simplest ones to the Hermitian elements and serendipities) is discussed. These opportunities are illustrated by the analysis of FEM schemes for Timoshenko shells and elasticity dynamic problems.

1 Conversion of Finite Element Numerical Schemes to Finite Difference Form

The investigation of difference schemes involves the standard scheme notation in the form of a grid operator equality. In the case of uniform difference grids the FEM scheme operator can be written in the final form, which is suitable for the theoretical analysis. The conversion procedure is similar to the constructing the system of Euler differential equations of the variational problem. Lets consider the construction method for the two-dimensional case, which can be naturally is generalized to the n- dimensional case. Let $R^2 = \{x\} = \{(x1, x2)\}$ set uniform (possibly oblique), the main grid coordinates of nodes

$$x_{ij} = \begin{bmatrix} x_{ij}^1 \\ x_{ij}^2 \end{bmatrix} = B_h \begin{bmatrix} i \\ j \end{bmatrix} + \begin{bmatrix} x_0^1 \\ x_0^2 \end{bmatrix} \qquad i, j \in Z, \tag{1}$$

where B_h – is real nonsingular matrix 2×2. To refine the concept of certainty uniform finite element mesh. The finite element mesh in R^2 is said to be uniform if the elements and their nodes are periodically with a period given by a grid form (1). Let a functional is given

$$W = \int_\Omega F(u, p_1, p_2) \, d\Omega, \tag{2}$$

where u is an unknown function satisfying the given boundary conditions; $p_1 = \frac{\partial u}{\partial x^1}$, $p_2 = \frac{\partial u}{\partial x^2}$.

© Springer International Publishing Switzerland 2015
I. Dimov et al. (Eds.): FDM 2014, LNCS 9045, pp. 153–160, 2015.
DOI: 10.1007/978-3-319-20239-6_14

It is necessary to find a function u, deliver functional W extreme or stationary value. Solution of this problem satisfies the differential equation of Euler

$$\frac{\partial F}{\partial u} - \frac{\partial}{\partial x^1}\frac{\partial F}{\partial p_1} - \frac{\partial}{\partial x^2}\frac{\partial F}{\partial p_2} = 0 \tag{3}$$

and boundary conditions. Constructing of FEM numerical scheme reduces to the partition of the region into finite elements (construction of the finite element mesh) and the choice of basis functions, then the FEM problem is defined. It further reduces by known methods [3,4] to an algebraic system.

We assume that the FEM problem is defined, i.e. functional (2) is given, built finite element mesh and selected basis functions. Consider the transformation scheme to FEM finite difference form in the simplest case of Lagrangian elements.

We assume that each cell is divided into r, finite elements, where $r = 1$ to quadrilateral elements and $r = 2$ for the triangular elements (three-dimensional case r can take values from 1 to 6). Then the functional (2) can be written as

$$W_h = \sum_{(i,j)\in \Omega_h} \sum_{k=1}^{r} \int_{E_{ijk}} F\left(u, p_1, p_2\right) d\Omega, \tag{4}$$

where E_{ijk} - k-th element in the (i, j)-th cell of the main foil. Assuming that the k-th element of the type comprising m_k node template has S_k. Unknown function u in the element is given as

$$u = \sum_{l=0}^{m_k-1} C_{lk}\varphi_{lk}, \tag{5}$$

where φ_{lk} $(l = 0,\ldots,m_k - 1)$- the basic function of the k-type element. For example, for a 4-node bilinear element

$$f(x^1, x^2) = c_0 + c_1(x^1 - x_c^1) + c_2(x^2 - x_c^2) + c_3(x^1 - x_c^1)(x^2 - x_c^2)$$

(here (x_c^1, x_c^2)- the coordinates of the element center). Coefficients c_m can be expressed through the function node values $c_m = \sum_{k=1}\beta_k^m f_k$ (it uses the local node numbering). As a result of the transition to a nodes global numbering for the l-type element the following formula we get:

$$c_m = d_{m,l}^{+}f_{ij} = \sum_{(p,q)\in S_l} \beta_{p,q}^{m,l} f_{i+p,j+q} . \tag{6}$$

This formula also defines the basic differential operators $d_{m,l}^{+}$. Coefficients $\beta_{p,q}^{m,l}$ depend from the element template and matrix B_h. We also introduce the characters taken from different operators adjoint to (6):

$$d_{m,l}^{-}f_{ij} = (-1)^{K_m} \sum_{(p,q)\in S_l} \beta_{p,q}^{m,l} f_{i-p,j-q} . \tag{7}$$

Here K_m- is order of derivative, which is approximated by the operator. C_{lk} coefficients can be expressed through the values of function u at the nodes of the element, and further through the difference operators $d_0^{k+}, \ldots, d_{m_k-1}^{k+}$. As a result, the function u in element E_{ijk} may be represented as

$$u = \sum_{l=0}^{m_k-1} \left(d_{l,k}^+ u \right)_{ij} \psi_{lk}, \tag{8}$$

where ψ_{lk}- some linear combinations of functions φ_{lk}.

After substituting (8) into (4) integration we obtain

$$W_h = \sum_{(i,j) \in \Omega_h} \sum_{k=1}^{r} \gamma_k \Phi_k \left(\xi_{ij0}, \ldots, \xi_{ijm_k-1} \right). \tag{9}$$

where $\xi_{ijl} = \left(d_l^+ u \right)_{ij}$.

We write the variation of the functional (9):

$$\delta W_h = \sum_{(i,j) \in \Omega_h} \sum_{k=1}^{r} \gamma_k \left(f_{k0} \delta \left(d_0^{k+} u \right) + \ldots + f_{km_k-1} \delta \left(d_{m_k-1}^{k+} u \right) \right)_{ij}. \tag{10}$$

Here we use the notation

$$(f_{kl})_{ij} = \frac{\partial \Phi_k}{\partial \xi_{ijl}}$$

Substituting (10) in the discrete variation equation

$$\delta W_h = 0$$

and applying grid integration by parts, we obtain

$$\sum_{(i,j) \in \Omega_h} \sum_{k=1}^{r} \gamma_k \sum_{l=0}^{m_k-1} (-1)^{K_l} d_{l,k}^- f_{kl} \delta u_l = 0.$$

The last equality is satisfied if the conditions

$$\sum_{k=1}^{r} \gamma_k \sum_{l=0}^{m_k-1} (-1)^{K_l} d_{l,k}^- f_{kl} = 0 \tag{11}$$

obey at all nodes of the difference grid. Equation (11) represents a difference scheme standard form. Thus, we obtain the final difference form of the FEM scheme.

This method of FEM schemes conversion to finite differences can be used for theoretical analysis of a wide FEM schemes class for both linear and nonlinear problems. The method of variational-difference schemes transformation to finite differences is similar with not great distinctions. Similar transformations for other types of finite elements (Hermitian, sirendipity etc.) were considered in [1].

2   Analysis of Approximation of Variational-Difference and Finite Element Schemes of Timoshenko Plates Theory

Consider the system of equations describing the transverse vibrations of the Tymoshenko plate [5] 1D case recorded in the dimensionless form. System has the form

$$a \left(\frac{\partial^2 w}{\partial x^2} + \frac{\partial \psi}{\partial x} \right) - \frac{\partial^2 w}{\partial t^2} = 0$$
$$\frac{\partial^2 w}{\partial x^2} - \frac{12a}{\eta^2} \left(\frac{\partial w}{\partial x} + \psi \right) - \frac{\partial^2 \psi}{\partial t^2} = 0 \tag{12}$$

It is equivalent to a single equation of the fourth order

$$\frac{\partial^4 w}{\partial x^4} + \frac{12a}{\eta^2} \frac{\partial^2 w}{\partial t^2} - \left(1 + \frac{1}{a} \right) \frac{\partial^4 w}{\partial x^2 \partial t^2} + \frac{1}{a} \frac{\partial^4 w}{\partial t^4} = 0. \tag{13}$$

Consider the difference schemes approximation of the equations system (12). Finite-difference scheme has the form

$$\begin{cases} a \left(D_{11}w + D_{01}\psi \right) - D_{tt}w = 0 \\ D_{11}\psi - \frac{12a}{\eta^2} \left(D_{01}w + \psi \right) - D_{tt}\psi = 0 \end{cases} \tag{14}$$

variational-difference scheme has the form

$$\begin{cases} a \left(D_{11}w + D_{01}\psi \right) - D_{tt}w = 0 \\ D_{11}\psi - \frac{12a}{\eta^2} \left(D_{01}w + D_{00}\psi \right) - D_{tt}\psi = 0 \end{cases} \tag{15}$$

linear finite element scheme has the form

$$\begin{cases} a \left(D_{11}w + D_{01}\psi \right) - D_{tt}w = 0 \\ D_{11}\psi - \frac{12a}{\eta^2} \left(D_{01}w + D_0\psi \right) - D_{tt}\psi = 0 \end{cases} \tag{16}$$

There in (14)–(16) $D_{11}f = \frac{1}{h^2} \left(f_{i+1} - 2f_i + f_{i-1} \right)$, $D_{00}f = \frac{1}{4} \left(f_{i+1} + 2f_i + f_{i-1} \right)$, $D_{01}f = \frac{1}{2h} \left(f_{i+1} - f_{i-1} \right)$, $D_0f = \frac{1}{6} \left(f_{i+1} + 4f_i + f_{i-1} \right)$, $D_{tt}f = \frac{1}{\tau^2} \left(f^{j+1} - 2f^j + f^{j-1} \right)$.

Schemes (14)-(16) differ by only one equations member approximation - function ψ. They all have second-order approximation. Transform them to the form similar to (13). Finite-difference scheme (14) takes the form:

$$\left(1 + 3a \left(\frac{h}{\eta} \right)^2 \right) D_{11}D_{11}w + \frac{12}{\eta^2} D_{tt}w - \left(1 + \frac{1}{a} \right) D_{11}D_{tt}w + \frac{1}{a} D_{tt}D_{tt}w = 0. \tag{17}$$

variational-difference scheme (15) takes the form:

$$D_{11}D_{11}w + \frac{12}{\eta^2} D_{00}D_{BB}w - \left(1 + \frac{1}{a} \right) D_{11}D_{tt}w + \frac{1}{a} D_{tt}D_{tt}w = 0. \tag{18}$$

linear finite element scheme (16) takes the form:

$$\left(1 + a\left(\frac{h}{\eta}\right)^2\right)D_{11}D_{11}w + \frac{12}{\eta^2}D_0^*D_{tt}w - \left(1 + \frac{1}{a}\right)D_{11}D_{tt}w + \frac{1}{a}D_{tt}D_{tt}w = 0.$$

(19)

Comparing (17)-(19) with the original differential equation (13), we conclude that at finite values of the quantity h/η (relationship grid spacing to plate thickness) difference equations (17) and (19) do not approximate equation (13) but, respectively, the equations

$$\left(1 + 3a\left(\frac{h}{\eta}\right)^2\right)\frac{\partial^4 w}{\partial x^4} + \frac{12}{\eta^2}\frac{\partial^2 w}{\partial t^2} - \left(1 + \frac{1}{a}\right)\frac{\partial^4 w}{\partial x^2 \partial t^2} + \frac{1}{a}\frac{\partial^4 w}{\partial t^4} = 0. \quad (20)$$

and

$$\left(1 + a\left(\frac{h}{\eta}\right)^2\right)\frac{\partial^4 w}{\partial x^4} + \frac{12}{\eta^2}\frac{\partial^2 w}{\partial t^2} - \left(1 + \frac{1}{a}\right)\frac{\partial^4 w}{\partial x^2 \partial t^2} + \frac{1}{a}\frac{\partial^4 w}{\partial t^4} = 0. \quad (21)$$

Thus, we can conclude that all schemes have the convergence in the usual sense, but the finite difference and finite element schemes do not have the uniform convergence by grid problem parameter. Scheme (15) has the uniform convergence of the parameter h/η. Quantitative analysis confirming the findings is given in [1]. A similar analysis was conducted and the approximation for two-dimensional schemes. Analysis results agree well with well known reduced integration technique by O. Zienkievich [6], due to which the scheme (16) becomes the scheme (15).

3   Influence of the Mutual Location of Finite Elements on the Accuracy of the Numerical Solution

Consider the effect of finite element mutual location influence to approximation and accuracy of schemes. We take the 3D elastic problem and 4-node linear finite element. Way of the base parallelepiped dividing into tetrahedrons defined by a set of templates elements. Pattern of each element contains four integer vector is a subset of $\{(000), (001), (010), (011), (100), (101), (110), (111)\}$.

Below you can see the types of partitions hexahedron.
(1) 5 tetrahedra (Fig. 1.):

$$S_1 = \{(000), (011), (101), (110)\}, \quad S_2 = \{(000), (011), (101), (001)\},$$
$$S_3 = \{(000), (100), (101), (110)\}, \quad S_4 = \{(000), (011), (010), (110)\},$$
$$S_5 = \{(111), (011), (101), (110)\}$$

(2) 6 tetrahedra with centrally symmetric partition (Fig. 2.):

$$S_1 = \{(010), (001), (101), (011)\}, \quad S_2 = \{(000), (001), (101), (010)\},$$
$$S_3 = \{(000), (100), (101), (010)\}, \quad S_4 = \{(101), (110), (010), (100)\},$$
$$S_5 = \{(111), (110), (010), (101)\}, \quad S_6 = \{(111), (011), (010), (101)\}.$$

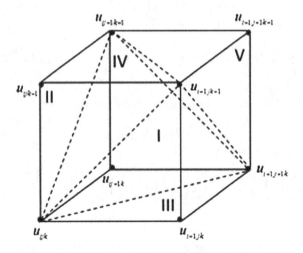

Fig. 1. 5 tetrahedra

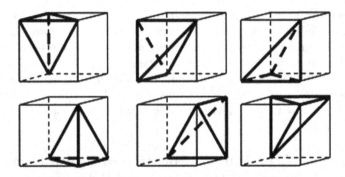

Fig. 2. 6 tetrahedra,

(3) 6 tetrahedra with rotational-symmetric partition:

$$S_1 = \{(010), (001), (101), (011)\}, \quad S_2 = \{(000), (001), (101), (010)\},$$
$$S_3 = \{(000), (100), (101), (010)\}, \quad S_4 = \{(100), (101), (011), (111)\},$$
$$S_5 - \{(100), (011), (110), (111)\}, \quad S_6 = \{(100), (011), (010), (110)\}.$$

(4) 6 tetrahedra with non-symmetric partition:

$$S_1 = \{(010), (001), (101), (011)\}, \quad S_2 = \{(000), (001), (101), (010)\},$$
$$S_3 = \{(000), (100), (101), (010)\}, \quad S_4 = \{(100), (010), (110), (111)\},$$
$$S_5 = \{(100), (010), (101), (111)\}, \quad S_6 = \{(010), (101), (011), (111)\}.$$

Unknown functions in linear element represented in the form

$$f(x^1, x^2, x^3) = c_0 + c_1(x^1 - x_c^1) + c_2(x^2 - x_c^2) + c_3(x^3 - x_c^3)$$

(here (x_c^1, x_c^3, x_c^3) - the coordinates of the center of the element).

We write the functional as energy internal of the linearly elastic body:

$$W = \frac{1}{2} \int_{\Omega} \sigma_{\alpha\beta} \varepsilon_{\alpha\beta} d\Omega$$

Further, according to the algorithm described above, we obtain the representation of FEM schemes in the traditional finite-difference form

$$(\lambda + \mu) \begin{vmatrix} D_{11} u_1 + D_{12} u_2 + D_{13} u_3 \\ D_{21} u_1 + D_{22} u_2 + D_{23} u_3 \\ D_{31} u_1 + D_{32} u_2 + D_{33} u_3 \end{vmatrix} + \mu \, D_\Delta \begin{vmatrix} u_1 \\ u_2 \\ u_3 \end{vmatrix} + \rho \begin{vmatrix} F_1 \\ F_2 \\ F_3 \end{vmatrix} = \rho D_{tt} \begin{vmatrix} u_1 \\ u_2 \\ u_3 \end{vmatrix}, \quad (22)$$

a similar system of Lame equations

$$(\lambda + \mu) \, grad \, div \, u + \mu \Delta u + \rho F = \rho \frac{\partial^2 u}{\partial t^2} \qquad (23)$$

where the operators D_{ij} approximate second derivatives, respectively, for the i-th and j-th coordinates, $D_{tt}f = \frac{1}{\tau^2} (f(t+\tau) - 2f(t) + f(t-\tau))$ approximates the second derivative with respect to time, $D_\Delta = D_{11} + D_{22} + D_{33}$ - the grid Laplace operator. D_{ij} operators have different specific form depending on the variant schemes investigated and may be either the first or second order approximation. Schemes for linear finite element we have $D_{ij} = \sum_{l=1}^{s} \gamma_l d_{i,l}^+ d_{j,l}^-$, where $s = 5, \gamma_1 = \frac{1}{3}, \gamma_2 = \gamma_3 = \gamma_4 = \gamma_5 = \frac{1}{6}$ the scheme with the partition parallelepiped 5 tetrahedron; $s = 6, \gamma_1 = \gamma_2 = \gamma_3 = \gamma_4 = \gamma_5 = \gamma_6 = \frac{1}{6}$ for schemes with the partition parallelepiped 6 tetrahedron. Analysis grid approximation equation (23) by equation (22) conducted a standard method for the case of orthogonal grid with the coordinates of the grid nodes are equal $x_{ijk}^1 = x_0^1 + h_1 i$, $x_{ijk}^2 = x_0^2 + h_2 j$, $x_{ijk}^3 = x_0^3 + h_3 k$) showed that one of this schemes (centrally symmetric partition) has second order approximation, and the other three - the first order approximation. The results of the test problem solutions also showed a different rate of schemes convergence

4   Variational-Difference and Finite Element Schemes on Rare Grids [7]

There in formulas (11) is an overall view of finite difference schemes represent FEM on uniform grids. They contain coefficients $\gamma_k = V_k/\Delta V$, where V_k - the volume (area) element k-type, ΔV - the basic unit of volume of a uniform grid of the form (1). Coefficients γ_k satisfy the obvious equality

$$\sum_{k=1}^{p} \gamma_k = 1, \qquad (24)$$

reflects the continuous filling elements of the computational domain (here p - the number of elements that make up the cell.) Varying set of coefficients γ_k, while maintaining this equality, we obtain new difference schemes, some of which can be quite successful. In particular, variation in the difference scheme similar to (15) on the coefficients of the triangular cells are equal $\gamma_1 = \gamma_2 = 1/2$. Substituting their values $\gamma_1 = 1, \gamma_2 = 0$, we obtain "rare mesh" variational- difference scheme. The scheme has a much better convergence than the original. Note it is also more economical, because it is actually two times less computational cells. A detailed analysis of this scheme are given in [1,2].

Further developing this method, we arrive at the idea of rare mesh schemes FEM. Under the rare mesh scheme we understand the scheme, in which some of the coefficients γ_k equal to zero. Relevant elements do not contribute to the numerical scheme and may be excluded from the calculations. This approach proved to be very productive in solving the three-dimensional elasticity problems. In particular, the scheme has been proposed on the basis of a linear 4 -node finite element, which for central tetrahedron (Fig. 1.) Ratio γ_k was 1, the remaining tetrahedra - zero. This scheme is significantly more economical than traditional and has better convergence. Also, it has no the drawback of numerical schemes on hexahedral elements - the "hourglass instability". Detailed description of the scheme, the results of its analysis and testing described are given in [7,8].

5   Conclusion

Method described in the study of numerical schemes based on the variational formulation of problems allows more deeply study their properties and to propose ways to improve, as evidenced by the examples discussed. He partly overcomes the gap between the theory of difference schemes and finite element method. This approach can be applied to the analysis of a wide variety of schemes FEM mathematical physics problems.

References

1. Bazhenov, V.G., Chekmarev, D.T.: Solving the problems of plates and shells dynamics by variational- difference method, Nizhny Novgorod (2000) (in Russian)
2. Bazhenov, V.G., Chekmarev, D.T.: On numerical differentiation index commutation. Comp. Math. and Math. Phys. **29**(5), 662–674 (1989). (in Russian)
3. Zienkiewicz, O.C., Morgan, K.: Finite Elements and Approximation. Wiley, New York (1983)
4. Strang, G., Fix, G.J.L: An Analysis of the Finite Element Method. Series in Aut. Comp. pp. XIV + 306 SM Fig. Prentice-Hall Inc, Englewood Clifs, NJ (1973)
5. Timoshenko, S.: Vibration problems in engineering. Van Nostrand, New York (1937)
6. Zienkiewich, O.C., Too, J., Taylor, R.L.: Reduced integration technique in general analysis of plates and shells. Int. J. Num. Meth. Engng. **3**(2), 275–290 (1971)
7. Chekmarev, D.T.: Finite element schemes on rare meshes. Probl. At. Sci. Techn. Ser. Math. Model. phys. proc. **2**, 49–54 (2009). (in Russian)
8. Zhidkov, A.V., Zefirov, S.V., Kastalkaya, K.A., Spirin, S.V., Chekmarev, D.T.: Rare mesh scheme for numerical solution of three-dimensional dynamic problems of elasticity and plasticity. Bull. Nizhny Novgorod Univ. **4**(4), 1480–1485 (2011). (in Russian)

Application of Finite Difference TVD Methods in Hypersonic Aerodynamics

Yury Dimitrienko$^{(\boxtimes)}$, Mikhail Koryakov, and Andrey Zakharov

Bauman Moscow State Technical University, 2-ya Baumanskay st., 5, Russia, Moscow
dimit@bmstu.ru
http://www.bmstu.ru/~fn11/english/echif.htm

Abstract. The numerical method for solving of the hypersonic nonequilibrium aerogasdynamics problems is suggested. The method is based on the full three-dimensional Navier-Stokes equations, supplemented by the equations of chemical kinetics and the finite difference TVD method. The developed algorithms are implemented in the computer-aided software package SIGMA. The results of simulation of the hypersonic flow about the spherical nose segment of a model hypersonic vehicle are presented.

Keywords: Computational fluid dynamics · Hypersonic flows · TVD schemes · Parallel processing

1 Introduction

TVD schemes [1] have recently become quite popular in solving of gas dynamics problems due to the higher accuracy and monotonic property of the numerical solutions. However, they are usually used in an orthogonal coordinate system, so experience with these schemes for geometrically adaptive meshes in domains of complex curvilinear shape while is not great. The aim of this work is the development of a variant of the method, based on the TVD Harten scheme of the second order accurate, for numerical solution of the equations of gas dynamics in arbitrary non-orthogonal curvilinear coordinate systems.

We consider the Navier-Stokes equations with chemical kinetics, describing high–speed flow around an aircraft. Integration of the Navier-Stokes equations is performed by the splitting into physical processes. At the first stage the members of the viscous terms are excluded from the consideration. The problem is solved for ideal gas by the developed TVD scheme. Then the viscous components are approximated using the implicit difference schemes with splitting method into the spatial directions, which are solved by the Thomas algorithm. Further the equations of chemical kinetics are solved in 3 stages. Firstly, an explicit-implicit scheme is used to solve the system of difference equations with source terms. Then the convection of the chemical components is taken into account. The TVD scheme is used to solve the equations similarly the step of the solving of the inviscid flow. And then the diffusion of the chemical components is taken

© Springer International Publishing Switzerland 2015
I. Dimov et al. (Eds.): FDM 2014, LNCS 9045, pp. 161–168, 2015.
DOI: 10.1007/978-3-319-20239-6_15

into account. The algorithm for solving the equations is similarly the step of the solving of the viscous components.

The results of a simulation of a flow about an spherical nose segment of a model hypersonic aircraft are presented. The chemical gas-phase model included all the main components of high temperature air for flight conditions in the Earth's atmosphere. The peak wavelengths of the spectral intensity of the body have been obtained.

2   The System of Equations

Consider the system of equations of a viscous heat-conducting gas (the Navier-Stokes equations) with chemical kinetics:

$$\frac{\partial \rho}{\partial t} + \boldsymbol{\nabla} \cdot \rho \boldsymbol{v} = 0, \tag{1}$$

$$\frac{\partial \rho \boldsymbol{v}}{\partial t} + \boldsymbol{\nabla} \cdot (\rho \boldsymbol{v} \otimes \boldsymbol{v} + p\boldsymbol{E} - \boldsymbol{T}_v) = \boldsymbol{0}, \tag{2}$$

$$\frac{\partial \rho \epsilon}{\partial t} + \boldsymbol{\nabla} \cdot ((\rho \epsilon + p)\,\boldsymbol{v} - \boldsymbol{T}_v \cdot \boldsymbol{v} + \boldsymbol{q}) = 0, \tag{3}$$

$$\frac{\partial \rho y_i}{\partial t} + \boldsymbol{\nabla} \cdot (\rho y_i \mathbf{v} - \rho D_{ij} \nabla y_i) = \omega_i, \tag{4}$$

where ρ — the gas density of the gas mixture, t — the time, $\boldsymbol{v}$ — the velocity vector of the center of mass of the mixture, p — the pressure, $\boldsymbol{E}$ — the identity tensor, ϵ — the total energy per unit volume, $y_i = \rho_i/\rho$ — the mass concentration of the i-th spice, ω_i — the source of generation of the i-th spice, D_{ij} — the diffusion coefficients.

This system adds relations for the perfect gas, viscous stress tensor and heat flux vector

$$p = \rho \frac{R_0}{M_0}\theta, \quad \frac{1}{M_0} = \sum_{i=1}^{7} \frac{y_i}{M_i}, \quad \epsilon = e + \frac{|\boldsymbol{v}|^2}{2}, \quad e = c_v\theta, \quad |\boldsymbol{v}|^2 = \boldsymbol{v} \cdot \boldsymbol{v},$$

$$c_v = \sum_{i=1}^{7} y_i c_{vi}, \quad \boldsymbol{T}_v = -\frac{2}{3}\mu(\boldsymbol{\nabla} \cdot \boldsymbol{v})\boldsymbol{E} + \mu(\nabla \otimes \boldsymbol{v} + \nabla \otimes \boldsymbol{v}^T), \quad \boldsymbol{q} = -\lambda \boldsymbol{\nabla}\theta,$$

where R — the universal gas constant, M_i — the molecular weight of the i-th spice, θ — the gas temperature, e — the internal energy per unit volume, c_{vi} — the specific heat at constant volume of the i-th spice, μ — the coefficient of viscosity, λ — the thermal conductivity of the gas.

We consider the binary diffusion model. The coefficients of diffusion, viscosity and thermal conductivity are given by the following functions [6]

$$D_{12} = 1.85 \cdot 10^{-7} \frac{\theta^{3/2}}{p\sigma_{12}^2 \Omega_{12}^{(1,1)}} \left(\frac{M_1 + M_2}{M_1 M_2}\right)^{1/2},$$

$$\mu = \sum_{i=1}^{6} \frac{\mu_i}{1 + \sum\limits_{j=1, j \neq i}^{6} G_{ij}^{\mu} \frac{\gamma_j}{\gamma_i}}, \qquad \lambda = \sum_{i=1}^{6} \frac{\lambda_i}{1 + \sum\limits_{j=1, j \neq i}^{6} G_{ij}^{\lambda} \frac{\gamma_j}{\gamma_i}}, \qquad \gamma_i = \frac{y_i}{M_i},$$

where σ_{12} — the characteristic distance, $\Omega_{12}^{(1,1)}$ — the collision integral, μ_i, λ_i — the viscosity and thermal conductivity of pure gases, G_{ij}^{μ}, G_{ij}^{λ} — the universal constants.

The gas mixture consists of seven spices and the possible chemical reactions that occur in shock layer at high temperatures are following

$$O_2 + M \leftrightarrow 2O + M \qquad N_2 + M \leftrightarrow 2N + M \qquad NO + M \leftrightarrow N + O + M$$
$$NO + O \leftrightarrow O_2 + N \qquad N_2 + O \leftrightarrow NO + N \qquad NO + O \leftrightarrow NO^+ + e^-,$$

where M is any of the six spices considered to be the catalyst, e^- — the electronic component. Associate the index $i = 1, 2, 3, 4, 5, 6$ with the components of O, N, NO, O_2, N_2, NO+, respectively.

Also the conditions of conservation of atomic composition

$$y_{O_2} = 0.21 - 0.5 \cdot (y_O + y_{NO} + y_{NO+}), \qquad y_{N_2} = 0.79 - 0.5 \cdot (y_N + y_{NO} + y_{NO+})$$

and mixture quasineutrality $y_{NO+} = y_{e-}$ are valid, thus only four spices are independent.

3   Numerical Method

The integration of the Navier-Stokes equations with chemical kinetics is performed by the method of the splitting into physical processes.

System of equations describing the ideal gas in the computational coordinates X^i [2,4] as follows

$$\frac{\partial \boldsymbol{U}}{\partial t} + \sum_{i,j=1}^{3} P_j^i \frac{\partial \boldsymbol{V}^j}{\partial X^i} = 0, \quad P_j^i = \frac{\partial X^i}{\partial x^j}. \tag{5}$$

where $\boldsymbol{U} = [\rho, \rho \bar{v}_x, \rho \bar{v}_y, \rho \bar{v}_z, \rho E]^T$, $\boldsymbol{F} = [\rho v, \rho v \bar{v}_x + p \delta_1^i, \rho v \bar{v}_y + p \delta_2^i, \rho v \bar{v}_z + p \delta_3^i, \rho (E + p) \boldsymbol{v}]^T$, P_j^i — the components of the inverse Jacobian matrix of the transformation of Cartesian coordinates x^j to the computational coordinates X^i, $\bar{v}^i$ — the Cartesian components of the velocity vector.

Consider the technique of obtaining the explicit second order accurate TVD scheme based on the Harten schemes [1], which is obtained by applying a nonoscillatory first order accurate scheme to an appropriately modified flux function $\tilde{f} = (f + \frac{1}{\lambda} g)$

$$\boldsymbol{U}_{\eta}^{n+1} = \boldsymbol{U}_{\eta}^{n} - \sum_{i,j=1}^{3} \lambda_{\eta}^{j} \left(\tilde{\boldsymbol{V}}_{\eta+1/2}^{i,j,n} - \tilde{\boldsymbol{V}}_{\eta-1/2}^{i,j,n} \right), \qquad \lambda_{\eta}^{j} = \frac{\Delta t}{X_{\eta+}^{j} - X_{\eta-}^{j}},$$

where $U^n_\eta = U(X_\eta, t^n)$, $\tilde{V}^{i,j,n}_{\eta+1/2}$ — the columns of the numerical flux function in the i-th direction defined by the following formula

$$\tilde{V}^{i,j,n}_{\eta+1/2} = 0{,}5 \cdot \left(V^{i,n}_\eta \cdot P^j_{i,\eta} + V^{i,n}_{\eta+} \cdot P^j_{i,\eta} + R^{j,n}_{\eta+1/2} \cdot \Phi^{i,j,n}_{\eta+1/2} \right).$$

Here $R^{i,n}_{\eta+1/2} = (R^{i,n}_{\eta+} + R^{i,n}_\eta)/2$, $R^{i,n}_\eta = R^i(X_\eta, t^n)$ — the matrix of the right eigenvectors of the matrix $G^i = \partial V^i / \partial U$.

Numerical viscosity vector $\Phi^{i,j,n}_{\eta+1/2}$ have the following components

$$\varphi^{i,j,m}_{\eta+1/2} = 0{.}5 \cdot \psi\left(\hat{a}^{i,j,m}_{\eta+1/2}\right) \cdot \left(g^{i,j,m}_\eta + g^{i,j,m}_{\eta+}\right) - \psi\left(\hat{a}^{i,j,m}_{\eta+1/2} + \gamma^{i,j,m}_{\eta+1/2}\right) \cdot \alpha^{j,m}_{\eta+1/2},$$

$$\gamma^{i,j,m}_{\eta+1/2} = \begin{cases} \dfrac{\psi\left(\hat{a}^{i,j,m}_{\eta+1/2}\right)\left(g^{i,j,m}_{\eta+} - g^{i,j,m}_\eta\right)}{2 \cdot \alpha^{i,m}_{\eta+1/2}}, & \alpha^{i,m}_{\eta+1/2} \neq 0; \\ 0, & \alpha^{i,m}_{\eta+1/2} = 0; \end{cases}$$

where $g^{i,j,m}_\eta = \text{minmod}\left(\alpha^{j,m}_{\eta-1/2}, \alpha^{j,m}_{\eta+1/2}\right)$, $\hat{a}^{i,j,m}_{\eta+1/2} = a^{i,m}_{\eta+1/2} \cdot P^j_{i,\eta}$, $a^{i,m}_{\eta+1/2} = a^m(X_\eta, t^n)$ — eigenvectors of the matrix G^i, $\alpha^j_{\eta+1/2} = \left(\alpha^{j,1}_{\eta+1/2}, \ldots, \alpha^{j,5}_{\eta+1/2}\right)^T = \left(R^{j,n}_{\eta+1/2}\right)^{-1} \cdot (U^n_{\eta+} - U^n_\eta)$, $\psi(z)$ — the numerical viscosity function, which is defined by the formula

$$\psi(z) = \begin{cases} |z|, & |z| \geq \varepsilon; \\ \left(z^2 + \varepsilon^2\right)/2\varepsilon, & |z| < \varepsilon. \end{cases}$$

Then the viscous components are taken into account

$$\rho \frac{\partial \bar{v}^k}{\partial t} = P^n_i \frac{\partial}{\partial X^n}\left(\bar{M}^{ikj}_l P^s_j \frac{\partial \bar{v}^l}{\partial X^s}\right),$$

$$\rho \frac{\partial e}{\partial t} = P^n_i \frac{\partial}{\partial X^n}\left(\lambda P^k_i \frac{\partial \theta}{\partial X^k}\right) + \bar{M}^{pqk}_l P^s_k P^m_p \frac{\partial \bar{v}^l}{\partial X^s}\frac{\partial \bar{v}_q}{\partial X^m},$$

$$M^{ijk}_l = \bar{M}^{mjsp} P^i_m P^k_s \delta_{pl}, \qquad \bar{M}^{mjsp} = \mu_1 \delta^{mj}\delta^{sp} + \mu_2(\delta^{ms}\delta^{jp} + \delta^{mp}\delta^{jp}).$$

Its are approximated using the implicit difference scheme with splitting method into the spatial directions

$$\text{Step 1:} \qquad \rho^{n+\frac{2}{6}}(\bar{U}^{n+\frac{3}{6}} - \bar{U}^{n+\frac{2}{6}}) = \frac{\Delta t}{2}\Lambda_1\bar{U}^{n+\frac{3}{6}},$$

where $\bar{U}^{n+\frac{2}{6}}$ — the values obtained after the solving of the Euler system.

$$\text{Step 2:} \qquad \rho^{n+\frac{3}{6}}(\bar{U}^{n+\frac{4}{6}} - \bar{U}^{n+\frac{3}{6}}) = \frac{\Delta t}{2}\Lambda_2\bar{U}^{n+\frac{4}{6}}.$$

$$\text{Step 3:} \qquad \rho^{n+\frac{4}{6}}(\bar{U}^{n+\frac{5}{6}} - \bar{U}^{n+\frac{4}{6}}) = \frac{\Delta t}{2}\Lambda_3\bar{U}^{n+\frac{5}{6}}.$$

These equations are solved sequentially and independently from each other by the Thomas algorithm. In step 4 the mixed derivatives are calculated by the explicit scheme

$$\rho^{n+\frac{2}{6}}(\bar{U}^{n+1} - \bar{U}^{n+\frac{5}{6}}) = \Delta t \left[\sum_{i=1}^{3} \Lambda_i \bar{U}^{n+\frac{5}{6}} + (\Lambda_{12} + \Lambda_{13} + \Lambda_{23})\bar{U}^{n+\frac{2}{6}} + \bar{F}^{n+\frac{2}{6}} \right].$$

$$\text{where} \quad \Lambda_i = \frac{\partial}{\partial X^i}\left(A\frac{\partial}{\partial X^i}\right), \quad \Lambda_{ij} = \frac{\partial}{\partial X^i}\left(A\frac{\partial}{\partial X^j}\right), i \neq j.$$

Here $U = [\bar{v}_x, \bar{v}_y, \bar{v}_z, \theta]^T$, $\bar{F} = [0, 0, 0, \bar{M}_l^{pqk}P_k^s P_p^m (\partial \bar{v}^l/\partial X^s)(\partial \bar{v}^q/\partial X^m)]^T$.

Further the equations of chemical kinetics are solved in 3 stages.

Firstly, the mass fluxes of each i-th spice due to chemical transformation are taken into account

$$\frac{\partial X_i}{\partial t} = \dot{\omega}_i, \qquad X_i = \frac{\rho y_i}{M_i}, \qquad \dot{\omega}_i = \frac{\omega_i}{M_i}. \tag{6}$$

The source of generation of the i-th spice is calculated using the law of mass action

$$\begin{aligned}
\dot{\omega}_1 &= \varphi_{11} + \varphi_{13} + \varphi_{14} + \varphi_{15} + \varphi_{16}, & \varphi_{11} &= 2k_1(K_1 X_4 - X_1^2), \\
\dot{\omega}_2 &= \varphi_{22} + \varphi_{13} - \varphi_{14} - \varphi_{15} + \varphi_{16}, & \varphi_{13} &= k_3(K_3 X_3 - X_1 X_2), \\
\dot{\omega}_3 &= -\varphi_{13} + \varphi_{14} - \varphi_{15}, & \varphi_{14} &= -k_4(K_4 X_1 X_3 - X_2 X_4), \\
\dot{\omega}_6 &= -\varphi_{16}, & \varphi_{15} &= -k_5(K_5 X_1 X_5 - X_2 X_3), \\
& & \varphi_{16} &= -k_6(K_6 X_1 X_2 - X_6 X_7), \\
& & \varphi_{22} &= 2k_2(K_2 X_5 - X_2^2),
\end{aligned}$$

where k_i, K_i — the specific rate of the reverse reactions and equilibrium constant, respectively, which are temperature-dependent [4].

The system (6) is solved by the explicit-implicit scheme [4]

$$\mathbf{X}^{n+1} = \tilde{\mathbf{X}}^{n+1} + \left[\mathbf{X}^n - \tilde{\mathbf{X}}^{n+1} + \Delta t \left(\alpha \dot{\omega}^n + (1-\alpha)\tilde{\dot{\omega}}^{n+1}\right)\right] \cdot$$
$$\cdot \left[\mathbf{E} - (1-\alpha)\Delta t \left[\frac{\partial \tilde{\dot{\omega}}^{n+1}}{\partial \mathbf{X}}\right]\right]^{-1}.$$

where $\tilde{X}$ — the intermediate value of the iterative process, α — the blending coefficient.

Then the convection of chemical components is taken into account

$$\frac{\partial \rho y_i}{\partial t} + \nabla \cdot \rho y_i \boldsymbol{v} = 0. \tag{7}$$

The system is solved by the TVD scheme similarly the step of the solving of the inviscid flow.

And then the diffusion of chemical components is taken into account

$$\frac{\partial \rho y_i}{\partial t} = \nabla \cdot (\rho D_{ij} \nabla y_i). \tag{8}$$

These equations are solved similarly the step of the solving of the viscous components.

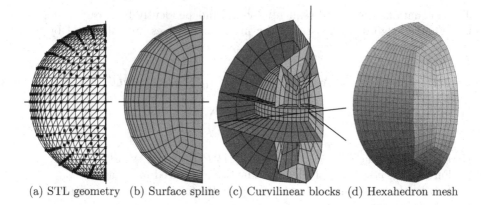

(a) STL geometry (b) Surface spline (c) Curvilinear blocks (d) Hexahedron mesh

Fig. 1. Import of geometry, domain construction and mesh generation

4   Description of Developed Software

The developed algorithms were implemented by the authors in their software package SIGMA [5]. It includes the complete set of modules that are required for the computer simulation. SIGMA includes a three–dimensional geometric modeling module, which allows to generate a wide range of geometric shapes; a module, which allows to set properties, parameters and initial conditions; an adaptive mesh generator (preprocessing components); a solver (processor) and a visualization tools (post–processor). Each SIGMA module is the detached cross–platform software, implemented using C++ with the ability to create extensions. Most of the iterative procedures for the mesh generation and solving are implemented using OpenMP and MPI libraries by means of geometric decomposition.

The preprocessor module has a graphical interface that allows you to create a solution domain visually. The domain is constructed from a set of initial hexahedral blocks (primitives) by their combining and subsequent deformation. The deformation is performed by changing of coordinates of control points of the domain by entering them or reading from a file. The control points of the domain are located on the boundary surfaces of the primitives, form a regular surface mesh and are the basis for the construction of the linear or cubic spline surfaces. It is possible to generate the curved blocks which are based on the geometries of the surfaces, imported from solid modeling software, see Fig. 1. In this case, the functions for generation of the points in given sections and along the lines between the two specified points on the surface are implemented for the construction of the regular mesh of the control points on the imported surfaces.

The mesh generator creates non-orthogonal block-structured grids (see Fig. 1) and uses explicit form of the algebraic transformation, which refers to the lagrangian coordinate transformations of transfinite interpolation methods [4]. Additional transformations of the grids are introduced to concentration the nodes near the boundaries. The preprocessor is able to construct the O-grid blocks for the certain types of curved domains like in ANSYS ICEM CFD.

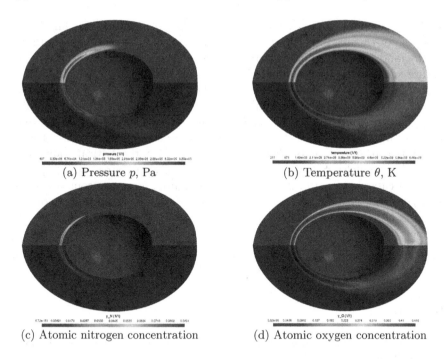

(a) Pressure p, Pa

(b) Temperature θ, K

(c) Atomic nitrogen concentration

(d) Atomic oxygen concentration

Fig. 2. Import of geometry, domains construction and meshes generation

5   Results

The simulation of the hypersonic flow about a spherical nose segment at Mach number $M = 20$ and altitude $h = 30\,\text{km}$ was considered. A solid ball had the following characteristics: radius $R = 30\,\text{cm}$, emissivity $\varepsilon = 1$. The main goal of the simulation was the calculation of the spectral radiant intensity of the body as a function of wavelength.

Figure 2 shows the three-dimensional distributions of gasdynamic parameters near the body. The coating materials of the ball should be able to withstand extremely high temperature up to $6500\,^\circ\text{K}$ in the stagnation point. At such temperatures the processes of dissociation and ionization of air molecules occur. The initial concentrations of a mixture of molecular oxygen and nitrogen are changed and the new concentrations of atomic nitrogen and oxygen are generated near the body. The heat transfer coefficient, heat flux and spectral irradiances are calculated using the distribution of the temperature in the flow and on the surface of the blackbody. And it was obtained that the peak wavelengths of the spectral intensity of the body is located in the near-infrared range.

6   Conclusion

A method based on the TVD scheme of the second order accurate for numerical solving of the full three-dimensional Navier-Stokes equations with chemical kinetics in domains with a complicated form has been proposed. The formulation allows to take into account the changes in the chemical composition of the flow and calculate more accurately the integral characteristics of the gas mixture such as thermal conductivity, viscosity, specific heat, which effect on the mechanical and thermal stresses near the critical components of aerospace vehicles. The developed algorithms have been implemented in the computer-aided software package SIGMA. SIGMA appropriates for simulation of supersonic and hypersonic flows, defining fields of mechanical and thermal stresses, as well as concentrations of chemical substances near the critical components of aerospace vehicles. SIGMA contains preprocessor, processor and postprocessor and is capable to perform calculations on high-performance computers. The developed numerical method and software can be applied to the analysis of aerodynamics of hypersonic aircrafts.

Acknowledgements. The reported study was supported by the Supercomputing Center of Lomonosov Moscow State University.

References

1. Harten, A.: High resolution schemes for hyperbolic conservation laws. J. Comp. Phys. **49**, 357–393 (1983)
2. Gil'manov, A.N.: Metody adaptivnyx setok v zadachax gazovoj dinamiki [Methods of Adaptive Meshes in Gas Dynamic Problems]. Fizmatlit, Moscow (2000)
3. Anderson, J.D.: Hypersonic and High-Temperature Gas Dynamics, 2nd edn. American Institute of Aeronautics and Astronautics, Reston (2006)
4. Dimitrienko, Y.I., Kotenev, V.P., Zakharov, A.A.: Metod lentochnyh adaptivnyh setok dlja chislennogo modelirovanija v gazovoj dinamike [Methods of Adaptive Band-Net Meshes for Numerical Simulation in Gas Dynamics]. Fizmatlit, Moscow (2011)
5. Dimitrienko, Y.I., Zakharov, A.A., Koryakov, M.N.: Razrabotka programmnogo obespechenija dlja chislennogo modelirovanija v zadachah giperzvukovoj ajerogazodinamiki perspektivnyh letatel'nyh apparatov [Software development for numeral modeling of hypersonic aerogasdynamics problem perspective aircrafts]. Programmnye sistemy: teorija i prilozhenija: jelektron. nauchn. zhurn. **3**(4), 17–26 (2012). http://psta.psiras.ru/read/psta2012_4_17-26.pdf
6. Hirschfelder, J.O., Curtiss, C.F., Bird, R.B.: Molecular Theory of Gases and Liquids. American Institute of Aeronautics and Astronautics, Wiley, New York (1954)

Finite Difference Equations for Neutron Flux and Importance Distribution in 3D Heterogeneous Reactor

A. Elshin$^{(\boxtimes)}$

Alexandrov Research Institute of Technology, Sosnovy Bor, Russia
elchine@niti.ru

Abstract. This paper describes an application of the surface harmonics method to derivation of few-group finite difference equations for neutron flux distribution in a 3D triangular-lattice reactor model. The Boltzmann neutron transport equation is used as the original equation. Few-group finite difference equations are derived, which describe the neutron importance distribution (the multiplication factor in the homogeneous eigenvalue problem) in the reactor core. The derived finite difference equations remain adjoint to each other like the original equation of neutron transport and its adjoint equation. Non-diffusion approximations apply to calculation of a whole reactor core if we increase the number of trial functions for describing the neutron flux distribution in each cell and the size of the matrices of the few-group coefficients for finite difference equations.

1 Introduction

Finite difference methods are used in many areas of nuclear engineering including nuclear reactor physics calculations. This paper addresses the issues of deriving finite difference equations for calculation of the space and energy distribution of neutrons within a heterogeneous reactor. Neutron flux distribution in the reactor is described by the transport equation for neutron density. Neutron density depends on coordinates, neutron energy, and $\boldsymbol{\Omega}$ - the unit vector of neutron direction ($\mathbf{v} = \{E, \boldsymbol{\Omega}\}$ is used for brevity). Since the operating reactor is almost always in the steady-state (critical) condition, the so-called conditionally critical equation (see [1]) can be used for calculating the steady-state neutron flux distribution

$$\hat{L}\Phi(\mathbf{r}, \mathbf{v}) - \frac{1}{k_{eff}}\hat{K}_f\,\Phi(\mathbf{r}, \mathbf{v}) = 0 \tag{1}$$

where k_{eff} is the eigenvalue of the problem, the so-called effective multiplication factor in the reactor physics,

$$\hat{L}\Phi(\mathbf{r}, \mathbf{v}) = \boldsymbol{\Omega}\nabla\,\Phi(\mathbf{r}, \mathbf{v}) + \Sigma_t(\mathbf{r}, E)\,\Phi(\mathbf{r}, \mathbf{v}) - \int\limits_0^\infty \int\limits_{4\pi} \Sigma_s(\mathbf{r}, \mathbf{v}' \to \mathbf{v})\,\Phi(\mathbf{r}, \mathbf{v}')\,dE'\,d\Omega'$$

is the neutron transport operator, $\hat{K}_f\Phi(\mathbf{r}, \mathbf{v}) = \frac{1}{4\pi}\int\limits_0^\infty \int\limits_{4\pi} \chi(\mathbf{r}, E, E')\nu_f\Sigma_f(\mathbf{r}, E')$

$\Phi(\mathbf{r}, \mathbf{v}')\,dE'\,d\Omega'$ is the fission neutron generation operator.

© Springer International Publishing Switzerland 2015
I. Dimov et al. (Eds.): FDM 2014, LNCS 9045, pp. 169–176, 2015.
DOI: 10.1007/978-3-319-20239-6_16

In reactor physics the homogenization approach is now most widely used to calculate neutron fluxes in nuclear reactor cores. This approach is based on a two-step calculation scheme: first, certain neutron flux distributions in cells of a reactor core model are calculated and used for "homogenization" of the cells and, second, the diffusion equation for the core with piecewise-constant properties is solved. The main approximations of the homogenization method are the use of neutron flux distribution in a closed cell for "homogenization" and the use of diffusion approximation for calculation of a whole reactor core. A number of attempts have been made to avoid these approximations. An example is the surface harmonics approach which has been known since 80-es of XX century. This paper develops principles of the surface harmonics method and presents its implementation for deriving finite difference equations that describe the neutron flux and neutron importance distributions in a 3D triangular-lattice heterogeneous reactor core model. The neutron importance obeys the following adjoint equation to (1):

$$\hat{L}^+ \Phi^+(\mathbf{r}, \mathbf{v}) - \frac{1}{k_{eff}} \hat{K}_f^+ \Phi^+(\mathbf{r}, \mathbf{v}) = 0. \tag{2}$$

The possibility of obtaining similar equations for 2D core simulations is demonstrated in [2]. This paper relies upon reciprocity relations for neutron flux and neutron importance distributions at cell faces given in [2,3] and applies the method to 3D geometry with the angular momenta of neutron flux distribution at the cell faces being of higher order than those in the diffusion approximation case.

2   Derivation of Finite Difference Equations

Following the surface harmonics method (or homogenization method), the reactor core with the reflector is divided in plan into unit cells depending on the lattice geometry. The choice of the cell heights is determined by the reactor core subdivision along the height assuming homogeneity of the physical properties of the cell materials along the height or symmetry of the properties relative to the central cross-section plane of the cell.

Next, the neutron flux distribution in each cell is written as a linear combination of certain trial functions (the cell index is omitted):

$$\Phi_N(\mathbf{r}, \mathbf{v}) = \sum_{n=0}^{N-1} \sum_{g=1}^{G} A_{ng} \Psi_{ng}(\mathbf{r}, \mathbf{v}). \tag{3}$$

Equation (3) distinguishes summation over energy groups of neutrons (which are used for reactor calculations) because trial functions of the same type and their functionals are most well suited for generating vectors of length G and matrices of size $G \times G$. Index n designates types of trial functions $\Psi_{ng}(\mathbf{r}, \mathbf{v})$ (the types of trial functions are ordered symmetrically, N types are used in this case), A_{ng} are amplitudes of trial functions.

It is assumed that the trial functions in (3) obey neutron transport Eq. (1) inside the cell and certain non-uniform boundary conditions. Various (symmetrically ordered) combinations of cell-inlet and cell-outlet neutron fluxes are considered as boundary conditions for trial functions. Substitution of (3) into (1) gives equation residual (as δ-function) localized only at the cell boundaries. The reason is that the continuity of neutron flux distribution at the cell boundaries cannot be obtained with a finite number of trial functions. The condition of zeroing this residual (with weight of neutron importance):

$$\sum_{k=0}^{K} \int_{S_k} \int_{0}^{\infty} \int_{4\pi} (\mathbf{\Omega}, \mathbf{n}) \Phi_{Nk-}^{+}(\mathbf{r}, \mathbf{v})[\Phi_{Nk+}(\mathbf{r}, \mathbf{v}) - \Phi_{Nk-}(\mathbf{r}, \mathbf{v})]dSdEd\Omega = 0 \qquad (4)$$

gives equations which link the trial function amplitudes of neighbor cells. Here $\mathbf{n}$ (normal to the cell boundary) is outward-directed, an integral over cell boundaries written as the sum of integrals over the surface S_k of each k-th cell (K is the total number of cells in the reactor), $\Phi_{N\pm}(\mathbf{r}, \mathbf{v})$ are neutron (or neutron importance) distributions to the right (+) and left (−) of the boundary. A sub integral expression for the boundary face of the cell in (4) is not considered here. The neutron importance is a reasonably smooth function of coordinates, therefore, the coefficients of spherical harmonics (see [4]) serial expansion for neutron importance at the cell boundaries should diminish fast. Inserting the expansion into (4), using a recurrent formula for associated Legendre polynomial, grouping even and odd angular momenta of neutron flux and neutron importance distributions, expressing the integral over the cell boundary as the sum of integrals over the x, y faces (S_{ki}, $i = 1, 2, ..., M$ the number of faces) and z faces (S_{kj}, $j = 1, 2$) of the k-th cell faces , we obtain the following expression for residual (to within numerical factors which are not important for equation derivation):

$$\sum_{k=1}^{K} \sum_{l=0}^{\infty} \sum_{m=-l}^{l} \Big\{ \sum_{i=1}^{M} \int_{S_{ki}} \int_{0}^{\infty} \{ U_{Nki-}^{2l,m+}(\mathbf{r}, E)[\Phi_{Nki+}^{2l+1,m}(\mathbf{r}, E) - \Phi_{Nki-}^{2l+1,m}(\mathbf{r}, E)]$$

$$+ \Phi_{Nki-}^{2l+1,m+}(\mathbf{r}, E)[U_{Nki+}^{2l,m}(\mathbf{r}, E) - U_{Nki-}^{2l,m}(\mathbf{r}, E)] \}dSdE$$

$$+ \sum_{j=1}^{2} \int_{S_{kj}} \int_{0}^{\infty} \{ U_{Nkj-}^{2l,m+}(\mathbf{r}, E)[\Phi_{Nkj+}^{2l+1,m}(\mathbf{r}, E) - \Phi_{Nkj-}^{2l+1,m}(\mathbf{r}, E)]$$

$$+ \Phi_{Nkj-}^{2l+1,m+}(\mathbf{r}, E)[U_{Nkj+}^{2l,m}(\mathbf{r}, E) - U_{Nkj-}^{2l,m}(\mathbf{r}, E)] \}dSdE \Big\} = 0 \qquad (5)$$

In this expression, $\Phi_{Nki\pm}^{2l+1,m}(\mathbf{r}, E)$, $(\Phi_{Nki\pm}^{2l+1,m+}(\mathbf{r}, E))$ are odd angular momenta of the neutron flux (importance) distribution, linear combinations of even angular momenta of the neutron flux distribution, $(\Phi_{Nki\pm}^{2l,m}(\mathbf{r}, E)$ and $\Phi_{Nki\pm}^{2l+2,m}(\mathbf{r}, E))$ are denoted by $U_{Nki\pm}^{2l,m}(\mathbf{r}, E)$ and similar combinations of even angular momenta of the neutron importance are denoted by $U_{Nki\pm}^{2l,m+}(\mathbf{r}, E)$, the equation residual is written separately for (x, y) and z faces of each cell. Quantities $U_{Nki\pm}^{2l,m}(\mathbf{r}, E)$

$(U_{Nki\pm}^{2l,m+}(\mathbf{r}, E))$ will be hereinafter referred as neutron (importance) levels similarly like N.I. Laletin called the combination of total neutron flux density and the second angular momentum as the neutron level in his early papers (see e.g. [5]).

Previous studies on the surface harmonics method (see e.g. [6,7]) demonstrate that limiting the summation in (5) to terms with $l = 0$ (and $m = 0$) leads to finite difference equations which look like a finite difference approximation of the few-group diffusion equation. There are no technical constraints on a larger number of terms. In that case we will have to additionally sew together odd angular momenta of higher order and combinations of even angular momenta. Sewing conditions determine the choice of trial functions for neutron flux distribution. It is convenient to use unit odd angular momenta as boundary conditions. Before defining the angular momenta, their energy and spatial dependence at the cell boundaries should be determined. Definition of G for linearly independent spectra of odd momenta $\theta_{rg}^{2l+1,m}(\mathbf{r}, E)$, $\theta_{zg}^{2l+1,m}(\mathbf{r}, E)$, $g=1, 2, \ldots, G$ will determine the number of groups for the reactor calculation. Sets of neutron spectra can certainly be different at the (x, y) and z faces. We here restrict to the case of the triangular lattice with hexagonal unit cells and write the neutron flux distribution in the k-th cell in a vector form as follows (also see [6]):

$$\Phi_N^k(\mathbf{r}, \mathbf{v}) = \left(\varphi_k^r(\mathbf{r}, \mathbf{v})\right)' \mathbf{I}_k^r + \left(\varphi_k^z(\mathbf{r}, \mathbf{v})\right)' \mathbf{I}_k^z + \left(\psi_k^x(\mathbf{r}, \mathbf{v})\right)' \mathbf{J}_k^x + \left(\psi_k^y(\mathbf{r}, \mathbf{v})\right)' \mathbf{J}_k^y$$
$$+ \left(\psi_k^z(\mathbf{r}, \mathbf{v})\right)' \mathbf{J}_k^z + \left(\xi_k^x(\mathbf{r}, \mathbf{v})\right)' \mathbf{P}_k^x + \left(\xi_k^y(\mathbf{r}, \mathbf{v})\right)' \mathbf{P}_k^y + \left(\tau_k^z(\mathbf{r}, \mathbf{v})\right)' \mathbf{T}_k \qquad (6)$$

In the above expression the length of vectors G is the number of groups used for reactor calculations; the amplitudes of trial functions are arranged as column-vectors; for preserving the laws of matrix multiplication, column-vectors of trial functions are transposed (transposition is labeled by the accent mark: vector elements will be arranged in a row). Terms in the sum (trial functions) represent various schemes of neutron currents to and out of the cell (through x, y, and z faces, see Fig. 1). Relation (5) for residual also includes neutron currents and importance levels as a weight function. Let us write neutron importance in each cell by analogy with the neutron flux distribution (3), (6) as a linear combination of trial functions (assuming vector elements with $+$ index being arranged in a row):

$$\Phi_N^{k+}(\mathbf{r}, \mathbf{v}) = \mathbf{I}_k^{r+}\left(\varphi_k^{r+}(\mathbf{r}, \mathbf{v})\right)' + \mathbf{I}_k^{z+}\left(\varphi_k^{z+}(\mathbf{r}, \mathbf{v})\right)' + \mathbf{J}_k^{x+}\left(\psi_k^{x+}(\mathbf{r}, \mathbf{v})\right)' \mid \mathbf{J}_k^{y+}\left(\psi_k^{y+}(\mathbf{r}, \mathbf{v})\right)'$$
$$+ \mathbf{J}_k^{z+}\left(\psi_k^{z+}(\mathbf{r}, \mathbf{v})\right)' + \mathbf{P}_k^{x+}\left(\xi_k^{x+}(\mathbf{r}, \mathbf{v})\right)' + \mathbf{P}_k^{x+}\left(\xi_k^{x+}(\mathbf{r}, \mathbf{v})\right)' + \mathbf{T}_k^+\left(\tau_k^+(\mathbf{r}, \mathbf{v})\right)'$$
$$(7)$$

It is assumed that trial functions in (7) satisfy adjoint equation of neutron transport (2) and certain non-uniform boundary conditions. Symmetrically ordered combinations of cell-inlet and cell-outlet importance currents are considered as boundary conditions for trial functions in (6). Then Eq. (5) can be rewritten as:

$$\sum_{k=1}^{K}\left\{S_{rk}\sum_{i=1}^{M}\left[\mathbf{u}_{ki}^+(\mathbf{j}_{ik}-\mathbf{j}_{ki})+\mathbf{j}_{ki}^+(\mathbf{u}_{ik}-\mathbf{u}_{ki})\right]+S_{zk}\sum_{j=1}^{2}\left[\mathbf{u}_{kj}^+(\mathbf{j}_{jk}-\mathbf{j}_{kj})+\mathbf{j}_{kj}^+(\mathbf{u}_{jk}-\mathbf{u}_{kj})\right]\right\}=0, \quad (8)$$

where S_{rk} and S_{zk} are areas of the (x, y) and z faces of the k-th cell, respectively; vectors $\mathbf{j}_{lm}$ ($\mathbf{j}_{lm}^+$) denote face average neutron (importance) currents at the m-th face of the l-th cell, vectors $\mathbf{u}_{lm}$ ($\mathbf{u}_{lm}^+$) denote face average neutron (importance) levels at the m-th face of the l-th cell. To satisfy (8), we now equate the expressions preceding the amplitudes of coupled functions to zero and obtain the following set of equations ($k=1, 2, \ldots, K$) for computing amplitudes of trial functions:

$$\frac{1}{M} \sum_{i=1}^{M} \left[\varphi_{\mathbf{rk}}^{\mathbf{r}}(\mathbf{j}_{ik} - \mathbf{j}_{ki}) + \mathbf{u}_{ki} - \mathbf{u}_{ik} \right] + \frac{1}{2} \sum_{j=1}^{2} \varphi_{\mathbf{zk}}^{\mathbf{r}}(\mathbf{j}_{jk} - \mathbf{j}_{kj}) = 0 \qquad (9)$$

$$\frac{1}{2} \sum_{j=1}^{2} \left[\varphi_{\mathbf{zk}}^{\mathbf{z}}(\mathbf{j}_{jk} - \mathbf{j}_{kj}) + \mathbf{u}_{kj} - \mathbf{u}_{jk} \right] + \frac{1}{M} \sum_{i=1}^{M} \varphi_{\mathbf{rk}}^{\mathbf{z}}(\mathbf{j}_{ik} - \mathbf{j}_{ki}) = 0 \qquad (10)$$

$$\sum_{i=1}^{M} \left[\psi_{\mathbf{k}}^{\mathbf{r}}(\mathbf{j}_{ik} - \mathbf{j}_{ki}) + \mathbf{u}_{ki} - \mathbf{u}_{ik} \right] \cos \alpha_i = 0$$

$$\sum_{i=1}^{M} \left[\psi_{\mathbf{k}}^{\mathbf{r}}(\mathbf{j}_{ik} - \mathbf{j}_{ki}) + \mathbf{u}_{ki} - \mathbf{u}_{ik} \right] \sin \alpha_i = 0 \qquad (11)$$

$$\sum_{j=2}^{M} \left[\psi_{\mathbf{k}}^{\mathbf{z}}(\mathbf{j}_{jk} - \mathbf{j}_{kj}) + \mathbf{u}_{kj} - \mathbf{u}_{jk} \right] \cos \alpha_j = 0 \qquad (12)$$

$$\sum_{i=1}^{M} \left[\xi_{\mathbf{k}}(\mathbf{j}_{ik} - \mathbf{j}_{ki}) + \mathbf{u}_{ki} - \mathbf{u}_{ik} \right] \cos 2\alpha_i = 0$$

$$\sum_{i=1}^{M} \left[\xi_{\mathbf{k}}(\mathbf{j}_{ik} - \mathbf{j}_{ki}) + \mathbf{u}_{ki} - \mathbf{u}_{ik} \right] \sin 2\alpha_i = 0 \qquad (13)$$

$$\sum_{i=1}^{M} \left[\tau_{\mathbf{k}}(\mathbf{j}_{ik} - \mathbf{j}_{ki}) + \mathbf{u}_{ki} - \mathbf{u}_{ik} \right] \cos 3\alpha_i = 0 \qquad (14)$$

Matrices $\varphi_{\mathbf{r}}^{\mathbf{r}}, \varphi_{\mathbf{r}}^{\mathbf{z}}, \varphi_{\mathbf{z}}^{\mathbf{r}}, \varphi_{\mathbf{z}}^{\mathbf{z}}, \psi^{\mathbf{r}}, \psi^{\mathbf{z}}, \xi, \tau$ in the last expressions are formed from few-group first face (perpendicular to axis) average neutrons levels (combinations of zero-order and second-order angular momenta) calculated for corresponding trial functions:

$$\mathbf{h}_r = \frac{1}{S_r} \int_0^\infty \int_{S_{r1}} \theta^r(E) \big(\mathbf{U}_r^h(\mathbf{r}, E) \big)' dSdE, h = \varphi^r, \varphi^z, \psi^r, \xi, \tau,$$

$$\mathbf{h}_z = \frac{1}{S_z} \int_0^\infty \int_{S_{z1}} \theta^z(E) \big(\mathbf{U}_z^h(\mathbf{r}, E) \big)' dSdE, h = \varphi^r, \varphi^z, \psi^z. \qquad (15)$$

Similar matrices

$$\mathbf{h}_r^+ = \frac{1}{S_r}\int\limits_0^\infty\int\limits_{S_{r1}}(\mathbf{U}_r^{h+}(\mathbf{r},E))'\theta^r(E)dSdE,\, \mathbf{h}_z^+ = \frac{1}{S_z}\int\limits_0^\infty\int\limits_{S_{z1}}(\mathbf{U}_z^{h+}(\mathbf{r},E))'\theta^z(E)dSdE,$$

will appear in expressions for u_k^+ - face average neutron importance levels at the $(x,\,y)$ and z faces of the k-th cell. Expressions (9–14) were obtain with using reciprocity relations (see e.g. [3]): $\varphi_r^{r+} = \varphi_r^r$, $\varphi_z^{z+} = \varphi_z^z$, $\psi^{r+} = \psi^r$, $\psi^{z+} = \psi^z$, $\xi^+ = \xi$, $\tau^+ = \tau$, $\varphi_z^{r+} = \frac{MS_r}{2S_z}\varphi_r^z$, $\varphi_r^{z+} = \frac{2S_z}{MS_r}\varphi_z^r$.

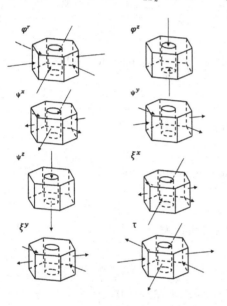

Fig. 1. Neutron currents for trial function

Generally speaking, set (9–14) is a closed algebraic system of equations which are solved to find amplitudes of all trial functions; knowing amplitudes and trial functions, the neutron flux distribution in each cell can be found. Writing the system (9–14) in a more convenient and clearer form is beyond the scope of our paper. However Eqs. (11), (12) can be expressed in the form that is close to the finite difference formulation of Ficks law. Eqs. (9), (10) can be used to construct a relation similar to the finite difference formulation of the diffusion equation for variables $\boldsymbol{\Phi}$ which are close to few-group cell average neutron flux densities, but with other formulas for calculating the diffusion coefficients, cross sections for generation, neutron absorption, and transition of neutrons from one energy group to another:

$$\frac{S_{rk}}{V_k}\hat{L}_k^r\boldsymbol{\Phi}_k + \frac{S_{zk}}{V_k}\hat{L}_k^z\boldsymbol{\Phi}_k - \boldsymbol{\Sigma}_k\boldsymbol{\Phi}_k = 0, \qquad (16)$$

where $\boldsymbol{\Sigma}_k$ - group cell macrosection matrix; $\hat{L}_k^r\boldsymbol{\Phi}_k = \sum (\psi_i + \psi_k)^{-1}(\boldsymbol{\Phi}_i - \boldsymbol{\Phi}_k)$, $\hat{L}_k^z\boldsymbol{\Phi}_k = \sum_{j=1}^2 (\psi_i^z + \psi_k^z)^{-1}(\mathbf{R}_i\boldsymbol{\Phi}_i - \mathbf{R}_k\boldsymbol{\Phi}_k)$ - finite-difference operators $\nabla D\nabla\boldsymbol{\Phi}_k$. A similar approach to the simplest case (2D geometry, symmetrical cells, one neutron group, neutron currents and neutron fluxes are sewed together at cell faces) is described in Chap. 9 of [9].

The above described approach can be applied to deriving finite difference equations for neutron importance in the reactor core. Transformations will yield a transposed system of equations of (9–14) type where square matrix multiplication by column–vector is replaced with raw–vector multiplication by the same square matrix. For example Eq. (16)-type is written as

$$\frac{S_{rk}}{V_k}\hat{L}_k^{r+}\boldsymbol{\Phi}_k^+ + \frac{S_{zk}}{V_k}\hat{L}_k^{z+}\boldsymbol{\Phi}_k^+ - \boldsymbol{\Phi}_k^+\boldsymbol{\Sigma}_k = 0,$$

where

$$\hat{L}_k^{r+}\Phi_k = \sum_{i=1}^{M}(\Phi_i^+ - \Phi_k^+)(\psi_i + \psi_k)^{-1},$$

$$\hat{L}_k^{z+}\Phi_k = \sum_{j=1}^{2}(\Phi_i^+ \mathbf{R}_i - \Phi_k^+ \mathbf{R}_k)(\psi_i^z + \psi_k^z)^{-1},$$

variables Φ are close to few-group cell average neutron importance.

3 Consideration of Higher Angular Momenta at Cell Faces

To construct finite difference equations, we used trial functions representing normal projections of neutron currents at the cell faces (when other odd angular momenta at the cell faces are equal to zero). Various schemes of higher, e.g. third-order angular momenta at the cell faces also can be used. With higher momenta, the size of matrices and vectors in (6), (7) is increased. A formal derivation and form of Eqs. (9–14) or (16) remains absolutely unchanged. When trial functions for defining third-order angular momenta such as $\Phi^{3,0}(\mathbf{r}_s, E)$ at the cell faces are added to description of neutron distribution in the cell, amplitudes of these trial functions are also added to corresponding vectors in (6), (7), thus increasing the length of the vectors of spectra and levels used in calculations of h-type matrices (15):

$$\mathbf{I} = \begin{pmatrix} \Phi_1^{1,0} \\ \Phi_2^{1,0} \\ \cdots \\ \Phi_G^{1,0} \\ \Phi_1^{3,0} \\ \Phi_2^{3,0} \\ \cdots \\ \Phi_G^{3,0} \end{pmatrix}, \quad \theta(E) = \begin{pmatrix} \theta_1^{1,0}(E) \\ \theta_2^{1,0}(E) \\ \cdots \\ \theta_G^{1,0}(E) \\ \theta_1^{3,0}(E) \\ \theta_2^{3,0}(E) \\ \cdots \\ \theta_G^{3,0}(E) \end{pmatrix}, \quad \mathbf{U}^h(\mathbf{r}, E) = \begin{pmatrix} U_1^{h1,0}(\mathbf{r}, E) \\ U_2^{h1,0}(\mathbf{r}, E) \\ \cdots \\ U_G^{h1,0}(\mathbf{r}, E) \\ U_1^{h3,0}(\mathbf{r}, E) \\ U_2^{h3,0}(\mathbf{r}, E) \\ \cdots \\ U_G^{h3,0}(\mathbf{r}, E) \end{pmatrix}.$$

As [8] shows, reciprocity relations also hold for type matrices of increased size and derivation of equation system (9–14) or (16) will not change in this case. So, system of Eqs. (9–14) or (16) avoids using diffusion approximations in calculation of a whole reactor core. It is advantageous that this avoidance is implemented by only increasing the matrix size with the equations remaining unchanged in form.

4 Conclusion

The present paper applies the surface harmonics method to obtaining a system of finite difference equations for description of neutron flux distribution in the

3D heterogeneous reactor with triangular lattice. The presented approach is also used to derive an adjoint system of finite difference equations for neutron importance distribution in the reactor (solution of an adjoint equation for neutron transport). The finite difference equations are derived by sewing together the linear combinations of angular momenta of neutron distribution or the angular momenta neutron importance distributions at the cell faces. Transition to higher-order approximations (no diffusion approximation needed) within the derived system of equations is implemented by only increasing the size of the coefficient matrices of equations and, consequently, the number of trial functions used for description of neutron distribution in the unit cells of the heterogeneous reactor. No homogenization is used because the method first solves a system of finite difference equations (with a spatial step corresponding to a lattice pitch) and then constructs a detailed neutron flux distribution in the reactor cells.

References

1. Bell, G.J., Glasstone, S.: Nuclear Reactor Theory. Atomizdat, Moscow (1974). in Russian
2. Elshin, A.: The evolution of the surface harmonic method to derive equations for a distribution of neutrons and their importance in a heterogeneous reactor. In: Proceeding of International Topical Meeting on Mathematics and Computations, Supercomputing, Reactor Physics and Nuclear and Biological Applications, Avignon, France (2005)
3. Elshin, A.: Obtaining finite difference equations for a heterogeneous reactor with spatial kinetics. AtomnayaEnergiya **103**(4), 222–232 (2007)
4. Corn, G.G., Corn, T.: Mathematical handbook for scientists and engineers. McGraw-Hill Book Company, Inc., New York, Toronto, London (1961)
5. Laletin, N.I.: Basic principles for developing equations for heterogeneous reactors a modification of the homogeneous method. Nucl. Sci. Eng. **85**, 133–138 (1983)
6. Laletin, N.I., Elshin, A.V.: System of refined finite difference equations for 3D heterogeneous reactor model. Atomnaya Energiya **60**(2), 96–99 (1986). (in Russian)
7. Boyarinov, V.F.: 3D equations of heterogeneous reactors in the method of surface harmonics with one unknown quantity per cell-group. Atomnaya Energiya **72**(2), 227–231 (1992). (in Russian)
8. Elshin, A.V., Abdullayev, A.M.: On reciprocity relations and high order approximations in the surface harmonics method. Proceeding of XXIII inter-departmental meeting: Neutronic problems of nuclear power plants with closed fuel cycle (Neutronics-2012), in two volumes. Obninsk, IPPE. 2, pp. 515–521 (2013) (in Russian)
9. Methods for calculation of thermal neutron fields in reactor lattices (ed. by Ya. V. Shevelev, PhD), Moscow, Atomizdat, 216–239 (1974)(in Russian)

Matrix-Free Iterative Processes for Implementation of Implicit Runge–Kutta Methods

Boris Faleichik$^{(\boxtimes)}$ and Ivan Bondar

Belarusian State University, 220030 Minsk, Belarus
faleichik@bsu.by

Abstract. In this work we present so-called generalized Picard iterations (GPI) – a family of iterative processes which allows to solve mildly stiff ODE systems using implicit Runge–Kutta (IRK) methods without storing and inverting Jacobi matrices. The key idea is to solve nonlinear equations arising from the base IRK method by special iterative process based on the idea of artificial time integration. By construction these processes converge for all asymptotically stable linear ODE systems and all A-stable base IRK methods at arbitrary large time steps. The convergence rate is limited by the value of "stiffness ratio", but not by the value of Lipschitz constant of Jacobian. The computational scheme is well suited for parallelization on systems with shared memory. The presented numerical results exhibit that the proposed GPI methods in case of mildly stiff problems can be more advantageous than traditional explicit RK methods.

Keywords: Runge–Kutta methods · Stiff problems · Parallel methods

1 Generalized Picard Iterations

Consider an initial value problem for the system of ordinary differential equations

$$y'(t) = f(t, y(t)), \quad y(t_0) = y_0, \tag{1}$$

where $y : \mathbb{R} \to \mathbb{R}^n$, $f : \mathbb{R} \times \mathbb{R}^n \to \mathbb{R}^n$, and some base s-stage implicit Runge–Kutta (RK) method for this problem:

$$y_1 = y_0 + \tau \sum_{i=0}^{s} b_i f(t_0 + c_i \tau, Y_i).$$

Here $y_1 \approx y(t_0 + \tau)$, and $Y_i \in \mathbb{R}^n$ are unknown stage values which satisfy the following system of nonlinear equations: $Y_i = y_0 + \tau \sum_{j=0}^{s} a_{ij} f(t_0 + c_j \tau, Y_j)$, $i = 1, \ldots, s$. We use standard notation for RK method coefficients $\left(a_{ij}\right)_{i,j=1}^{s} = A$,

© Springer International Publishing Switzerland 2015
I. Dimov et al. (Eds.): FDM 2014, LNCS 9045, pp. 177–184, 2015.
DOI: 10.1007/978-3-319-20239-6_17

$(b_1, \ldots, b_s)^T = b$, $(c_1, \ldots, c_s)^T = c$. In practice it is handy to perform a standard change of variables to minimize roundoff issues: $Z_i = Y_i - y_0$:

$$r_i(Z) = -Z_i + \tau \sum_{j=1}^{s} a_{ij} f(t_0 + c_j\tau, y_0 + Z_j) = 0, \quad i = 1, \ldots, s, \qquad (2)$$

or simply

$$r(Z) = (r_1(Z), \ldots, r_s(Z)) = 0, \quad Z = (Z_1, \ldots, Z_s)^T. \qquad (3)$$

Our goal is to construct a method for matrix-free solution of (3), i.e. without storing and inverting Jacobian matrix of f, which is usually done when Newton's or similar methods are used. In [1,2] we proposed a family of such methods, which is called generalized Picard iterations. Let's give a brief description of the approach.

The idea is to use artificial time integration of an 'embedding' differential equation $Z' = r(Z)$ by some *auxiliary* explicit one-step method of RK type with constant artificial time step ω. This results in the process which we shall call the *generalized Picard iteration* (GPI). Its general form is simply

$$Z^{[m+1]} = \Phi(Z^{[m]}), \quad m = 1, 2, \ldots \qquad (4)$$

where Φ is the time-stepping mapping of the auxiliary method. The key task now is to define this mapping, i.e. to determine the coefficients of the auxiliary method. We perform this by optimizing the convergence behavior of (4) on linear problems. To make this precise we shall give the following definition.

Definition 1. *Consider the linear model ODE $y'(t) = \lambda y(t)$, $\lambda \in \mathbb{C}$, and corresponding GPI process (4) for the solution of induced RK equation of the form (3). A region $D \subset \mathbb{C}$, such that (4) converges for all $\lambda\tau \in D$ is called the linear convergence region of (4).*

By substituting $f(t, y) = \lambda y$ in (2) we have

$$r(Z) = (\lambda\tau A - I)Z + g_0, \quad Z \in \mathbb{R}^s, \qquad (5)$$

where $g_0 = \lambda\tau A(y_0 1_s)$, $1_s = (1, \ldots 1)^T \in \mathbb{R}^s$. In this case we have

$$\Phi(Z) = R(\omega(\lambda\tau A - I))Z + Q(\omega, \lambda\tau A - I)g_0, \qquad (6)$$

where R is the stability polynomial of the auxiliary method, $Q(\omega, z) = (R(\omega z) - 1)/z$. According to the convergence criterion of linear fixed-point iterations, the linear convergence region of GPI (4) in this case will be

$$D = \bigcap_{i=1}^{s} \mu_i^{-1}(\omega^{-1}S + 1), \qquad (7)$$

where $\{\mu_i\} = \Sigma(A)$, $\Sigma(\cdot)$ is spectrum of a matrix, and S is the stability region of the auxiliary method: $S = \{z \in \mathbb{C} : |R(z)| < 1\}$. Furthermore, the convergence

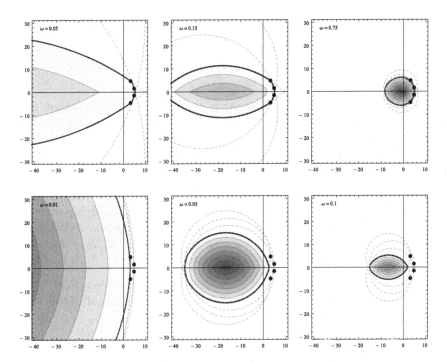

Fig. 1. Convergence regions of ordinary (upper) and 'preconditioned' (lower) GPI methods for base RadauIIA 4-stage method and auxiliary explicit Euler method. Black points are μ_i^{-1}. Contour lines correspond to the constant values of convergence factor $K(z)$ (8), (10).

factor of GPI process for (6) is determined by the spectral radius of $R(\omega(\lambda\tau A - I))$, which is equal to $K(\lambda\tau)$, where

$$K(z) = \max_i |R(\omega(z\mu_i - 1))|. \tag{8}$$

The examples of convergence regions for GPI based on (2) with 7th order RadauIIA base method [3, Sect. IV.5] and explicit Euler method being the auxiliary method are shown in the first row of Fig. 1. We see that generally as $\omega \to 0$ the area of D increases, but the overall convergence factor grows significantly.

In order to improve the situation instead of (2) we consider the 'preconditioned' RK system

$$r_i(Z) = -\sum_{j=1}^{s} \tilde{a}_{ij} Z_j + \tau f(t_0 + c_i\tau, y_0 + Z_i) = 0, \quad i = 1, \ldots, s, \tag{9}$$

where $(\tilde{a}_{ij}) = \tilde{A} = A^{-1}$. If A is not singular this system is equivalent to (2), but the corresponding GPI process (4) behaves much better for $\lambda\tau \ll 0$ (in stiff case). Indeed, in scalar linear case instead of (5) now we have

$$r(Z) = (\lambda\tau I - \tilde{A})Z + \lambda\tau \mathbf{1}_s y_0,$$

and the convergence region and convergence factor become respectively

$$D = \bigcap_{i=1}^{s}(\mu_i^{-1} + \omega^{-1}S) \quad \text{and} \quad K(z) = \max_i |R(\omega(z - \mu^{-1}))|, \qquad (10)$$

see the second row in Fig. 1. We see that the preconditioned equation (9) is better to use in stiff case, but for $\lambda\tau \approx 0$ the ordinary RK equation (2) should be used.

The simple analysis of the preconditioned GPI process allows to prove the next important property.

Proposition 1. *Let the base IRK method be A-stable and there exists $r_0 > 0$ such that the open disk of radius r_0 and center in $(-r_0, 0)$ is entirely covered by the stability region of the auxiliary RK method. Then for any linear ODE system $y' = Jy$ with $\Sigma(J) \subset \mathbb{C}^-$ and any time step $\tau > 0$ there exists $\omega_0 > 0$ such that the preconditioned GPI iterations (4), (9) converge for all $\omega \in (0, \omega_0)$.*

As we see, in order to achieve faster convergence of GPI we need $|R(z)|$ to take minimal possible values over the whole stability region S. In light of this we use the following scheme of auxiliary method construction:

1. Select the desired shape of stability region $\Omega \approx S$ taking the condition $\Sigma\left(\frac{\partial f}{\partial y}(t_0, y_0)\right) \subset D$ as a reference point, see (7) and (10).
2. Choose σ – the desired number of stages for the auxiliary method, and construct a stability polynomial R of degree σ basing on the condition[1] $\iint_{\Omega} |R(z)|^2 dz \to \min$. In our experiments we use minimization over an angular sector in the left halfplane:

$$\int_0^1 \int_{\pi-\theta}^{\pi+\theta} |R(\rho e^{i\varphi})|^2 d\varphi d\rho \to \min.$$

We solve this problem numerically with higher-precision arithmetic using *Mathematica* system. For example, if the spectrum of Jacobian is close to the real axis we take $\theta = \pi/180$ and for $\sigma = 7$ get a stability polynomial which stability region is shown in Fig. 2.
3. Build an explicit RK scheme which implements the constructed stability polynomial. This step can be performed in variety of ways. We use Lebedev's approach briefly described in [3], see also [1]: the factorized stability polynomial $R(z) = R_1(z)R_2(z)\ldots R_M(z)$ yields the representation of Φ as a composition of one and two-stage methods: $\Phi = \Phi_1 \circ \Phi_2 \circ \ldots \circ \Phi_M$, where

$$R_k(z) = \begin{cases} 1 + \delta z & \text{for } odd\ \sigma \text{ and } k = 1, \\ (1 + \delta_k z)(1 + \delta_k' z), & \text{in quadratic case;} \end{cases}$$

---

[1] We use this kind of optimization mostly for simplicity reasons. Of course, in general case this condition does not imply $|R(z)| < 1\ \forall z \in \Omega$, so special care should be taken here.

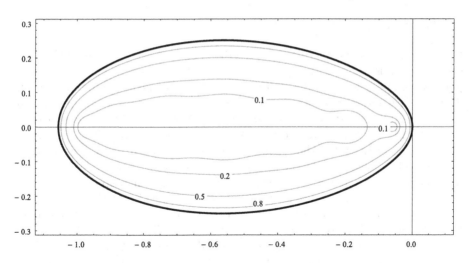

Fig. 2. Stability region of 7-stage auxiliary method optimized with $\theta = \pi/180$

$$
\begin{aligned}
&\Phi_k(X) = X + \omega\delta r(X), && \text{if } \deg R_k = 1; \\
&\Phi_k(X) = g_{2,k} - h\alpha_k\gamma_k(r_{2,k} - r_{1,k}), && \text{if } \deg R_k = 2, \text{where} \\
& && r_{1,k} = r(X), \\
& && g_{1,k} = X + h\alpha_k r_{1,k}, \\
& && r_{2,k} = r(g_{1,k}), \\
& && g_{2,k} = g_{1,k} + h\alpha_k r_{2,k}.
\end{aligned}
$$

Here $\alpha_k = (\delta + \delta')/2$, $\gamma_k = 1 - \delta\delta'/\alpha_k^2$, $k = 1,\dots,M$, and δ are the coefficients which uniquely determine the auxiliary method mapping Φ.

In practice we perform iterations of GPI process (4) until the estimated error $\|Z^{[m]} - Z^*\|$, where $r(Z^*) = 0$, is less than $0.05 \times Atol$. Here $Atol$ is, as usual, the required tolerance for the local error $y_1 - y(t_0 + \tau)$. The error estimation technique is based on the estimation of the convergence factor Θ as described in [3, Section IV.8].

It is important to mention that the resulting method $y_1^{[m]} = y_0 + \tau\sum_{i=1}^{s} f(t_0 + c_i\tau, y_0 + Z_i^{[m]})$ is equivalent to some explicit RK method of order one at least. Though instead of this form we use

$$
y_1^{[m]} = y_0 + \sum_{i=1}^{s} d_i Z_i^{[m]},
$$

where $(d_1,\dots,d_s)^T = b^T A^{-1}$, which gives method of only order zero, but performs better on stiff problems (see [1] for details).

2   Numerical Experiments

Our experimental code based on GPI is written in C++ and has a parallelization option, which is implemented using OpenMP. If this option is enabled the independent components r_i of the residual function r (3) are evaluated in parallel. The step size and error control is implemented in a standard way by using two methods of different order. In our case these are 4-stage RadauIIA and Gaussian methods of order 7 and 8 respectively. Since both of them are collocation methods, we effectively exploit the continuous polynomial approximation which they provide: this polynomial is used for predicting the initial approximation $Z^{[0]}$ for the error controller method and for the main method on new steps.

The Jacobian spectral radius estimate should be provided by the user in order to properly select the value of auxiliary time step ω. In our tests we compute this estimate on each step using Gershgorin theorem. This estimate is also used for switching between 'stiff' and 'non-stiff' GPI methods. In stiff case we use the 'preconditioned' residual function (9) with 7-stage auxiliary method from Fig. 2. In non-stiff case we use ordinary residual (2) with explicit Euler auxiliary method which linear convergence regions have been shown in the first row of Fig. 1.

Further details of the implementation the interested reader can find in [1].

Since GPI methods are actually explicit, in its current state our code can not compete with implicit methods in cases when Newton's method is applicable. That's why we compare the performance of our code with highly-regarded DOP853 code, which implements explicit Dormand-Prince RK methods with variable order and is applicable in case of mildly stiff problems. We used C language version of this code[2]. Each diagram shows results of the solvers with required absolute tolerance settings $Atol = 10^{-i}$, $i = 2, 3, \ldots, 10$. The actual absolute error at the endpoint and elapsed CPU time measured in seconds are depicted in logarithmic scales.

The experiment was performed on a machine with 4-core Intel Core 2 Quad Q6600 2.4 GHz processor and Linux operating system.

2.1   HIRES Problem

The first test problem is the well-known HIRES problem which describes a chemical reaction of photomorphogenesis [3, Section IV.10]. This is a system of 8 nonlinear ODEs. The endpoint of integration is 421.8122, the reference solution was downloaded from E. Hairer's webpage[3]. The results of the experiment are shown in Fig. 3. We see that for moderate tolerances the serial GPI code outperforms DOP853, which is quite surprising. The parallel version works much slower, which is expected, since the dimension of the system is too small and thus the parallelization overhead is higher than the speed-up.

One may also note the unnatural behavior of GPI codes for $Atol = 0.01$, which means that the error controlling mechanism needs to be tweaked.

---

[2] http://www.unige.ch/~hairer/prog/nonstiff/cprog.tar.

[3] http://www.unige.ch/~hairer/testset/stiff/hires/res_exact_pic.

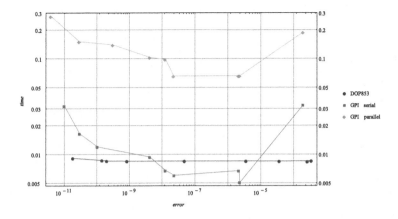

Fig. 3. HIRES problem test results.

2.2   BRUSS2D Problem

The second is another classic test problem BRUSS-2D which is a method-of-lines discretisation of two-dimensional parabolic reaction-diffusion PDE [3, Section IV.10]. We solved this problem on two spatial grids: 64×64 (Fig. 4) and 128×128 (Fig. 5). The value of the diffusion coefficient α is 1 in both cases.

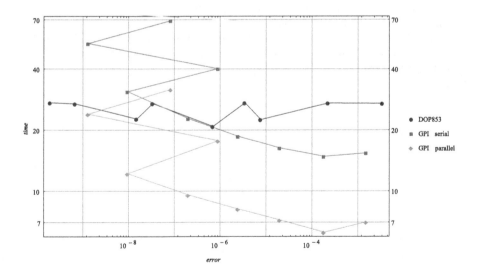

Fig. 4. BRUSS-2D problem with $N = 64$, $\alpha = 1$. The dimension of ODE is 8192.

For both of these tests the parallel version of GPI (running on 4 processors) was approximately 2.5 times faster than the serial.

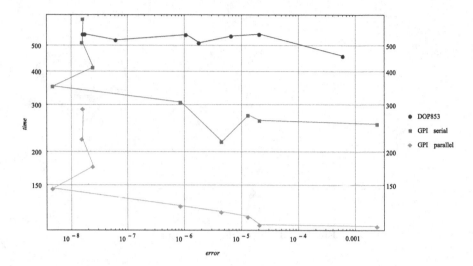

Fig. 5. BRUSS-2D problem with $N = 128$, $\alpha = 1$. The dimension of ODE is 32768.

References

1. Faleichik, B., Bondar, I., Byl, V.: Generalized Picard iterations: a class of iterated Runge-Kutta methods for stiff problems. J. Comp. Appl. Math. **262**, 37–50 (2013)
2. Faleichik, B.V.: Explicit implementation of collocation methods for stiff systems with complex spectrum. J. Numer. Anal. **5**(1–2), 49–59 (2010)
3. Hairer, E., Wanner, G.: Solving ordinary differential equations II. In: Hairer, E., Wanner, G. (eds.) Stiff and Differential-Algebraic Problems, 2nd edn. Springer, Heidelberg (1996)

Simulation of Technogenic and Climatic Influences in Permafrost for Northern Oil Fields Exploitation

M.Yu. Filimonov[1,2(✉)] and N.A. Vaganova[1]

[1] Institute of Mathematics and Mechanics of Ural Branch of Russian Academy of Sciences, Ekaterinburg, Russia
{fmy,vna}@imm.uran.ru
[2] Ural Federal University, Ekaterinburg, Russia

Abstract. In this paper a mathematical model for simulation of thermal fields from wells located in permafrost area is considered, which takes into account basic physical, technological, and climatic factors that lead to a nonlinear boundary condition on the surface of the soil. To find the thermal fields a finite-difference method is used to solve the problem of Stefan type, and solvability of the corresponding difference problem is proved. Possibilities of the developed software are presented to carry out various numerical experiments and make long-term forecasts in simulations of thermal fields in the system "well – permafrost" with annual cycle of thawing/freezing the upper layers of the soil due to seasonal temperature changes, intensity of solar radiation and technical parameters of the wells. Comparison of numerical and experimental data are in good agreement (difference is about 5 %) due to, in particular, that the software adapts to the geographic location by using special iterative algorithm of determination of the parameters, included in the non-linear boundary condition on the soil surface.

1 Introduction

In Russia, permafrost, the soils which are conserved with frozen state during long time, occupies a total area of 10 million km^2, up to 65 % of the territory. These areas are extremely important for Russian economy, as here there are produced about 93 % of Russian natural gas and 75 % of oil, which in monetary terms, provides up to 70 % of exports. Thawing of ice-rich soils will be followed by the earth's surface subsidence and activation of cryogenic hazardous geological processes, called thermokarst. It is found that not only climate change, but human activities lead to permafrost thawing.

According to Russian Constructive Standards, before developing an oil and gas field it is necessary that the soil surface at the work site must be first covered by a riprap of about two meters thickness, which usually consists of penoplex, sand layer and a concrete slabs.

Then on the prepared work site, a number of wells may be placed with dependence of production rate. The following restriction is used: two wells cannot

© Springer International Publishing Switzerland 2015
I. Dimov et al. (Eds.): FDM 2014, LNCS 9045, pp. 185–192, 2015.
DOI: 10.1007/978-3-319-20239-6_18

be drilled at a distance lower than two radii of thawing from each other (i.e., lower than the distance which will be covered by zero isotherm for 25–30 years of operation of a single well).

Thus, the problem of the optimal planning and designing work sites is very important, as to deliver materials for construction is expensive by flights, because of absence of land-ways in northern and polar regions of oil and gas fields.

2   Mathematical Model

We consider a problem of Stefan type for heat propagation in frozen soil from heated well(tube) which may be insulated, taking into account phase transition. Numerical methods for solving problems of heat conductivity are the most effective and universal methods of researching such processes. A large number of works are devoted to development of numerical methods for solving boundary value problems of heat conduction. Basics of finite difference methods are detailed in the works [1,2]. It is known that the difference methods for solving mathematical problems provide highly accurate results, take into account a large number of parameters and do not require rigid restrictions and assumptions. However, as a rule, computational codes are cumbersome, and analysis of results is difficult. On the other hand, the existing analytical solutions (for example, [3]), and a numerically–analytical methods [4,5] cannot usually consider real boundary conditions and can be used only for testing numerical methods.

At present there are the following difference methods for solving problems of Stefan type: a method of front catching by difference grid node, a method of fronts straightening, a method of smoothing factors and through calculation methods [6], and others. The method of front catching by grid node is used only for one-dimensional single-front problems, and straightening method only for multiple-front problems. A essential feature of these methods is that the finite difference schemes are constructed with explicit front of phase transition. It should be noted that the methods with explicit fixing of phase transition are not suitable in the case of temperature cycling on the boundary, because there may be several non-monotonic moving fronts, and some of them may merge with each other or disappear. In [7] a through calculation scheme developed with smoothing of discontinuous phase transition coefficients in heat conduction equation. Through calculation scheme is characterized by the phase transition boundary is not determined explicitly, and homogeneous difference schemes may be used. The heat of phase transformation is introduced with using Dirac δ-function as a concentrated heat of phase transition in the specific heat ratio. The obtained discontinuous function then "shared" with respect to temperature, and does not depend on the number of measurements, phases, and fronts.

The models which are most similar to the results of this work, we have to mention ones with using a one-dimensional heat equation taking into account a variety of factors: snow cover, vegetation etc. A detailed review of these models is presented in [8], but it is not assumed that there is any engineering systems installed in the permafrost zone. When taking into account solar radiation and the presence of snow cover it was thought that short-wave part of the radiation can penetrate through snow cover to a considerable depth, varying by Bouguer–Lambert law. In our three-dimensional model snow cover, vegetation, solar radiation per year, as well as other climatic factors are taken into consideration by using a special iterative algorithm of some coefficients variating, which are a terms of non-linear boundary condition on the surface of the soil. However, these parameters can be found more accurately, if there is additional information about temperature distribution up to depth 10–11 m for a season for this temperature. This approach allows to makes it easier to compose an initial data that are often cannot be accurately, or, for example, it is impossible to measure the thickness of snow cover, and determine the snow properties depending on solar radiation, etc.

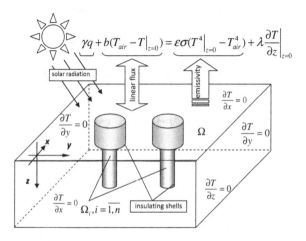

Fig. 1. The main heat flows and boundary conditions

In this paper the results of numerical simulations of thermal fields from two wells are presented. The computational domain is a three-dimensional box in which the vertical wells are considered as inner boundaries. The considered soil is possible to be non-homogenous and may include a number of elements, such as ice-rock lenses, different layers, engineering constructions (riprap), and insulating layers around the wells.

As a basic mathematical model with including localized heat of phase transition is considered – an approach to solve the problem of Stefan type, without the explicit separation of the phase transition [6]. The heat of phase transformation is introduced with using Dirac δ-function as a concentrated heat of phase transition in the specific heat ratio.

First let consider heat exchange on a flat ground surface directly illuminated by the sun. Let the initial time be $t_0 = 0$, and the ground is a box Ω and $T_0(x, y, z)$ is initial temperature. The computational domain is a three-dimensional box, where x and y axes are parallel to the ground surface and the

z axis is directed downward. We assume that the size of the box Ω is defined by positive numbers L_x, L_y, L_z: $-L_x \leq x \leq L_x$, $-L_y \leq y \leq L_y$, $-L_z \leq z \leq 0$. Let $T = T(t, x, y, z)$ be soil temperature at the point (x, y, z) at the time moment t. The main heat flow associated with climatic factors on the surface $z = 0$ is shown in Fig. 1.

The ground surface z=0 is a main zone of formation of natural thermal fields. On this surface the equation of balance of flows is used as a boundary condition, with taking into account the main climate factors: air temperature and solar radiation. $T_{air} = T_{air}(t)$ denotes the temperature in the surface layer of air, which varies from time to time in accordance with the annual cycle of temperature; $\sigma = 5,67 \cdot 10^{-8}$ $W/(m^2 K^4)$ is Stefan–Boltzmann constant; $b = b(t, x, y)$ is heat transfer coefficient; $\lambda = \lambda(T)$ is thermal conductivity coefficient; $\varepsilon = \varepsilon(t, x, y)$ is the coefficient of emissivity. The coefficients of heat transfer and emissivity depend on the type and condition of the soil surface. Ω can include a number of engineering structures. Suppose that in Ω there are 2 objects that are heat sources (for example producing insulated wells). We denote the surface of these objects by $\Omega_i = \Omega_i(x, y, z)$, $i = 1, 2$, (see Fig. 1).

Thus, the modeling of thawing in the soil is reduced to the solution in Ω of the following heat equation [6]:

$$\rho\big(c_\nu(T) + k\delta(T - T^*)\big)\frac{\partial T}{\partial t} = \nabla\left(\lambda(T)\Delta T\right), \tag{1}$$

where ρ is density $[kg/m^3]$, T^* is temperature of phase transition $[K]$,

$$c_\nu(T) = \begin{cases} c_1(x, y, z), T < T^*, \\ c_2(x, y, z), T > T^*, \end{cases} \text{is specific heat [J/kg K],}$$

$$\lambda(T) = \begin{cases} \lambda_1(x, y, z), T < T^*, \\ \lambda_2(x, y, z), T > T^*, \end{cases} \text{is thermal conductivity coefficient [W/m K],}$$

$k = k(x, y, z)$ is specific heat of phase transition, δ is Dirac delta function. Thus, it is necessary to solve equation (1) in the area Ω with initial condition

$$T(0, x, y, z) = T_0(x, y, z) \tag{2}$$

and boundary conditions [9,10]

$$\gamma q + b(T_{air} - T(x, y, 0, t)) = \varepsilon\sigma(T^4(x, y, 0, t) - T_{air}^4) + \lambda\frac{\partial T(x, y, 0, t)}{\partial z}, \tag{3}$$

$$T\Big|_{\Omega_{10}} = T_1(t), \quad T\Big|_{\Omega_{20}} = T_2(t), \tag{4}$$

$$\frac{\partial T}{\partial x}\Big|_{x=\pm L_x} = \frac{\partial T}{\partial y}\Big|_{y=\pm L_y} = \frac{\partial T}{\partial z}\Big|_{z=-L_z} = 0. \tag{5}$$

Boundary condition (2) determines an initial distribution of soil temperature at the time moment from which we plan to start the numerical calculation. Condition (3) is obtained from the balance the heat fluxes at the ground surface $z = 0$. Here total solar radiation $q(t)$ is the sum of direct solar radiation. Soil is absorbed only part of the total radiation which equal to $\gamma q(t)$, where $\gamma = \gamma(t, x, y)$ is the part of energy that is formed to heat the soil, which in general depends on atmospheric conditions, angle of incidence of solar radiation, i.e. latitude and time.

Thus, the simulation of heat transfer in three-dimensional domain with the phase transition is reduced to solving the initial-boundary value problem (1)–(5).

3   Numerical Results

The base of this numerical method is an algorithm with good reliability in finding thermal fields of underground pipelines [14–16], but in view of specificity, related to the possible phase transitions in the soil [9–12].

On the base of ideas in [6] a finite difference method is used with splitting by the spatial variables in three-dimensional domain to solve the problem for Eq. (1) in Ω. We construct an orthogonal grid, uniform, or condensing near the ground surface or to the surfaces of Ω_1 and Ω_2.

In computations the nonlinear boundary condition is approximated, as a rule, by a linear one or with using an iterative process of calculations. At present implicit finite difference schemes, which leads to solving nonlinear equation, are not used in investigations devoted to direct numerical simulation of heat propagation problems with nonlinear boundary conditions.

To solve a difference system for Eq. (1) a sweep method is considered [13]. In [15] solvability of the system with nonlinear difference boundary condition was proved.

With using the model (1)-(5) numerical calculations were carried out for Western Siberia region (Suzun oil and gas field). Monthly average solar radiation and air temperature used in computations are consistent with the data in Atmospheric Science Data Center (ASDC) at NASA Langley Research Center.

As a basic soil we will use a loam with the following parameters. Thermal conductivity: frozen — 1.82 W/(m K), melted — 1.58 W/(m K), volumetric heat: frozen — 2130 kJ/(m^3 K), melted — 3140 kJ/(m^3 K), volumetric heat of phase transition — 1.384·10^5 kJ/(m^3 K). The background temperature of permafrost is -0.7°C, except for the layer of seasonal thawing (freezing) of soil.

There is a layer of riprap of 2.5 m. The riprap consists of three layers: penoplex (0.2 m), sand (2.0 m) and the concrete slab on the top (0.3 m). Parameters: concrete slab with density 2500.0 kg/m^3, thermal conductivity 1.69 W/(m K), specific heat 0.84 kJ/(kg K); sand with density 1600.0 kg/m^3, thermal conductivity 0.47 W/(m K), specific heat 0.84 kJ/(kg K), penoplex with density 35.0 kg/m^3, thermal conductivity 0.031 W/(m K), specific heat 1.53 kJ/(kg K).

Let consider two wells at the distance of 10 m from each other. Radius of the wells is 0.089 m. Temperature in the wells is 45 °C basic temperature of

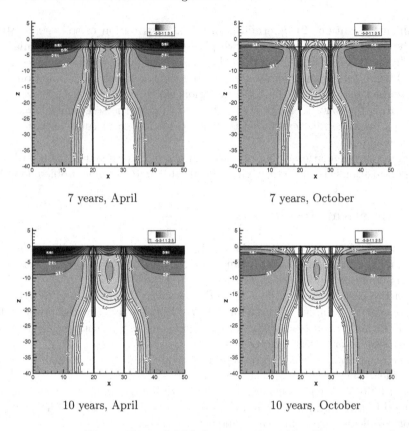

7 years, April   7 years, October

10 years, April   10 years, October

Fig. 2. Thermal fields from two identical wells.

permafrost is $-0.7\,°C$. The wells have a cement shell with thickness of 0.176 m, and upper part up to 22 m has two additional insulating shells: up to radius 0.410 m a penoplex layer is inserted, up to radius 0.5 m — cement.

Further the results of simulations of different regimes of exploitation are shown.

In Fig. 2 thermal fields for 7 and 10 years of exploitation of wells are shown. Upper layers in April and October are different because of seasonal changes. We may see thermal fluxes which propagate from non-insulated lower part and interaction with fluxes from upper boundary. Joining two fronts of thawing is observed on 3rd year of exploitation.

In Fig. 3 the left well has been stopped after 7 years of exploitation and soil around the well began to cool. It doesn't return to the initial temperature due to the molten state of the soil and the influence of the second well still being in operation. But the upper layers of the soil are frozen in winter and these processes may leads to thermokarst processes which may destroy the well equipping.

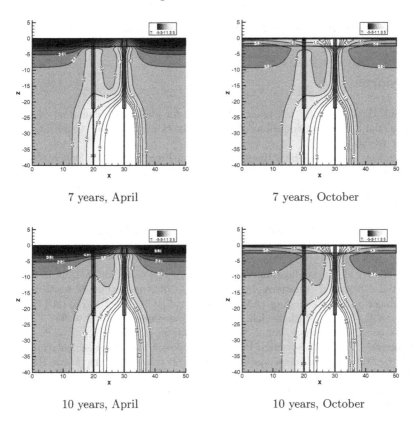

7 years, April

7 years, October

10 years, April

10 years, October

Fig. 3. Thermal fields from two wells, when one is turned off.

4 Conclusion

Based on a mathematical model numerical algorithm and software package Wellfrost is developed. Solvability of the corresponding difference equations with nonlinear boundary condition (3) is proved, which is not commonly used by researchers when considering similar problems. The difference between the numerical and experimental data was about 5 %, in particular, due to boundary condition (3) allowed to describe natural thermal fields.

Developed software complex has been tested for eight oil and gas fields in permafrost zone. Numerical simulation using and analysis of radii of thawing allowed to choose optimal parameters and to reduce costs in field development at the design stage. In the numerical calculations it were observed some patterns in increasing speed of propagation of permafrost thawing between two wells, depending on various parameters. These results allow the standards in distance between wells to be corrected.

Acknowledgement. Supported by Programs of UD RAS "Arktika" and Program 15–16–1–10, and by Russian Foundation for Basic Research 13–01–00800.

References

1. Crank, J., Nicolson, P.: A practical method for numerical evaluation of solutions of partial differential equations of heat-conduction type. Proc. Cambridge Philos. Soc. **43**, 50–67 (1947)
2. Patankar, S.V.: Numerical Heat Transfer and Fluid Flow. Hemisphere, New York (1980)
3. Javierre, E., Vuik, C., Vermolen, F.J., Zwaag, S.V.: A comparison of numerical models for one-dimensional Stefan problems. J. Comput. Appl. Math. **192**(2), 445–459 (2006)
4. Filimonov, M.Y.: Representation of solutions of initial-boundary value problems for nonlinear partial differential equations by the method of special series. Differ. Equ. **39**(8), 1159–1166 (2003)
5. Filimonov, M.Y.: Application of method of special series for solution of nonlinear partial differential equations. In: Proceedings of Applications of Mathematics in Engineering and Economics (AMEE 2014), vol. 1631, pp. 218–223 (2014)
6. Samarsky, A.A., Vabishchevich, P.N.: Computational Heat Transfer. The Finite Difference Methodology, vol. 2. Wiley, Chichester (1995)
7. Samarskii, A.A., Moiseyenko, B.D.: An economic continuous calculation scheme for the Stefan multidimensional problem. USSR Comput. Math. Math. Phys. **5**(5), 43–58 (1965)
8. Zhang, Y., Chen W., Cihlar, J.: A process-based model for quantifying the impact of climate change on permafrost thermal regimes. J. Geophys. Res. **108**(D22) (2003)
9. Filimonov, M.Y., Vaganova, N.A.: Simulation of thermal fields in the permafrost with seasonal cooling devices. In: Proceedings of ASME 45158, Volume 4: Pipelining in Northern and Offshore Environments; Strain-Based Design; Risk and Reliability; Standards and Regulations, pp. 133–141 (2012)
10. Filimonov, M.Y., Vaganova, N.A.: Simulation of thermal stabilization of soil around various technical systems operating in permafrost. Appl. Math. Sci. **7**(144), 7151–7160 (2013)
11. Vaganova, N.A., Filimonov, M.Y.: Simulation of Engineering Systems in Permafrost. Vestnik Novosibirskogo Gosudarstvennogo Universiteta. Seriya Matematika, Mekhanika, Informatika **13**(4), 37–42 (2013). (in Russian)
12. Filimonov, M.Y., Vaganova, N.A.: Prediction of changes in permafrost as a result technogenic effects and climate. Acad. J. Sci. **3**(1), 121–128 (2014)
13. Samarsky, A.A.: The Theory of Difference Schemes. Nauka, Moscow (1983)
14. Bashurov, V.V., Vaganova, N.A., Filimonov, MYu.: Numerical simulation of thermal conductivity processes with fluid filtration in soil. J. Comput. Technol. **16**(4), 3–18 (2011). (in Russian)
15. Vaganova, N.A.: Existence of a solution of an initial-boundary value difference problem for a linear heat equation with a nonlinear boundary condition. Proc. Steklov Inst. Math. **261**(1), 260–271 (2008)
16. Vaganova, N.A.: Mathematical model of testing of pipeline integrity by thermal fields. In: Proceedings of Applications of Mathematics in Engineering and Economics (AMEE 2014), vol. 1631, pp. 37–41 (2014)

Iterative Implicit Methods for Solving Nonlinear Dynamical Systems: Application of the Levitron

Jürgen Geiser$^{(\boxtimes)}$, Karl Felix Lüskow, and Ralf Schneider

Institute of Physics, Ernst Moritz Arndt University of Greifswald,
Felix-Hausdorff-Str. 6, 17489 Greifswald, Germany
{juergen.geiser,lueskow,schneider}@uni-greifswald.de

Abstract. In this paper we apply modified implicit methods for nonlinear dynamical systems related to constrained and non-separable Hamiltonian problems. The application of well-known standard Runge-Kutta integrator methods based on splitting schemes failed, while the energy conservation is no longer guaranteed. We propose a novel class of iterative implicit method that resolves the nonlinearity and achieve an asymptotic symplectic behavior. In comparison to explicit symplectic methods we achieve more accurate results for 5–10 iterations for only double computational time.

Keywords: Semi-implicit integrators · Levitron problem · Iterative Euler method · Crank-Nicolson methods · Symplectic splitting · Long time computations

1 Introduction

Nonlinear dynamical systems with non-separable Hamiltonian are important in real-life applications, by the way, they are delicate to solve. While standard integrators fail, see [1,3,4,8], it is important to consider new iterative solver techniques, that combine the ideas of decomposing into simpler equation parts and resolving the nonlinearity. The main challenge is to design implicit integrators, which are cheap in computational costs and fulfill the physical constraints of energy conservation, see [2,5,6].

We deal with the nonlinear evolution equation of the dynamical variable $u(\mathbf{q}, \mathbf{p})$ (including $\mathbf{q}$ and $\mathbf{p}$ themselves) is given by the Poisson bracket,

$$\partial_t u(\mathbf{q}, \mathbf{p}) = \left(\frac{\partial u}{\partial \mathbf{q}} \cdot \frac{\partial H}{\partial \mathbf{p}} - \frac{\partial u}{\partial \mathbf{p}} \cdot \frac{\partial H}{\partial \mathbf{q}} \right) = (A(\mathbf{q}, \mathbf{p}) + B(\mathbf{q}, \mathbf{p})) u(\mathbf{q}, \mathbf{p}). \quad (1)$$

where A can be an unbounded and nonlinear operator. For solving Hamiltonian problems, we can apply the structure of decomposing $L = A + B$, where A is the kinetic and B is the potential operator, depending on the variables $(\mathbf{q}, \mathbf{p})$.

To study the trajectories of the dynamical system, we deal with the following system of nonlinear ordinary differential equation:

$$\frac{d\mathbf{c}}{dt} = \dot{\mathbf{c}}(t) = \mathbf{f}(\mathbf{c}, t), \ t \in [0, T], \text{ and } \mathbf{c}(0) = \mathbf{c}_0, \quad (2)$$

© Springer International Publishing Switzerland 2015
I. Dimov et al. (Eds.): FDM 2014, LNCS 9045, pp. 193–200, 2015.
DOI: 10.1007/978-3-319-20239-6_19

where $\mathbf{c} = (\mathbf{p}, \mathbf{q})^t$ and the nonlinear operator $\mathbf{f} = (\mathbf{f}_p, \mathbf{f}_q)^t$ is given as

$$\mathbf{f}(\mathbf{c}, t) = \mathbf{f}(\mathbf{p}, \mathbf{q}, t), \tag{3}$$

where $\mathbf{f}_p(\mathbf{p}, \mathbf{q}, t) = -\frac{\partial H}{\partial \mathbf{q}}(\mathbf{p}, \mathbf{q})$, $\mathbf{f}_q(\mathbf{p}, \mathbf{q}, t) = \frac{\partial H}{\partial \mathbf{p}}(\mathbf{p}, \mathbf{q})$. We assume a system with m differential equations, where m is even with $\mathbf{c} = (c_1, \ldots, c_m)^t = (p_1, \ldots, p_{m/2}, q_1, \ldots, q_{m/2})^t$ and in detail the nonlinear operators are componentwise given as:

$$\mathbf{f}(\mathbf{c}) = (-\frac{\partial H}{\partial q_1}(\mathbf{p}, \mathbf{q}), \ldots, -\frac{\partial H}{\partial q_{m/2}}(\mathbf{p}, \mathbf{q}), \frac{\partial H}{\partial p_1}(\mathbf{p}, \mathbf{q}), \ldots, \frac{\partial H}{\partial p_{m/2}}(\mathbf{p}, \mathbf{q}))^t. \tag{4}$$

For such a nonlinear dynamical equation system, we have to apply time-integrators, that can resolve this nonlinear behavior. Further, we need energy conservation for long times and fast algorithms to study the stability of the trajectories. In the following, we propose a novel class of asymptotic symplectic integrators.

The paper is outlined in the following:

The time-integrators are discussed in Sect. 2. Application of the numerical schemes to the test problem of a Levitron are done in Sect. 3. In the conclusions, that are given in Sect. 4, we summarize our results.

2 Time-Integrator Methods

We compare the standard Runge-Kutta integrator with a new class of iterative semi-implicit solvers with respect to long-term stability of the Levitron. In the following, different methods will be compared.

2.1 Implicit Methods Embedded to Waveform-Relaxation Schemes

To circumvent the expensive computations of implicit methods, we use the property of the Hamiltonian, that the nonlinear function depends only on the recent variable in the function producing a decoupling with respect to Waveform-relaxation schemes. Therefore, we can use the implicit behavior of the multistep methods without any fixpoint schemes as an explicit method. In the asymptotic behavior of the iterative scheme for $i \to \infty$, we obtain a semi-implicit Euler method, which is symplectic, see [5].

The initial value of (1) is rewritten in the following form:

$$\mathbf{u}' = \mathbf{f}(\mathbf{u}, \mathbf{u}, t), \quad \mathbf{u}(0) = \mathbf{u}_0, \tag{5}$$

where we have the special structure of the Hamiltonian problem:

$$\mathbf{u} = \begin{pmatrix} \mathbf{p} \\ \mathbf{q} \end{pmatrix}, \quad \mathbf{f}(\mathbf{u}, \mathbf{u}, t) = \begin{pmatrix} -\frac{\partial H}{\partial \mathbf{q}}(\mathbf{p}, \mathbf{q}) \\ \frac{\partial H}{\partial \mathbf{p}}(\mathbf{p}, \mathbf{q}) \end{pmatrix}. \tag{6}$$

The well-known Picard or Waveform-relaxation scheme, see [7], for system (5) has the form:

$$\mathbf{u}'^{i+1} = \mathbf{f}(\mathbf{u}^{i+1}, \mathbf{u}^i, t), \ \mathbf{u}^{i+1}(0) = \mathbf{u}_0, \tag{7}$$

where $x^0(t)$ is an initial iteration and the nonlinear splitting function $\mathbf{f} : (\mathbb{R}^m)^2 \times [0, T] \to \mathbb{R}^m$.

The semi-implicit Euler scheme, see [5], is applied with the difference approximation and is given as:

$$\mathbf{u}^{n+1} = \mathbf{u}^n + \Delta t \ \mathbf{f}(\mathbf{u}^{n+1}, \mathbf{u}^n, t^n). \tag{8}$$

The exact integrator is given as: $(J\mathbf{u}')(t) = u(0) + \int_0^t \mathbf{u}'(s) \, ds$.

The semi-implicit integrators, with respect to the Hamiltonian structure (1) are given as

- Semi-implicit Euler Integrator: $(J_{\text{implicit Euler}} \mathbf{u}')(t) = t\mathbf{u}(t)$.
- Fractional-Step Integrator: $(J_{\text{FS}} \mathbf{u}')(t) = t(\theta \mathbf{u}(t) + (1 - \theta)\mathbf{u}(0))$, $\theta \in [0, 1]$.
- Crank-Nicolson Integrator: $(J_{\text{CN}} \mathbf{u}')(t) = \frac{t}{2}(\mathbf{u}(t) + \mathbf{u}(0))$.
- Leap-Frog Integrator:
  $(J_{\text{Leap}} \mathbf{u}')(t) = (J_{\text{implicit Euler}}) \circ (J^*_{\text{implicit Euler}})(\mathbf{u}')(t) = t \ \mathbf{u}(t/2)$.

Corollary 1. *The composition of two symplectic one-step integrators, in our case, e.g., $(J_{\text{implicit Euler}})$ and the conjugate integrator $J^*_{\text{implicit Euler}}$, is also symplectic in the composition, see [2].*

Corollary 2. *The system (5) has a unique solution $\mathbf{u}'^*$ and the sequence $\{\mathbf{u}^i\}$ applied by the algorithm (7) converge to $\mathbf{u}^*$, where $\mathbf{u}^* = J\mathbf{u}'^*$ is the unique solution of (5).*

Definition 1. *We apply the norm $|| \cdot ||_m$ in $\mathbb{R}^m$ such that:*

$$||\mathbf{f}(\mathbf{u}_1, \mathbf{u}_2)(t) - \mathbf{f}(\mathbf{v}_1, \mathbf{v}_2)(t)||_{L^2} \leq \sum_{l=1}^{2} a_l ||\mathbf{u}_l - \mathbf{v}_l||, \tag{9}$$

a_1, a_2 *are Lipschitz constants.*

We have $\tilde{\mathbf{u}} = \mathbf{u}'$ for $t \in [0, T]$ and

$$(J\tilde{\mathbf{u}})(t) = \mathbf{u}(0) + \int_0^t \tilde{\mathbf{u}} \, ds, \tag{10}$$

where $\mathbf{u}(0) = \mathbf{u}_0$ is the initial condition.

For the implicit Euler integrator we apply:

$$(J_{implicit\ Euler} \tilde{\mathbf{u}})(t) = t\tilde{\mathbf{u}}(t), \tag{11}$$

where we have a first order scheme and the error is given as $||(J\tilde{\mathbf{u}})(t) - (J_{implicit\ Euler} \tilde{\mathbf{u}})(t)|| \leq \mathcal{O}(t^2)$. Analogue this was done for the fractional step scheme for $\theta \in [0, \frac{1}{2})$ and $\theta \in (\frac{1}{2}, 1]$.

For the Crank-Nicolson or trapezoidal integrator we apply:

$$(J_{CN}\tilde{\mathbf{u}})(t) = \frac{t}{2}(\tilde{\mathbf{u}}(t) + \tilde{\mathbf{u}}(0)), \tag{12}$$

where we have a second order scheme and the error is given as $\|(J\tilde{\mathbf{u}})(t) - (J_{CN}\tilde{\mathbf{u}})(t)\| \leq \mathcal{O}(t^3)$.

Corollary 3. *The system (5) has a unique solution* $\tilde{\mathbf{u}}^*$ *and the sequence* $\{\mathbf{u}^i\}$ *applied by the algorithm (7) converge to* $\mathbf{u}^*$, *where* $\mathbf{u}^* = J\tilde{\mathbf{u}}^*$ *is the unique solution of (5).*

Proof. The proof idea is also described in [7].
We apply the following details:

$$\|\mathbf{f}(J(\tilde{\mathbf{u}}_1, J(\tilde{\mathbf{u}}_1, t) - \mathbf{f}(J(\tilde{\mathbf{u}}_2, J(\tilde{\mathbf{u}}_2, t))\| \leq \sum_{l=1}^{2} a_l \|J(\tilde{\mathbf{u}}_1) - J(\tilde{\mathbf{u}}_2)\|$$

$$\leq \sum_{l=1}^{2} a_l \int_0^t \|\tilde{\mathbf{u}}_1) - \tilde{\mathbf{u}}_2\| \, ds \leq \sum_{l=1}^{2} a_l \, \mathcal{V}_c \|\tilde{\mathbf{u}}_1) - \tilde{\mathbf{u}}_2\|, \tag{13}$$

where $\mathcal{V}_c : C([0,T], \mathbb{R}^m) \to C([0,T], \mathbb{R}^m)$ *is a linear Voltera integral operator with* $(\mathcal{V}_c\mathbf{p})(t) = \int_0^t \mathbf{p}(s) \, ds$. *Further* $\mathcal{L} = \sum_{l=1}^{2} a_l \mathcal{V}_c$ *is a compact and non-negative operator.*

The iterative scheme can be estimated as:

$$\|\tilde{\mathbf{u}}^{i+k+1}(t) - \tilde{\mathbf{u}}^i(t)\| \leq \sum_{j=0}^{k-1} \|\tilde{\mathbf{u}}^{i+j+1}(t) - \tilde{\mathbf{u}}^{i+j}(t)\|$$

$$\leq \sum_{j=0}^{k-1} \mathcal{L}^j \|\tilde{\mathbf{u}}^{i+1}(t) - \tilde{\mathbf{u}}^i(t)\| \leq (\mathcal{I} - \mathcal{L})^{-1}\mathcal{L}^k \|\tilde{\mathbf{u}}^1(t) - \tilde{\mathbf{u}}^0(t)\| = \Lambda^{(i)}(t). \tag{14}$$

Therefore, the inequality (14) converges uniformly, $\{\tilde{\mathbf{u}}^i(t)\}$ is a Cauchy sequence in the function space $C([0,T], \mathbb{R}^m)$ as long as $\{\Lambda^{(i)}\}$ uniformly converges in $C([0,T], \mathbb{R}^{m \times m})$,
While $\rho(\mathcal{L}) < 1$, we find a norm $\|\mathcal{L}\|_\epsilon \leq \rho(\mathcal{L}) + \epsilon < 1$ and

$$\|\Lambda^{(i)}\|_\epsilon \leq \|\mathcal{L}\|_\epsilon^i \|\phi_0\|_\epsilon, \tag{15}$$

where $\phi_0 = (\mathcal{I} - \mathcal{L})^{-1}\|\tilde{\mathbf{u}}^1(t) - \tilde{\mathbf{u}}^0(t)\|$.
Therefore the sequence $\{\tilde{\mathbf{u}}^i(t)\}$ converge to a unique solution $\tilde{\mathbf{u}}^*(t)$ and we obtain a unique solution $\mathbf{u}^*(t)$ of system (5).

2.2 Algorithmic Structures of the Iterative Semi-implicit Integrators

The iterative semi-implicit methods are given in Algorithm 1. We deal with the following superscripts of the solutions **u**:

- $\mathbf{u}^n$ is defined as the solution at the time-point t^n,
- $\mathbf{u}^{i,n}$ is defined as the iterative solution of $\mathbf{u}$ in the i-th iterative step at the time-point t^n.
- If we apply $\mathbf{u}^{n+1} = \mathbf{u}^{i,n+1}$, we replace the solution in the next time point t^{n+1} with the iterative solution in the i-th iterative step.
- If we apply $\mathbf{u}^{0,n+1} = \mathbf{u}^n$, we initialize the iterative scheme with the solution in time point t^n.

Algorithm 1. *We compute the solution* $\mathbf{u}(t^n)$ *at the time points* $n = 1, 2, 3, \ldots, N$, *the initialization of the iterative scheme is given with the initial condition of the Eq. (5) as* $u^{0,1} = u^0$. *The time-step is given with* $\Delta t = t^n - t^{n-1}$ *and the error bound is given as* $\epsilon = 10^{-5}$. *We start with* $n = 1$:

1. Initialization $i = 0$:

$$\mathbf{u}^{0,n+1} = \mathbf{u}^n, \tag{16}$$

2. Iterative Steps:
   – Semi-implicit Euler Step:

$$\mathbf{u}^{i,n+1} = \mathbf{u}^n + \Delta t\, \mathbf{f}(\mathbf{u}^{i-1,n+1}, \mathbf{u}^{i,n+1}, t^{n+1}). \tag{17}$$

   – Fractional Step:

$$\mathbf{u}^{i,n+1} = \mathbf{u}^n + \Delta t\, \left(\theta \mathbf{f}(\mathbf{u}^{i-1,n+1}, \mathbf{u}^{i,n+1}, t^{n+1}) \right.$$
$$\left. + (1 - \theta)\mathbf{f}(\mathbf{u}^n, \mathbf{u}^n, t^n) \right), \ \theta \in [0, 1]. \tag{18}$$

   – Crank-Nicolson Step:

$$\mathbf{u}^{i,n+1} = \mathbf{u}^n + \frac{\Delta t}{2}\, \left(\mathbf{f}(\mathbf{u}^{i-1,n+1}, \mathbf{u}^{i,n+1}, t^{n+1}) + \mathbf{f}(\mathbf{u}^n, \mathbf{u}^n, t^n) \right). \tag{19}$$

   – Composition of two symplectic Integrators, e.g. Leap-Frog Integrator:

$$\mathbf{p}^{i,n+1/2} = \mathbf{p}^n + \frac{\Delta t}{2}\, \mathbf{f}_p(\mathbf{u}^{i-1,n+1/2}, \mathbf{u}^{i,n+1/2}, t^{n+1/2}), \tag{20}$$

$$\mathbf{q}^{i,n+1/2} = \mathbf{q}^{n+1/2} + \frac{\Delta t}{2}\, \mathbf{f}_q(\mathbf{u}^{i-1,n+1/2}, \mathbf{u}^{i,n+1/2}, t^{n+1/2}), \tag{21}$$

$$\mathbf{q}^{i,n+1} = \mathbf{q}^{n+1/2} + \frac{\Delta t}{2}\, \mathbf{f}_q(\mathbf{u}^{i-1,n+1}, \mathbf{u}^{i,n+1}, t^{n+1}), \tag{22}$$

$$\mathbf{p}^{i,n+1} = \mathbf{p}^{n+1/2} + \frac{\Delta t}{2}\, \mathbf{f}_p(\mathbf{u}^{i-1,n+1}, \mathbf{u}^{i,n+1}, t^{n+1}). \tag{23}$$

3. Stopping Criterion: If $i = I$ or the error is given as

$$||\mathbf{u}^{i,n+1} - \mathbf{u}^{i-1,n+1}|| \leq \epsilon, \tag{24}$$

we have $\mathbf{u}^{n+1} = \mathbf{u}^{i,n+1}$.
Else $n = n + 1$,
   If $n = N + 1$, we are done and have computed all N solutions and stop,
   Else go to step 1.

We obtain a new class of asymptotic symplectic integrators, which is symplectic only for separable Hamiltonians, using e.g. splitting, and embed the method into a waveform-relaxation scheme, see [7]. The waveform-relaxation scheme resolves iteratively the nonlinearity of the dynamical system with a convergence order of $\mathcal{O}(\Delta t^i)$, where i is the number of iterations and Δt is the time-step, see [7].

The order of the semi-implicit Euler integrator can be improved by applying two semi-implicit Euler integrators (positive and negative time step) and one gets a Leap-Frog integrator of second order, see [5].

The semi-implicit Crank-Nicolson method, based on the trapezoidal rule, while the Trapezoidal rule is symplectic, see [8], we could embed such an asymptotic symplecticity.

Remark 1. We have derived asymptotic symplectic methods, based on the symplectic kernel-functions. Such methods can be improved by compositions of symmetric integrators, see [2]. An alternative multistep method is not symplectic, see [2] and such methods can not be applied. We have applied BDF2 methods and obtained less stable results as for example with semi-implicit Euler methods.

3   Numerical Example

The numerical example is based on a Levitron, which is well-studied, see for example [1,4]. A Levitron is a spinning gyroscope which levitates in a magnetic field. To analyze the trajectories of the gyroscope it is important to energy preserving integrators, see [1].

The model equations are the following [4]:

$$H = \frac{1}{2m}\left(p_x^2 + p_y^2 + p_z^2\right) + \frac{p_\theta^2}{2A} + \frac{p_\psi^2}{2C} + \frac{(p_\phi - p_\psi \sin\theta)^2}{2A\cos^2\theta} + mgz - \mu\rho\left(\frac{1}{2}\Phi_2(z)\right.$$

$$\left.\cdot(x\,\sin\theta + y\cos\theta\sin\phi) + (-\Phi_1(z) + \frac{1}{4}(x^2+y^2)\Phi_3(z))\cos\theta\cos\phi)\right), \qquad (25)$$

where the higher order $\Phi_i(z)$ are defined as $\Phi_i(z) = \frac{\partial \Phi_{i-1}(z)}{\partial z}$.

For the Levitron test case we compare standard explicit integrators with the novel iterative semi-implicit integrators. At least only sufficient energy conservation can guarantee realistic numerical solutions of the Levitron, see [4].

Our parameters of the model Eq. (1) are given as: $a = 5\,cm$ (radius of the base plate), $b = 1.5\,cm$ (radius of the Levitron), $m = 20.0\,g$ (mass of the top), $A = \frac{mb^2}{4}$ (first moments of inertia), $C = \frac{mb^2}{2}$ (second moment of inertia), $\sigma = 20 \cdot 2\pi\,\frac{1}{s}$ (spin rate), $\mu \cdot \rho = -0.000095\,V^2s^2$ (dipole strength of the top $\times$ dipole density of the base), $(x,y,z) = (1,0,31.3)$mm (initial point of the Levitron), $\phi = \psi = \theta = 0°$ (initial angles) $(p_x, p_y, p_z) = (0,0,0)$kg m^2 s^{-1} (initial momenta) and $(p_\theta, p_\phi, p_\psi) = (0,0,\sigma C)$kg m^2 s^{-1}.

The time step is $\Delta t = 10^{-6}$s and the maximum number of time-steps which was tested is $N = 10^8$, representing a time of 100 s.

We see, that the energy conservation for long-time computations $t \geq 80$s is fulfilled only for the iterative semi-implicit integrators, i.e. IEU10 (implicit Euler with $i = 10$, CN10 (Crank-Nicolson with $i = 10$), Leap Frog with $i = 10$. Here, we resolve the nonlinearity of the Hamiltonian and gain asymptotic symplecticity, due to the fact of the symplectic kernel. While shorter times in the range of $20 \leq t \leq 80$s can be reached with sufficient energy conservation by explicit schemes (e.g. Runge-Kutta 4th order schemes), the overall benefit of the novel schemes are long-time conservations, see Fig. 1. For long-time studies, the Levitron moved into the stable attractor $x = 0, y = 0, z = z_s$, which is the stable point and we see only small perturbations.

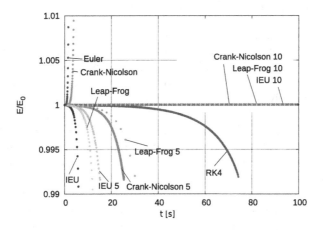

Fig. 1. Long energy computations with semi-implicit methods in the time-interval $t = [0, 100]$s (10^6 timesteps).

The computational time of such schemes increases, see Table 1, such that an application of the iterative semi-implicit schemes is more efficient for long-time studies.

Table 1. Computational time for 10^6 time-steps of explicit and implicit schemes.

Schemes	Comput. Time [s]	Schemes	Comput. Time [s]
expl. Euler	20.6	it. CN $i = 1$	40.9
expl. RK (Ord. 4)	141.0	it. CN $i = 5$	121.2
it. Euler $i = 1$	21.4	it. CN $i = 10$	222.0
it. Euler $i = 5$	101.4	Leapfrog $i = 1$	40.3
it. Euler $i = 10$	201.4	Leapfrog $i = 5$	201.5
		Leapfrog $i = 10$	404

In summary, the iterative semi-implicit schemes are more effective, because they can resolve the nonlinear Hamiltonian structure and gain asymptotic symplecticity. We overcome restriction of the CFL (Courant-Friedrich-Levy) condition, which is known for the explicit schemes. In real-life applications, the choice of the integrator are given by the dilemma: if the problem does not require long times to be studied a relatively simple integrator can be the most efficient, e.g. explicit symplectic Runge-Kutta methods. For long-time analysis like the stability problem of the Levitron it is necessary to have very small error in energy conservation for this system. Therefore, only more complex composite iterative solvers are important.

4   Conclusions and Discussions

We present a new class of iterative implicit time-integrators for non-linear problems with non-separable Hamiltonians. We could prove asymptotic symplecticity for the iterative implicit schemes. We see the benefit of iterative semi-implicit methods, which produce stable results for long time computations. Further, they are cheaper than fully implicit schemes, while they only update explicit the next iterative solution. Nevertheless explicit methods are fast but they failed in energy conservation for longer time intervals, e.g. $t \geq 50\,\text{s}$. Based on this fact, iterative semi-implicit methods are more attractive, while combining resolution of the nonlinearity and the asymptotic symplecticity with additional iterative cycles. We could demonstrate long-time stability of about 20 h for a test trajectory, which have at least only twice computational effort compared with standard 4th order schemes. In the future, we will concentrate on analyzing and designing this composition types of iterative semi-implicit integrators for nonlinear and non-separable Hamiltonian systems.

References

1. Dullin, H.R.: Poisson integrator for symmetric rigid bodies. Regul. Chaotic Dyn. **9**, 255–264 (2004)
2. Fenga, Q.-D., Jiaoa, Y.-D., Tanga, Y.-F.: Conjugate symplecticity of second-order linear multi-step methods. J. Comput. Appl. Math. **203**, 6–14 (2007)
3. Geiser, J.: Nonlinear extension of multiproduct expansion schemes and applications to rigid bodies. Int. J. Differ. Equ. **2013**, 10 (2013). Article ID. 681575
4. Geiser, J., Lüskow, K.F., Schneider, R.: Levitron: multiscale analysis of stability. Dyn. Syst. **29**(2), 208–224 (2014)
5. Hairer, E., Lubich, C., Wanner, G.: Geometric numerical integration illustrated by the StörmerVerlet method. Acta Numerica **12**, 399–450 (2003)
6. Hairer, E., Lubich, C., Wanner, G.: Geometric Numerical Integration: Structure-Preserving Algorithms for Ordinary Differential Equations. Springer Series in Computational Mathematics, 2nd edn. Springer, Heidelberg (2006)
7. Jiang, Y.-L., Wing, O.: A note on convergence conditions of waveform relaxation algorithms for nonlinear differential–algebraic equations. Appl. Numer. Math. **36**(2–3), 281–297 (2001)
8. McLachlan, R.I., Atela, P.: The accuracy of symplectic integrators. Nonlinearity **5**, 541–562 (1992)

Modeling Textile Fabric Used in Pest Control with a 3 Scale Domain Decomposition Method

Tineke Goessens[1]([⊠]), Benny Malengier[2], and Lieva Van Langenhove[2]

[1] Department of Mathematical Analysis, Research Group NaM²,
Faculty of Engineering and Architecture, Ghent University, Ghent, Belgium
tineke.goessens@ugent.be
http://www.nam2.ugent.be/
[2] Department of Textiles, Faculty of Engineering and Architecture,
Ghent University, Ghent, Belgium
benny.malengier@ugent.be

Abstract. In this paper we present a model to simulate textile as used in pest control. For this application, textile is coated with a repellent, protecting the user from insect bites, and one wants to determine optimal material properties. The model extends an existing 3 scale method to allow for simulations in saturated conditions. This is achieved with the addition of an overlapping domain decomposition approach for the fiber-yarn interaction.

With the model we present how the performance of a coating can be determined: how much material is required, what evaporative properties are needed, how can the coating be replenished? Furthermore, the model can be used to evaluate the effects of the used textile substrate, like the type and number of fibers or the weaving structure. Lastly, it can be used to validate simple first-order models of coated textile.

Numerical results indicate the 3 scale approach is valid. The influence of different textile properties on the effectiveness of the resulting textile component is presented.

Keywords: Diffusion · Textile modeling · Upscaling · Multi-scale modeling · Domain decomposition

1 Introduction

Typical textiles used in pest control are treated scrims. These are open structures. One can however also consider treated shirts and trousers, as an alternative to smearing products onto exposed skin, or spraying products onto textile. Textile can be treated in different ways: printed film coating and bath coating. We consider the case of a bath coating that results into coated fibers as opposed to coated yarns or fabric. This method retains maximally the original properties of the textile. As the coating is on the fiber level, it is useful to include a meso-model which describes the interaction from fiber to yarn and from yarn to fabric, [1].

© Springer International Publishing Switzerland 2015
I. Dimov et al. (Eds.): FDM 2014, LNCS 9045, pp. 201–208, 2015.
DOI: 10.1007/978-3-319-20239-6_20

To model the application in mind we thus make a distinction on three levels of the scrim. First we model the fiber with one or more coatings containing an active component (AC). To this end the fiber will be seen as a cylindrical object. The boundary conditions depend on the chosen textile and the void space characteristics. Secondly we model the yarn, a porous structure built out of fibers, upscaling the outcome of the fiber model using an overlapping domain decomposition method. The third model represents the scrim or fabric itself, with its environment, again using overlapping domain decomposition to calculate the overall properties of the fabric using the yarn properties.

The aim of the model is to develop a textile fabric that can maximally protect the user from harmful insect bites. To achieve this, the textile must spatially repel the insects. This will be possible with repellents when their vapour concentration is higher than the minimum effective concentration (MEC). For example, for DEET which repels mosquitoes, this is around $2\mu g/L$ air (depending on the species), [2]. As a consequence, pure DEET evaporation could be reduced a factor 22 and still give rise to repellency. Applying the DEET to textile via a coating which reduces the evaporation rate would hence be beneficial. Moreover, as the total fiber surface is larger than the fabric surface, a further reduction in DEET evaporation is possible.

We start with an outline of the three scale model in the next Section, and then show in Sect. 3 numerical examples of how the model can aid in the design of protective textiles.

2   Multiscale Model

2.1   The Micro-level

We consider the cylindrical diffusion equation in [1]. However, to upscale the fiber concentration to the yarn level of the model we now use the overlapping domain decomposition technique. Therefore the original domain $[0, R_f]$ (with R_f the radius of the fiber cross section) is extended with a zone $\Omega_o^f = [R_f, kR_f]$ overlapping with the yarn model domain, where k is determined in such a way that after upscaling to the yarn level the overlap domain corresponds to the available air space on the yarn level. In this zone the governing PDE is given by

$$\frac{\partial C_f(r,t)}{\partial t} = \frac{1}{r}\frac{\partial}{\partial r}\left(rD_f\frac{\partial C_f(r,t)}{\partial r}\right) - \Gamma_f(\Omega_o^f, t), \quad R_f \leq r \leq kR_f, \quad (1)$$

with homogeneous Neumann BC

$$\frac{\partial C_f}{\partial r}(kR_f, t) = 0,$$

where D_f is the diffusion coefficient of the AC in air. The extra sink term $\Gamma_f(\Omega_o^f, t)$ is the concentration of AC that is removed from the micro-scale due to diffusion to the meso-level, see Sect. 2.2.

Equation 1 can be combined with the equation in $[0, R_f]$ by setting Γ_f to zero there, and considering D_f as the diffusion in the fiber, instead of in air. At the fiber surface, $r = R_f$, the evaporation condition, [3], is imposed with an evaporative flux

$$\mathcal{F}_{\text{evap}}(t) = Sh_{lg}\left(C_s(T) - C_B\right) \cdot \mathcal{H}\left(C_f(R_f, t) - C_{\min}, C_s(T) - C_B\right). \tag{2}$$

In this, S is the effective area fraction where evaporation or condensation takes place, h_{lg} the mass transfer coefficient from liquid to gas [mm/s], $C_s(T)$ the saturated concentration in air (temperature depending), C_B the concentration of the component in the air at the boundary, $C_{\min}$ a minimum amount of component in the fibers which relates to the absorption isotherm and chemically bound component, and $\mathcal{H}$ defined as

$$\mathcal{H}(v, c) = \begin{cases} 1, & c \leq 0 \\ 0, & c > 0 \ \& \ v < 0 \ . \\ 1, & c > 0 \ \& \ v > 0 \end{cases} \tag{3}$$

2.2 The Meso-level with Overlapping Domain Decomposition Technique

Based on the model described in [1,3], the governing model for the concentration of the AC on the yarn level for a yarn of radius R_y is

$$\epsilon \frac{\partial C_y(r, t)}{\partial t} = \frac{1}{r} \frac{\partial}{\partial r}\left(\epsilon r \frac{D_y}{\tau_y} \frac{\partial C_y(r, t)}{\partial r}\right) + \Gamma_{in}(\Omega_o^f, t), \quad r \in [0, R_y], \tag{4}$$

where D_y is the diffusion coefficient of the AC in the yarn air gaps, ϵ is the porosity depending on position and τ_y is the tortuosity of the yarn.

The term $\Gamma_{in}(\Omega_o^f, t)$ in the equation above is a source term that describes the upscaled amount of AC change in the overlap zone of the fiber level due to release from the fiber. It is calculated by upscaling the concentration change in the overlapping zone of the fiber during the current time step, i.e. by upscaling $C_f(\Omega_o^f, t_j) - C_f(\Omega_o^f, t_{j-1})$ representing the concentration coming from one single fiber to the overlap zone. Upscaling is done by multiplying with the total number of fibers in a yarn cross section, taking possible blends of different kinds of fibers into account. Note that the yarn domain will be split up in zones, and in every zone a fiber model will be solved. The value of C_B used in the fiber model will be the average of the computed C_y over the zone. We call these fiber models the representative fiber models. As the fibers are present everywhere for $r \in [0, R_y]$, the fiber overlap zone is implicit in the yarn model.

After solving the yarn model for t_j we know the concentration change in the implicit overlap zones $[R_f, kR_f]$ for the fibers in the yarn zone considered (new C_B value) and downscale it again to the concentration change for one single fiber. This will be used in the next time step as $\Gamma_f(\Omega_o^f, t_{j+1})$ for the fiber model.

We did not consider boundary conditions yet. Due to the radial symmetry, at $r = 0$ we have a homogeneous Neumann BC. At $r = R_y$, we again consider the domain overlapping technique. Hence, to upscale to the macro-level model containing the total fabric and its environment we extend the original domain $[0, R_y]$ with an overlapping domain Ω_o^y at the interval $[R_y, 2R_y]$ with the PDE (4) slightly adapted to

$$\epsilon \frac{\partial C_y(r,t)}{\partial t} = \frac{1}{r} \frac{\partial}{\partial r} \left(\epsilon r \frac{D_y}{\tau_y} \frac{\partial C_y(r,t)}{\partial r} \right) + \Gamma_{in}(\Omega_o^f, t) - \Gamma_{out}(\Omega_o^y, t), \qquad (5)$$

with a homogeneous Neumann BC

$$\frac{\partial C_y}{\partial r}(2R_y) = 0.$$

The sink term $\Gamma_{out}(\Omega_o^y, t)$ is the concentration of AC that is removed from the meso-scale due to diffusion to the macro-level, see Sect. 2.3. No special treatment at $r = R_y$ is needed, as that boundary is only characterized by ϵ and τ_y becoming unity. Furthermore, Γ_{in} is zero for $r > R_y$, and Γ_{out} is zero for $r < R_y$.

2.3   The Macro-level and Domain Decomposition Method

As mentioned above we will use an overlap zone $\Omega_o^y = [R_y, 2R_y]$, to upscale from the meso-level to the fabric level.

The room 1D diffusion equation for a room of length L with textile placed at $x = 0$ is

$$\partial_t C = \partial_x (D \partial_x C) + \Gamma_s(x, t), \quad x \in [R_y, L], \quad \partial_x C(R_y) = 0,$$

with D the diffusion of the AC in air and $\Gamma_s(x, t)$ the concentration per time unit added/removed at x and only non-zero for $x < 2R_y$. We situate the overlap zone in the first cell for numerical integration so there the source term Γ_s is corresponding to the upscaled concentration change in the overlap zone of the yarns $C_y(\Omega_o^y, t_j) - C_y(\Omega_o^y, t_{j-1})$ according to the number of yarns in the fabric. Then the fabric model is solved to obtain a new concentration value near a yarn. Because we model only half the room while the full yarn is considered, the calculated mass loss should be multiplied by 2. Next the correct $\Gamma_{out}(\Omega_o^y, t_j)$ term on the yarn level is set, downscaling the mass calculated from the room model in this time step to keep mass balance. This yarn model sink term is based on the concentration change occurring in the fabric overlap zone, i.e. the mass removed from the fabric overlap zone, downscaled to one yarn.

As BC at $x = L$ we can consider a homogeneous Neumann condition for simulations in a closed environment, or a homogeneous Dirichlet BC to take basic ventilation into account (open cup test).

2.4   The Complete Three Step Model

For the numerical scheme of the algorithm we make the distinction between the r-coordinate of the fiber level, r_i, $1 \le i \le I$ and the r-coordinate of the yarn level, r'_k, $1 \leq k \leq K$. On each time step one complete three step model is solved using on every scale a finite volume discretization in space, and a ODE Backward Differentiation Formula solver in time (method of lines).

We initiate the numerical computations with the fiber level with initial concentrations $C_f(r_i, r'_k, t_0) = C_{init}(r'_k)$ for a fiber at radial position r'_k in the yarn corresponding to the amount of AC in the coating after production of the fabric. The sink term $\Gamma_f(t)$ is zero for this first step. Next a yarn model is solved using the change in concentration in the overlap zone of the fiber over the current time step calculated in the fiber model and upscaled to the yarn source term Γ_{in}. The sink term Γ_{out} is the concentration change in the overlap zone of the yarn due to the fabric model diffusion in the previous time step. For the first time step Γ_{out} is zero.

Afterwards a fabric model is solved now upscaling the concentration change in the overlap zone of the yarn to the source term Γ_s.

A typical fabric can consist of different types of yarns. In this case two different yarn models are used. Likewise, a yarn can consist of a blend of different fibers. In that case, for every yarn zone we consider as many fiber models as there are fibers types present. Per construction, the upscaling and downscaling of mass changes in the overlap zones are fully mass conservative. This requires a correct characterization of the fabric, yarns and fibers used.

3   Experiments

We consider the problem of having the best possible protection. For this the concentration of AC must be sufficiently high for the evaporation to cause a build up of component sufficient to repel insects. From (2) we can deduce that the evaporation speed for textiles reduces over its lifetime, as the effective area S reduces with available compound. That is, we have that evaporation surface S is equal to $nC_f(R_f, t)/\rho$, where ρ is the density of the compound, and n the porosity available for the compound. As a consequence, a minimum effective dose of a compound can be determined from dose studies. For DEET absorbed by cotton muslin cloth this is 20 nmol/cm^2, [4].

The above reasoning leads one to conclude that all concentrations above the MED are evaporating more compound than needed for repellency. As for our example cotton muslin cloth can absorb up to 7500 nmol/cm^2, a lot of material is wasted to achieve long duration protection.

We propose and investigate two techniques to overcome this drawback. We will concentrate on DEET as a compound, as all material properties of this repellent are well known.

3.1   Extending Lifetime

Consider the following production method.

1. A finished cotton product (scrim or apparel) needs to obtain repellent properties.

2. The product is drenched with DEET so as to absorb a pre-set mass of DEET.
3. The product is dried to remove surface DEET.
4. The product is coated with a polymer blocking layer that reduces the diffusion of DEET to the surface.
5. The product is coated with a second polymer layer that has normal diffusion properties, but reduced evaporation properties. This layer can optionally also be enriched with DEET.

To simulate this, we consider an initial condition of the fiber consisting of 3 zones: the cotton zone with a high DEET concentration, a blocking zone with no DEET content and low diffusion coefficient, and a surface layer zone. Typical simulation results for 4 cases are shown in Figs. 1 and 2. Case 1 has a blocking layer combined with a pure DEET mass transfer coefficient ($h_{lg} = 0.897$mm/s). Case 2 is like Case 1 but with a blocking layer that has it's diffusion coefficient an order smaller. Case 3 is like Case 1 but with mass transfer coefficient 4 orders smaller. Case 4 is like Case 1 but with a mass transfer coefficient for DEET as typical with a polymer coating technique ($S = 1$ combined with $h_{lg} = 8 \, 10^{-5}$mm/s).

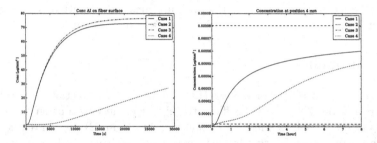

Fig. 1. Left: evolution of the concentration of active component (AC) at the fiber surface (Case 3 & 4 overlap). Right: resulting concentration in the air 4 mm away from the fabric, bottom dashed line indicates the minimum amount needed for repellency of DEET, the top dashed line the saturation concentration.

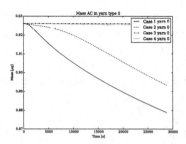

Fig. 2. Total mass in a single yarn of the fabric.

The results give insight in how material properties can be adapted to maximize the effectiveness of the textile product. The threshold for repellency is the bottom horizontal dashed line in Fig. 1. It can be seen that Case 1 and 2 give excellent protection at the expense of overusing AC. After 8 hours also Case 4 offers protection, and this with a minimal loss of AC mass, as evident in Fig. 2. The model allows to investigate the material properties needed to obtain an optimal lifetime. The model next can be used to investigate the effects of changing the fabric types: mixing in polyester fibers, changing yarn densities, ...

3.2   Replenishing Compound

An alternative approach to reduce the use of repellent is reactivating the textile after use. In this case the textile is folded in a closed container, at high initial temperature. DEET is released in the container up to saturation concentration, after which the temperature is reduced, leading to oversaturation and condensation of DEET, according to Eq. (2) with $S = n$. In this case the textile will absorb the repellent. As no liquid DEET is used, no unwanted side effects due to over-use should be present.

We perform a simplified simulation, in which we don't take temperature effects into account and assume an oversaturation can be maintained for a long time. This will sufficiently correspond to the actual behaviour.

The result of a typical investigation of this use case is given in Fig. 3. In this figure the concentration at the surface of a fiber is given during the activation of the fiber. As the diffusion into a cotton fiber is fast, this will closely match the concentration over the entire fiber. Note that the activation is a slow process, over 8 hours, the concentration in the fiber is only augmented to 2 $\mu g/mm^3$. This however is enough to give protection for 10 hours, as is evident from the release profile shown in Fig. 4. Due to the low initial content, the protection at 4 mm only starts after 30 min however.

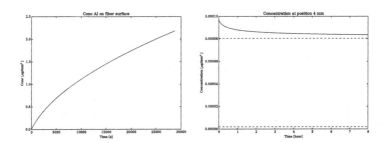

Fig. 3. Left: Evolution of the concentration of active component (AC) at the fiber surface during replenishing a fabric in a closed container. Right: reduction of compound in a closed container due to absorption by the fabric.

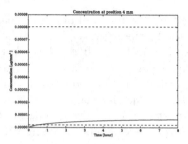

Fig. 4. Release of a replenished fabric over time, when used in a large room

4   Conclusion and Future Work

Our interest is the development of protective clothing. To this aim a textile model has been developed to investigate the influence of characteristics on 3 different scales: fiber properties, yarn properties and finally how yarns are combined into a garment or scrim. We have shown that the model can be used in the design phase of the textile. As future work we will validate the model with laboratory tests, to show that the model results are in accordance with experimental results of lifetime and effectiveness. In a later stadium we plan to develop optimization tools that allow guided design of protective clothing based the constructed model.

Acknowledgements. The authors gratefully acknowledge the support of the European Commission, FP7, project number 228639.

References

1. Goessens, T., Malengier, B., Constales, D., Staelen, R.D.: A volume averaging and overlapping domain decomposition technique to model mass transfer in textiles. J. Comput. Appl. Math. **79**, 163–174 (2014)
2. Hoffmann, E.J., Miller, J.R.: Reduction of mosquito (diptera: Culicidae) attacks on a human subject by combination of wind and vapor-phase deet repellent. J. Med. Entomol. **39**(6), 935–938 (2002)
3. Ye, C., Huang, H., Fan, J.: Numerical study of heat and moisture transfer in textile materials by a finite volume method. Commun. Comput. Phy. **4**(4), 929–948 (2008)
4. Ali, A., Cantrell, C.L., Bernier, U.R., Duke, S.O., Schneider, J.C., Agramonte, N.M., Khan, I.: Aedes aegypti (diptera: Culicidae) biting deterrence: Structure-activity relationship of saturated and unsaturated fatty acids. J. Med. Entomol. **49**(6), 1370–1378 (2012)

The Theory and Applications of the SMIF Method for Correct Mathematical Modeling of the Incompressible Fluid Flows

Valentin Gushchin and Pavel Matyushin$^{(\boxtimes)}$

Institute for Computer Aided Design of the Russian Academy of Sciences,
Moscow, Russia
gushchin@icad.org.ru, pmatyushin@mail.ru

Abstract. For solving of the Navier-Stokes equations describing the incompressible viscous fluid flows the Splitting on physical factors Method for Incompressible Fluid flows (SMIF) with hybrid explicit finite difference scheme (second-order accuracy in space, minimum scheme viscosity and dispersion, capable for work in the wide range of Reynolds (Re) and internal Froude (Fr) numbers and monotonous) based on the Modified Central Difference Scheme (MCDS) and the Modified Upwind Difference Scheme (MUDS) with a special switch condition depending on the velocity sign and the signs of the first and second differences of the transferred functions has been developed and successfully applied. At the present paper the description of the numerical method SMIF and its applications for simulation of the 3D separated homogeneous and density stratified fluid flows around a sphere and a circular cylinder are demonstrated.

Keywords: Direct numerical simulation · Viscous fluid · Visualization of the vortex structures · Flow regime · Sphere · Cylinder

1 Introduction

Many phenomena in the nature may be considered in the frame of the incompressible fluid flows. Such flows are described by the Navier-Stokes equations. As usually we have deal with the flows with large gradient of hydrodynamic parameters (flows with free surface, stratified fluid flows, separated flows, etc.). For the direct numerical simulation of such flows the finite difference schemes should possess by the following properties: high order of accuracy, minimum scheme viscosity and dispersion and monotonisity. The Splitting on the physical factors Method for Incompressible Fluid flows (SMIF) with hybrid explicit finite difference scheme (second-order accuracy in space, minimum scheme viscosity and dispersion, capable for work in wide range of Reynolds and Froude numbers and monotonous) has been developed [1,2] and successfully applied for solving of the different problems: 3D separated homogeneous and stratified viscous fluid

© Springer International Publishing Switzerland 2015
I. Dimov et al. (Eds.): FDM 2014, LNCS 9045, pp. 209–216, 2015.
DOI: 10.1007/978-3-319-20239-6_21

flows around a sphere and a circular cylinder including transitional regimes [3–12]; the flows with free surface including regimes with broken surface wave [1,2]; the air, heat and mass transfer in the clean rooms [13]. At the present paper the classifications of the regimes of the 3D separated homogeneous viscous fluid flows around a circular cylinder (at $Re < 400$) and the 3D density stratified viscous fluid flows around a sphere (at $Re < 700$) will be demonstrated.

2   Numerical Method SMIF

Let $\rho(x, y, z) = 1 - x/(2A) + S(x, y, z)$ is the non-dimensional density of the linearly stratified fluid where x, y, z are the Cartesian coordinates; z, x, y are the streamwise, lift and lateral directions (x, y, z have been non-dimensionalized by $d/2, d$ is a diameter of the body); $A = \Lambda/d$ is the scale ratio, Λ is the buoyancy scale, which is related to the buoyancy frequency N and period $T_b(N = 2\pi/T_b, N^2 = g/\Lambda)$; g is the scalar of the gravitational acceleration; S is a dimensionless perturbation of salinity. The density stratified viscous fluid flows have been simulated on the basis of the Navier-Stokes equations in the Boussinesq approximation (1)-(2) (including the diffusion equation (1) for the stratified component (salt)) with four dimensionless parameters: $Fr = U/(N \cdot d), Re = U \cdot d/\nu, A \gg 1, Sc = \nu/\kappa = 709.22$, where U is the scalar of the body velocity, ν is the kinematical viscosity, κ is the salt diffusion coefficient.

$$\frac{\partial S}{\partial t} + (\mathbf{v} \cdot \nabla) S = \frac{2}{Sc \cdot Re} \Delta S + \frac{v_x}{2A} \tag{1}$$

$$\frac{\partial \mathbf{v}}{\partial t} + (\mathbf{v} \cdot \nabla) \mathbf{v} = -\nabla p + \frac{2}{Re} \Delta \mathbf{v} + \frac{A}{2Fr^2} S \frac{\mathbf{g}}{g} \qquad \nabla \cdot \mathbf{v} = 0 \tag{2}$$

In (1)-(2) $\mathbf{v} = (v_x, v_y, v_z)$ is the velocity vector (non-dimensionalized by U), p is a perturbation of pressure (non-dimensionalized by $\rho_0 U^2$).

For solving of the Navier-Stokes equations (1)-(2) the Splitting on physical factors Method for Incompressible Fluid flows (SMIF) has been used [1,2].

Let the velocity, the perturbation of pressure and the perturbation of salinity are known at some moment $t_n = n \cdot \tau$, where τ is time step, and n is the number of time-steps. Then the calculation of the unknown functions at the next time level $t_{n+1} = (n + 1) \cdot \tau$ for equations (1)-(2) can be presented in the following four-step form:

$$\frac{S^{n+1} - S^n}{\tau} = -(\mathbf{v}^n \cdot \nabla) S^n + \frac{2}{Sc \cdot Re} \Delta S^n + \frac{v_x^n}{2A} \tag{3}$$

$$\frac{\tilde{\mathbf{v}} - \mathbf{v}^n}{\tau} = -(\mathbf{v}^n \cdot \nabla) \mathbf{v}^n + \frac{2}{Re} \Delta \mathbf{v}^n + \frac{A}{2Fr^2} S^{n+1} \frac{\mathbf{g}}{g} \tag{4}$$

$$\tau \Delta p = \nabla \cdot \tilde{\mathbf{v}} \qquad \frac{\mathbf{v}^{n+1} - \tilde{\mathbf{v}}}{\tau} = -\nabla p \tag{5}$$

The Poisson equation for the pressure (5) has been solved by the diagonal Preconditioned Conjugate Gradients Method.

In order to understand the finite-difference scheme for the convective terms of the equations (1)-(2) let us consider the linear model equation and a finite-difference approximation of this equation:

$$f_t + uf_x = 0, \quad u = const; \qquad \frac{f_i^{n+1} - f_i^n}{\tau} + u\frac{f_{i+1/2}^n - f_{i-1/2}^n}{h} = 0 \qquad (6)$$

Let us investigate the class of the difference scheme which can be written in the form of the two-parameter family which depends on the parameters α and β in the following manner:

$$f_{i+1/2}^n = \alpha f_{i-1}^n + (1 - \alpha - \beta)f_i^n + \beta f_{i+1}^n, \quad if \quad u \geq 0,$$
$$f_{i+1/2}^n = \alpha f_{i+2}^n + (1 - \alpha - \beta)f_{i+1}^n + \beta f_i^n, \quad if \quad u < 0. \qquad (7)$$

In this case the first differential approximation for equation (6) has the form

$$f_t + uf_x = [\alpha - \beta + \gamma] \cdot h \cdot |u| \cdot f_{xx}, \qquad (8)$$

where $\gamma = 0.5 \cdot (1 - C)$ and $C = \tau|u|/h$ is the Courant number.

If we put $\alpha = \beta = 0$ in (7) we'll obtain usual first order monotonic scheme which is stable when $0 \leq C \leq 1$.

It is known that it is impossible to construct a homogeneous monotonic difference scheme of higher order than the first order of the approximation for equation (6). A monotonic scheme of higher order can therefore only be constructed either on the basis of second-order homogeneous scheme using smoothing operators, or on the basis of the hybrid schemes using different switch conditions from one scheme to another (depending on the nature of the solution), possibly with the use of smoothing. Here we are going to consider a hybrid monotonic difference scheme.

Let us investigate schemes with upwind differences (UD), i.e. $\beta = 0$. The requirement that the scheme viscosity should be a minimum, as can readily be seen from equation (8), impose the following condition on $\alpha = -\gamma$ (the modified upwind difference scheme (MUDS)).

For schemes with $\alpha = 0$, the analogous condition is $\beta = \gamma$ (the modified central difference scheme (MCDS)).

Since an explicit finite difference scheme considered, we shall restrict the subsequent analysis to the necessary condition $0 \leq C \leq 1$ for stability.

Let us assume that there is a monotonic net function f_i^n , for example, $\Delta f_{i+1/2}^n \equiv f_{i+1}^n - f_i^n \geq 0$ at any i.

The function f_i^{n+1} will also be monotonic when the following conditions are satisfied: (a) $\Delta f_{i+1/2}^n \geq \zeta(C) \cdot \Delta f_{i-1/2}^n$ (for MUDS); b) $\Delta f_{i+1/2}^n \leq \sigma(C) \cdot \Delta f_{i-1/2}^n$ (for MCDS), where $\zeta(C) = 0.5 \cdot (1 - C)/(2 - C)$; $\sigma(C) = 2 \cdot (1 + C)/C$.

It can be seen from this that the domains of monotonicity of the homogeneous schemes being considered have a non-empty intersection. Hence, a whole class of hybrid schemes is distinguished by the condition of switching from one homogeneous scheme to another. The general form of this condition is as follows: $\Delta f_{i+1/2}^n = \delta \cdot \Delta f_{i-1/2}^n$, where $\zeta(C) \leq \delta \leq \sigma(C)$.

The choice of $\delta = 1$ corresponds to the points of the interchange of the sign of the second difference f_i^n and makes it possible to obtain the estimation $f_{xx} = O(h)$ for the required function f at the intersection points, by means of which a second-order approximation is retained with respect to the spatial variables of smooth solutions. We used the following switching condition: if $(u \cdot \Delta f \Delta^2 f)_{i+1/2}^n \geq 0$, then MUDS is used; if $(u \cdot \Delta f \Delta^2 f)_{i+1/2}^n < 0$, then MCDS is used; where $\Delta^2 f_{i+1/2}^n = \Delta f_{i+1}^n - \Delta f_i^n$.

On smooth solutions this scheme has a second order of approximation with respect to the time and spatial variables. It is stable when the Courant criterion $0 \leq C \leq 1$ is satisfied and monotonic. More over it was shown that this hybrid scheme comes nearest to the third order schemes. The generalization of the considered finite-difference scheme for 2D and 3D problems is easily performed for convective terms in equations (1) - (2). For the approximation of other space derivatives in equations (1) - (2) the central differences are used.

The efficiency of the method SMIF and the greater power of supercomputers make it possible adequately to model the 3D separated incompressible viscous flows past a sphere and a circular cylinder at moderate Reynolds numbers [3–12] and the air, heat and mass transfer in the clean rooms [13].

3   The Visualization Techniques

For the visualization of the 3D vortex structures in the fluid flows the isosurfaces of the streamwise component of vorticity w_x (Fig. 1, $\mathbf{w} = rot\mathbf{v}$) and the isosurfaces of β and λ_2 have been drawing, where β is the imaginary part of the complex-conjugate eigen-values of the velocity gradient tensor $\mathbf{G}$ [14] (Fig. 2b), λ_2 is the second eigen-value of the $\mathbf{S}^2 + \Omega^2$ tensor, where $\mathbf{S}$ and Ω are the symmetric and antisymmetric parts of $\mathbf{G}$ [4,15] (Fig. 2a). The good efficiency of this β-visualization technique has been demonstrated in [5,6,9].

4   Some Fluid Flow Regimes Around a Circular Cylinder

Let us consider the homogeneous viscous fluid flow regimes around a circular cylinder. At $Re > 40$ the periodical formation of vortex tubes is simulated in

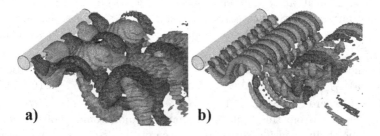

a)   **b)**

Fig. 1. The isosurfaces of the streamwise component of vorticity w_x ($\mathbf{w} = rot\mathbf{v}$) for the circular cylinder with length $L = 7.5 \cdot d$: a) $Re = 240$ (mode A); b) $Re = 320$ (mode B).

Fig. 2. Vortex structures of the sphere wake at $Fr > 10$: a-c - $Re = 200, 250, 350$; a) $\lambda_2 = -10^{-6}$ and -0.16; b) $\beta = 0.04$; c) $\lambda_2 = -2 \cdot 10^{-5}$.

the wake. At $Re = 191$ 2D-3D transition is observed in the wake. It means that at $Re > 191$ there is a periodicity of the flow along the circular cylinder axis. At $191 < Re < 300$ and $300 \leq Re \leq 400$ the periodicity scales are equal to $3.5 \cdot d \leq \lambda \leq 4 \cdot d$ (mode A) and $0.8 \cdot d \leq \lambda \leq 1.0 \cdot d$ (mode B) correspondingly (Fig. 1). Owing to our investigations it was found that the values of the maximum phase difference along the circular cylinder axis are approximately equal to $0.1 - 0.2 \cdot T$ (for mode A) and $0.015 - 0.030 \cdot T$ (for mode B), where the time T is the period of the flow [3].

5 The Classification of Fluid Flow Regimes Around a Sphere

For $Fr > 10$ the homogeneous viscous fluid flow regime **I** is observed in the sphere wake. The following classification of the flow subregimes of **I** [5] has been obtained by SMIF at $Re \leq 700, Fr > 10$: **I** – **1**) $Re \leq 20.5$ - a steady axisymmetrical flow without separation; **I** – **2**) $20.5 < Re \leq 200$ - a steady axisymmetrical wake with a vortex ring in the recirculation zone (**RZ**) and a vortex sheet (**VSh**) surrounding **RZ** (Fig. 2a); **I** – **3**) $200 < Re \leq 270$ a steady double-thread wake with a deformed vortex ring in **RZ** (Fig. 2b); **I** – **4**) $270 < Re \leq 400$ a periodical generation of the vortex loops (**VLs**) (facing upwards), the periodical separation of the one edge of **VSh** (Fig. 2c) (at $270 < Re \leq 290, 290 < Re \leq 320$ and $320 < Re \leq 400$ the different chains of the basic formation mechanisms of vortices (FMV) are realized in **RZ**); **I** – **5**) $400 < Re \leq 700$ - the periodical separation of the opposite edges of the irregularly rotating **VSh**.

For $Fr \leq 10$ let us first consider the flow regimes at $100 < Re < 700$.

For **flow regimeII**($1.5 \leq Fr \leq 10$ - the quasi-homogeneous case (with four additional threads connected with **VSh** surrounding the sphere, Fig. 3a)) the boundaries of the analogous sub-regimes II-1, II-2, II-3, II-4, II-5 are slightly shifted. The vortex structures are flattened in vertical direction and combined with four additional vortex threads. With decreasing of Fr from 10 to 2 the vortex ring is deformed in an oval. In the vertical plane the part of fluid is supplied in RZ. Then this fluid goes through the core of the vortex oval and is emitted downstream in the horizontal plane. The 3D instantaneous stream lines which are going near the sphere surface go around this vortex oval and form the four vortex threads (see Fig. 3a for the sub-regime **II** – **2**).

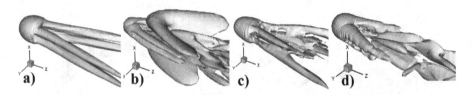

Fig. 3. The vortex structures of the stratified fluid around a moving sphere at $Re = 100$ (a-b) and $Re = 250$ (c-d): a-d- $Fr = 2, 1, 2, 1$; a-d- the isosurfaces of $\beta = 0.055; 0.02; 0.04; 0.08$.

For subregime **II-3** (at $Re = 250, Fr = 2$) the six-threads wake is observed (Fig. 3c). The domination of the four additional threads is obvious here. Unlike $Fr > 10$ at $Fr = 2, Re = 250$ the unsteadiness in the form of the periodical fluctuation of the rear stagnation point around axis z is observed.

At $Fr < 1.5 (200 < Re \leq 700)$ the big initial vertical flattening of the flow prevent the vortex formation mechanisms typical for the homogeneous fluid. At $Fr < 1.5$ the new vortex formation mechanisms (which are typical for the stratified fluid) are realized with increasing of Re. With decreasing of Fr the fluid structures around the sphere are slowly flattened both along the vertical axis x and along the line of the sphere motion (along axis z) (the length of the internal waves (**IWs**) in the vertical plane is $\lambda/d \approx 2 \cdot \pi \cdot Fr$). For example two free foci in $y - z$ plane are approaching to the sphere with decreasing of Fr. The length of four threads (connected with the vortex sheet surrounding the sphere) is also diminished with reducing of Fr and at $Fr \leq 0.1$ these threads are transformed in the sheets of density (**DSh**) before the sphere (Fig. 4b).

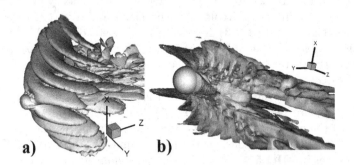

Fig. 4. The vortex structures of the stratified fluid around a moving sphere at $Re = 100$: a-b- $Fr = 0.5, 0.08$; a-b- the isosurfaces of $\beta = 0.02; 0.005$.

Thus the following classification of the viscous fluid flow regimes around a sphere at $100 < Re < 700$ and $Fr \leq 10$ [6,9] has been obtained by SMIF: **II**)$1.5 \leq Fr \leq 10$ the quasi-homogeneous case (with four additional threads connected with VSh surrounding the sphere, Fig. 3a); **III – 2**)$0.9 < Fr < 1.5$ - the non-axisymmetric attached vortex in RZ, Fig. 3b; **IV – 2**)$0.6 < Fr \leq 0.9$

the two symmetric vortex loops in RZ; $\mathbf{V} - \mathbf{2})0.4 \leq Fr \leq 0.6$ the absence of RZ (Fig. 4a); $\mathbf{VI} - \mathbf{2})0.25 < Fr < 0.4$ - a new RZ; $\mathbf{VII} - \mathbf{2})Fr \leq 0.25$ - the two vertical vortices in new RZ (bounded by IWs) (Fig. 4b). At $Fr \leq 0.3, Re > 120$ a periodical generation of the vortex loops (facing left or right) has been observed. The corresponding Strouhal numbers $0.19 < St = f \cdot d/U < 0.24$ (where f is the frequency of shedding) and horizontal and vertical separation angles are in a good agreement with the experiment [16]. The increments of the drag coefficient $\Delta \ C_d \ (Fr) = C_d(Fr) - C_d(\infty)$ are in a good agreement with the experiment [17,18] conducted at $Re > 150$.

Let us now consider the flow regimes at $Re = 10, Fr \leq 10$. At $Fr < 1$ the increments of the drag coefficient $\Delta \ C_d(Fr)$ for $Re = 10$ are $2 \div 6$ times greater than for $100 < Re < 700$ (for $Fr = 0.004, 0.05, 0.1, 0.2, 0.3, \ 0.4, 0.6, 0.8, 1.0, 2.0$ $\Delta C_d \ = \ 1.461, 1.546, 1.626, 1.816, \ 1.783, 1.481, 0.628, 0.076, 0.179, 0.164$ correspondingly). At $Re = 10$ the following classification of viscous fluid flow regimes around a sphere has been obtained by SMIF: $\mathbf{II} - \mathbf{1})0.32 \leq Fr < 10$ - the quasi-homogeneous case - the flow without separation with the vertical and horizontal symmetry planes; $\mathbf{V} - \mathbf{1})0.25 < Fr < 0.32$ - the nearest wave crest on axis z (in wake) is very close to the sphere; $\mathbf{VI} - \mathbf{1})1.6 < Fr \leq 0.25$ - a new RZ (generated from the nearest wave crest) - the two vertical vortices in new RZ (bounded by IWs); $\mathbf{VII} - \mathbf{1})Fr \leq 0.16$ - the absence of RZ.

Acknowledgments. This work has been partly supported by Russian Foundation for Basic Research (grants No. $13 - 01 - 92696, 14 - 01 - 00428$), by the programs of the Presidium of RAS No. $15, 18$ and by the program No. 3 of the Department of Mathematical Sciences of RAS.

References

1. Belotserkovskii, O.M., Gushchin, V.A., Konshin, V.N.: Splitting method for studying stratified fluid flows with free surfaces. USSR Comput. Math. Math. Phys. **27**(2), 181–196 (1987)
2. Gushchin, V.A., Konshin, V.N.: Computational aspects of the splitting method for incompressible flow with a free surface. J. Comput. Fluids **21**(3), 345–353 (1992)
3. Gushchin, V.A., Kostomarov, A.V., Matyushin, P.V., Pavlyukova, E.R.: Direct numerical simulation of the transitional separated fluid flows around a sphere and a circular cylinder. J. Wind Eng. Ind. Aerodyn. **90**(4–5), 341–358 (2002)
4. Gushchin, V.A., Kostomarov, A.V., Matyushin, P.V.: 3D visualization of the separated fluid flows. J. Vis. **7**(2), 143–150 (2004)
5. Gushchin, V.A., Matyushin, P.V.: Vortex formation mechanisms in the wake behind a sphere for $200 < \ Re \ < 380$. Fluid Dynamics **41**(5), 795–809 (2006)
6. Gushchin, V.A., Matyushin, P.V.: Numerical simulation and visualization of vortical structure transformation in the flow past a sphere at an increasing degree of stratification. Comput. Math. Math. Phys. **51**(2), 251–263 (2011)
7. Gushchin, V.A., Rozhdestvenskaya, T.I.: Numerical study of the effects occurring near a circular cylinder in stratified fluid flows with short buoyancy periods. J. Appl. Mech. Tech. Phys. **52**(6), 905–911 (2011)

8. Gushchin, V.A., Mitkin, V.V., Rozhdestvenskaya, T.I., Chashechkin, Y.D.: Numerical and experimental study of the fine structure of a stratified fluid flow over a circular cylinder. J. Appl. Mech. Tech. Phys. **48**(1), 34–43 (2007)

9. Matyushin, P.V., Gushchin, V.A.: Transformation of Vortex Structures in the wake of a sphere moving in the stratified fluid with decreasing of internal Froude Number. J. Phys.Conf. Ser. **318**, 062017 (2011)

10. Matyushin, P.V., Gushchin, V.A.: Direct numerical simulation of the 3D separated viscous fluid flows around the horizontally moving blunt bodies.In: ECCOMAS 2012, e-Book Full Papers, pp. 5232–5241 (2012)

11. Gushchin, V.A., Matyushin, P.V.: Mathematical modeling of the 3D separated viscous fluid flows. AIP Conf. Proc. **1487**, 22–29 (2012)

12. Gushchin, V., Matyushin, P.: Method SMIF for incompressible fluid flows modeling. In: Dimov, I., Faragó, I., Vulkov, L. (eds.) NAA 2012. LNCS, vol. 8236, pp. 311–318. Springer, Heidelberg (2013)

13. Gushchin, V.A., Narayanan, P.S., Chafle, G.: Parallel computing of industrial aerodynamics problems: clean rooms. In:Schiano, P., Ecer, A., Periaux, J., Satofuka, N. (eds.) Parallel CFD 1997, ELSEVIER Science B.V. (1997)

14. Chong, M.S., Perry, A.E., Cantwell, B.J.: A general classification of three-dimensional flow field. Phys. Fluids **A2**(5), 765–777 (1990)

15. Jeong, J., Hussain, F.: On the identification of a vortex. J. Fluid Mech. **285**, 69–94 (1995)

16. Lin, Q., Lindberg, W.R., Boyer, D.L., Fernando, H.J.S.: Stratified flow past a sphere. J. Fluid Mech. **240**, 315–354 (1992)

17. Lofquist, K.E.B., Purtell, L.P.: Drag on a sphere moving horizontally through a stratified liquid. J. Fluid Mech. **148**, 271–284 (1984)

18. Mason, P.J.: Forces on spheres moving horizontally in a rotating stratified fluid. Geophys. Astrophys. Fluid Dyn. **8**, 137–154 (1977)

Determination of the Time-Dependent Thermal Conductivity in the Heat Equation with Spacewise Dependent Heat Capacity

M.S. Hussein[1,2(✉)] and D. Lesnic[1]

[1] Department of Applied Mathematics, University of Leeds, Ls2 9jt, Leeds, UK
mmmsh@leeds.ac.uk, amt5ld@maths.leeds.ac.uk
[2] Department of Mathematics, University of Baghdad, College of Science, Baghdad, Iraq

Abstract. In this paper, we consider an inverse problem of determining the time-dependent thermal conductivity from Cauchy data in a one-dimensional heat equation with space-dependent heat capacity. The parabolic partial differential equation is discretised using the finite - difference method and the inverse problem is recast as a nonlinear least-squares minimization. This is solved using the *lsqnonlin* routine from the MATLAB toolbox. Numerical results are presented and discussed showing that accurate and stable numerical solutions are achieved.

Keywords: Inverse problem · Finite-difference method · Thermal conductivity

1   Introduction

The scope of inverse problems has existed in various branches of physics, engineering and mathematics for a long time. The theory of inverse problems has been extensively developed within the past decade due partly to its importance in applications; on the other hand the numerical solutions to such problems need huge computations and also reliable numerical methods. For instance, deconvolution in seismic exploration, image reconstruction and parameter identification all require high performance computers and reliable solution methods to carry out the computation [7].

Parameter identification problems consists in using the input of actual observation or indirect measurement, contaminated with noise, to infer the values of the parameters characterizing the system under investigation. Often, these inverse problems are ill-posed according to the Hadamard concept which is: if the solution does not exist or, is not unique or, if it violates the continuous dependence upon input data. Most identification problems satisfy the first two conditions and violate the third one which is the stability.

Determination of leading coefficient or, the coefficient of highest-order derivative in the parabolic heat equation has been investigated widely. For example,

© Springer International Publishing Switzerland 2015
I. Dimov et al. (Eds.): FDM 2014, LNCS 9045, pp. 217–224, 2015.
DOI: 10.1007/978-3-319-20239-6_22

in [2] the problem of space-dependent diffusivity identification has been studied, while the time-dependent case has been investigated in [4]. Also, for the temperature-dependent case we refer to [1,8].

In this paper, we consider obtaining the numerical solution of inverse time-dependent multiplier of the highest-order derivative in the parabolic heat equation for which its unique solvability has previously been established/proved by Ivanchov [3, Sect. 4.3]. Physically, this unknown thermal property coefficient corresponds to the thermal conductivity of the heat conducting system which has also a space-varying known heat capacity.

The mathematical formulation for the inverse nonlinear problem is described in Sect. 2. The numerical method is based on the Crank-Nicholson finite-difference scheme used as direct solver in a least-squares minimization, as described in Sects. 3 and 4, respectively. This combination yields an accurate and stable numerical solution, as it is discussed in Sect. 5. Finally, the conclusions of this research are highlighted in Sect. 6.

2   Mathematical Formulation

Let $L > 0$ and $T > 0$ be fixed numbers and consider the inverse problem of finding the time-dependent thermal conductivity $C[0,T] \ni a(t) > 0$ for $t \in [0,T]$ and the temperature $u(x,t) \in C^{2,1}(Q_T) \cap C^{1,0}(\overline{Q}_T)$, which satisfy the heat equation

$$c(x)\frac{\partial u}{\partial t}(x,t) = a(t)\frac{\partial^2 u}{\partial x^2}(x,t) + F(x,t), \qquad (x,t) \in Q_T := (0,L) \times (0,T), \quad (1)$$

where $c(x) > 0$ is the heat capacity and F is a heat source, the initial condition

$$u(x,0) = \phi(x), \qquad\qquad x \in [0,L], \qquad (2)$$

the Dirichlet boundary conditions

$$u(0,t) = \mu_1(t), \qquad\qquad u(h,t) = \mu_2(t), \qquad\qquad t \in [0,T], \quad (3)$$

and the heat flux additional measurement

$$- a(t)u_x(0,t) = \mu_3(t), \qquad\qquad t \in [0,T]. \qquad (4)$$

Dividing Eq. (1) by $c(x)$ and denoting

$$b(x) = \frac{1}{c(x)}, \quad f(x,t) = \frac{F(x,t)}{c(x)} \qquad\qquad (5)$$

we obtain

$$\frac{\partial u}{\partial t}(x,t) = a(t)b(x)\frac{\partial^2 u}{\partial x^2}(x,t) + f(x,t), \qquad\qquad (x,t) \in Q_T. \qquad (6)$$

This inverse problem was previously investigated theoretically by Ivanchov [3] who established its unique solvability as follows:

Theorem 1. (Existence)

Suppose that the following conditions hold:

1. $b \in C^1[0, L]$, $\phi \in C^1[0, L]$, $\mu_i \in C^1[0, T]$, $i = 1, 2$, $\mu_3 \in C[0, T]$, $f \in C^{1,0}(\overline{Q}_T)$;
2. $b(x) > 0$, $\phi'(x) > 0$, $b'(x) \leq 0$ for $x \in [0, L]$; $\mu_3(t) < 0$, $\mu_1'(t) - f(0, t) \leq 0$,
* $\mu_2'(t) - f(L, t) \geq 0$ for $t \in [0, T]$, $f_x(x, t) \geq 0$ for $(x, t) \in \overline{Q}_T$;*
3. $\phi(0) = \mu_1(0)$, $\phi(L) = \mu_2(0)$.

Then there exists a solution to the inverse problem (2)–(4) and (6).

Theorem 2. (Uniqueness)

If $b \in C^1[0, L]$, $b(x) > 0$ for $x \in [0, L]$, $\mu_3(t) \neq 0$ for $t \in [0, T]$, then the solution of the inverse problem (2)–(4) and (6) is unique.

3   Solution of Direct Problem

In this section, we consider the direct initial boundary value problem given by equations (2), (3) and (6), where $a(t)$, $b(x)$, $f(x, t)$, $\phi(x)$ and $\mu_i(t)$, $i = 1, 2$, are known and the temperature $u(x, t)$ is the solution to be determined. We use the finite-difference method (FDM) with a Crank-Nicholson scheme [5], which is unconditionally stable and second-order accurate in space and time.

The discrete form of direct problem is as follows. We subdivided the domain $Q_T = (0, L) \times (0, T)$ into $M \times N$ subintervals of equal step length $\Delta x = L/M$ and $\Delta t = T/N$, respectively. At the node (i, j) we denote $u_{i,j} = u(x_i, t_j)$, $a(t_j) = a_j$, $b(x_i) = b_i$ and $f(x_i, t_j) = f_{i,j}$, where $x_i = i\Delta x$, $t_j = j\Delta t$, for $i = \overline{0, M}$, $j = \overline{0, N}$.

Considering the general partial differential equation

$$u_t = G(x, t, u_{xx}), \tag{7}$$

the Crank-Nicolson method is basically based on central finite-difference approximations for space and forward finite-difference approximations for time which gives second-order convergence rate. This method is equivalent to take average of forward and backward Euler schemes in time, hence Eq. (7) can approximated as:

$$\frac{u_{i,j+1} - u_{i,j}}{\Delta t} = \frac{1}{2}\left(G_{i,j} + G_{i,j+1}\right), i = \overline{1, (M-1)}, \ j = \overline{0, (N-1)}, \tag{8}$$

$$u_{i,0} = \phi(x_i), i = \overline{0, M}, \tag{9}$$

$$u_{0,j} = \mu_1(t_j), u_{M,j} = \mu_2(t_j), j = \overline{0, N}, \tag{10}$$

where

$$G_{i,j} = G\left(x_i, t_j, \frac{u_{i+1,j} - 2u_{i,j} + u_{i-1,j}}{(\Delta x)^2}\right),$$
$$i = \overline{1, (M-1)}, \ j = \overline{0, (N-1)}. \tag{11}$$

For our problem, Eq. (1) can be discretised in the form of (8) as

$$-A_{i,j+1}u_{i-1,j+1} + (1 + B_{i,j+1})u_{i,j+1} - C_{i,j+1}u_{i+1,j+1} =$$
$$A_{i,j}u_{i-1,j} + (1 - B_{i,j})u_{i,j} + C_{i,j}u_{i+1,j} + \frac{\Delta t}{2}\left(f_{i,j} + f_{i,j+1}\right) \tag{12}$$

for $i = \overline{1, (M-1)}$, $j = \overline{0, N}$, where

$$A_{i,j} = C_{i,j} = \frac{(\Delta t)a_j b_i}{2(\Delta x)^2} \quad B_{i,j} = \frac{(\Delta t)a_j b_i}{(\Delta x)^2}.$$

At each time step t_{j+1}, for $j = \overline{0, (N-1)}$, using the Dirichlet boundary conditions (10), the above difference equation can be reformulated as a $(M-1) \times (M-1)$ system of linear equations of the form,

$$D\mathbf{u_{j+1}} = E\mathbf{u_j} + \mathbf{b}, \tag{13}$$

where

$$\mathbf{u_{j+1}} = (u_{1,j+1}, u_{2,j+1}, ..., u_{M-1,j+1})^{tr},$$

$$D = \begin{pmatrix} 1 + B_{1,j+1} & -C_{1,j+1} & 0 & \cdots & 0 & 0 & 0 \\ -A_{2,j+1} & 1 + B_{2,j+1} & -C_{2,j+1} & \cdots & 0 & 0 & 0 \\ \vdots & \vdots & \vdots & \ddots & \vdots & \vdots & \vdots \\ 0 & 0 & 0 & \cdots & -A_{M-2,j+1} & 1 + B_{M-2,j+1} & -C_{M-2,j+1} \\ 0 & 0 & 0 & \cdots & 0 & -A_{M-1,j+1} & 1 + B_{M-1,j+1} \end{pmatrix},$$

$$E = \begin{pmatrix} 1 - B_{1,j} & C_{1,j} & 0 & \cdots & 0 & 0 & 0 \\ A_{2,j} & 1 - B_{2,j} & C_{2,j} & \cdots & 0 & 0 & 0 \\ \vdots & \vdots & \vdots & \ddots & \vdots & \vdots & \vdots \\ 0 & 0 & 0 & \cdots & A_{M-2,j} & 1 - B_{M-2,j} & C_{M-2,j} \\ 0 & 0 & 0 & \cdots & 0 & A_{M-1,j} & 1 - B_{M-1,j} \end{pmatrix},$$

and

$$\mathbf{b} = \begin{pmatrix} \frac{\Delta t}{2}(f_{1,j} + f_{1,j+1}) + A_{1,j+1}\mu_1(t_j) \\ \frac{\Delta t}{2}(f_{2,j} + f_{2,j+1}) \\ \vdots \\ \frac{\Delta t}{2}(f_{M-2,j} + f_{M-2,j+1}) \\ \frac{\Delta t}{2}(f_{M-1,j} + f_{M-1,j+1}) + C_{M-1,j+1}\mu_2(t_j) \end{pmatrix}.$$

4   Numerical Approach to the Inverse Problem

In the inverse problem, we wish to obtain stable and accurate reconstructions of the time-dependent thermal conductivity $a(t)$ and the temperature $u(x,t)$ satisfying the Eqs. (2)–(4) and (6). The most common approach based on imposing the measurement (4) in a least-squares sense, namely, is minimizing

$$F(a) := \|a(t)u_x(0,t) + \mu_3(t)\|^2, \tag{14}$$

where the norm is usually the $L^2[0, T]$-norm. The discretization of (14) yields

$$F(\underline{a}) = \sum_{j=0}^{N} \left[a_j u_x(0, t_j) + \mu_3(t_j) \right]^2. \tag{15}$$

The minimization of the objective function (15) subject to the physical simple lower bound constraints $\underline{a} > \underline{0}$ is accomplished using the MATLAB toolbox routine *lsqnonlin*, which does not require supplying by the user the gradient of the objective function, [6].

This routine attempts to find a minimum of a scalar function of several variables, starting from an initial guess, subject to constraints and this generally is referred to as a constrained nonlinear optimization. We use the Trust-Region-Reflective (TRR) algorithm from *lsqnonlin*, [6], and the positive components of the vector $\underline{a}$ are sought in the interval $(10^{-10}, 10^3)$.

We also take the parameters of the routine as follows:

- Number of variables $M = N = 40$.
- Maximum number of iterations $= 10^2 \times$ (number of variables).
- Maximum number of objective function evaluations
  $= 10^3 \times$ (number of variables).
- Solution Tolerance (aTol) $= 10^{-20}$.
- Object function Tolerance (FunTol) $= 10^{-20}$.
- Nonlinear constraint tolerance $= 10^{-6}$.

In our problem, we take the initial guess as $\underline{a}^{(0)} = \underline{1}$. It is worth mentioning that at the first time step, i.e. $j = 0$, the derivative $u_x(0, 0)$ is obtained from the initial condition as

$$u_x(0, 0) = \frac{4\phi_1 - \phi_2 - 3\phi_0}{2(\Delta x)}, \tag{16}$$

where $\phi_i = \phi(x_i)$ for $i = \overline{0, M}$.

The inverse problem under investigation is solved subject to both exact and noisy heat flux measurements, $\mu_3(t)$. The noisy data is numerically simulated as

$$\mu_3^\epsilon(t_j) = \mu_3(t_j) + \epsilon_j, \quad j = \overline{0, N}, \tag{17}$$

where ϵ_j are random variables generated from a Gaussian normal distribution with mean zero and standard deviation σ given by

$$\sigma = p \times \max_{t \in [0,T]} |\mu_3(t)|, \tag{18}$$

where p represents the percentage of noise. We use the MATLAB function *norm-rnd* to generate the random variables $\underline{\epsilon} = (\epsilon_j)_{j=\overline{0,N}}$ as follows:

$$\underline{\epsilon} = normrnd(0, \sigma, N + 1). \tag{19}$$

5   Numerical Results and Discussion

In this section, we present a test example to illustrate the accuracy and stability of the numerical scheme based on the FDM combined with the minimization of the least-squares functional (15), as described in Sect. 4.

We take $L = T = 1$ and present the numerical results obtained with $M = N = 40$ for the inverse problem (2)–(4) and (6) with the input data

$$\phi(x) = u(x,0) = x + \sin(x), \quad b(x) = 2 - x^2,$$
$$\mu_1(t) = u(0,t) = 8t, \quad \mu_2(t) = u(1,t) = 1 + \sin(1) + 8t,$$
$$f(x,t) = 8 + (1+t)(2-x^2)\sin(x), \quad \mu_3(t) = -a(t)u_x(0,t) = -2 - 2t.$$

One can observe that the conditions of Theorems 1 and 2 are satisfied hence the problem is uniquely solvable. The analytical solution is given by

$$a(t) = 1 + t, \quad u(x,t) = x + \sin(x) + 8t. \tag{20}$$

Although not illustrated, it is reported that excellent agreement between the exact and numerical solutions for $(a(t), u(x,t))$ has been obtained for exact data, i.e. $p = 0$.

We next investigate the stability of the numerical solution with respect to the noise in the data (4), defined by Eq. (17). Although not illustrated, it is reported that a decreasing convergence of the objective functional (15) is achieved for $p \in \{2\%, 20\%\}$ noise in 8 iterations each, to reach a stationary value of $O(10^{-24})$.

Figure 1 shows the numerical solutions for the thermal conductivity $a(t)$, for $p \in \{2\%, 20\%\}$ noise. From this figure, it can be seen that the numerical solution for the thermal conductivity coefficient $a(t)$ converges to the exact solution $a(t) = 1 + t$, as the percentage of noise p decreases from 20 % to 2 % and then to 0. The nonlinear least-squares minimization produces good and consistent retrievals of the solution even for a large amount of noise such as 20 %. That is to say, our inverse problem is rather stable and in fact no regularization was needed to be included in the least-squares functional (15).

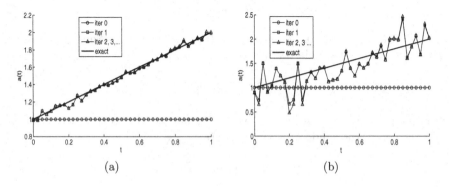

(a)   (b)

Fig. 1. The thermal conductivity $a(t)$ for (a) $p = 2\%$ and (b) $p = 20\%$ noise.

Figure 2 shows the exact solution, the numerical solution for the temperature $u(x,t)$ and the relative error between them. From this figure it can be seen that the numerical solution is stable and furthermore, its accuracy is consistent with the amount of noise included into the input data (4).

Numerical outputs such as the number of iterations and function evaluations, as well as the final convergent value of the objective function are provided in Table 1.

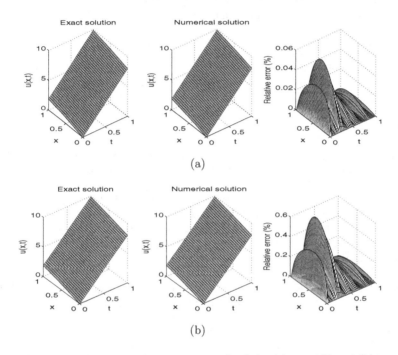

Fig. 2. The exact and numerical temperature $u(x,t)$ for (a) $p = 2\%$ and (b) $p = 20\%$ noise. The relative error between them is also included.

Table 1. Number of iterations, number of function evaluations, and value of objective function (15) at final iteration, for $p \in \{0, 2\%, 20\%\}$ noise.

	$p = 0$	$p = 2\%$	$p = 20\%$
No. of iterations	10	8	8
No. of function evaluations	451	328	328
Function value	$1.7E - 24$	$1.1E - 24$	$1.6E - 24$

6 Conclusions

The inverse problem which requires determining a time-dependent thermal conductivity when the spacewise dependent heat capacity is given for the heat equation under an overspecified heat flux boundary condition has been investigated. The direct solver based on a Crank-Nicolson finite difference scheme has been developed. The inverse problem has been reduced to a nonlinear least-squares minimization which has been solved using the MATLAB toolbox routine *lsqnonlin*. Numerical results show that an accurate and stable solution has been obtained.

Acknowledgments. M.S. Hussein would like to thank the Higher Committee of Education Development in Iraq (HCEDiraq) for their financial support.

References

1. Cannon, J.R., DuChateau, P.: Determining unknown coefficients in a nonlinear heat conduction problem. SIAM J. Appl. Math. **24**, 298–314 (1973)
2. Doris, H.G., Peralta, J., Luis, E.O.: Regularization algorithm within two parameters for the identification of the heat conduction coefficient in the parabolic equation. Math. Comput. Model. **57**, 1990–1998 (2013)
3. Ivanchov, M.I.: Inverse Problems for Equations of Parabolic Type. VNTL Publications, Lviv, Ukraine (2003)
4. Lesnic, D., Yousefi, S.A., Ivanchov, M.: Determination of a time-dependent diffusivity from nonlocal conditions. J. Appl. Math. Comput. **41**, 301–320 (2013)
5. Smith, G.D.: Numerical Solution of Partial Differential Equations: Finite Difference Methods, Oxford Applied Mathematics and Computing Science Series, Third Edition (1985)
6. Mathworks R2012 Documentation Optimization Toolbox-Least Squares (Model Fitting) Algorithms. www.mathworks.com/help/toolbox/optim/ug/brnoybu.html
7. Wang, Y., Yang, C., Yagola, A.: Optimization and Regularization for Computational Inverse Problems and Applications. Springer-Verlag, Berlin (2011)
8. Wang, P., Zheng, K.: Determination of an unknown coefficient in a nonlinear heat equation. J. Math. Anal. Appl. **271**, 525–533 (2002)

Inverse Problems of Simultaneous Determination of the Time-Dependent Right-Hand Side Term and the Coefficient in a Parabolic Equation

Vitaly L. Kamynin[✉]

National Research Nuclear University MEPhI (Moscow Engineering Physics Institute), 31, Kashirskoe Shosse, 115409 Moscow, Russia
vlkamynin2008@yandex.ru

Abstract. This work deals with the inverse problem of simultaneous determination of two unknown time-dependent terms in a one-dimensional parabolic equation. The additional information is given by two integral observations. We prove theorems of existence and uniqueness of solution. We also give estimates of maximum modulus of unknown right-hand side and unknown coefficient of the equation with constants derived explicitly in terms of input data.

1 Introduction

In present paper we study the existence and the uniqueness of the solution $\{u(t,x), d(t), f(t)\}$ of the inverse problem

$$u_t - a(t,x)u_{xx} + b(t,x)u_x + d(t)u = f(t)g(t,x), \ (t,x) \in Q; \tag{1}$$

$$u(0,x) = u_0(x), \ x \in [-l,l]; \quad u(t,-l) = u(t,l) = 0, \ t \in [0,T]; \tag{2}$$

$$\int_{-l}^{l} u(t,x)\omega(x)dx = \varphi(t), \quad \int_{-l}^{l} u(t,x)\chi(x)dx = \psi(t), \ t \in [0,T]. \tag{3}$$

Here $Q = [0,T] \times [-l,l]$, $a(t,x), b(t,x), g(t,x), u_0(x), \omega(x), \chi(x), \varphi(t), \psi(t)$ are some known functions.

Additional information is given in the form of integral observations (3) which can physically mean, say, a measurement of function $u(t,x)$ by sensor averaging over the segment $[-l,l]$ of space variable.

Inverse problems of determination of two coefficients in parabolic equations in the various posing other than (1)–(3) were considered in several papers [1–3], etc. However there the inverse problems contained additional conditions different from (3).

The spaces $L_p(Q), L_p([-l,l]), L_\infty([0,T]), W_2^{1,2}(Q), C^{0,\alpha}(Q)$ with corresponding norms are defined as usual (e.g. see [4]). We also denote $L_\infty^+([0,T]) = \{h(t) \in L_\infty([0,T]) : h(t) \geq 0\}$, $B_R = \{h(t) \in L_\infty([0,T]) : |h(t)| \leq R\}$, $B_R^+ = \{h(t) \in L_\infty([0,T]) : 0 \leq h(t) \leq R\}$, $R = \text{const} > 0$.

© Springer International Publishing Switzerland 2015
I. Dimov et al. (Eds.): FDM 2014, LNCS 9045, pp. 225–232, 2015.
DOI: 10.1007/978-3-319-20239-6_23

Throughout the paper we assume that all functions occurring in the input data of problem (1)–(3) are measurable and satisfy the following conditions:

(A) $a_1 \le a(t,x) \le a_2, |a_x(t,x)| \le K_a^*, |a_{xx}(t,x)| \le K_a^{**}, (t,x) \in Q;$

(B) $|b(t,x)| \le K_b, |b_x(t,x)| \le K_b^*, |g(t,x)| \le K_g, (t,x) \in Q;$

(C) $|\omega(x)| \le K_\omega, |\omega'(x)| \le K_\omega^*, |\omega''(x)| \le K_\omega^{**}, |\chi(x)| \le K_\chi,$

   $|\chi'(x)| \le K_\chi^*, |\chi''(x)| \le K_\chi^{**}, x \in [-l, l]; \omega(\pm l) = \chi(\pm l) = 0;$

(D) $u_0(x) \in \overset{0}{W}{}_2^1[-l, l]), |u_0(x)| \le M_0, x \in [0, T];$

(E) $|\varphi(t)| \le K_\varphi, |\varphi'(t)| \le K_\varphi^*, |\psi(t)| \le K_\psi, |\psi'(t)| \le K_\psi^*, t \in [0, T];$

$$\varphi(0) = \int_{-l}^{l} u_0(x)\omega(x)dx, \ \psi(0) = \int_{-l}^{l} u_0(x)\chi(x)dx.$$

Here $a_1, a_2, M_0, K_g, K_\omega, K_\chi = \text{const} > 0, K_a^*, K_a^{**}, K_b, K_b^*, K_\omega^*, K_\omega^{**}, K_\chi^*, K_\chi^{**} = \text{const} \ge 0.$

Let us denote

$$G_\omega(t) = \int_{-l}^{l} g(t,x)\omega(x)dx, \ G_\chi(t) = \int_{-l}^{l} g(t,x)\chi(x)dx,$$

$$\Delta_1(t) = \psi(t)G_\omega(t) - \varphi(t)G_\chi(t), \ \Delta_2(t) = G_\chi(t)\varphi'(t) - G_\omega(t)\psi'(t),$$

$$Q_\omega(t;u) = \int_{-l}^{l} [(a\omega)_{xx} + (b\omega)_x]u(t,x)dx, \ Q_\chi(t;u) = \int_{-l}^{l} [(a\chi)_{xx} + (b\chi)_x]u(t,x)dx,$$

$$J_\omega(t;u) = -\varphi'(t) + Q_\omega(t;u), \ J_\chi(t;u) = -\psi'(t) + Q_\chi(t;u),$$

$$\Delta_d(t;u) = G_\omega(t)J_\chi(t;u) - G_\chi(t)J_\omega(t;u), \ \Delta_f(t;u) = \varphi(t)J_\chi(t;u) - \psi(t)J_\omega(t;u),$$

where $u(t,x)$ is a function from $L_\infty(Q)$.

Let us also put $K_1 = K_a^{**}K_\omega + 2K_a^*K_\omega^* + a_2K_\omega^{**} + K_b^*K_\omega + K_bK_\omega^*, K_2 = K_a^{**}K_\chi + 2K_a^*K_\chi^* + a_2K_\chi^{**} + K_b^*K_\chi + K_bK_\chi^*, M(R) = M_0 + K_gRT, R = \text{const} > 0.$

Definition. A generalized solution of the problem (1)–(3) is a triplet of functions $\{u(t,x), d(t), f(t)\}, u(t,x) \in W_2^{1,2}(Q) \bigcap C^{0,\alpha}(Q), \alpha = \text{const} \in (0,1), d(t) \in L_\infty^+([0,T]), f(t) \in L_\infty([0,T])$, satisfying Eq. (1) almost everywhere in Q and such that the function $u(t,x)$ satisfies conditions (2) and (3).

Remark. In view of assumptions (A), (B), (C) for each $d(t) \in L_\infty^+([0,T]), f(t) \in L_\infty([0,T])$ there exists a solution $u(t,x)$ of the direct problem (1)–(2) (see [4]). Moreover, if $|f(t)| \le R_f, t \in [0,T]$, then by the maximum principle we have

$$|u(t,x)| \le M_0 + K_gR_fT \equiv M(R_f). \tag{4}$$

2   Uniqueness of the Solution of the Inverse Problem

Theorem 1. Let conditions $(A) - (E)$ be satisfied. Suppose that there exists a constant $\delta_1 > 0$ that $\forall t \in [0, T]$

$$\Delta_1(t) \geq \delta_1 > 0. \tag{5}$$

Then there exists at most one solution of the inverse problem (1)–(3).

Proof. Suppose that there exist two distinct solutions $\{u^{(1)}(t, x), d^{(1)}(t), f^{(1)}(t)\}$ and $\{u^{(2)}(t, x), d^{(2)}(t), f^{(2)}(t)\}$ of the problem (1)–(3). Set

$$\hat{u}(t, x) = u^{(1)}(t, x) - u^{(2)}(t, x), \ \hat{d}(t) = d^{(1)}(t) - d^{(2)}(t), \ \hat{f}(t) = f^{(1)}(t) - f^{(2)}(t).$$

Then

$$\hat{u}_t - a(t, x)\hat{u}_{xx} + b(t, x)\hat{u}_x + d^{(1)}(t)\hat{u}$$

$$= -\hat{d}(t)u^{(2)}(t, x) + \hat{f}(t)g(t, x), \ (t, x) \in Q, \tag{6}$$

$$\hat{u}(0, x) = 0, \ x \in [-l, l], \quad \hat{u}(t, -l) = \hat{u}(t, l) = 0, \ t \in [0, T], \tag{7}$$

$$\int_{-l}^{l} \hat{u}(t, x)\omega(x)dx = 0, \ \int_{-l}^{l} \hat{u}_x(t, x)\omega(x)dx = 0, \quad t \in [0, T]. \tag{8}$$

We multiply (6) by $\omega(x)$ and integrate the resulting relation over $[-l, l]$. Then integrating by parts and taking into account (8) and $(A) - (C)$ we obtain

$$\hat{d}(t)\varphi(t) - \hat{f}(t)G_\omega(t) = Q_\omega(t; \hat{u}). \tag{9}$$

Similarly, multiplying (6) by $\chi(x)$ and integrating the resulting relation over $[-l, l]$ we obtain

$$\hat{d}(t)\psi(t) - \hat{f}(t)G_\chi(t) = Q_\chi(t; \hat{u}). \tag{10}$$

From (9) and (10), via the condition (5), we obtain

$$\hat{d}(t) = \Delta_d^*(t; \hat{u})/\Delta_1(t), \ \hat{f}(t) = \Delta_f^*(t; \hat{u})/\Delta_1(t), \tag{11}$$

where

$$\Delta_d^*(t; \hat{u}) = G_\omega(t)Q_\chi(t; \hat{u}) - G_\chi(t)Q_\omega(t; \hat{u}),$$

$$\Delta_f^*(t; \hat{u}) = \varphi(t)Q_\chi(t; \hat{u}) - \psi(t)Q_\omega(t; \hat{u}).$$

We substitute (11) into (6) and obtain the following integro-differential relation:

$$\hat{u}_t - a(t, x)\hat{u}_{xx} + b(t, x)\hat{u}_x + d^{(1)}(t)\hat{u}$$

$$= -\frac{\Delta_d^*(t; \hat{u})}{\Delta_1(t)}u^{(2)}(t, x) + \frac{\Delta_f^*(t; \hat{u})}{\Delta_1(t)}g(t, x), \ (t, x) \in Q. \tag{12}$$

Let us multiply (12) by $-\hat{u}_{xx}$ and integrate over $Q_\tau \equiv [0,\tau] \times [-l,l]$, where $\tau \in [0,t_1]$ and $t_1 > 0$ will be chosen below. By integrating by parts in the resulting relation and by using conditions $(A) - (E)$, Poincaré inequality and the well-known inequality

$$|ab| \leq \frac{\varepsilon}{2}a^2 + \frac{1}{2\varepsilon}b^2, \ \varepsilon > 0,$$

we obtain

$$\frac{1}{2}\|\hat{u}_x(\tau,\cdot)\|^2_{L_2([-l,l])} + \frac{a_0}{2}\|\hat{u}_{xx}\|^2_{L_2(Q_\tau)}$$

$$\leq c_1\|\hat{u}_x\|^2_{L_2(Q_\tau)} + c_2\int_0^\tau |\Delta_d^*(t;\hat{u})|^2 dt + c_3\int_0^\tau |\Delta_f^*(t;\hat{u})|^2 dt, \quad (13)$$

where $c_1, c_2, c_3 = \text{const} > 0$ depend on l, T, δ_1, the constants from $(A) - (E)$ and maximum modulus of functions $d^{(i)}(t), f^{(i)}(t), u^{(i)}(t,x), i = 1,2$, but do not depend on t_1.

From the definition of $\Delta_d^*(t;\hat{u})$ and $\Delta_f^*(t;\hat{u})$ it is easy to prove that

$$\int_0^\tau |\Delta_d^*(t;\hat{u})|^2 dt \leq c_4\|\hat{u}\|^2_{L_2(Q_\tau)} \leq c_5\|\hat{u}_x\|^2_{L_2(Q_\tau)},$$

$$\int_0^\tau |\Delta_f^*(t;\hat{u})|^2 dt \leq c_6\|\hat{u}\|^2_{L_2(Q_\tau)} \leq c_7\|\hat{u}_x\|^2_{L_2(Q_\tau)},$$

where $c_4, c_5, c_6, c_7 = \text{const} > 0$ do not depend on t_1; here we also use Poincaré inequality.

Then from (13) we have

$$\frac{1}{2}\|\hat{u}_x(\tau,\cdot)\|^2_{L_2([-l,l])} + \frac{a_0}{2}\|\hat{u}_{xx}\|^2_{L_2(Q_\tau)}$$

$$\leq c_8\|\hat{u}_x\|^2_{L_2(Q_\tau)} \leq c_8 \cdot t_1 \cdot \sup_{0 \leq \tau \leq t_1} \|\hat{u}_x(\tau,\cdot)\|^2_{L_2([-l,l])}, \quad (14)$$

where the positive constant c_8 does not depend on t_1.

Now we choose a $t_1 > 0$ such that $c_8 \cdot t_1 < 1/2$. Then from (14) it follows that

$$\sup_{0 \leq \tau \leq t_1} \|\hat{u}_x(\tau,\cdot)\|^2_{L_2([-l,l])} = 0.$$

This together with relation (7) implies that $\hat{u}(t,x) = 0$ a.e. in $Q_{t_1} \equiv [0,t_1] \times [-l,l]$.

By reproducing the above arguments for the rectangles $[t_1, 2t_1] \times [-l,l]$, $[2t_1, 3t_1] \times [-l,l]$ and so on, in finitely many steps we find that $\hat{u}(t,x) = 0$ a.e. in Q. Then from relation (11) we obtain that $\hat{d}(t) = \hat{f}(t) = 0$, $t \in [0,T]$.

The proof of Theorem 1 is complete.

3   Existence of the Solution of the Inverse Problem

In this section in addition to conditions (A)–(E) and (5) we make the following assumptions: there exists a positive constant δ_2 that

$$\Delta_2(t) \equiv G_\chi(t)\varphi'(t) - G_\omega\psi'(t) \geq \delta_2 > 0, \tag{15}$$

and also

$$2lK_gT(K_\varphi K_2 + K_\psi K_1) < \delta_1. \tag{16}$$

Then we put

$$R_f = \frac{K_\varphi K_\psi^* + K_\psi K_\varphi^* + 2lM_0(K_\varphi K_2 + K_\psi K_1)}{\delta_1 - 2lK_gT(K_\varphi K_2 + K_\psi K_1)} \tag{17}$$

and suppose that

$$4l^2 K_g(M_0 + K_gTR_f)(K_\chi K_1 + K_\omega K_2) \leq \delta_2. \tag{18}$$

Let us derive a system of operator equations for the functions $d(t)$ and $f(t)$. Let $d(t) \in L_\infty^+([0,T]), f(t) \in B_{R_f}$ and let $u(t,x)$ be a solution of direct problem (1)–(2) with selected functions $d(t)$ and $f(t)$ in the Eq. (1).

We multiply (1) by $\omega(x)$ and integrate the resulting relation over $[-l,l]$. By taking into account conditions $(A) - (E)$, (2), (3) we obtain

$$d(t)\varphi(t) - f(t)G_\omega(t) = J_\omega(t; u). \tag{19}$$

Similarly, multiplying (1) by $\chi(x)$, integrating the resulting relation over $[-l,l]$ and taking into account conditions $(A) - (E)$, (2), (3) we obtain

$$d(t)\psi(t) - f(t)G_\chi(t) = J_\chi(t; u). \tag{20}$$

Due to condition (5), the algebraic system (19)–(20) has a unique solution $\{d(t), f(t)\}$ given by formulas

$$d(t) = \Delta_d(t; u)/\Delta_1(t), \; f(t) = \Delta_f(t; u)/\Delta_1(t). \tag{21}$$

We introduce a nonlinear operator $\mathcal{A} \equiv (\mathcal{A}_1, \mathcal{A}_2) : L_\infty^+([0,T]) \times B_{R_f} \longrightarrow L_\infty([0,T]) \times L_\infty([0,T])$ by setting

$$\mathcal{A}_1(d, f) = \Delta_d(t; u)/\Delta_1(t), \; \mathcal{A}_2(d, f) = \Delta_f(t; u)/\Delta_1(t), \tag{22}$$

where $u(t,x)$ is a solution of direct problem (1)–(2) with selected functions $d(t), f(t)$ in Eq. (1).

By the definition of $\mathcal{A}_1$ and $\mathcal{A}_2$, relations (21) can be represented in the form

$$d(t) = \mathcal{A}_1(d, f), \; f(t) = \mathcal{A}_2(d, f). \tag{23}$$

Lemma 1. Let conditions $(A)-(E)$, (5), (15), (16) and (18) (with R_f from (17)) be satisfied. Then, for any $d(t) \in L_\infty^+([0,T])$, $f(t) \in B_{R_f}$ we have $\mathcal{A}_1(d, f) \geq 0$.

Proof. Since the function $u(t, x)$ satisfies the estimate (4), then, from (18) and (15) together with the definitions of $\Delta_d(t; u)$ and the constants K_1, K_2 the inequality $\Delta_d(t; u) \geq 0$ holds. Then, by (5) and the definition of $\mathcal{A}_1(d, f)$ we have $\mathcal{A}_1(d, f) \geq 0$. Lemma 1 is proved.

Lemma 2. Let conditions $(A) - (E)$, (5), (15), (16) hold. Then the triplet of functions $\{u(t, x), d(t), f(t)\}$, $u(t, x) \in W_2^{1,2}(Q) \bigcap C^{0,\alpha}(Q), \alpha = \text{const} \in (0, 1)$, $d(t) \in L_\infty^+([0, T])$, $f(t) \in B_{R_f}$, is a generalized solution of problem (1)–(3) if and only if this triplet of functions satisfies the relations (1), (2) and (23).

The proof of this Lemma is standard (see, for example, [5]).

Lemma 3. Let conditions of Lemma 2 hold. Then for all $d(t) \in L_\infty^+([0, T])$, $f(t) \in B_{R_f}$

$$\|\mathcal{A}_2(d, f)\|_{L_\infty([0,T])} \leq R_f. \tag{24}$$

Proof. By the definition of $\mathcal{A}_2$ and inequality (4) we have

$$\|\mathcal{A}_2(d, f)\|_{L_\infty([0,T])} \leq \frac{1}{\delta_1} \|\Delta_f(t; u)\|_{L_\infty([0,T])}$$

$$\leq \frac{1}{\delta_1} \left(K_\varphi K_\psi^* + K_\psi K_\varphi^* \right) + \frac{2l}{\delta_1} \left(K_\varphi K_2 + K_\psi K_1 \right) \cdot \left(M_0 + K_g R_f T \right). \tag{25}$$

Now taking into account the definition of R_f in (17) we deduce from (25) the inequality (24). Lemma 3 is proved.

Lemma 4. Let conditions of Lemma 2 hold and $f(t) \in B_{R_f}$. Then

$$\|\mathcal{A}_1(d, f)\|_{L_\infty([0,T])} \leq R_d, \tag{26}$$

where

$$R_d = \frac{2l}{\delta_1} K_g \left(K_\omega K_\psi^* + K_\chi K_\varphi^* \right) + \frac{4l^2}{\delta_1} K_g \left(K_\omega K_2 + K_\chi K_1 \right) \cdot \left(M_0 + K_g R_f T \right). \tag{27}$$

Proof. The assertion of the lemma is an immediate consequence of the definition of the $\mathcal{A}_2$, conditions $(A) - (E)$, (5) and inequality (4).

Lemma 5. Let conditions of Lemma 2 hold. Then the operator $\mathcal{A} \equiv (\mathcal{A}_1, \mathcal{A}_2)$ given by (22) is continuous on the set $B_{R_d}^+ \times B_{R_f}$.

Proof. In the proof of this lemma by c_i we denote positive constants depending on l, T, constants from conditions $(A) - (E)$, as well as R_d and R_f.

Let $\{d^{(i)}(t), \{f^{(i)}(t)\} \in B_{R_d}^+ \times B_{R_f}$, $i = 1, 2$, and $u^{(i)}(t, x)$ be generalized solutions of corresponding direct problem (1)–(2). Set

$$\hat{u}(t, x) = u^{(1)}(t, x) - u^{(2)}(t, x), \ \hat{d}(t) = d^{(1)}(t) - d^{(2)}(t), \ \hat{f}(t) = f^{(1)}(t) - f^{(2)}(t).$$

Then, these functions satisfy the relations (6) and (7).

By virtue of the maximum principle we have the estimate

$$|u^{(2)}(t,x)| \le M(R_f) \equiv M_0 + K_g R_f T, \ (t,x) \in Q.$$

Once again using the maximum principle for the problem (6)–(7) we obtain

$$|\hat{u}(t,x)| \le T[M(R_f)\|\hat{d}\|_{L_\infty([0,T])} + K_g\|\hat{f}\|_{L_\infty([0,T])}]. \tag{28}$$

By the definition of $J_\omega(t;u), J_\chi(t;u), \Delta_d(t;u), \Delta_f(t;u)$ and conditions $(A) - (E)$ we have

$$\|\Delta_d(t;u^{(1)}) - \Delta_d(t;u^{(2)})\|_{L_\infty([0,T])} \le c_1\|\hat{u}\|_{L_\infty(Q)},$$

$$\|\Delta_f(t;u^{(1)}) - \Delta_f(t;u^{(2)})\|_{L_\infty([0,T])} \le c_2\|\hat{u}\|_{L_\infty(Q)}.$$

Then, the definition of the operators $\mathcal{A}_i(d,f)$, $i = 1,2$, and condition (7) imply that

$$\|\mathcal{A}_i(d^{(1)},f^{(1)}) - \mathcal{A}_i(d^{(2)},f^{(2)})\|_{L_\infty([0,T])} \le c_3\|\hat{u}\|_{L_\infty(Q)}, \ i = 1,2.$$

From the last estimate using (28) we find that

$$\|\mathcal{A}_i(d^{(1)},f^{(1)}) - \mathcal{A}_i(d^{(2)},f^{(2)})\|_{L_\infty([0,T])} \to 0$$

as

$$\|d^{(1)} - d^{(2)}\|_{L_\infty([0,T])} + \|f^{(1)} - f^{(2)}\|_{L_\infty([0,T])} \to 0.$$

Thus, the operator $\mathcal{A}$ is continuous on the set $B_{R_d}^+ \times B_{R_f}$. Lemma 5 is proved.

Lemma 6. Let conditions of Lemma 1 hold. Then $\mathcal{A}$ is compact operator mapping $B_{R_d}^+ \times B_{R_f}$ into itself.

Proof. Let $d(t) \in B_{R_d}^+$, $f(t) \in B_{R_f}$, and let $u(t,x)$ be the generalized solution of direct problem (1)–(2) with chosen $d(t)$ and $f(t)$ in the Eq. (1). Then by [4]

$$|\hat{u}|_{C^{0,\alpha}(Q)} \le R_0, \tag{29}$$

where $R_0 = $ const > 0 and $\alpha = $ const $\in (0,1)$ depend on the input data of problem (1)–(2) and on R_d and R_f.

Since the space $C^{0,\alpha}(Q)$ is compactly embedded in the space $C(Q)$, then from the definition of the operator $\mathcal{A}$ and Lemmas 1, 3, 4 it follows that $\mathcal{A}$ is a compact operator mapping set $B_{R_d}^+ \times B_{R_f}$ into itself. By Lemma 5 it is also continuous operator on this set. Consequently, it is a completely compact operator mapping $B_{R_d}^+ \times B_{R_f}$ into itself. The proof of Lemma 6 is complete.

Theorem 2. Let conditions of Lemma 1 hold. Then there exists a generalized solution $\{u(t,x), d(t), f(t)\}$ of the inverse problem (1)–(3). Moreover, $0 \le d(t) \le R_d$, $|f(t)| \le R_f$ and $u(t,x)$ satisfies the estimate (4).

Proof. By Lemma 6, the operator $\mathcal{A}$ is a completely compact operator mapping the bounded convex closed set $B^+_{R_d} \times B_{R_f}$ into itself. By the Shauder fixed point theorem (e.g. see [6], p. 193) the system of Eq. (23) has a solution $\{d(t), f(t)\} \in B^+_{R_d} \times B_{R_f}$ Now the assertion of the theorem follows from Lemma 2.

Acknowledgements. This work was supported by the Program for improving the competitiveness of universities (NRNU MEPHI).

References

1. Ivanchov, N.I.: On the inverse problem of simultaneous determination of the coefficients of heat conductivity and heat capacity. Siberian Math. J. **35**(3), 612–621 (1994)
2. Anikonov, Y.E., Belov, Y.Y.: Determining two unknown coefficients of parabolic type equations. J. Inverse Ill-Posed Prob. **8**(1), 1–19 (2000)
3. Ivanchov, N.I., Pobyrivska, N.V.: On determination of two coefficients in parabolic equation depending on time. Siberian Math. J. **43**(2), 402–413 (2002)
4. Kruzhkov, S.N.: Quasilinear parabolic equations and systems with two independent variables. Trudy Sem. im. I. G. Petrovskogo. **5**, 217–272 (1979)
5. Kamynin, V.L.: Unique solvability of the inverse problem of determination of the leading coefficient in a parabolic equation. Diff. Equat. **47**(1), 91–101 (2011)
6. Lyusternik, L.A., Sobolev, V.I.: Kratkii Kurs Functcional'nogo analiza (Brief Course of Functional Analysis). Vyssh. Shkola, Moscow (1982)

Effectiveness of the Parallel Implementation of the FEM for the Problem of the Surface Waves Propagation

Evgeniya Karepova[✉] and Ekaterina Dementyeva

Institute of Computational Modelling of SB RAS, Siberian Federal University, 660036
Akademgorodok, Krasnoyarsk, Russia
e.d.karepova@icm.krasn.ru
http://icm.krasn.ru

Abstract. In this paper effectiveness of several parallel implementations
of the finite element method is investigated for an algorithm of a numerical solution of the boundary function problem for the shallow water
equations. The parallel technologies MPI, OpenMP and MPI+OpenMP
are used.

Keywords: data assimilation problem · finite element method and high
performance computation

1 Introduction

The shallow water equations are used for the numerical modeling of the surface
waves in a large area taking into account the sphericity of the Earth and the Corio-
lis acceleration [1–4]. In [5] the finite element method (the FEM) was constructed
for this differential problem and the problem is reduced to the vector-matrix form.
The system of linear algebraic equations is solved with a Jacobi-type iterative
method which can be parallelized well. The diagonal dominance of its convergence
is provided by choosing the time step.

The using of the FEM for spatial discretization of the problem has a number
of advantages. The main one is the possibility to use unstructured nonuniform
grids for numerical domains of complex shapes. At the same time, the FEM is
more compute-intensive [6] than the finite difference method. Grids with a large
number of finite elements are often used for physical problems. Therefore, usage
of high-performance computing systems for the FEM is urgent [7–12].

Nowadays SMP-nodes clusters are widespread. The hybrid MPI+OpenMP
approach is a natural parallel programming paradigm for this architecture [11,12].
In this case the FEM parallelizing with MPI is used for the distributed memory
architectures. Then OpenMP can be employed within each MPI process. In order
the hybrid program to be effective it is necessary to make an efficient MPI and

The work was supported by Russian Science Foundation (grant 14-11-00147).

© Springer International Publishing Switzerland 2015
I. Dimov et al. (Eds.): FDM 2014, LNCS 9045, pp. 233–240, 2015.
DOI: 10.1007/978-3-319-20239-6_24

OpenMP implementations of the FEM. In the case OpenMP is used it is important to avoid the data dependency in the loops concerning the FEM.

In the present work for the shallow water equations inverse problem we study the effectiveness of several parallel implementations of the FEM based on technologies MPI, OpenMP and MPI+OpenMP. For development of the effective parallel programs some features of FEM implementation on high-performance systems with shared and distributed memory are investigated theoretically and numerically.

We consider the following problem (for details see [13,14]). Let (r, λ, φ) be spherical coordinates with the origin at the south pole, $0 \leq \lambda \leq 2\pi$, $0 \leq \varphi < \pi$, $r = R_E$, where R_E is the radius of the Earth which is assumed to be constant. Let Ω be a domain in the plain (λ, φ) with the piecewise smooth Lipchitz boundary $\Gamma = \Gamma_1 \cup \Gamma_2$ of the class $C^{(2)}$, where Γ_1 is the coastline and $\Gamma_2 = \Gamma \setminus \Gamma_1$ is the sea boundary. The points $\varphi = 0$ and $\varphi = \pi$ (poles) are not involved in Ω. Divide the segment $[0, T]$ into K intervals by points: $0 = t_0 < t_1 < \cdots < t_K = T$ with the step $\tau = T/K$. Write in Ω the vertically averaged equations of motion and continuity [1,4] for the unknown functions u, v and ξ at the time instant t_{k+1}, $k = 0, 1, \ldots, K - 1$:

$$
\left(\frac{1}{\tau} + R_f\right) u - lv - mg \frac{\partial \xi}{\partial \lambda} = f_1 + \frac{1}{\tau} u^k \quad \text{in} \quad \Omega,
$$

$$
\left(\frac{1}{\tau} + R_f\right) v + lu - ng \frac{\partial \xi}{\partial \varphi} = f_2 + \frac{1}{\tau} v^k \quad \text{in} \quad \Omega, \tag{1}
$$

$$
\frac{1}{\tau}\xi - m \left(\frac{\partial}{\partial \lambda}(Hu) + \frac{\partial}{\partial \varphi}\left(\frac{n}{m}Hv\right)\right) = f_3 + \frac{1}{\tau}\xi^k \quad \text{in} \quad \Omega,
$$

where u and v are components of the velocity vector $\mathbf{U}$ in λ and φ directions, respectively; ξ is a deviation of a free surface from the nonperturbed level; $H(\lambda, \varphi) > 0$ is a depth of a water area at a point (λ, φ); the function $R_f = r_*|\mathbf{U}^k|/H$ takes into account the base friction force, r_* is the friction coefficient; $l = -2\omega \cos \varphi$ is the Coriolis parameter; $m = 1/(R_E \sin \varphi)$; $n = 1/R_E$; g is the acceleration of gravity; $f_1 = f_1(t, \lambda, \varphi)$, $f_2 = f_2(t, \lambda, \varphi)$ and $f_3 = f_3(t, \lambda, \varphi)$ are given functions of external forces. For an arbitrary function $f(t, \lambda, \phi)$ we use $f^k = f(t_k, \lambda, \phi)$, $f = f(t_{k+1}, \lambda, \phi) = f^{k+1}$.

We consider boundary conditions in the following form:

$$
HU_n + \beta \chi_2 \sqrt{gH}\xi = \chi_2 \sqrt{gH}d \quad \text{on} \quad \Gamma, \tag{2}
$$

where $U_n = \mathbf{U} \cdot \mathbf{n}$, $\mathbf{n} = (n_1, \frac{n}{m}n_2)$ is the vector of an outer normal to the boundary; $\beta \in (0, 1]$ is a given parameter; $\chi_2 = 0$ on Γ_1, $\chi_2 = 1$ on Γ_2; $d = d(t, \lambda, \varphi)$ is an unknown boundary function on the boundary Γ_2 and equal to zero on Γ_1.

To close the problem (1)–(2) we consider the following closed condition:

$$
\xi = \xi_{obs} \quad \text{on} \quad \Gamma_0, \tag{3}
$$

where $\xi_{obs} \in L_2(\Gamma_0)$ is a given function (for example, from observation data) on some part of the boundary $\Gamma_0 \subset \Gamma$.

Problem 1 (Inverse Problem). *Let for a fixed time instant* t_{k+1}, $k = 0, 1, ..., K - 1$, *the observation function* ξ_{obs} *be given on* Γ_0, *the function* d *be unknown on* Γ_2 *and equal to zero on* Γ_1. *At the time instant* t_{k+1} *find* u, v, ξ, d *satisfying system* (1), *boundary condition* (2) *and closure condition* (3).

In [13,14] an iterative algorithm is proposed to find the numerical solution of Problem 1. It consists of an alternate solving of the following coupled equations: direct, adjoint and the boundary function refinement equation. The algorithm is compute-intensive on each iteration. At the same time, recovery of the boundary function requires a lot of iterations in case the quality observation data is poor, i.d., for instance, with white noise or with gaps. Therefore, the development of the effective parallel implementation of the algorithm is urgent.

2   The Parallel Implementation on SMP-node Cluster

For the numerical modeling of long waves propagation in a large water area the finite element method with linear triangular finite elements is used [5].

The Jacobi-type iterative process is used when solving the linear algebraic system generated by the FEM. Notice some features of implementation of the algorithm related to the FEM. The global stiffness matrix depends on time and must be recalculated at each time step. However, when implementing the Jacobi-type method on finite elements there is no need to store the global stiffness matrix explicitly. Only elements of local stiffness matrices are stored in the program. Moreover, only their diagonal elements depend on time and are recalculated at each time step. A residual is assembled with the use of finite elements local stiffness matrices. An assembling residual process includes computing a running sum with use of finite elements local stiffness matrices.

To investigate a speedup of several parallel implementations of the algorithm the model problem is considered for the square domain on the sphere. A regular grid with consistent triangulation is used in the computational domain. In the numerical experiments the grid of 1600×1600 points in space directions is used and 100 time steps are done. The experiments are carried out using up to 100 computational cores of a cluster and demonstrate a good scalability.

The most time-consuming operation in the FEM is the residual assembling using finite elements local stiffness matrices. There are at least two ways to compute a residual:

(1) by elements (Fig. 1 to the right) (the traditional way realizing the most profitable memory allocation for storage of information about triangulation);
(2) by grid points (Fig. 1 to the right) (the way which requires creation irregular structures in the memory for storing the information on triangulation).

The numerical experiments on studying effectiveness of several parallel implementations of our FEM are performed on high-performance clusters of SFU and SSCC SB RAS [15]. Consider the following theoretical and numerical results.

Fig. 1. The scheme of residual assembling by grid points (to the left) and by elements (to the right)

1. *Sequential implementation.* More or less, the execution time T_1 of the sequential program consists of the time T_1^{calc} of calculations and the time T_1^{mem} of memory operations:

$$T_1 = T_1^{calc} + T_1^{mem}.$$

The numerical experiments shows that the execution time for the program with the residual assembling by elements is 1.5 times less than the execution time for program with the residual assembling by grid points (Fig. 2). This effect can be explained by more favourable memory distribution causing a decrease of T_1^{mem} value when a residual is assembled by elements.

While compiling the sequential versions of the program we use a set of compiler keys reducing the execution time T_1. This value of T_1 is used to calculate a speedup of the parallel versions. For calculating a speedup of each parallel version the same way of the residual assembling must be used for both the sequential and the parallel versions.

2. *OpenMP-implementations for shared memory architecture.* In this case the following time overheads arise: (1) the time for threads creation and exit; (2) the time for OpenMP loop scheduling (*static, dynamic, guided*, etc.); (3) the time for threads synchronization in the main loop to the residual assembling by elements; (4) the time for the global reduction operation calculating the stop criterion of the iterative process.

Consider in detail item 3. A complete residual in a point is assembled by running sum using several triangle elements (Fig. 1). In case of the residual assembling by elements different threads may simultaneously handle the same grid point while bypassing the finite elements in the main loop. Hence, in this case it is extremely necessary to synchronize the threads additionally.

Estimate the execution time T_{th} of OpenMP-program with th threads. The execution time T_{th} consists of the time T_{th}^{calc} of performing calculations, the time T_{th}^{mem} of memory operation and the time T_{th}^{synch} of overheads specified in points (1)–(4): $T_{th} = T_{th}^{calc} + T_{th}^{mem} + T_{th}^{synch}.$

Moreover, when using OpenMP the time T_1^{calc} of calculations usually decreases proportionally to the number of OpenMP-threads involved. The time of memory operations by th threads is also reduced as compared to T_1^{mem} but possible simultaneous reading and writing impair parallelism [16]. Then, the execution time of OpenMP-program with th threads is the following:

$$T_{th} = \frac{T_1^{calc}}{th} + \frac{\gamma(th)}{th}T_1^{mem} + T_{th}^{synch}, \quad 1 < \gamma(th) < th.$$

Thus, to obtain a good speedup of OpenMP-program the multithreading overheads are to be small as compared to the computational effort.

The numerical results show that when a residual is assembled by elements the threads synchronization in the main loop is the most time-consuming operation. The synchronization overheads take up to 40 % of the execution time T_{th} of the program. In this case efficiency of the paralleling is about 25 % (Fig. 2). When a residual is assembled by grid points the supplementary threads synchronization is not required which makes this approach advantageous (Fig. 2). If up to 30 threads are used efficiency runs up to 90 % while for more than 30 threads it reaches about 80 % (a thread onto a core).

3. *MPI-implementations for distributed memory architecture.* In this case data parallelism and decomposition of the computational domain without shadow lines (overlaps) are used. For both ways of the residual assembling the following overheads [14] arise inevitably at each iteration: (1) the time required to point-to-point exchanges between adjacent processes for computation of the complete residual in a cut line of the computational domain; (2) the time for the global reduction operation calculating the stop criterion of the iterative process.

The execution time T_p of an MPI-program with p MPI-processes (one MPI-process per a core) consists of the time T_p^{calc} of performing calculations, the time T_p^{mem} of memory operations and the overhead time T_p^{comm} for communication: $T_p = T_p^{calc} + T_p^{mem} + T_p^{comm}$. As unblocked point-to-point communication is used and the number of neighbours in item 1 is independent of p so the time T_p^{comm} depends on p weakly.

Moreover, all MPI-processes perform their parts of calculations and memory operations simultaneously. Then, the execution time of an MPI-program with p processes is the following:

$$T_p = \frac{T_1^{calc} + T_1^{mem}}{p} + T_p^{comm}. \tag{4}$$

From (4) follows that the way of the residual assembling has no significant effects on a speedup of an MPI-program. It can be explained by the fact that the execution time difference in these two cases of the residual assembling is determined by the difference in their time T_p^{mem} of memory operations. The numerical experiments validate these conclusions. They show the advantage of the residual assembling by elements. Figures 2 and 3 demonstrate that this version of the program has efficiency of about 80 % and is performed faster than an MPI-program with the residual assembling by grid points. The reason is more profitable memory distribution in the first case. It is necessary to note that the execution time of the best MPI-program is less than the execution time of the fastest OpenMP-program (Fig. 2a).

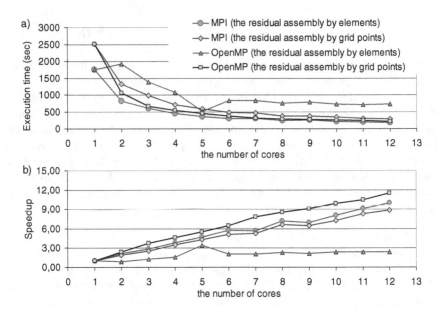

Fig. 2. Graphs of dependence of execution time (a) and a speedup (b) on the number of cores for MPI and OpenMP programs

4. *Joint usage of MPI and OpenMP technologies for a SMP-node cluster* (Fig. 3). The execution time $T_{p \times th}$ of an MPI+OpenMP-program with p MPI-processes and th OpenMP-threads per a MPI-process is specified as

$$T_{p \times th}^{SMP} = \frac{T_1^{calc}}{p \times th} + \frac{\gamma(th) T_1^{mem}}{p \times th} + T_{th}^{synch} + T_p^{comm}, \quad 1 < \gamma < th. \quad (5)$$

From (4)-(5) it is concluded that the benefits from joint usage of OpenMP and MPI technologies compared to the straight MPI approach can be derived only if the overhead generated by th OpenMP-threads and p MPI-processes is less than the overhead of $p \times th$ MPI-processes:

$$T_{p \times th}^{comm} > \frac{\gamma(th) - 1}{p \times th} T_1^{mem} + T_{th}^{synch} + T_p^{comm}, \quad 1 < \gamma < th.$$

In the numerical experiments (Fig. 3) there are two MPI-processes per node with five OpenMP-threads per each MPI-process. Figure 3 shows that in the case of the residual assembling by grid points the speedup of the MPI+OpenMP-program almost coincides with the linear speedup, and the efficiency is about 100 %. In the case of the residual assembling by elements the MPI+OpenMP-program efficiency is only 40 %, which should be explained by a high overhead of OpenMP-threads synchronization.

Figure 3 demonstrates that the MPI+OpenMP-program with the residual assembling by grid points is the most effective among the examined parallel implementations. At the same time, this parallel version is most difficult to implement as it requires creation, storage and processing of additional irregular

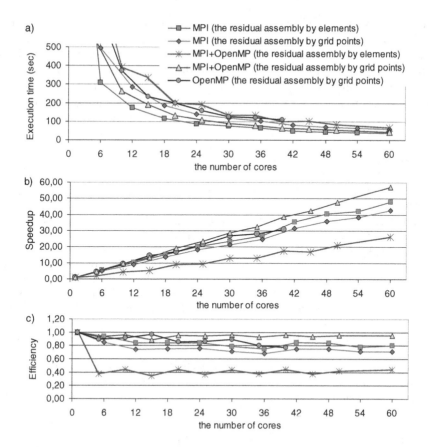

Fig. 3. Graphs of dependence of execution time (a), a speedup (b) and efficiency (c) on the number of cores for MPI, OpenMP and MPI+OpenMP programs

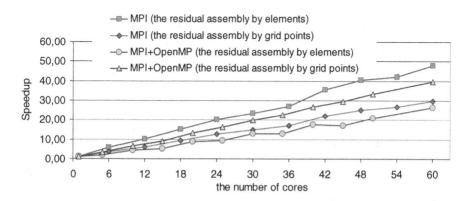

Fig. 4. Graphs of dependence of a speedup on the number of cores for MPI and MPI+OpenMP programs. The speedup is calculated with respect to the time T_1 of the sequential program witn the residual assembly by elements

structures. However, the MPI-program with the residual assembling by elements has the least execution time (Fig. 3a).

Thus far a speedup was considered as $S_p = T_1/T_p$, where the algorithms for sequential and parallel versions coincide. However, the best time T_1^{best} of the sequential program is achieved in the case of the residual assembling by elements. Figure 4 shows that the MPI-program with the residual assembling by elements has the best speedup calculated as $S_p = T_1^{best}/T_p$ for all the programs.

References

1. Marchuk, G.I., Kagan, B.A.: Dynamics of Ocean Tides. Leningrad, Gidrometizdat (1983)
2. Gill, A.E.: Atmosphere-Ocean Dynamics. Academic Press, New York (1982)
3. Kowalik, Z., Polyakov, I.: Tides in the Sea of Okhotsk. J. Phys. Oceanogr. **28**(7), 1389–1409 (1998)
4. Agoshkov, V.I.: Inverse problems of the mathematical theory of tides: boundary-function problem. Russ. J. Numer. Anal. Math. Modelling **20**(1), 1–18 (2005)
5. Kamenshchikov, L.P., Karepova, E.D., Shaidurov, V.V.: Simulation of surface waves in basins by the finite element method. Russ. J. Numer. Anal. Math. Model. **21**(4), 305–320 (2006)
6. Ilin, V.P.: Methods and technologies of finite elements. ICMMG SB RAS Press, Novosibirsk (2007)
7. Choporov, S.: Parallel Computing Technologies in the Finite Element Method. In: Third International Conference "High Performance Computing" HPC-UA 2013, pp. 85–91. Kyiv (2013)
8. Vutov, Y.: Parallel incomplete factorization of 3D NC FEM elliptic systems. In: Boyanov, T., Dimova, S., Georgiev, K., Nikolov, G. (eds.) NMA 2006. LNCS, vol. 4310, pp. 114–121. Springer, Heidelberg (2007)
9. Jimack, P.K., Touheed, N.: Developing parallel finite element soft-ware using MPI. In: High Performance Computing for Computational Mechanics, pp. 15–38 (2000)
10. Pantale, O.: Parallelization of an object-oriented FEM dynamics code: influence of the strategies on the Speedup. Adv. Eng. Softw. **36**(6), 361–373 (2005)
11. Mahinthakumar, G., Saied, F.: A hybrid MPI-openMP implementation of an implicit finite-element code on parallel architectures. Int. J. High Perform. Comput. Appl. **16**(4), 371–393 (2002)
12. Vargas-Felix, M., Botello-Rionda, S.: Solution of finite element problems using hybrid parallelization with MPI and openMP. Acta Universitaria **22**(7), 14–24 (2012)
13. Karepova, E., Dementyeva, E.: The numerical solution of the boundary function inverse problem for the tidal models. In: Dimov, I., Faragó, I., Vulkov, L. (eds.) NAA 2012. LNCS, vol. 8236, pp. 345–354. Springer, Heidelberg (2013)
14. Dementeva, E., Karepova, E., Shaidurov, V.: Recovery of a boundary function by observation data in a problem for the shallow water model. AIP Conf. Proc. **1629**, 373–380 (2014)
15. Hybrid cluster of SSCC SB RAS. Available via DIALOG. http://www2.sscc.ru/HKC-30T/HKC-30T.htm
16. Gergel, V.P.: High-performance computing for the multicore multiprocessor systems. NNSU Publ, Nizhny Novgorod (2010)

Splitting Scheme for Poroelasticity and Thermoelasticity Problems

Alexandr E. Kolesov[1]([✉]), Petr N. Vabishchevich[1,2],
Maria V. Vasilyeva[1], and Victor F. Gornov[3]

[1] North-Eastern Federal University, Yakutsk, Russia
kolesov.svfu@gmail.com
[2] Nuclear Safety Institute, RAS, Moscow, Russia
[3] JSC Insolar-Invest, Moscow, Russia

Abstract. We consider an unconditionally stable splitting scheme for solving coupled systems of equations arising in poroelasticity and thermoelasticity problems. The scheme is based on splitting the systems of equation into physical processes, which means the transition to the new time level is associated with solving separate sub-problems for displacement and pressure/temperature. The stability of the scheme is achieved by switching to three-level finite-difference scheme with weight. We present stability estimates of the scheme based on Samarskii's theory of stability for operator-difference schemes. We provide numerical experiments supporting the stability estimates of the splitting scheme.

1 Introduction

The poroelasticity and thermoelasticity problems play important role in applied mathematical modeling [12]. The basic mathematical model includes the Lame elliptic equation for the displacement vector and non-stationary parabolic equation for the fluid pressure/temperature. The most important feature of mathematical models of poroelasticity and thermoelasticity consists in that these two equations are strongly tied together. On the one hand, the equation for displacements contains the body force, which is proportional to the pressure/temperature gradient. On the other hand, the equation for pressure includes the term, which describes the compressibility of porous medium (the divergence of displacement velocity).

Computational algorithms for numerical solving poroelasticity and thermoelasticity problems are mostly based on the finite element discretization in space [1,5] and the standard two-level schemes for time discretization. In work [4] the stability and convergence of two-level schemes for poroelasticity problem with the finite-difference discretization are studied on the basis of the general theory of stability (correctness) for operator-difference schemes [9,10].

For poroelasticity and thermoelasticity problems, it is essential to construct schemes with splitting into physical processes, when the transition to a new time level is associated with the sequential solution of separate sub-problems for the displacement and pressure/temperature. Various classes of such schemes are

I. Dimov et al. (Eds.): FDM 2014, LNCS 9045, pp. 241–248, 2015.
DOI: 10.1007/978-3-319-20239-6_25

based on the additive splitting of operator [8,11]. For poroelasticity problem, splitting schemes are constructed and used in works [6].

In this work, on the basis of Samarskii's regularization principle [9] we construct the splitting scheme for poroelasticity and thermoelasticity problems. The stability of the scheme is achieved by switching to a three-level finite-difference scheme with weight.

2 Mathematical Model

Lets u be the displacement vector and p be the fluid pressure. For homogeneous, isotropic porous medium the poroelasticity equations can be expressed as

$$\operatorname{div} \boldsymbol{\sigma}(\boldsymbol{u}) - \alpha \operatorname{grad} p = 0, \tag{1}$$

$$\alpha \frac{\partial \operatorname{div} \boldsymbol{u}}{\partial t} + \frac{1}{M} \frac{\partial p}{\partial t} - \operatorname{div}\left(\frac{k}{\eta} \operatorname{grad} p\right) = f(\boldsymbol{x}, t). \tag{2}$$

Here, $\boldsymbol{\sigma}(\boldsymbol{u})$ is the stress tensor:

$$\boldsymbol{\sigma} = 2\mu\varepsilon(\boldsymbol{u}) + \lambda \operatorname{div} \boldsymbol{u} \, \boldsymbol{I},$$

where μ, λ is the Lame coefficients, $\boldsymbol{I}$ is the identity tensor, and ε is the strain stress:

$$\varepsilon(\boldsymbol{u}) = \frac{1}{2}\left(\operatorname{grad} \boldsymbol{u} + \operatorname{grad} \boldsymbol{u}^T\right).$$

Other notations: M is the Biot modulus, k is the permeability, η is the viscosity, α is the Biot-Willis coefficient and $f(\boldsymbol{x}, t)$ is the function that represents the source term.

The problem for the system of Eqs. (1), (2) is considered in a bounded domain Ω with a boundary Γ, where we set the following boundary conditions. For the displacement ($\Gamma = \Gamma_D^u + \Gamma_N^u$) we set

$$\boldsymbol{u} = 0, \quad \boldsymbol{x} \in \Gamma_D^u, \quad \boldsymbol{\sigma}\boldsymbol{n} = 0, \quad \boldsymbol{x} \in \Gamma_N^u, \tag{3}$$

where $\boldsymbol{n}$ is the unit normal to the boundary. Similarly, boundary conditions for the pressure are:

$$p = 0, \quad \boldsymbol{x} \in \Gamma_N^u, \quad -\frac{k}{\eta} \frac{\partial p}{\partial n} = 0, \quad \boldsymbol{x} \in \Gamma_D^u. \tag{4}$$

In addition, we set the following initial condition for the pressure:

$$p(\boldsymbol{x}, 0) = s(\boldsymbol{x}), \quad \boldsymbol{x} \in \Omega. \tag{5}$$

Similarly, we set the initial-boundary value problem for thermoelasticity. In this case, the increment θ with respect to an initial temperature T is the unknown instead of the pressure p. The thermoelasticity equations have the following form:

$$\operatorname{div} \boldsymbol{\sigma}(\boldsymbol{u}) - \alpha \operatorname{grad} \theta = 0,$$

$$\alpha T \frac{\partial \operatorname{div} \boldsymbol{u}}{\partial t} + c \frac{\partial \theta}{\partial t} - \operatorname{div}(k \operatorname{grad} \theta) = f(\boldsymbol{x}, t).$$

In this case, α is proportional to the coefficient of volume expansion α_T ($\alpha = \alpha_T(3\lambda + 2\mu)$), k is the thermal conductivity, and c is heat capacity in the absence of deformation.

3 Finite Element Discretization

For the discretization in space using the finite-element method we first come to a variational formulation of problem (1)–(5). We define the Hilbert space for scalar values $L_2(\Omega)$ with the following scalar product and norm, respectively,

$$< u, v >= \int_\Omega u(\boldsymbol{x})\, v(\boldsymbol{x})\, dx, \quad ||u|| =< u, u >^{1/2}.$$

For vector values we use $\boldsymbol{L}_2(\Omega) = [L_2(\Omega)]^m$, where $m = 2, 3$ is the dimension of the domain Ω. Let $H^1(\Omega)$ and $\boldsymbol{H}^1(\Omega)$ be the Sobolev spaces. We define spaces for scalar and vector functions:

$$Q = \{q \in H^1(\Omega): q(\boldsymbol{x}) = 0, \ \boldsymbol{x} \in \Gamma_N^u\},$$

$$V = \{\boldsymbol{v} \in \boldsymbol{H}^1(\Omega): \boldsymbol{v}(\boldsymbol{x}) = 0, \ \boldsymbol{x} \in \Gamma_D^u\}.$$

Multiplying Eqs. (1) and (2) by test functions $\boldsymbol{v}$ and q, respectively, and integrating by parts to eliminate the second order derivatives, we obtain the variational problem: Find $\boldsymbol{u} \in \boldsymbol{V}$, $p \in Q$ such that

$$a(\boldsymbol{u}, \boldsymbol{v}) + g(p, \boldsymbol{v}) = 0, \quad \boldsymbol{v} \in \boldsymbol{V}, \tag{6}$$

$$d\left(\frac{d\boldsymbol{u}}{dt}, q\right) + c\left(\frac{dp}{dt}, q\right) + b(p, q) =< f, q >, \quad q \in Q. \tag{7}$$

The bilinear forms in (6), (7) are defined as follows:

$$a(\boldsymbol{u}, \boldsymbol{v}) = \int_\Omega \boldsymbol{\sigma}(\boldsymbol{u})\, \varepsilon(\boldsymbol{v})\, dx, \quad g(p, \boldsymbol{v}) = \alpha \int_\Omega \operatorname{grad} p\, \boldsymbol{v}\, dx,$$

$$c(p, q) = \frac{1}{M} \int_\Omega p\, q\, dx, \quad d(\boldsymbol{u}, q) = \alpha \int_\Omega \operatorname{div} \boldsymbol{u}\, q\, dx,$$

$$b(p, q) = \int_\Omega \frac{k}{\eta} \operatorname{grad} p\, \operatorname{grad} q\, dx.$$

The bilinear forms $a(\cdot, \cdot)$, $b(\cdot, \cdot)$, and $c(\cdot, \cdot)$ are symmetric and positive-defined, and $d(\boldsymbol{v}, q) = -g(q, \boldsymbol{v})$ for the boundary conditions (3), (4).

The discrete problem is obtained by restricting the variational problem to discrete spaces: Find $u_h \in V_h \subset V$, $p_h \in Q_h \subset Q$ such that

$$a(u_h, v_h) + g(p_h, v_h) = 0, \quad v_h \in V_h \subset V, \tag{8}$$

$$d\left(\frac{du_h}{dt}, q_h\right) + c\left(\frac{dp_h}{dt}, q_h\right) + b(p_h, q_h) = <f_h, q_h>, \quad q_h \in Q_h \subset Q. \tag{9}$$

The system (8), (9) is completed by the initial condition

$$<p_h(0), q_h> = <s, q_h>, \quad q_h \in Q_h \subset Q. \tag{10}$$

For stability analysis we come to an operator-difference formulation. We link operators A_h, B_h, C_h, D_h, G_h with the corresponding bilinear forms, for example:

$$<A_h u_h, v> = a(u_h, v), \quad \forall u_h, v \in V_h,$$

It allows to switch from the problem (8)–(10) to the following Cauchy problem:

$$A_h u_h + G_h p_h = 0, \tag{11}$$

$$D_h \frac{du_h}{dt} + C_h \frac{dp_h}{dt} + B_h p_h = f_h, \tag{12}$$

$$p_h(0) = s_h, \tag{13}$$

where $<f_h, q> = <f, v>$, $<s_h, q> = <s, v>$, $\forall q \in V_h$. The finite-dimensional operator A_h, B_h, C_h are stationary, self-adjoint, and positive-defined:

$$A_h = A_h^* > 0, \quad B_h = B_h^* > 0, \quad C_h = C_h^* > 0,$$

and $D_h = -G_h^*$.

4   Splitting Scheme

For the time discretization we use a uniform grid with step-size $\tau > 0$. Let $u^n = u_h(x, t^n)$, $p^n = p_h(x, t^n)$, where $t^n = n\tau$, $n = 0, 1, ...,$. Then, the standard implicit two-level scheme reads as

$$A_h u_h^{n+1} + G_h p_h^{n+1} = 0, \tag{14}$$

$$D_h \frac{u_h^{n+1} - u_h^n}{\tau} + C_h \frac{p_h^{n+1} - p_h^n}{\tau} + B_h p_h^{n+1} = f_h^{n+1}. \tag{15}$$

The computational realization of the implicit scheme (13)–(15), is associated with simultaneous calculation of u^{n+1}, p^{n+1} at every time level $n = 0, 1, ...$ This coupled system of equations requires special computational algorithms [7]. Therefore, splitting schemes for the poroelasticity and thermoelasticity problems are attracting increasing interest [6]. The simplest splitting scheme is an explicit-implicit scheme:

$$A_h u_h^{n+1} + G_h p_h^n = 0, \tag{16}$$

$$D_h \frac{u_h^{n+1} - u_h^n}{\tau} + C_h \frac{p_h^{n+1} - p_h^n}{\tau} + B_h p_h^{n+1} = f_h^{n+1}. \tag{17}$$

In this case, the transition to a next time level is associated with solving the separate sub-problems for the displacement (16) and pressure (17).

Expressing u_h from (11) and inserting it into (12), we obtain the single equation for the pressure p_h:

$$- D_h A_h^{-1} G_h \frac{dp_h}{dt} + C_h \frac{dp_h}{dt} + B_h p_h = f_h. \tag{18}$$

We write (18) as the standard Cauchy problem for a homogeneous evolutionary equation:

$$\widetilde{B} \frac{dp_h}{dt} + \widetilde{A} p_h = f_h, \tag{19}$$

$$p(0) = s_h. \tag{20}$$

where $\widetilde{A} = B_h$ and the operator $\widetilde{B}$ is the sum of two self-adjoint, positive-define operators

$$\widetilde{B} = \widetilde{B}_0 + \widetilde{B}_1, \quad \widetilde{B}_0 = C_h, \quad \widetilde{B}_1 = -D_h A_h^{-1} G_h. \tag{21}$$

The operators $\widetilde{A}, \widetilde{B}$ are stationary, self-adjoint, and positive-defined.

Under the additional constraint

$$\widetilde{B}_1 \leq \gamma \widetilde{B}_0, \quad \gamma > 0, \tag{22}$$

for the numerical solution of (19), (20), we can use a three-level explicit-implicit scheme

$$\widetilde{B}_0 \left(\theta \frac{p^{n+1} - p^n}{\tau} + (1 - \theta) \frac{p^n - p^{n-1}}{\tau} \right) + \widetilde{B}_1 \frac{p^n - p^{n-1}}{\tau} + \widetilde{A} p^{n+1} = f^{n+1}. \tag{23}$$

The equation is supplemented by the initial condition:

$$p^0 = s, \quad \widetilde{B} \frac{p^1 - p^0}{\tau} + \widetilde{A} p^1 = f^1. \tag{24}$$

The value of the weight θ must be chosen from a stability condition of the scheme (23), (24) [3].

Theorem 1. *For $2\theta \geq 1 + \gamma$ the scheme (23), (24) is unconditionally stable and its solution satisfies a priori estimate*

$$\mathcal{E}^{n+1} \leq \mathcal{E}^n + \frac{\tau}{2} \left\| f^{n+1} \right\|_{C^{-1}}^2, \tag{25}$$

where

$$\mathcal{E}^n = \left\| \frac{u^n + u^{n-1}}{2} \right\|_A^2 + \left\| \frac{u^n - u^{n-1}}{\tau} \right\|_{D - \frac{\tau^2}{4} A}^2,$$

and the operators C, D are

$$C = \widetilde{B} + \tau \widetilde{A}, \quad D = \frac{\tau}{2} \left((2\theta - 1) \widetilde{B}_0 - \widetilde{B}_1 \right) + \frac{\tau^2}{2} \widetilde{A}. \tag{26}$$

Combining (21) and (23), we get

$$-D_h A_h^{-1} G_h \frac{p^n - p^{n-1}}{\tau} + C_h \left(\theta \frac{p^{n+1} - p^n}{\tau} + (1 - \theta) \frac{p^n - p^{n-1}}{\tau} \right) \\ + B_h p^{n+1} = f^{n+1}. \tag{27}$$

The stability condition of scheme (27) (theorem 1) is $2\theta \geq 1 + \gamma_h$, where γ_h is defined by (see (22)) the condition

$$-D_h A_h^{-1} G_h \leq \gamma_h C_h.$$

For the poroelasticity problem (1)–(5), we have

$$\gamma_h = M \bar{\lambda}_{max} \tag{28}$$

where $\bar{\lambda}$ is the largest eigenvalue of the following eigenproblem:

$$G_h D_h \boldsymbol{u} = \bar{\lambda} A_h \boldsymbol{u}. \tag{29}$$

Taking into account (16), we get

$$D_h \frac{\boldsymbol{u}^{n+1} - \boldsymbol{u}^n}{\tau} + C_h \left(\theta \frac{p^{n+1} - p^n}{\tau} + (1 - \theta) \frac{p^n - p^{n-1}}{\tau} \right) \\ + B_h p^{n+1} = f^{n+1}. \tag{30}$$

Thus, the natural approximation (17) corresponds to the weight $\theta = 1$.

The main result of our analysis is the following theorem

Theorem 2. *The splitting scheme (16), (30) is unconditionally stable for $2\theta \geq 1 + \gamma_h$, where γ_h is the constant in (28).*

5   Numerical Experiments

We consider a poroelasticity problem in a unit square domain with an applied pressure $p = f(t)$ on some centered upper boundary as in Fig. 1. The applied pressure is $f(t) = \beta \sin(\beta t)$, where $\beta = (\lambda + 2\mu)k/\eta$ and βt is the dimensionless time. The rest of boundary is drained $p = 0$. Also, we set a free boundary condition for displacement $\boldsymbol{u} \times \boldsymbol{n} = 0$, $(\boldsymbol{\sigma} \cdot \boldsymbol{n}) \cdot \boldsymbol{n} = 0$. The initial pressure is zero $p = 0$. This problem is similar to the well-known Barry-Mercer problem widely used to test the accuracy of numerical methods [2]. The following set of material parameters are chosen: $\lambda = 15$ GPa, $\mu = 10$ GPa, $M = 100$ GPa, $\alpha = 0.9$, $k = 10^{-13}$ m^2, $\eta = 0.001$ Pa/s.

For the numerical solution we use two computational meshes: the coarse mesh with 6600 cells and the fine mesh with 26400 cells. The time step is chosen as $\tau = 0.01\beta t$. We find the largest eigenvalue $\bar{\lambda}_{max}$ of the eigenproblem (29) to calculate γ_h. For our problem, $\gamma_h = 3.4$. Note that γ_h hardly depends on the mesh size h.

Figure 2 shows the dimensionless pressure $\bar{p} = p/(2\lambda + \mu)$ (left) and displacement (right) at time $\beta t = \pi/2$, which are obtained using the implicit scheme (14)–(15) and the fine mesh. The displacement is presented on the deformed domain. These results will be used as "etalon" solutions $\bar{p}_e$ and $\boldsymbol{u}_e$ for evaluating the errors of the pressure $\varepsilon_p = ||\bar{p}_e - \bar{p}||$ and displacement $\varepsilon_u = ||\boldsymbol{u}_e - \boldsymbol{u}||$, respectively.

Figure 3 illustrates the errors of the dimensional pressure ε_p and displacement ε_u obtained using the implicit scheme and splitting scheme for several values of weights θ on the coarse mesh. We see that the splitting scheme (16), (30) is unstable for $\theta = 1$. The increase in the value of θ improves the stability of the scheme. For $\theta = 0.5(\gamma_h + 1) = 2.2$, the scheme is stable. This confirms the proposed stability condition of our scheme.

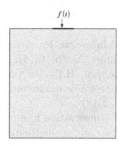

Fig. 1. Computational domain

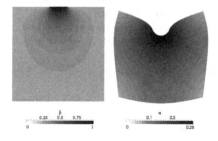

Fig. 2. Dimensionless pressure (left) and displacement distributions (right)

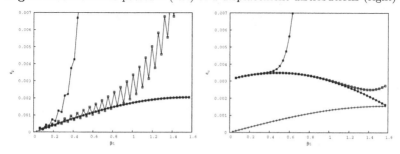

Fig. 3. Errors of the dimensional pressure (left) and displacement (right) obtained using the implicit scheme $(-\!\!+\!\!-)$ and the splitting scheme for $\theta = 1.0$ $(-\!\!\blacksquare\!\!-)$, $\theta = 1.4$ $(-\!\!\boxminus\!\!-)$, $\theta = 1.8$ $(-\!\!\ominus\!\!-)$, and $\theta = 2.2$ $(-\!\!*\!\!-)$

Acknowledgements. This work is supported by CJSC OptoGan (contract N02.G25.31.0090); RFBR (project N13-01-00719A); The Ministry of Education and Science of Russian Federation (contract RFMEFI5791X0026).

References

1. Armero, F.: Formulation and finite element implementation of a multiplicative model of coupled poro-plasticity at finite strains under fully saturated conditions. Comput. Methods Appl. Mech. Eng. **171**(3), 205–241 (1999)
2. Barry, S., Mercer, G.: Exact solutions for two-dimensional time-dependent flow and deformation within a poroelastic medium. J. Appl. Mech. **66**, 1–5 (1999)
3. Gaspar, F., Grigoriev, A., Vabishchevich, P.: Explicit-implicit splitting schemes for some systems of evolutionary equations. Int. J. Numer. Anal. Model. **11**(2), 346–357 (2014)
4. Gaspar, F., Lisbona, F., Vabishchevich, P.: A finite difference analysis of biot's consolidation model. Appl. Numer. Math. **44**(4), 487–506 (2003)
5. Haga, J.B., Osnes, H., Langtangen, H.P.: On the causes of pressure oscillations in low-permeable and low-compressible porous media. Int. J. Numer. Anal. Meth. Geomech. **36**(12), 1507–1522 (2012)
6. Jha, B., Juanes, R.: A locally conservative finite element framework for the simulation of coupled flow and reservoir geomechanics. Acta Geotech. **2**(3), 139–153 (2007)
7. Lisbona, F., Vabishchevich, P., Gaspar, F., Oosterlee, C.: An efficient multigrid solver for a reformulated version of the poroelasticity system. Comput. Methods Appl. Mech. Eng. **196**(8), 1447–1457 (2007)
8. Marchuk, G.I.: Splitting and alternating direction methods. In: Ciarlet, P.G., Lions, J.L. (eds.) Handbook of Numerical Analysis, vol. I, pp. 197–462. North-Holland, Amsterdam (1990)
9. Samarskii, A.A.: The Theory of Difference Schemes. Marcel Dekker, New York (2001)
10. Samarskii, A.A., Matus, P.P., Vabishchevich, P.N.: Difference Schemes with Operator Factors. Kluwer Academic Publisher, The Netherlands (2002)
11. Vabishchevich, P.N.: Additive Operator-Difference Schemes: Splitting schemes. de Gruyter, Berlin (2013)
12. Wang, H.: Theory of Linear Poroelasticity with Applications to Geomechanics and Hydrogeology. Princeton University Press, Princeton (2000)

Numerical Investigation of Adaptive Immune Response to Viral Infection

Mikhail Kolev[1]([✉]), Ana Markovska[2], and Boiana Garkova[2]

[1] Faculty of Mathematics and Computer Science, University of Warmia and Mazury,
Słoneczna 54, 10-710 Olsztyn, Poland
`kolev@matman.uwm.edu.pl`
[2] Faculty of Mathematics and Natural Sciences, South-West University
"Neofit Rilski", Ivan Mihajlov 66, 2700 Blagoevgrad, Bulgaria

Abstract. In this paper we present a new mathematical model describing acquired immune response to viral infection. The model is formulated as a system of six ordinary differential equations (ODE). Conditions for existence, uniqueness and non-negativity of the solutions are studied. Numerical simulations for the case of dominating cellular immunity and various initial values of concentrations of virus particles are presented and discussed.

Keywords: Numerical simulations · Differential equations · Nonlinear dynamics · Kinetic model · Virus · Immune system

1 Introduction

The use of mathematical models for investigations of the behavior of immune system of organisms infected by pathogens such as viruses or organisms suffering from cancer, can be effective tool for determining the tendencies of the disease under medical treatments or without them [4–6, 8–10, 12].

An organism that meets a specific antigen for the first time possesses only a small amount of lymphocytes able to recognize and neutralize the pathogen. That is why the acquired immune system needs at least several days while bigger amount of specific lymphocytes are produced and activated. During this period of time the fight against the pathogen is performed by the innate immunity, which functions quickly but does not possess specificity and efficiency. As a result the infection can become strong and difficult to eradicate [1]. That is why the acquired (or adaptive) immune mechanisms are often needed in order to clean the infection.

When foreign antigens enter an organism, both humoral and cellular types of acquired immunity start to function. Their mechanisms of functioning are different. The humoral immunity applies antibodies, which neutralize free viral particles. The cellular immunity system employs cytotoxic T lymphocytes (CTL), which destroy infected host cells [1].

© Springer International Publishing Switzerland 2015
I. Dimov et al. (Eds.): FDM 2014, LNCS 9045, pp. 249–256, 2015.
DOI: 10.1007/978-3-319-20239-6_26

In our paper we present a model, which is a generalization of a basic model proposed by G. Marchuk [10] and a model proposed by D. Wodarz [12]. In our model we assume that the growth of the virus depends on the amount of the infected cells as well as on the amount of the free viral particles that have entered the organism. Additionally, we suppose that the production of antibodies and CTL depends on the degree of the damage of the target organ: the higher is the damage, the weaker is the production of antibodies and CTL.

The purpose of this paper is to illustrate the application of mathematical and computational methods to immunology. The contents of our work are organized as follows. In Sect. 2 we describe our mathematical model of acquired immune response to viral infection. The model is a complicated system of ordinary differential equations. Theorem for existence, uniqueness and non-negativity of its solution is proved. In Sect. 3 we present some results of our simulations and comment their biological meaning.

2   Mathematical Model

The interacting populations included in our model and their notations are the following:

- $x(t)$ - concentration of the susceptible uninfected cells of the target organ;
- $y(t)$ - concentration of the infected cells;
- $v(t)$ - concentration of the free virus particles;
- $z(t)$ - concentration of CTL specific for the virus;
- $w(t)$ - concentration of antibodies (immunoglobulins) specific for the virus;
- $m(t)$ - degree of the target organ damage.

The proposed model describing the time dynamics of the considered variables consists of the following six ordinary differential equations (ODE):

$$\dot{x}(t) = L - dx(t) - \beta x(t)v(t), \tag{1}$$

$$\dot{y}(t) = \beta x(t)v(t) - ay(t) - py(t)z(t), \tag{2}$$

$$\dot{v}(t) = ky(t)v(t) - qv(t)w(t), \tag{3}$$

$$\dot{z}(t) = c\xi(m)v(t)z(t) - \delta(z(t) - \bar{z}) - by(t)z(t), \tag{4}$$

$$\dot{w}(t) = \gamma\xi(m)v(t)w(t) - h(w(t) - \bar{w}) - rv(t)w(t), \tag{5}$$

$$\dot{m}(t) = sv(t) - nm(t). \tag{6}$$

We suppose that the parameters of the model (1)-(6) are non-negative constants and parameters $L, \bar{z}$ and $\bar{w}$ are positive. We look for solution such that the unknown functions are continuously differentiable with non-negative initial conditions.

Equation (1) describes the dynamics of the population of the susceptible uninfected cells. The meaning of its parameters is the following: L describes the production of uninfected cells; d - the rate of decrease of uninfected cells due to their natural death; β - the rate of decrease of uninfected cells due to their infection by virus.

Equation (2) describes the dynamics of the population of the infected cells. Parameter a characterizes the decrease of concentration of infected cells due to their natural death; p denotes the rate of decrease of infected cells due to their destruction by CTL.

Equation (3) describes the time dynamics of the concentration of free virus particles. The viruses are produced inside the infected cells. The meaning of its parameters is the following: k denotes the rate of production of virus particles inside the infected cells; q - the decrease of virus particles due to their neutralization by antibodies.

Equation (4) describes the dynamics of CTL. The meaning of its parameters is the following: c characterizes the production of CTL; δ - the natural death of CTL; b - the decrease of concentration of CTL due to their killing activity against infected cells; $\bar{z}$ is the amount of CTL circulating in a healthy organism.

Equation (5) describes the dynamics of the concentration of antibodies. Their production depends on the amount of viruses and on the degree of target organ damage. The meaning of its parameters is the following: γ characterizes the production of antibodies; h - the rate of their natural death; r - the decrease of concentration of antibodies due to their antiviral activity; $\bar{w}$ is the amount of antibodies circulating in a healthy organism.

Equation (6) describes the degree of target organ damage. The damage depends on the amount of virus particles and can decrease due to repair processes in the organism. The meaning of its parameters is the following: s denotes the rate of damage of the target organ by viruses; n - the rate of recovery of the target organ.

$\xi(m)$ (participating in Eqs. (4) and (5)) is assumed to be non-increasing non-negative continuous function that accounts for the violation of the normal functioning of the immune system due to the damage of the target organ [1]. We assume that there exists its limit value $\bar{m} \in (0, 1)$. If the value of m is less than $\bar{m}$ we suppose that the damage of the infected organ is small and it does not affect the efficiency of the immune system. On the other hand, if m is greater than $\bar{m}$, we suppose that the damage of the infected organ is considerable and the immune response is weakened. The form of the function $\xi(m)$ is not uniquely defined. It can be chosen such that $\xi(m) = 1$ for $m \leq \bar{m}$ and $\xi(1) = 0$. For example it can be defined as follows:

$$\xi(m) = \begin{cases} 1 & \text{for } 0 \leq m \leq \bar{m}, \\ \frac{m-1}{\bar{m}-1} & \text{for } \bar{m} < m \leq 1. \end{cases} \tag{7}$$

Now we formulate theorems for non-negativity, existence and uniqueness of the solution to model (1) - (6).

Theorem 1. *If the system (1) - (6) with initial conditions $x(0) = x_0 > 0$, $y(0) = y_0 \geq 0$, $v(0) = v_0 \geq 0$, $z(0) = z_0 = \bar{z} > 0$, $w(0) = w_0 = \bar{w} > 0$, $m(0) = m_0 \geq 0$ possesses solution then this solution is non-negative for every $t \geq 0$.*

Proof. Consider Eq. (1). Let us assume that that there exist values of $t > 0$ such that $x(t) < 0$. From the initial condition $x(0) > 0$ and the continuity of the function $x(t)$ it follows that there exists an instant in time t_1 at which $x(t)$ changes its sign (i.e. $x(t) > 0$ for $t < t_1$, $x(t_1) = 0$ and $x(t) < 0$ for $t > t_1$). From here we would have $\dot{x}(t_1) < 0$. This would be a contradiction with Eq. (1) giving $\dot{x}(t_1) = L > 0$. Therefore the assumption about the possible negativity of $x(t)$ is incorrect.

The solution to Eq. (3) can be written in the form:

$$v(t) = v(0)e^{\int_0^t [ky(u)-qw(u)]du} \geq 0 \text{ for } t \geq 0.$$

From the non-negativity of $x(t)$ and $v(t)$ the non-negativity of $y(t)$ follows, since from Eq. (2) we obtain:

$$y(t) = e^{-\int_0^t [pz(u)+a]du}\left[y(0) + \int_0^t \beta x(u)v(u)e^{\int_0^t [pz(u)+a]du}du\right].$$

From Eq. (4) the non-negativity of $z(t)$ follows, since

$$z(t) = e^{\int_0^t [c\xi(m)v(u)-\beta-by(u)]du}\left[z(0) + \int_0^t \delta\bar{z}e^{-\int_0^t [c\xi(m)v(u)-\beta-by(u)]du}du\right].$$

Similarly, from Eq. (5) the non-negativity of $w(t)$ follows, since

$$w(t) = e^{\int_0^t [\gamma\xi(m)v(u)-h-rv(u)]du}\left[w(0) + \int_0^t h\bar{w}e^{-\int_0^t [\gamma\xi(m)v(u)-h-rv(u)]du}du\right].$$

Finally, from Eq. (6) the non-negativity of $m(t)$ follows, since

$$m(t) = e^{-nt}\left[m(0) + \int_0^t sv(u)e^{nu}du\right].$$

Theorem 2. *For every $T > 0$ on the interval $[0, T]$ there exists a unique continuously differentiable solution to the system (1) – (6) with initial conditions $x(0) = x_0 > 0$, $y(0) = y_0 \geq 0$, $v(0) = v_0 \geq 0$, $z(0) = z_0 = \bar{z} > 0$, $w(0) = w_0 = \bar{w} > 0$, $m(0) = m_0 \geq 0$.*

Proof. The local existence of the solution follows from the continuity of the right-hand sides (Peano theorem [7]). The uniqueness of the solution follows from the continuity of the partial derivatives of the right-hand sides with respect to the unknown functions [7].

It can be shown that the functions $x(t)$, $y(t)$ and $v(t)$ are bounded on $[0, T]$. Let us denote their maximal values with X, Y and V respectively. The following a priory bounds can be established for the solution on $[0, T]$:

$$\dot{x}(t) \leq L - dx(t), \tag{8}$$

$$\dot{y}(t) \leq \beta Xv(t) - ay(t), \tag{9}$$

$$\dot{v}(t) \leq kYv(t), \tag{10}$$

$$\dot{z}(t) \leq cVz(t) - \delta(z(t) - \bar{z}), \tag{11}$$

$$\dot{w}(t) = \gamma Vw(t) - h(w(t) - \bar{w}), \tag{12}$$

$$\dot{m}(t) = sv(t) - nm(t). \tag{13}$$

By the estimations (8) -(13) we see that, the nonlinear system (1) – (6) behaves not worse than a linear system. Therefore a global solution on $[0, T]$ exists.

3   Numerical Experiments and Discussion

The Cauchy problem (1)–(6) consisting of six nonlinear ODE is solved numerically. The system is solved by using the code `ode15s` from the Matlab ODE suite with $RelTol = 10^{-3}$ and $AbsTol = 10^{-4}$. `ode15s` is a multistep solver using numerical differential formulae (see, e.g. [11]).

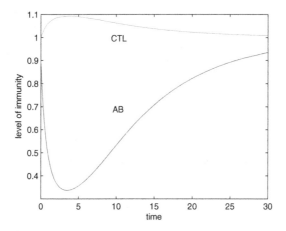

Fig. 1. Concentrations of antibodies (AB) and CTL at $v_0 = 0.1$ (low initial virus load).

The aim of our numerical experiments is to study the role of the magnitude v_0 of the initial virus load for the outcome of the competition between the immune system and the viral infection in the case when both parts of the immune systems (the cellular and the humoral immunity) are strong.

The initial conditions and parameters of the model have been set to simulate a full adaptive (humoral and cellular) immune response to viral infection. We have assumed the initial presence of susceptible uninfected cells, virus particles, antibodies and precursor CTL, as well as the initial absence of infected cells and effector CTL. The initial values for populations and the parameters have been set as follows:

$$x(0) = 1, \quad y(0) = 0, \quad z(0) = 0.9, w(0) = 1, m(0) = 0,$$

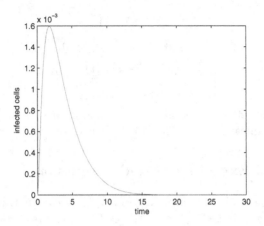

Fig. 2. Concentration of infected cells at $v_0 = 0.1$ (low initial virus load).

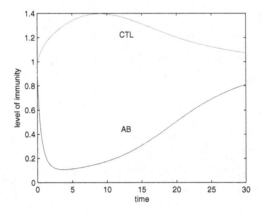

Fig. 3. Concentrations of antibodies and CTL at $v_0 = 0.2$ (high initial virus load).

$$L = 10, \quad d = 2, \beta = 0.01, \quad a = 0.1,$$

$$p = 1, \quad k = 1, \quad q = 1,$$

$$c = 1, \quad \delta = 0.1, \quad b = 10, \quad \gamma = 0.5, \quad h = 0.1,$$

$$r = 10, \quad s = 10, \quad n = 1.5.$$

Additionally, we assume that $\xi(m) = 1$ for every value of m and chose various values of the parameter v_0 specified in the captions of the figures.

Results of our numerical experiments are presented in Figs. 1–4. In the first part of our numerical experiments we consider the case of low initial virus load ($v_0 = 0.1$). The results for the dynamics of concentrations of antibodies and CTL as well as of infected cells are presented in Figs. 1 and 2 respectively.

In the second part of our numerical experiments we consider the case of high initial virus load ($v_0 = 0.2$). The results are presented in Figs. 3 and 4.

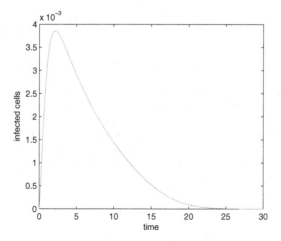

Fig. 4. Concentration of infected cells at $v_0 = 0.2$ (high initial virus load).

Our results show that in the presence of both cellular and humoral types of immunity, the strength and prolongation of infection does not depend significantly of the initial virus load. Strong cellular immunity kill the infected cells, which are needed for the viral replication. On the other hand, strong humoral response is able to destroy the free viral particles. Working in cooperation, the both adaptive immune mechanisms are able to eradicate the infection very effectively. Thus, the enhancement of the immune system is of crucial importance in the fight against viral infections.

We conclude that numerical simulations utilizing mathematical models may lead to a reduction in the quantity of experimental studies performed in virology. One of our future aims includes the determination of the parameters of the system (1)–(6) in order to fit existing experimental and clinical data. Another future plan is investigation of the role of function $\xi(m)$.

References

1. Abbas, A., Lichtman, A.: Basic Immunology: Functions and Disorders of the Immune System. Elsevier, Philadelphia (2009)
2. Arlotti, L., Bellomo, N., Lachowicz, M.: Kinetic equations modelling population dynamics. Transp. Theory Statist. Phys. **29**, 125–139 (2000)
3. Belleni-Morante, A.: Applied Semigroups and Evolution Equations. Oxford University Press, Oxford (1979)
4. Bellomo, N., Carbonaro, B.: Toward a mathematical theory of living systems focusing on developmental biology and evolution: a review and perspectives. Phys. Life Rev. **8**, 1–18 (2011)
5. Bellomo, N., Bianca, C.: Towards a Mathematical Theory of Complex Biological Systems. World Scientific, Singapore (2011)

6. Bianca, C.: Mathematical modelling for keloid formation triggered by virus: malignant effects and immune system competition. Math. Models Methods Appl. Sci. **21**, 389–419 (2011)
7. Hartman, P.: Ordinary Differential Equations. John Wiley and Sons, New York (1964)
8. Kolev, M., Korpusik, A., Markovska, A.: Adaptive immunity and CTL differentiation - a kinetic modeling approach. Math. Eng. Sci. Aerosp. **3**, 285–293 (2012)
9. Kolev, M., Garkova, B.: Numerical implementation of reaction-diffusion-chemotaxis model of cancer invasion using nonstandard finite difference method. In: Proceedings of the XIX National Conference on Applications of Mathematics in Biology and Medicine, Jastrzebia Gora, Poland, pp. 60–65 (2013)
10. Marchuk, G.: Mathematical modeling of Immune Response in Infections Diseases. Kluwer Academic Publishers, Dordrecht (1997)
11. Shampine, L., Reichelt, M.: The Matlab ODE suite. SIAM J. Sci. Comput. **18**, 1–22 (1997)
12. Wodarz, D.: Killer Cell Dynamics. Springer, New York (2007)

Efficient Application of the Two-Grid Technique for Solving Time-Fractional Non-linear Parabolic Problem

Miglena N. Koleva[✉]

FNSE, University of Rousse, 8 Studentska Str., 7017 Rousse, Bulgaria
mkoleva@uni-ruse.bg

Abstract. In this paper we present numerical methods for solving a non-linear time-fractional parabolic model. To cope with non-local in time nature of the problem, we exploit the idea of the two-grid method and develop fast numerical algorithms. Moreover, we show that suitable modifications of the standard two-grid technique lead to significant reduction of the computational time. Numerical results are also discussed.

1 Introduction

Fractional derivative have broad applications in mathematics, physics and engineering [5,15], such as anomalous transport in disordered systems, some percolations in porous media, and the diffusion of biological populations. The most important advantage of using fractional differential equations is their non-local property. This means that the next state of a system depends not only upon its current state but also upon all of its historical states. Thus the fractional-order models are more realistic and it is one reason why fractional calculus has been extensively investigated.

However the non-linear fractional differential equations are difficult to get their exact solutions [8,13]. An effective method for solving such equations is needed.

Consider the following non-linear time-fractional equation

$$D_c^\alpha u_t = (u^m)_{xx} + f(u), \quad 0 < \alpha < 1, \quad m > 1, \quad m \in \mathbb{R}, \quad (x,t) \in \mathbb{R} \times (0,T], \quad (1)$$

with given initial condition $u(x,0) = u^0(x)$, where D_c^α is the differential operator in the sense of Caputo

$$D_c^\alpha u_t = \frac{1}{\Gamma(1-\alpha)} \int_0^t \frac{\partial u(x,s)}{\partial s} (t-s)^{-\alpha} ds, \quad 0 < \alpha < 1. \quad (2)$$

If $m = 2$, Eq. (1) is the generalized non-linear biological population equation [4], where u denotes the population density and f represents the population supply due to births and deaths.

© Springer International Publishing Switzerland 2015
I. Dimov et al. (Eds.): FDM 2014, LNCS 9045, pp. 257–265, 2015.
DOI: 10.1007/978-3-319-20239-6_27

An example of constitutive equation for f is $f(u) = c_1 u^p (1 - c_2 u^q)$, where c_1, c_2, p, q are real numbers. More precisely, if $c_2 = 0$, $p = 1$ we have Malthusian Law [6], the case $p = q = 1$, $c_1 > 0$, $c_2 > 0$ corresponds to Verhulst Law [6] and $c_2 = 0$, $0 < p < 1$, $c_1 \leq 0$ is Porous Media model [2,16].

For $\alpha \to 1$, some properties of (1) such as Hölder estimates of its solutions are studied in [14]. Different iterative methods to construct approximate analytical solutions to the generalized non-linear biological population equation are developed: new iterative method [10], homotopy perturbation method [13,18], Adomian decomposition method [4].

For brevity, we shall concentrate on the following boundary value problem

$$D_c^\alpha u_t = (u^m)_{xx}, \quad m > 1, \quad 0 < \alpha < 1, \quad (x,t) \in (a,b) \times (0,T],$$
$$u(a,t) = u_a(t), \quad u(b,t) = u_b(t), \quad t \in (0,T], \tag{3}$$
$$u(x,0) = u^0(x), \quad x \in [a,b].$$

The main difficulty in the numerical computation of (3) is in the treatment of the memory integral $\int_0^t v(t-s)w(s)ds$, which arise from the discretization of Caputo time derivative. The solution process involves, at any given time step, the history of all computed solutions at each previous time levels. Thus, there is a potential need to store and operate on the entire history of the numerical solution. To cope with this problem in many articles (see [17,19], for example) is obtained a recursive formula, which involves only values at the current and previous time step, and not the entire history. Then this formula is incorporated into a numerical scheme, which then become local in time. This approach can be applied for known functions v and w or in the case of specific form of one of these functions - for example exponential, which are not our case. Moreover, the approach presented in [19] involves infinity series, which have to be computed at each grid node, and therefore undesired truncation error appears.

The aim of this paper is to develop *efficient numerical method*, based on the two-grid technique, for solving (3). The two-grid idea is to solve non-linear problem (time-fractional in our case) on the coarse mesh and linear problem (time-fractional in our case) on the fine mesh. Thus we attain the same precision as with one-grid procedure on the fine mesh, but saving computational time. We extend this approach and develop numerical algorithm (MTGM) which is equivalent of solving a non-linear time-fractional differential model on a coarse mesh and *linear classical (integer)* parabolic problem on a fine mesh. We also propose another technique (PTGM), where the two-grid idea is implemented as postprocessing procedure. The presented strategies lead additionally to significant reduction of the computational time and can be easily implemented for the initial value problem (1). In fact, we emphasize on the rapid computation of the memory integral.

The rest of the paper is organized as follows. In the next section, implementing the two-grid idea in different manner, we develop three numerical algorithms. In Sect. 3 we analyze numerically accuracy, convergence and efficiency of the proposed methods. Comparison results are also available. Finally, we provide some concluding remarks.

2   Numerical Method

In this section we shall implement the the two-grid methodology in different manner in order to save computational effort and accelerate the convergence.

The two grid approach was proposed in [1,20], for a linearization of non-linear elliptic problems. The two-grid finite element method was also used by Xu and many other scientists (see the reference in [9]) for discretizing non-symmetric indefinite elliptic and parabolic equations. By employing two finite element spaces of different scales, one coarse and one fine space, the method was used for symmetrization of nonsymmetric problems, which reduces the solution of a nonsymmetric problem on a fine grid to the solution of a corresponding (but much smaller) nonsymmetric problem, discretized on the coarse grid and the solution of a symmetric positive definite problem on the fine grid. Many different applications of the two-grid approach for *classical* differential problems are developed, see [7,11,12,21] among others.

In this section we shall present the following two-grid approaches:

- Application of the *classical two-grid method* (CTGM) for the model problem (3).
- *Modified two-grid method* (MTGM). We differ from the CTGM in the treatment of memory integral.
- Application of the *two-grid technique* as postprocessing procedure (PTGM).

All these algorithms are based on Newton's iteration procedure.

2.1   Classical Newton's Method

The computational domain $[a, b] \times [0, T]$ is discretized by uniform mesh ω:

$$\omega = \omega_h \times \omega^\tau, \quad \omega_h = \{x_i = a + ih, \ i = 0, \ldots, N, \ h = (b - a)/N\},$$
$$\omega^\tau = \{t_n = n\tau, \ n = 0, \ldots, M, \ M\tau = T\}.$$

Next, the numerical solution at point (x_i, t_n) will be denoted by $u_i^n = u(x_i, t_n)$ and $u^n = [u_0^n, u_1^n, \ldots, u_N^n]$.

The time-fractional derivative is replaced by simple quadrature formula [3]:

$$D_c^\alpha u_t(x_i, t_n) = \frac{\tau^{-\alpha}}{\Gamma(2 - \alpha)} \left[\lambda_0 u_i^n - \sum_{\substack{l=1 \\ (n>1)}}^{n-1} (\lambda_{n-l-1} - \lambda_{n-l}) u_i^l - \lambda_{n-1} u_i^0 \right] + \mathcal{O}(\tau^{2-\alpha}),$$

$$(4)$$

where $\lambda_j = (j + 1)^{1-\alpha} - j^{1-\alpha}$.

Now, using the standard central difference quotient, from (4), at each time level $n = 0, 1, \ldots$, we get the following discretization of the model problem (3)

$$\frac{\tau^{-\alpha}}{\Gamma(2 - \alpha)} \lambda_0 u_i^{n+1} + \Lambda u_i^{n+1} = \frac{\tau^{-\alpha}}{\Gamma(2 - \alpha)} F(u_i^n, \ldots, u_i^0), \ i = 1, \ldots, N - 1,$$

$$u_0^{n+1} = u_a(t_{n+1}), \quad u_N^{n+1} = u_b(t_{n+1}), \qquad u^0(x_i) = u_i^0, \quad i = 0, \ldots, N,$$

where $\Lambda u_i^n := [(u_{i+1}^n)^m - 2(u_i^n)^m + (u_{i-1}^n)^m]/h^2$ and $F(u_i^n, \ldots, u_i^0) := \sum_{l=1}^{n} (\lambda_{n-l-1} - \lambda_{n-l})u_i^l + \lambda_{n-l}u_i^0$.

Applying Newton's method, the solution u^{n+1} at each time layer $n = 0, 1, \ldots$, is a result of the iteration process $u^{k+1,n+1} = u^{k,n+1} + \triangle^{k+1,n+1}$, $k = 0, 1, \ldots$

$$\mathcal{L}(u_i^{k,n+1}, \triangle_i^{k+1,n+1}) = \frac{\tau^{-\alpha}}{\Gamma(2-\alpha)} \left[-u_i^{k,n+1} + F(u_i^n, \ldots, u_i^0) \right] + \Lambda u_i^{k,n+1},$$

$$\triangle_0^{k+1,n+1} = u_a(x_0) - u_0^{k,n+1}, \quad \triangle_N^{k+1,n+1} = u_b(x_N) - u_N^{k,n+1}, \tag{5}$$

$$\mathcal{L}(v_i, w_i) := \left(\frac{\tau^{-\alpha}}{\Gamma(2-\alpha)} + \frac{2mv_i^{m-1}}{h^2} \right) w_i - \frac{mv_{i+1}^{m-1}}{h^2} w_{i+1} - \frac{mv_{i-1}^{m-1}}{h^2} w_{i-1},$$

where $i = 1, \ldots, N-1$ and $u_i^{0,n+1} = u_i^n$ for $i = 0, \ldots, N$.

2.2 Two-Grid Algorithms

We define two uniform meshes in space: *coarse mesh* ω_H and *fine mesh* ω_h, $h << H$:

$$\omega_H = \{x_i = a + iH, \ i = 0, \ldots, N_c, \ H = (b-a)/N_c\},$$
$$\omega_h = \{x_i = a + ih, \ i = 0, \ldots, N_f, \ h = (b-a)/N_f\}.$$

Following the idea of the two-grid approach, we construct the algorithm

CTGM At each time level $n = 0, 1 \ldots$ perform the following two steps:

Step 1. Solve the non-linear fractional order problem (3) on the coarse mesh ω_H, applying Newton's iteration procedure (5) to obtain the solution u^H.

Step 2. Perform *only one* Newton's iteration (5) on the mesh ω_h with initial guess u^H

$$\mathcal{L}(u_i^H, \triangle_i^h) = \frac{\tau^{-\alpha}}{\Gamma(2-\alpha)} \left[-u_i^H + \mathbf{F}(\mathbf{u_i^n}, \ldots, \mathbf{u_i^0}) \right] + \Lambda u_i^H, \quad i = 1, \ldots, N_f - 1,$$

$$\triangle_0^h = u_a(x_0) - u_0^H, \quad \triangle_{N_f}^h = u_b(x_{N_f}) - u_{N_f}^H,$$

to obtain the solution $u^{n+1} = u^H + \triangle^h$. Actually, u^H is interpolated solution u^H on the fine mesh ω_h.

Next, we modify the second step of this algorithm

MTGM At each time level $n = 0, 1 \ldots$ perform the following two steps:

Step 1. Solve the non-linear fractional order problem (3) on the coarse mesh ω_H, applying Newton's iteration procedure (5) to obtain the solution u^H.

Step 2. Perform *only one* Newton's iteration (5) on the mesh ω_h with initial guess u^H

$$\mathcal{L}(u_i^H, \triangle_i^h) = \frac{\tau^{-\alpha}}{\Gamma(2-\alpha)} \left[-u_i^H + \mathbf{F_i^H} \right] + \Lambda u_i^H, \quad i = 1, \ldots, N_f - 1,$$

$$\triangle_0^h = u_a(x_0) - u_0^H, \quad \triangle_{N_f}^h = u_b(x_{N_f}) - u_{N_f}^H,$$

to obtain the solution $u^{n+1} = u^H + \triangle^h$. Here F^H and u^H are interpolated *coarse* values of u^H and $F(\cdot)$, respectively on the fine mesh ω_h.

With CTGM we need to store the solution history, obtained from the second step of the algorithm, i.e. on the fine mesh, and at each time level we have to recompute the summation also at step 2, while in MTGM we use interpolated value of F, obtained on the coarse mesh. Thus, at each time level, the memory integral computes only once - on the coarse grid for $k = 0$ and we need the history only for the solution at coarse grid nodes. Namely, on the fine mesh with MTGM we avoid the non-local nature of the problem.

The next algorithm can be interpret as postprocessing procedure on a substantially refined grid, because Step 2 performs only at the final time $t = T$.

PTGM

Step 1. While $t \leq T$, solve the non-linear fractional order problem (3) on the coarse mesh ω_H, applying Newton's iteration procedure (5). Denote the solution at final time by $u^{H,T}$

Step 2. At the final time $t = T$, continue Newton iteration process (5) on the fine mesh ω_h, starting with initial guess $u^{0,n+1} := u^{H,T}$ (interpolated on the fine mesh), obtained in Step 1 at time $t = T$. As in MTGM we use the interpolated value $\mathbf{F_i^H}$ of F on the fine mesh ω_h.

Remark 1. The non-linear term $f(u)$ in (1) can be easy incorporated in the two-grid Newton iteration procedure. To solve the initial value problem (1) in desired computational interval, a second order left and right finite difference can be used in order to approximate the second spatial derivative at both ends of the interval. This affects the accuracy, but *not* the order of convergence.

3 Numerical Experiments

In this section we shall verify accuracy, convergence rate and performance of the presented two-grid algorithms: CTGM, MTGM, PTGM. We deal with exact solution $u_{ex}(x,t) = t^{1+\alpha} e^{Cx}$ [3] adding residual term in the right-hand side of the differential equation in the model. Error (E^N) and convergence rate (CR) are computed in maximal discrete norm $(\| \cdot \|)$, using the grid size reduction $CR = \log_2[E^{N/2}/E^N]$, $E^N = \|u_{ex}(x,t_n) - u^n\|$, where u^n is computed on uniform mesh with N grid nodes in space.

Due to quadratic convergence of the Newton iterations, the expected order of convergence for CTGM and MTGM is $\mathcal{O}(\tau^{2-\alpha} + H^4 + h^2)$. This confirms from the computations. Thus the *optimal* choice for the fine mesh size is $h = H^2$. For PTGM we observe $\mathcal{O}(\tau^{2-\alpha} + H^3 + h^2)$ convergence rate for $\alpha \geq 0.5$ and $\mathcal{O}(\tau^{2-\alpha} + H^{5/2} + h^2)$ for $\alpha < 0.5$. Again the selection of fine mesh step size is small enough in order to capture the rate of convergence on the coarse mesh, i.e. $h = H^{3/2}$. Similarly, the time step τ is fixed in agreement with the ratio $\tau = h^{2/(2-\alpha)}$.

Newton's iterations process continue until $\|u^{k+1,n+1} - u^{k,n+1}\| \leq 10^{-10}$. Computations are performed in the interval $[0,1]$ up to final time T and $m = 2$.

Table 1. Errors, convergence rates and CPU time for different α, Example 1

α	N_c	N_f	τ	CTGM			MTGM		
				E^{N_f}	CR	CPU	E^{N_f}	CRs	CPU
0.8	5	25	4.678e-3	5.899e-4		0.58	6.461e-4		0.42
0.8	10	100	4.642e-4	3.780e-5	3.96	78.30	4.251e-5	3.93	9.43
0.8	20	400	4.605e-5	2.371e-6	3.99	25931.42	2.512e-6	4.08	1431.92
0.5	5	25	1.368e-2	6.681e-4		0.36	6.763e-4		0.29
0.5	10	100	2.154e-3	4.758e-5	3.81	4.40	4.769e-5	3.83	1.15
0.5	20	400	3.393e-4	3.097e-6	3.94	545.45	3.093e-6	3.95	35.05
0.2	5	25	2.797e-2	1.796e-4		0.26	1.794e-4		0.23
0.2	10	100	5.995e-3	1.497e-5	3.58	0.95	1.496e-5	3.58	0.50
0.2	20	400	1.285e-3	1.079e-6	3.79	40.95	1.078e-6	3.79	4.90
0.2	40	1600	2.753e-4	7.321e-8	3.88	3305.10	7.312e-8	3.88	157.80

Table 2. Errors, convergence rates and CPU time for different α, PTGM, Example 1

α	N_c	N_f	τ	E^{N_f}	CR	CPU
0.8	5	11	1.789e-2	2.671e-3		0.179
0.8	10	32	3.162e-3	3.655e-4	2.87	0.418
0.8	20	90	5.590e-4	4.708e-5	2.96	10.402
0.8	40	254	9.882e-5	5.961e-6	2.98	609.470
0.5	5	11	4.000e-2	2.743e-3		0.176
0.5	10	32	1.000e-2	4.428e-4	2.63	0.218
0.5	20	90	2.500e-3	6.051e-5	2.87	0.867
0.5	40	254	6.250e-4	8.039e-6	2.91	16.521
0.2	5	11	6.840e-2	6.248e-4		0.169
0.2	10	32	2.154e-2	1.280e-4	2.29	0.196
0.2	20	90	6.786e-3	2.100e-5	2.61	0.345
0.2	40	254	2.137e-3	3.564e-6	2.56	2.103

In order to compare the results at one and the same final time, we use a linear interpolation in time. To interpolate the coarse mesh solution on the fine mesh we use a cubic spline interpolation, while for the opposite action - from fine to coarse mesh solution we apply a linear interpolation.

Example 1 (*Exact solution test for* (3)). In this example we will compare the accuracy, convergence rate and computational cost (CPU time) of the presented methods for $C = 1$. The results for CTGM, MTGM, PTGM at $T = 0.1$ and different values of α are given in Tables 1, 2. It become clear that MTGM attains the same accuracy and convergence rate as CTGM, but MTGM is much more fast in comparison with CTGM. Although, PTGM is also cheap (in the sense of

Table 3. Errors and CPU time for different algorithms, $\alpha = 0.5$, Example 1

METHOD	T	N_c	N_f	τ	E^{N_f}	CPU
CTGM	1	5	25	0.01	2.48406e-4	5.124
MTGM	1	5	25	0.01	2.25752e-4	1.584
PTGM	1	14	52	0.01	2.29260e-4	3.159
CTGM	2	5	25	0.005	6.96817e-4	71.843
MTGM	2	5	25	0.005	6.61889e-4	15.983
PTGM	2	12	42	0.005	6.55361e-4	28.930
PTGM	2	11	36	0.005	7.74930e-4	26.313

Table 4. Errors and convergence rate in maximal and L_2 discrete norms, CPU time, MTGM, $\alpha = 0.5$, Example 2

N_c	τ	E^{N_f}	CR	CPU
25	9.685e-2	1.81611e-2		0.407
50	1.368e-2	1.40391e-3	*3.6933*	0.923
100	2.154e-3	9.34142e-5	*3.9097*	17.498

computational effort) algorithm, we observe slower convergence in comparison with MTGM, especially for $\alpha < 0.5$.

Next we will show how one and the same accuracy can be attained with two-grid algorithms CTGM, MTGM, PTGM for an *optimal choice* of the fine mesh step size for $\alpha = 0.5$. The results are listed in Table 3. It was found that for a long time, MTGM is even more precise than CTGM and PTGM is more stable, but needs more time than MTGM to reach one and the same precision.

Example 2 (*Initial value biological model*). Consider the problem (1), $f(u) = u(1 - 2u)$. As it was mentioned in Remark 1, the presented algorithms can be applied also in this case, using left and right second-order difference for the space derivative at both ends of the computational interval. We repeat the convergence test from Example 1 in the interval $[-5, 5]$ with MTGM, $\alpha = 0.5$, $C = 0.2$, such that the maximal discrete norm of the solution in the computational interval to be equal to the one in Example 1, see Table 4.

4   Conclusions

The main advantage of the proposed methods is that we attain the same accuracy as applying Newton method on the fine mesh, but saving *computational cost*. The efficiency of the two-grid approach is obvious: we reach a high accuracy, solving non-linear equations on a coarse grid and linear equations on a fine mesh. Moreover, the coarse grid can be very coarse and the results are still precise. Applying this idea in a different manner we obtain new algorithms

(MTHM, PTGM), which performance is very fast and leads to significant reduction of the computational cost. Moreover, they are more stable in time.

Acknowledgement. This research is supported by the Bulgarian National Fund of Science under the Project I02/20 - 2014.

References

1. Axelsson, O.: On mesh independence and Newton methods. Appl. Math. **38**(4–5), 249–265 (1993)
2. Bear, J.: Dynamics of Fluids in Porous Media. American Elsevier, New York (1972)
3. Cui, M.: Convergence analysis of high-order compact alternating direction implicit schemes for the two-dimensional time fractional diffusion equation. Numer. Algor. **62**(3), 383–409 (2013)
4. El-Sayed, A.M.A., Rida, S.Z., Arafa, A.A.M.: Exact solutions of fractional-order biological population model. Commun. Theor. Phys. **52**, 992–996 (2009)
5. Hilfer, R.: Applications of Fractional Calculus in Physics. World Scientific, Singapore (2000)
6. Gurtin, M.E., Maccamy, R.C.: On the diffusion of biological population. Math. Biosci. **33**, 35–49 (1977)
7. Ishimura, N., Koleva, M.N., Vulkov, L.G.: Numerical solution via transformation methods of nonlinear models in option pricing. Am. Inst. Phys. CP **1301**, 387–394 (2010)
8. Jiang, X.Y., Xu, M.Y.: Analysis of fractional anomalous diffusion caused by an instantaneous point source in disordered fractal media. Int. J. Nonlinear Mech. **41**, 156–165 (2006)
9. Jin, J., Shu, S., Xu, J.: A two-grid discretization method for decoupling systems of partial differential equations. Math. Comput. **75**, 1617–1626 (2006)
10. Koçak, H., Yildirim, A.: An efficient new iterative method for finding exact solutions of nonlinear time-fractional partial differential equations. Nonlinear Anal. Model. Control **16**(4), 403–414 (2011)
11. Koleva, M.N., Vulkov, L.G.: A two-grid approximation of an interface problem for the nonlinear Poisson-Boltzmann equation. In: Margenov, S., Vulkov, L.G., Waśniewski, J. (eds.) NAA 2008. LNCS, vol. 5434, pp. 369–376. Springer, Heidelberg (2009)
12. Koleva, M.N., Vulkov, L.G.: Two-grid quasilinearization approach to ODEs with applications to model problems in physics and mechanics. Comput. Phys. Commun. **181**, 663–670 (2010)
13. Liu, Y., Li, Z., Zhang, Y.: Homotopy perturbation method to fractional biological population equation. Fract. Diff. Calc. **1**, 117–124 (2011)
14. Lu, Y.G.: Hölder estimates of solutions of biological population equations. Appl. Math. Lett. **13**, 123–126 (2000)
15. Metzler, R., Klafter, J.: The random walks guide to anomalous diffusion: a fractional dynamics approach. Phys. Rep. **339**, 1–77 (2000)
16. Okubo, A.: Diffusion and Ecological Problem: Mathematical Models. Biomathematics 10. Springer, Berlin (1980)
17. Patlashenko, I., Givoli, D., Barbone, P.: Time-stepping schemes for systems of Volterra integrodifferential equations. Comput. Methods Appl. Mech. Eng. **190**, 5691–5718 (2001)

18. Roul, P.: Application of homotopy perturbation method to biological population model. Appl. Appl. Math. **5**(10), 1369–1378 (2010)
19. Wang, P., Zheng, C., Gorelick, S.: A general approach to advective-dispersive transport with multirate mass transfer. Adv. Water. Res. **28**, 33–42 (2005)
20. Xu, J.: A novel two-grid method for semilinear elliptic equations. SIAM J. Sci. Comput. **15**(1), 231–237 (1994)
21. Xu, J.: Two-grid discretization techniques for linear and nonlinear PDEs. SIAM J. Numer. Anal. **33**, 1759–1777 (1996)

Error Estimates of Four Level Conservative Finite Difference Schemes for Multidimensional Boussinesq Equation

Natalia Kolkovska[✉]

Institute of Mathematics and Informatics, Bulgarian Academy of Sciences,
Acad. Bonchev Str. Bl.8, 1113 Sofia, Bulgaria
`natali@math.bas.bg`

Abstract. A family of *four level conservative finite difference schemes* (FDS) for the multidimensional Boussinesq Equation is constructed and studied theoretically. A preservation of the discrete energy for this approach is established. We prove that the discrete solution of the FDS converges to the exact solution with a second order of convergence with respect to space and time mesh steps in the first discrete Sobolev norm and in the uniform norm. The numerical experiments for the one-dimensional problem confirm the theoretical rate of convergence and the preservation of the discrete energy in time.

1 Introduction

In this paper we consider the Cauchy problem for the Boussinesq type equation (BE)

$$\frac{\partial^2 u}{\partial t^2} = \Delta u + \beta_1 \Delta \frac{\partial^2 u}{\partial t^2} - \beta_2 \Delta^2 u + \Delta f(u), \quad x \in \mathbb{R}^d, \quad 0 < t \le T, \quad T < \infty \quad (1)$$

on the unbounded region $\mathbb{R}^d$ with asymptotic boundary conditions

$$u(x,t) \to 0, \quad \Delta u(x,t) \to 0, \quad |x| \to \infty, \quad (2)$$

and initial conditions

$$u(x,0) = u_0(x), \quad \frac{\partial u}{\partial t}(x,0) = u_1(x). \quad (3)$$

Here Δ is the Laplace operator and the constants β_1 and β_2 are positive. The nonlinearity function f can be in the form $f(u) = u^p, p = 2, 3, ..., f(u) = au^p + bu^{2p+1}$, etc.

BE occurs in a number of physical systems, for example in the modeling of surface waves in shallow waters (in this case $f(u) = \alpha u^2$ with $\alpha > 0$), see [6]. The form (1) of BE is derived from the original Boussinesq system in [6] and is referred in the literature as the 'generalized double dispersion equation' or the 'Boussinesq paradigm equation'.

© Springer International Publishing Switzerland 2015
I. Dimov et al. (Eds.): FDM 2014, LNCS 9045, pp. 266–273, 2015.
DOI: 10.1007/978-3-319-20239-6_28

BE (1)–(3) may have either a globally defined bounded solution or a blowing up solution. Sufficient conditions for the global existence or for the blow up of the weak solutions to (1)–(3) are given in [13,14,17,19] in terms of the initial data u_0, u_1, the non-linearity f and the initial energy. In this paper we assume that the solution to (1)–(3) exists and is smooth enough. We note also that the solution to (1)–(3) satisfies the energy conservation law, see e.g. [9,19]

It is of great importance to preserve the energy conservation law of (1)–(3) under the discretization method for solving (1), especially when solving problem (1)–(3) over a long time interval. The numerical methods that satisfy discrete conservation identity are called conservative methods.

Several numerical methods – finite difference methods, finite element methods, spectral and pseudo-spectral methods, vector additive schemes – have been proposed for the BE, see e.g. [1–5,8,11,15,16,18]. Only certain numerical schemes for approximation of BE conserve the discrete energy, see [6,9,12].

In this paper we analyze theoretically the family of new conservative four level finite difference schemes, presented in [10]. These new schemes are explicit with respect to the non-linearity in the sense that no iterations are needed for evaluation of the discrete solution on the highest time level (the three level schemes are implicit in the same sense - a system of nonlinear equations has to be solved for evaluation of the discrete solution on the highest time level). Section 3 contains the main results - the proof of the discrete conservation law and convergence theorems of the method. We establish a second order of convergence in the first discrete Sobolev mesh norm and in the uniform norm. This rate of convergence is compatible with the rate of convergence of the similar linear problem. The convergence of the schemes is demonstrated numerically in Sect. 4 for the one dimensional problem. Algorithmic aspects of the scheme's implementation and extensive numerical experiments can also be found in [10].

2   Finite Difference Scheme

For simplicity of the presentation we deal with the two dimensional case, i.e. $d = 2$ in (1). We discretize BPE (1)–(3) on a sufficiently large space domain $\Omega = [-L_1, L_1] \times [-L_2, L_2]$ assuming that the solution and its first and second derivatives are negligible outside Ω. For integers N_1, N_2 set the space steps $h_i = L_i/N_i, i = 1, 2$ and $h = (h_1, h_2)$. Let $\Omega_h = \{(x_i, y_j) : x_i = ih_1, i = -N_1, \dots, N_1, y_j = jh_2, j = -N_2, \dots, N_2\}$. Next, for integer N we denote the time step by $\tau = T/N$. For each of the time levels $t^k = k\tau$, $k = 0, 1, 2, \dots, N$ we consider a mesh function $v_{i,j}^{(k)}$ defined on $\Omega_h \times \{t^k\}$. Whenever possible the sub-indexes i, j of the mesh functions are omitted.

The discrete scalar product $\langle v, w \rangle = \sum_{i,j} h_1 h_2 v_{i,j} w_{i,j}$ is associated with the space of mesh functions v, w, which vanish on the boundary of Ω_h. Denote by Δ_h the standard 5-point discrete Laplacian. We use also the following notation $v_t^{(k)} = \left(v^{(k+1)} - v^{(k)}\right)\tau^{-1}$.

The standard approximation to the second time derivative $\frac{\partial^2 u}{\partial t^2}(\cdot, t^k)$ uses three time levels of the discrete functions and is defined as

$$v_{\bar{t}t}^{(k)} = \left(v^{(k+1)} - 2v^{(k)} + v^{(k-1)}\right)\tau^{-2}.$$

In this paper the time derivative $\frac{\partial^2 u}{\partial t^2}(\cdot, t^k + \tau/2)$ is approximated using four consecutive time levels $(k+2)$, $(k+1)$, (k) and $(k-1)$ by the expression

$$v_{\hat{t}\hat{t}}^{(k)} = 0.5(v^{(k+2)} - v^{(k+1)} - v^{(k)} + v^{(k-1)})\tau^{-2}.$$

Spatial derivatives included in (1) can be evaluated on the four time levels. Thus we introduce two symmetric approximations to $u(\cdot, t^k + \tau/2)$ with real parameters θ and μ:

$$v^{\theta(k)} = \theta v^{(k+2)} + (0.5 - \theta)v^{(k+1)} + (0.5 - \theta)v^{(k)} + \theta v^{(k-1)},$$

$$v^{\mu(k)} = \mu v^{(k+2)} + (0.5 - \mu)v^{(k+1)} + (0.5 - \mu)v^{(k)} + \mu v^{(k-1)}.$$

The approximation $v^{\theta(k)}$ with parameter θ will be applied when constructing approximations to the Laplacian Δv and the approximation $v^{\mu(k)}$ with parameter μ- for the bi-Laplacian $(\Delta)^2 v$.

The nonlinear term f in (1) can be treated in different ways, see [7,12,15,16]. Here we approximate $f(u)$ using the gradient form of f, $f(s) = F'(s)$, $F(u) = \int_0^u f(s)ds$ and the values of v on time levels k and $k + 1$:

$$\frac{F(v^{(k+1)}) - F(v^{(k)})}{v^{(k+1)} - v^{(k)}}. \tag{4}$$

Note that in the case of a polynomial function f the expression (4) can be evaluated easily, without division (if $f(u) = \alpha u^2$, then the function in (4) is equal to $\alpha(v^{(k+1)2} + v^{(k+1)}v^{(k)} + v^{(k)2})/3$).

Our four-time-level finite difference scheme defines the approximate solution $v_{i,j}^{(k)}$ as the solution of

$$v_{\hat{t}\hat{t}}^{(k)} - \beta_1 \Delta_h v_{\hat{t}\hat{t}}^{(k)} - \Delta_h v^{\theta(k)} + \beta_2 (\Delta_h)^2 v^{\mu(k)} = \Delta_h \frac{F(v^{(k+1)}) - F(v^{(k)})}{v^{(k+1)} - v^{(k)}} \tag{5}$$

at the internal mesh points of Ω_h, i.e. $|i| < N_1$ and $|j| < N_2$.

Initial values $v^{(0)}$ and $v^{(1)}$ on time levels $t = 0$ and $t = \tau$ are evaluated by formulas

$$v_{i,j}^{(0)} = u_0(x_i, y_j), \tag{6}$$

$$v_{i,j}^{(1)} = u_0(x_i, y_j) + \tau u_1(x_i, y_j)$$
$$+ 0.5\tau^2(I - \beta_1\Delta_h)^{-1}\left(\Delta_h u_0 - \beta_2(\Delta_h)^2 u_0 + \Delta_h f(u_0)\right)(x_i, y_j),$$

where I stands for the identity operator. The third initial value $v^{(-1)}$ at time level $t = -\tau$ is found from the equation

$$v_{\bar{t}t(i,j)}^{(0)} = \left(v_{(i,j)}^{(1)} - 2v_{(i,j)}^{(0)} + v_{(i,j)}^{(-1)}\right)\tau^{-2} \tag{7}$$

$$= (I - \beta_1\Delta_h)^{-1}\left(\Delta_h u_0 - \beta_2\Delta_h^2 u_0 + \Delta_h f(u_0)\right)(x_i, y_j).$$

The boundary conditions at the boundary mesh points, i.e. $|i| = N_1$ or $|j| = N_2$, of Ω_h are

$$v_{i,j}^{(k)} = 0, \quad \Delta_h v_{i,j}^{(k)} = 0, \quad k = 1, 2, \ldots, N. \tag{8}$$

In order to implement the second boundary condition in (8), the grid overlaps the domain Ω_h by one line at each boundary.

Equations (5)–(8) constitute a family of finite difference schemes depending on the parameters θ and μ. By Taylor series expansion of the solution u at point $(x_i, y_j, t^k + \tau/2)$ one concludes that for each fixed θ and μ the local approximation error of the finite difference scheme is $O(|h|^2 + \tau^2)$.

3   Convergence of the Family of Finite Difference Schemes

In the space of functions which vanish on the boundary of Ω_h we define operators $A = -\Delta_h$ and $B = I - \beta_1 \Delta_h - 2\tau^2 \theta \Delta_h + 2\tau^2 \beta_2 \mu (\Delta_h)^2$. Note that these operators are self-adjoint and positive definite (the additional requirements $\theta \geq 0$ and $\mu \geq 0$ are needed for $B > 0$). For the analysis of the error we rewrite (5) in the operator form

$$Bv_{\bar{t}\bar{t}}^{(k)} + 0.5(A + \beta_2 A^2)(v^{(k+1)} + v^{(k)}) + A \frac{F(v^{(k+1)}) - F(v^{(k)})}{v^{(k+1)} - v^{(k)}} = 0. \tag{9}$$

and introduce the discrete energy functional $E_h(v^{(k)})$ as

$$E_h(v^{(k)}) = 0.5 \left\langle A^{-1} B v_t^{(k)}, v_t^{(k-1)} \right\rangle + 0.5 \left\langle v^{(k)} + \beta_2 A v^{(k)}, v^{(k)} \right\rangle + \left\langle F(v^{(k)}), 1 \right\rangle. \tag{10}$$

We multiply (9) by $A^{-1} \left(v^{(k+1)} - v^{(k)} \right)$, use summation by parts and the boundary conditions (8). In this way we obtain the following

Theorem 1 (Discrete Conservation Law). *The discrete energy (10) of the solution to the difference schemes (5)–(8) is conserved in time:*

$$E_h(v^{(k)}) = E_h(v^{(0)}), \quad k = 1, 2, \ldots, N.$$

Theorem 1 states that, as in the continuous case, the conservation property holds for the solution to the finite difference scheme (5)–(8), i.e. the proposed finite difference schemes are conservative.

The key result of the paper is the following theorem for the convergence of the solution v of the finite difference scheme to the exact solution u to BPE. Denote by $z = v - u$ the error of the numerical solution.

Theorem 2 (Convergence of the Method). *Assume that f is a polynomial of u of degree m, $u \in C^{4,4}(\mathbb{R}^2 \times [0, T])$ and that:*

(i) *the solution v to the finite difference scheme (5)–(8) is bounded in the maximal norm;*

(ii) *the parameters θ and μ satisfy the operator inequality*

$$A^{-1} + \beta_1 I + \tau^2(2\theta - 0.5)I + \tau^2\beta_2(2\mu - 0.5)A > \epsilon I \tag{11}$$

with some positive real number ϵ.

Let M be a constant such that $M \geq \max_{i,j,k} \left(|u(x_i, y_j, t_k)|, |v_{i,j}^{(k)}|\right)$ and $\tau < C_1 M^{-1}$ (the constant C_1 is independent of h, τ, u and v). Then the discrete solution v to (5)–(8) converges to the exact solution u as $|h|, \tau \to 0$ and there exists a constant C (independent of h, τ and u) such that the following estimate holds for the error $z = u - v$ at every time level $k = 1, 2, ..., N$

$$0.25\|z^{(k)} + z^{(k+1)}\|^2 + 0.25\beta_2\|A^{1/2}(z^{(k)} + z^{(k+1)})\|^2 \leq Ce^{Mt^k}\left(|h|^2 + \tau^2\right)^2. \tag{12}$$

Proof. We sketch the proof because it is too lengthy. First, we substitute $v = z + u$ into the finite difference equation (9) and obtain the following equation for the error z:

$$Bz_{\bar{t}t}^{(k)} + 0.5(A + \beta_2 A^2)(z^{(k+1)} + z^{(k)}) = \psi^{(k)}, \tag{13}$$

where

$$\psi^{(k)} = -A\frac{F(v^{(k+1)}) - F(v^{(k)})}{v^{(k+1)} - v^{(k)}} - Bu_{\bar{t}t}^{(k)} - 0.5(A + \beta_2 A^2)(u^{(k+1)} + u^{(k)}).$$

We use Taylor series for the function u about the point $(x_i, y_j, t^k + \tau/2)$ and represent the error $\psi^{(k)}$ as $\psi^{(k)} = \psi_1^{(k)} + A\psi_2^{(k)}$ with $\psi_1^{(k)} = O(|h|^2 + \tau^2)$ and $|\psi_2^{(k)}| \leq (|z^{(k)}| + |z^{(k+1)}|)(\max(|u^{(k)}|, |u^{(k+1)}|, |v^{(k)}|, |v^{(k+1)}|))^{m-1}$. The initial conditions (3) are approximated locally by (6) and (7) with $O(|h|^2 + \tau^2)$ error.

By multiplication of the difference equation (13) by $A^{-1}(z^{(k+2)} + z^{(k+1)} - z^{(k)} - z^{(k-1)})$ we get for $Z^{(k)} = 0.5(z^{(k+1)} + z^{(k)})$ the main inequality

$$\left\langle (A^{-1}B - 0.5\tau^2 I - 0.5\tau^2\beta_2 A)Z_t^{(k)}, Z_t^{(k)}\right\rangle + 0.5\left\langle|Z^{(k+1)}|\right\rangle^2 + 0.5\left\langle|Z^{(k)}|\right\rangle^2$$

$$+ 0.5\beta_2\left\langle|A^{0.5}Z^{(k+1)}|\right\rangle^2 + 0.5\beta_2\left\langle|A^{0.5}Z^{(k)}|\right\rangle^2$$

$$\leq \epsilon_1\tau\left\langle|\psi_1^{(k)}|\right\rangle^2 + \epsilon_1\tau\left\langle|\psi_2^{(k)}|\right\rangle^2 + \frac{\tau}{4\epsilon_1}\left\langle Z_t^{(k)}, Z_t^{(k)}\right\rangle + \frac{\tau}{4\epsilon_1}\left\langle Z_t^{(k-1)}, Z_t^{(k-1)}\right\rangle$$

$$+ 0.5\left\langle|Z^{(k)}|\right\rangle^2 + 0.5\left\langle|Z^{(k-1)}|\right\rangle^2 + 0.5\beta_2\left\langle|A^{0.5}Z^{(k)}|\right\rangle^2$$

$$+ 0.5\beta_2\left\langle|A^{0.5}Z^{(k-1)}|\right\rangle^2 + \left\langle(A^{-1}B - 0.5\tau^2 I - 0.5\tau^2\beta_2 A)Z_t^{(k-1)}, Z_t^{(k-1)}\right\rangle.$$

Then we sum the above inequalities over $k = 1, 2, \cdots N$, use Gronwall's inequality, the properties of ψ_1 and ψ_2 and obtain the desired estimate (12).

Theorem 2 gives second order of convergence of the FDS in the discrete W_2^1 norm, which is compatible with the rate of convergence of the similar linear problem. Thus, the non-linearity does not deteriorate the rate of convergence.

The assumptions for the boundedness of the exact and discrete solutions in Theorem 2 could not be dropped. Note that the initial differential problem may have either bounded on the time interval $[0, T)$ solutions or blowing up solutions.

Corollary 1. *If $\theta \geq 0.25$ and $\mu \geq 0.25$ then the convergence of the discrete solution to the exact solution is of second order when $|h|$ and τ go independently to zero.*

If $\mu = 0$ then the discrete solution converges to the exact solution provided
$$\tau^2 < h^2 \frac{\beta_1 - \epsilon + \tau^2(2\theta - 0.5)}{2d\beta_2}.$$

Combining Theorem 2 with the embedding theorems we get error estimates in the uniform norm:

Corollary 2. *Under the assumptions of Theorem 2 the finite difference scheme (5)–(8) admits the following error estimate*

$$\max_i |z_i^{(k)} + z_i^{(k+1)}| \leq C e^{Mt^k} \left(|h|^2 + \tau^2 \right), \qquad d = 1;$$

$$\max_{i,j} |z_i^{(k)} + z_i^{(k+1)}| \leq C e^{Mt^k} \sqrt{\ln(\max\{N_1, N_2\})} \left(|h|^2 + \tau^2 \right), \qquad d = 2.$$

The above estimates are optimal for the 1D case and almost optimal (up to a logarithmic factor) for the 2D case.

4   Numerical Results

The numerical procedure for computation of the discrete solutions at the higher time level is very efficient, especially when $\mu = 0$. In this case a system of linear equations (5) with the discrete Laplacian equations in a rectangular domain has to be solved. The restriction (11) on the time step is mild (see Corollary 1). Note also that no iterations are needed for evaluation of the discrete solution on the highest time level nevertheless the considered method has a nonlinear term and is conservative! This is because the non-linearity is approximated on the previous time levels.

Now we present some numerical results concerning the proposed finite difference scheme in the one-dimensional case - for $d = 1$ and $\beta_1 = 1.5$, $\beta_2 = 0.5$, $f(u) = 3u^2$, $\mu = 0$, $\theta = 0.25$, $\tau = h\sqrt{(\beta_1/(8\beta_2))}$. It is well known that in this case BE possesses exact solutions - waves of permanent form, which propagate in time with the velocity c:

$$\tilde{u}(x, t; x_0, c) = \frac{3}{2} \frac{c^2 - 1}{\alpha} \text{sech}^2 \left(\frac{x - x_0 - ct}{2} \sqrt{\frac{c^2 - 1}{\beta_1 c^2 - \beta_2}} \right)$$

These solutions, called solitons, are analytical functions with maximum located at the initial moment at the point x_0.

Example 1: Single solitary wave (1 soliton)
We solve the problem (1)–(3) using following initial data

$$u(x,0) = \tilde{u}(x,0;0,2), \quad \frac{\partial u}{\partial t}(x,0) = \frac{\partial \tilde{u}(x,0;0,2)}{\partial t}.$$

We compare the numerical solution to the exact one and evaluate the error E in the discrete uniform norm. The results are shown in the first two columns of Table 1.

Example 2: Colliding solitary waves (2 solitons)
The initial conditions in this case are:

$$u(x,0) = \tilde{u}(x,0;-40,2) + \tilde{u}(x,0;50,-1.5),$$

$$\frac{du}{dt}(x,0) = \frac{d\tilde{u}}{dt}(x,0;-40,2) + \frac{d\tilde{u}}{dt}(x,0;50,-1.5).$$

In this case no explicit solution to (1)–(3) is available. Thus to illustrate the method's convergence we relay on grid-doubling. For every h the error is calculated by the Runge method as $E_1^2/(E_1 - E_2)$ with $E_1 = \|u_{[h]} - u_{[h/2]}\|$, $E_2 = \|u_{[h/2]} - u_{[h/4]}\|$, where $u_{[h]}$ is the calculated solution with step h for $T = 20$. The numerical rate of convergence is $(\log E_1 - \log E_2)/\log 2$.

Table 1. Errors in uniform norm and rate of convergence for one and two solitons, $T = 20$

h	1 soliton		2 soliton	
	L_∞ Error	Rate	L_∞ Error	Rate
0.08	0.0029301			
0.04	0.0007326	1.9998523	0.1193204	
0.02	0.0001836	1.9963024	0.0298805	1.9975659
0.01	4.9437 e-005	1.8930601	0.0008015	1.8984112

The calculations shown on Table 1 confirm that the schemes are of order $O(h^2 + \tau^2)$. Our calculations also show that the discrete energy functional $E_h(v^{(k)})$ is preserved in time with high accuracy - for $t \in (0,20]$ the relative error in calculation of $E_h(v^{(k)})$ is $7.0407e - 08$.

Acknowledgments. This work is partially supported by the Bulgarian Science Fund under grant DDVU 02/71.

References

1. Bratsos, A.: A predictor-corrector scheme for the improved Boussinesq equation. Chaos, Solitons Fractals **40**, 2083–2084 (2009)
2. Chertock, A., Christov, C., Kurganov, A.: Central-upwind schemes for the boussinesq paradigm equation. Comp. Sci. High Perform. Comp IV NNFM **113**, 267–281 (2011)
3. Choo, S.M.: Pseudo spectral methods for the damped Boussinesq equation. Comm. Korean Math. Soc. **13**, 889–901 (1998)
4. Choo, S.M., Chung, S.K.: Numerical solutions for the damped Boussinesq equation by FD-FFT-perturbation method. Comp. Math. with Appl. **47**, 1135–1140 (2004)
5. Christou, M., Christov, C.I.: Galerkin spectral method for the 2D solitary waves of Boussinesq paradigm equation. AIP **1186**, 217–224 (2009)
6. Christov, C.I.: An energy-consistent dispersive shallow-water model. Wave motion **34**, 161–174 (2001)
7. Dimova, M., Kolkovska, N.: Comparison of some finite difference schemes for Boussinesq paradigm equation. In: Adam, G., Buša, J., Hnatič, M. (eds.) MMCP 2011. LNCS, vol. 7125, pp. 215–220. Springer, Heidelberg (2012)
8. El-Zoheiry, H.: Numerical study of the improved Boussinesq equation. Chaos, Solitons and Fractals **14**, 377–384 (2002)
9. Kolkovska, N.T.: Convergence of finite difference schemes for a multidimensional Boussinesq equation. In: Dimov, I., Dimova, S., Kolkovska, N. (eds.) NMA 2010. LNCS, vol. 6046, pp. 469–476. Springer, Heidelberg (2011)
10. Kolkovska, N.: Four level conservative finite difference schemes for Boussinesq paradigm equation. AIP **1561**, 68–74 (2013)
11. Kolkovska, N., Angelow, K.: A multicomponent alternating direction method for numerical solution of Boussinesq paradigm equation. In: Dimov, I., Faragó, I., Vulkov, L. (eds.) NAA 2012. LNCS, vol. 8236, pp. 371–378. Springer, Heidelberg (2013)
12. Kolkovska, N., Dimova, M.: A new conservative FDS for Boussinesq paradigm equation. Cent. Eu. J. Math. **10**, 1159–1171 (2012)
13. Kutev, N., Kolkovska, N., Dimova, M.: Global existence of Cauchy problem for Boussinesq paradigm equation. Comp. and Math. with Appl. **65**(3), 500–511 (2013)
14. Kutev, N., Kolkovska, N., Dimova, M.: Global existence to generalized Boussinesq equation with combined power-type nonlinearities. J. Math. Anal. Appl. **410**, 427–444 (2014)
15. Ortega, T., Sanz-Serna, J.M.: Nonlinear stability and convergence of finite-difference methods for the "good" Boussinesq equation. Numer. Math. **58**, 215–229 (1990)
16. Pani, A., Saranga, H.: Finite element galerkin method for the "Good" Boussinesq equation. Nonlinear Anal. **29**, 937–956 (1997)
17. Polat, N., Ertas, A.: Existence and blow-up of solution of Cauchy problem for the generalized damped multidimensional Boussinesq equation. J. Math. Anal. Appl. **349**, 10–20 (2009)
18. Shokri, A., Dehghan, M.: A Not-a-Knot meshless method using radial basis functions and predictorcorrector scheme to the numerical solution of improved Boussinesq equation. Comp. Phys. Commun. **181**, 1990–2000 (2010)
19. Runzhang, X., Yacheng, L.: Global existence and nonexistence of solution for Cauchy problem of multidimensional double dispersion equation. J. Math. Anal. Appl. **359**, 739–751 (2009)

Positive Solutions for Boundary Value Problem of Nonlinear Fractional Differential Equation with p-Laplacian Operator

Hongling Lu, Zhenlai Han$^{(\boxtimes)}$, Chao Zhang, and Yan Zhao

School of Mathematical Sciences, University of Jinan,
Jinan 250022, People's Republic of China
lhl4578@126.com, hanzhenlai@163.com, {ss_zhangc,ss_zhaoy}@ujn.edu.cn

Abstract. In this paper, we deal with the following p-Laplacian fractional boundary value problem: $\phi_p(D_{0+}^{\alpha}u(t)) + f(t, u(t)) = 0$, $0 < t < 1$, $u(0) = u'(0) = u'(1) = 0$, where $2 < \alpha \leqslant 3$ is a real number. D_{0+}^{α} is the standard Riemann–Liouville differentiation, and $f : [0, 1] \times [0, +\infty) \to [0, +\infty)$ is continuous. By the properties of the Green function and some fixed-point theorems on cone, some existence and multiplicity results of positive solutions are obtained. As applications, examples are presented to illustrate the main results.

Keywords: Fractional differential equation · Boundary value problem · P-Laplacian operator · Green's function · Fixed-point theorem

1 Introduction

Fractional differential equations have been of great interest recently. It is caused both by the intensive development of the theory of fractional calculus itself and by the applications, see [1,2]. Many people pay attention to the existence and multiplicity of solutions or positive solutions for boundary value problems of nonlinear fractional differential equations by means of some fixed-point theorems, such as the Schauder fixed-point theorem, the Leggett-Williams fixed-point theorem, the Guo–Krasnosel'skii fixed-point theorem, and the upper and lower solutions method, see [3–14]. However, there are only a few papers devoted to the study of fractional differential equations with p-Laplacian operator, and the theories and applications seem to be just being initiated.

Bai and Lü [13] studied the following two-point boundary value problem of fractional differential equations $D_{0+}^{\alpha}u(t) + f(t, u(t)) = 0, 0 < t < 1, u(0) = u(1) = 0$, where $1 < \alpha \leqslant 2$ is a real number and D_{0+}^{α} is the standard Riemann–Liouville fractional derivative. They obtained the existence of positive solutions by means of Guo–Krasnosel skii fixed-point theorem and Leggett–Williams fixed-point theorem.

Zhao and Sun etc. [14] considered the existence of multiple positive solutions for the nonlinear fractional differential equation boundary value problem

© Springer International Publishing Switzerland 2015
I. Dimov et al. (Eds.): FDM 2014, LNCS 9045, pp. 274–281, 2015.
DOI: 10.1007/978-3-319-20239-6_29

$D_{0+}^{\alpha}u(t) + f(t, u(t)) = 0,\ 0 < t < 1, u(0) = u'(0) = u'(1) = 0,$ where $2 < \alpha \leqslant 3$ is a real number and D_{0+}^{α} is the Riemann–Liouville fractional differentiation. Using lower and upper solution method and Leggett–Williams fixed point theorem, some new existence criteria are obtained for the above boundary value problems.

From the above works, we can see a fact, although the fractional boundary value problems have been studied by some authors, to the best of our knowledge, the fractional differential equation with p-Laplacian operator is seldom considered. In this paper, we investigate the following existence of positive solutions of fractional differential equation with p-Laplacian operator:

$$\phi_p(D_{0+}^{\alpha}u(t)) + f(t, u(t)) = 0, \qquad 0 < t < 1, \tag{1}$$
$$u(0) = u'(0) = u'(1) = 0, \tag{2}$$

where $2 < \alpha \leqslant 3, \phi_p(s) = |s|^{p-2}s, p > 1, (\phi_p)^{-1} = \phi_q, 1/p + 1/q = 1$ and D_{0+}^{α} is the standard Riemann-Liouville differentiation, and $f : [0, 1] \times [0, +\infty) \to [0, +\infty)$ is continuous.

In this paper, we firstly derive the corresponding Green's function. Consequently, problem (1) and (2) is deduced to a equivalent Fredholm integral equation of the second kind. Finally, by the means of Guo–Krasnosel'skii fixed-point theorem and Leggett–Williams fixed-point theorem, the existence and multiplicity of positive solutions are obtained.

The plan of this paper is as follows. In Sect. 2, we will present some definitions and lemmas. In Sect. 3, some results are given. In Sect. 4, we present examples to demonstrate our results.

2   Background Materials and Preliminaries

In this section, we present some necessary definitions and lemmas.

Definition 1. *([13]) The fractional integral of order $\alpha > 0$ of a function $y : (0, +\infty) \to \mathbb{R}$ is given by $I_{0+}^{\alpha}y(t) = \dfrac{1}{\Gamma(\alpha)} \displaystyle\int_0^t (t - s)^{\alpha-1}y(s)ds$ provided the right side is pointwise defined on $(0, +\infty)$.*

Definition 2. *([13]) The fractional derivative of order $\alpha > 0$ of a continuous function $y : (0, +\infty) \to \mathbb{R}$ is given by $D_{0+}^{\alpha}y(t) = \dfrac{1}{\Gamma(n - \alpha)}(\dfrac{d}{dt})^n \displaystyle\int_0^t \dfrac{y(s)}{(t - s)^{\alpha-n+1}}ds$, where $n = [\alpha] + 1$, provided that the right side is point wise defined on $(0, +\infty)$.*

Remark 1. ([13]) As a basic example, we quote for $\lambda > -1, D_{0+}^{\alpha}t^{\lambda} = \dfrac{\Gamma(\lambda+1)}{\Gamma(\lambda-\alpha+1)}t^{\lambda-\alpha}$, giving in particular $D_{0+}^{\alpha}t^{\alpha-m} = 0, m = 1, 2, \cdots, N$, where N is the smallest integer greater than or equal to α.

In fact $\lambda > -1, D_{0+}^{\alpha}t^{\lambda} = \frac{1}{\Gamma(n-\alpha)}(\frac{d}{dt})^n \int_0^t \frac{s^{\lambda}}{(t-s)^{\alpha-n+1}}ds = \frac{1}{\Gamma(n-\alpha)}(\frac{d}{dt})^n t^{n-\alpha+\lambda}$ $\int_0^1 z^{\lambda}(1 - z)^{n-\alpha-1}dz = \frac{\Gamma(\lambda+1)}{\Gamma(\lambda+1+n-\alpha)}(\frac{d}{dt})^n t^{n-\alpha+\lambda}$. So, $D_{0+}^{\alpha}t^{\alpha-m} = \frac{\Gamma(\alpha-m+1)}{\Gamma(n-m+1)}$ $(\frac{d}{dt})^n t^{n-m} = 0,\ for\ m = 1, 2, \cdots, N$. From Definition 2 and Remark 1, we then obtain follow Lemma.

Lemma 1. *([13]) Let $\alpha > 0$. If we assume $u \in C(0,1) \cap L(0,1)$. Then the fractional deferential equation $D_{0+}^{\alpha} u(t) = 0$ has $u(t) = c_1 t^{\alpha-1} + c_2 t^{\alpha-2} + \cdots + c_n t^{\alpha-n}, c_i \in \mathbb{R}, i = 1, 2, \cdots, n$, as unique solutions.*

As $D_{0+}^{\alpha} I_{0+}^{\alpha} u = u$ for all $u \in C(0,1) \cap L(0,1)$. From Lemma 1 we deduce the following law of composition.

Lemma 2. *([13]) Assume that $u \in C(0,1) \cap L(0,1)$, with a fractional derivative of order $\alpha > 0$ that belongs to $C(0,1) \cap L(0,1)$. Then $I_{0+}^{\alpha} D_{0+}^{\alpha} u(t) = u(t) + c_1 t^{\alpha-1} + c_2 t^{\alpha-2} + \cdots + c_n t^{\alpha-n}$, for some $c_i \in \mathbb{R}, i = 1, 2, \cdots, n$.*

Lemma 3. *Given $y \in C[0,1]$ and $2 < \alpha \leqslant 3$, the unique solution of*

$$\phi_p(D_{0+}^{\alpha} u(t)) + y(t) = 0, \qquad 0 < t < 1, \tag{3}$$

$$u(0) = u'(0) = u'(1) = 0, \tag{4}$$

is $u(t) = \int_0^1 G(t,s)\phi_q(y(s))ds$, where

$$G(t,s) = \begin{cases} \dfrac{t^{\alpha-1}(1-s)^{\alpha-2} - (t-s)^{\alpha-1}}{\Gamma(\alpha)}, & 0 \leqslant s \leqslant t \leqslant 1, \\ \dfrac{t^{\alpha-1}(1-s)^{\alpha-2}}{\Gamma(\alpha)}, & 0 \leqslant t \leqslant s \leqslant 1. \end{cases} \tag{5}$$

Proof. We may apply Lemma 2 to reduce Eq. (3) to an equivalent integral equation $u(t) = -I_{0+}^{\alpha} \phi_q(y(t)) + c_1 t^{\alpha-1} + c_2 t^{\alpha-2} + c_3 t^{\alpha-3}$ for some $c_1, c_2, c_3 \in \mathbb{R}$. Consequently, the general solution of Eq. (3) is

$$u(t) = -\int_0^t \frac{(t-s)^{\alpha-1}}{\Gamma(\alpha)} \phi_q(y(s))ds + c_1 t^{\alpha-1} + c_2 t^{\alpha-2} + c_3 t^{\alpha-3}.$$

By (4), $c_3 = 0, c_2 = 0, c_1 = \int_0^1 \dfrac{(1-s)^{\alpha-2}}{\Gamma(\alpha)} \phi_q(y(s))ds$. Therefore, the unique solution of problem (3) and (4) is

$$\begin{aligned} u(t) &= -\int_0^t \frac{(t-s)^{\alpha-1}}{\Gamma(\alpha)} \phi_q(y(s))ds + \int_0^1 \frac{(1-s)^{\alpha-2} t^{\alpha-1}}{\Gamma(\alpha)} \phi_q(y(s))ds \\ &= \int_0^t \frac{t^{\alpha-1}(1-s)^{\alpha-2} - (t-s)^{\alpha-1}}{\Gamma(\alpha)} \phi_q(y(s))ds + \int_t^1 \frac{t^{\alpha-1}(1-s)^{\alpha-2}}{\Gamma(\alpha)} \phi_q(y(s))ds \\ &= \int_0^1 G(t,s)\phi_q(y(s))ds. \end{aligned}$$

The proof is complete.

Lemma 4. *([13]) The function $G(t,s)$ defined by (5) satisfies the following conditions: (1) $G(t,s) > 0, for\ t, s \in (0,1)$;*

$$(2)\ q(t)G(1,s) \leqslant G(t,s) \leqslant G(1,s), for\ t, s \in [0,1], where\ q(t) = t^{\alpha-1}. \tag{6}$$

Definition 3. *([13]) The map θ is said to be a nonnegative continuous concave functional on a cone P of a real Banach space E provided that $\theta : P \to [0, +\infty)$ is continuous and $\theta(tx + (1-t)y) \geqslant t\theta(x) + (1-t)\theta(y)$ for all $x, y \in P$ and $0 \leqslant t \leqslant 1$.*

Lemma 5. *([13]) Let E be a Banach space, $P \subseteq E$ a cone, and Ω_1, Ω_2 two bounded open balls of E centered at the origin with $\bar{\Omega}_1 \subset \Omega_2$. Suppose that $A : P \cap (\bar{\Omega}_2 \setminus \Omega_1) \to P$ is a completely continuous operator such that either*

(i) $||Ax|| \leqslant ||x||, x \in P \cap \partial\Omega_1$ and $||Ax|| \geqslant ||x||, x \in P \cap \partial\Omega_2$, or
(ii) $||Ax|| \geqslant ||x||, x \in P \cap \partial\Omega_1$ and $||Ax|| \leqslant ||x||, x \in P \cap \partial\Omega_2$

holds. Then A has a fixed point in $P \cap (\bar{\Omega}_2 \setminus \Omega_1)$.

Lemma 6. *([13]) Let P be a cone in a real Banach space E, $P_c = \{x \in P| ||x|| \leqslant c\}, \theta$ a nonnegative continuous concave functional on P such that $\theta(x) \leqslant ||x||$, for all $x \in \bar{P}_c$, and $P(\theta, b, d) = \{x \in P| b \leqslant \theta(x), ||x|| \leqslant d\}$. Suppose $A : \bar{P}_c \to \bar{P}_c$ is completely continuous and there exist constants $0 < a < b < d \leqslant c$ such that $(C1)$ $\{x \in P(\theta, b, d)|\theta(x) > b\} \neq \emptyset$ and $\theta(Ax) > b$ for $x \in P(\theta, b, d)$; $(C2)$ $||Ax|| < a$ for $x \leqslant a$; $(C3)$ $\theta(Ax) > b$ for $x \in P(\theta, b, c)$ with $||Ax|| > d$. Then A has at least three fixed points x_1, x_2, x_3 with*

$$||x_1|| < a, \quad b < \theta(x_2), \quad a < ||x_3|| \quad with \quad \theta(x_3) < b.$$

Remark 2. ([13]) If there holds d=c, then condition (C1) of Lemma 6 implies condition (C3) of Lemma 6.

3   Main Results

In this section, we impose growth conditions on f which allow us to apply Lemmas 5 and 6 to establish some results of existence and multiplicity of positive solutions for problem (1) and (2). Let $E = C[0, 1]$ be endowed with the ordering $u \leqslant v$ if $u(t) \leqslant v(t)$ for all $t \in [0, 1]$, and the maximum norm $||u|| = \max_{0 \leqslant t \leqslant 1} |u(t)|$. Define the cone $P \subset E$ by $P = \{u \in E| u(t) \geqslant 0\}$.

Let the nonnegative continuous concave functional θ on the cone P be defined by $\theta(u) = \min_{1/4 \leqslant t \leqslant 3/4} |u(t)|$.

Lemma 7. *Let $T : P \to P$ be the operator defined by $Tu(t) := \int_0^1 G(t, s)\phi_q (f(s, u(s)))ds$. Then $T : P \to P$ is completely continuous.*

Proof. The operator $T : P \to P$ is continuous in view of nonnegativeness and continuity of $G(t, s)$ and $f(t, u)$. Let $\Omega \subset P$ be bounded, i.e., there exists a positive constant $M > 0$ such that $||u|| \leqslant M$, for all $u \in \Omega$. Let $L = \max_{0 \leqslant t \leqslant 1, 0 \leqslant u \leqslant M} |\phi_q(f(t, u))| + 1$, then, for $u \in \Omega$, we have $|Tu(t)| = \int_0^1 G(t, s)\phi_q (f(s, u(s)))ds \leqslant L \int_0^1 G(1, s)ds$. Hence, $T(\Omega)$ is bounded.

On the other hand, given $\epsilon > 0$, setting $\delta = \min\{\frac{1}{2}, \frac{1}{2}(\frac{\Gamma(\alpha)\epsilon}{L})\}$, then, for each $u \in \Omega, t_1, t_2 \in [0,1], t_1 < t_2$, and $t_2 - t_1 < \delta$, we have $|Tu(t_2) - Tu(t_1)| < \epsilon$. That is to say, $T(\Omega)$ is equicontinuity. In fact,

$$|Tu(t_2) - Tu(t_1)| = |\int_0^1 G(t_2, s)\phi_q(f(s, u(s)))ds - \int_0^1 G(t_1, s)\phi_q(f(s, u(s)))ds|$$

$$= \int_0^{t_1} [G(t_2, s) - G(t_1, s)]\phi_q(f(s, u(s)))ds + \int_{t_2}^1 [G(t_2, s) - G(t_1, s)]\phi_q(f(s, u(s)))ds$$

$$+ \int_{t_1}^{t_2} [G(t_2, s) - G(t_1, s)]\phi_q(f(s, u(s)))ds < \frac{L}{\Gamma(\alpha)}[\int_0^{t_1} (1 - s)^{\alpha-2}(t_2^{\alpha-1} - t_1^{\alpha-1})ds$$

$$+ \int_{t_2}^1 (1 - s)^{\alpha-2}(t_2^{\alpha-1} - t_1^{\alpha-1})ds + \int_{t_1}^{t_2} (1 - s)^{\alpha-2}(t_2^{\alpha-1} - t_1^{\alpha-1})ds]$$

$$< \frac{L}{\Gamma(\alpha)}(t_2^{\alpha-1} - t_1^{\alpha-1}).$$

In the following, we divide the proof into two cases.

Case 1. $\delta \leqslant t_1 < t_2 < 1$,

$$|Tu(t_2) - Tu(t_1)| < \frac{L}{\Gamma(\alpha)}(t_2^{\alpha-1} - t_1^{\alpha-1}) \leqslant \frac{L}{\Gamma(\alpha)}(\alpha - 1)(t_2 - t_1) \leqslant \frac{L}{\Gamma(\alpha)}(\alpha - 1)\delta \leqslant \epsilon.$$

Case 2. $0 \leqslant t_1 < \delta, t_2 < 2\delta$,

$$|Tu(t_2) - Tu(t_1)| < \frac{L}{\Gamma(\alpha)}(t_2^{\alpha-1} - t_1^{\alpha-1}) \leqslant \frac{L}{\Gamma(\alpha)}t_2^{\alpha-1} \leqslant \frac{L}{\Gamma(\alpha)}t_2 < \frac{L}{\Gamma(\alpha)}(2\delta) \leqslant \epsilon.$$

By the means of the Arzela–Ascoli theorem, we have $T : P \to P$ is completely continuous. The proof is complete. Denote

$$M = (\int_0^1 G(1, s)ds)^{-1}, \quad \sigma = \min_{1/4 \leqslant t \leqslant 3/4} q(t) = (\frac{1}{4})^{\alpha-1}, \quad N = (\int_{1/4}^{3/4} \sigma G(1, s)ds)^{-1}.$$

Theorem 1. *Let $f(t, u)$ is continuous on $[0,1] \times [0, +\infty)$. Assume that there exist two positive constants $r_2 > r_1 > 0$ such that*

(A1) $\phi_q(f(t, u)) \leqslant Mr_2, for (t, u) \in [0, 1] \times [0, r_2]$;
(A2) $\phi_q(f(t, u)) \geqslant Nr_1, for (t, u) \in [0, 1] \times [0, r_1]$.

Then problem (1) and (2) has at least one positive solution u such that $r_1 \leqslant ||u|| \leqslant r_2$.

Proof. By Lemmas 3 and 7, we know $T : P \to P$ is completely continuous and problem (1) and (2) has a solution $u = u(t)$ if and only if u solves the operator equation $u = Tu$. In order to apply Lemma 5, we separate the proof into the following two steps.

Step 1. Let $\Omega_2 := \{u \in P|\ ||u|| < r_2\}$. For $u \in \partial\Omega_2$, we have $0 \leqslant u(t) \leqslant r_2$ for all $t \in [0,1]$. It follows form (A1) that for $t \in [0,1]$,

$$||Tu|| = \max_{0\leqslant t\leqslant 1} \int_0^1 G(t,s)\phi_q(f(s,u(s)))ds \leqslant Mr_2 \int_0^1 G(1,s)ds = r_2 = ||u||.$$

Step 2. Let $\Omega_1 := \{u \in P|\ ||u|| < r_1\}$. For $u \in \partial\Omega_1$, we have $0 \leqslant u(t) \leqslant r_1$ for all $t \in [0,1]$. It follows form (A2) that for $t \in [1/4, 3/4]$,

$$Tu(t) = \int_0^1 G(t,s)\phi_q(f(s,u(s)))ds \geqslant \int_0^1 q(t)G(1,s)\phi_q(f(s,u(s)))ds$$

$$\geqslant Nr_1 \int_{1/4}^{3/4} \sigma G(1,s)ds = r_1 = ||u||.$$

So, $||Tu|| \geqslant ||u||$, *for* $u \in \partial\Omega_1$. Therefore, by (ii) of Lemma 5, we complete the proof.

Theorem 2. *Let* $f(t,u)$ *is continuous on* $[0,1] \times [0, +\infty)$ *and there exist constants* $0 < a < b < c$ *such that the following assumptions hold:*

(B1) $\phi_q(f(t,u)) < Ma$, *for* $(t,u) \in [0,1] \times [0,a]$;

(B2) $\phi_q(f(t,u)) \geqslant Nb$, *for* $(t,u) \in [1/4, 3/4] \times [b,c]$;

(B3) $\phi_q(f(t,u)) \leqslant Mc$, *for* $(t,u) \in [0,1] \times [0,c]$. *Then, the boundary value problem* (1) *and* (2) *has at least three positive solutions* $u_1, u_2,$ *and* u_3 *with*

$$\max_{0\leqslant t\leqslant 1} |u_1(t)| < a, \qquad b < \min_{1/4\leqslant t\leqslant 3/4} |u_2(t)| < \max_{0\leqslant t\leqslant 1} |u_2(t)| \leqslant c,$$

$$a < \max_{0\leqslant t\leqslant 1} |u_3(t)| \leqslant c, \qquad \min_{1/4\leqslant t\leqslant 3/4} |u_3(t)| < b.$$

Proof. We show that all the conditions of Lemma 6 are satisfied.

If $u \in \bar{P}_c$, then $||u|| \leqslant c$. Assumption (B3) implies $\phi_q(f(t,u(t))) \leqslant Mc$ for $0 \leqslant t \leqslant 1$. Consequently,

$$||Tu|| = \max_{0\leqslant t\leqslant 1} |\int_0^1 G(t,s)\phi_q(f(s,u(s)))ds| \leqslant \int_0^1 G(1,s)\phi_q(f(s,u(s)))ds$$

$$\leqslant \int_0^1 G(1,s)Mcds \leqslant c.$$

Hence, $T : \bar{P}_c \to \bar{P}_c$. In the same way, if $u \in \bar{P}_a$, then assumption (B1) yields, $\phi_q(f(t,u(t))) < Ma, 0 \leqslant t \leqslant 1$. Therefore, condition (C2) of Lemma 6 satisfied.

To check condition (C1) of Lemma 6, we choose $u(t) = (b+c)/2, 0 \leqslant t \leqslant 1$. It is easy to see that $u(t) = (b+c)/2 \in P(\theta,b,c), \theta(u) = \theta((b+c)/2) > b$, consequently, $\{u \in P(\theta,b,c)|\ \theta(u) > b\} \neq \emptyset$. Hence, if $u \in P(\theta,b,c)$, then $b \leqslant u(t) \leqslant c$ for $1/4 \leqslant t \leqslant 3/4$. Form assumption (B2), we have $\phi_q(f(t,u(t))) \geqslant Nb$, for $1/4 \leqslant$

$t \leqslant 3/4$. So $\theta(Tu) = \min\limits_{1/4 \leqslant t \leqslant 3/4} |(Tu)(t)| \geqslant \int_0^1 q(t)G(1,s)\phi_q(f(s,u(s)))ds >$
$\int_{1/4}^{3/4} \sigma G(1,s)Nbds = b$, i.e., $\theta(Tu) > b$, for all $u \in P(\theta, b, c)$. This shows that
condition (C1) of Lemma 6 is also satisfied.

By Lemma 6 and Remark 2, the boundary value problem (1.1) and (1.2) has
at least three positive solutions u_1, u_2, and u_3 satisfying

$$\max_{0 \leqslant t \leqslant 1} |u_1(t)| < a, b < \min_{1/4 \leqslant t \leqslant 3/4} |u_2(t)|, a < \max_{0 \leqslant t \leqslant 1} |u_3(t)|, \min_{1/4 \leqslant t \leqslant 3/4} |u_3(t)| < b.$$

The proof is complete.

4   Examples

In this section, we will give some examples to illustrate our main results.

Example 4.1. Consider the following problem $\phi_{3/2}(D_{0+}^{3/2}u(t)) + u^2 + \frac{\sin t}{4} + \frac{1}{4} = 0$, $0 < t < 1, u(0) = u'(0) = u'(1) = 0$. We have $M = (\int_0^1 G(1,s)ds)^{-1} = \frac{5\sqrt{\pi}}{4} \approx 2.2156, N = (\int_{1/4}^{3/4} \sigma G(1,s)ds)^{-1} \approx 28.2479$.
Choosing $r_1 = 1/500, r_2 = 1$, then

$$\phi_3(f(t,u)) = (u^2 + \frac{\sin t}{4} + \frac{1}{4})^2 \leqslant 2.13268 \leqslant Mr_2, \ for \ (t,u) \in [0,1] \times [0,1],$$

$$\phi_3(f(t,u)) = (u^2 + \frac{\sin t}{4} + \frac{1}{4})^2 \geqslant 0.0625 \geqslant Nr_1, \ for \ (t,u) \in [0,1] \times [0,1/500].$$

From Theorem 1, the problem has at least one solution u such that $1/500 \leqslant \|u\| \leqslant 1$.

Example 4.2. Consider the following problem $\phi_{3/2}(D_{0+}^{3/2}u(t)) + f(t,u) = 0$, $0 < t < 1, u(0) = u'(0) = u'(1) = 0$. where

$$f(t,u) = \begin{cases} \frac{t}{20} + \frac{37}{9}u, & for \ u \leqslant 1, \\ \frac{t}{20} + \frac{u}{9} + 4, & for \ u > 1. \end{cases}$$

We have $M = (\int_0^1 G(1,s)ds)^{-1} = \frac{5\sqrt{\pi}}{4} \approx 2.2156, N = (\int_{1/4}^{3/4} \sigma G(1,s)ds)^{-1} \approx 28.2479$.
Choosing $a = 1/10, b = 1/2, c = 18$, then

$$\phi_3(f(t,u)) = (\frac{t}{20} + \frac{37}{9}u)^2 \leqslant 0.2125 \leqslant Ma \approx 0.2216, \ for \ (t,u) \in [0,1] \times [0,1/10],$$

$$\phi_3(f(t,u)) = (\frac{t}{20} + \frac{u}{9} + 4)^2 \geqslant 16.97 \geqslant Nb \approx 14.1239, \ for \ (t,u) \in [1/4, 3/4] \times [1/2, 18],$$

$$\phi_3(f(t,u)) = (\frac{t}{20}+\frac{u}{9}+4)^2 \leqslant 36.1562 \leqslant Mc \approx 39.8808, \; for \; (t,u) \in [0,1]\times[0,18].$$

By Theorem 2, the problem has at least three positive solutions u_1, u_2 and u_3 with $\max_{0\leqslant t\leqslant 1} |u_1(t)| < 1/10$, $1/2 < \min_{1/4\leqslant t\leqslant 3/4} |u_2(t)| < \max_{0\leqslant t\leqslant 1} |u_2(t)| \leqslant 18$, $1/10 < \max_{0\leqslant t\leqslant 1} |u_3(t)| \leqslant 18$, $\min_{1/4\leqslant t\leqslant 3/4} |u_3(t)| < 1/2$.

Acknowledgements. Corresponding author: Zhenlai Han. This research is supported by the Natural Science Foundation of China (61374074), Natural Science Outstanding Youth Foundation of Shandong Province (JQ201119) and supported by Shandong Provincial Natural Science Foundation (ZR2012AM009, ZR2013AL003).

References

1. Machado, J.T., Kiryakova, V., Mainardi, F.: Recent history of fractional calculus. Commun. Nonlinear Sci. Numer. Simul. **16**, 1140–1153 (2011)
2. Agarwal, R.P., Zhou, Y., Wang, J., Luo, X.: Fractional functional differential equations with causal operators in Banach spaces. Math. Comput. Modell. **54**, 1440–1452 (2011)
3. Wang, J., Zhou, Y.: Existence of mild solutions for fractional delay evolution systems. Appl. Math. Comput. **218**, 357–367 (2011)
4. Jiao, F., Zhou, Y.: Existence of solutions for a class of fractional boundary value problems via critical point theory. Comput. Math. Appl. **62**, 1181–1199 (2011)
5. Chen, F., Nieto, J., Zhou, Y.: Global attractivity for nonlinear fractional differential equations. Nonlinear Anal. RWA **13**, 287–298 (2012)
6. Zhou, Y., Jiao, F.: Nonlocal cauchy problem for fractional evolution equations. Nonlinear Anal. RWA **11**, 4465–4475 (2010)
7. Zhou, Y., Jiao, F., Li, J.: Existence and uniqueness for p-type fractional neutral differential equations. Nonlinear Anal. TMA **71**, 2724–2733 (2009)
8. Zhao, Y., Sun, S., Han, Z., Zhang, M.: Positive solutions for boundary value problems of nonlinear fractional differential equations. Appl. Math. Comput. **217**, 6950–6958 (2011)
9. Han, Z., Lu, H., Sun, S., Yan, D.: Positive solutions to boundary value problems of p-Laplacian fractional differential equations with a parameter in the boundary conditions. Electron. J. Differ. Equat. **2012**(213), 1–14 (2012)
10. Zhao, Y., Sun, S., Han, Z., Li, Q.: Positive solutions to boundary value problems of nonlinear fractional differential equations. Abstract Appl. Anal. Art. ID **390543**(2011), 1–16 (2011)
11. Feng, W., Sun, S., Han, Z., Zhao, Y.: Existence of solutions for a singular system of nonlinear fractional differential equations. Comput. Math. Appl. **62**, 1370–1377 (2011)
12. Lu, H., Han, Z., Sun, S.: Multiplicity of positive solutions for Sturm-Liouville boundary value problems of fractional differential equations with p-Laplacian. Bound. Value Prob. **2014**(26), 1–17 (2014)
13. Bai, Z., Lü, H.: Positive solutions for boundary value problem of nonlinear fractional differential equation. J. Math. Anal. Appl. **311**, 495–505 (2005)
14. Zhao, Y., Sun, S., Han, Z., Li, Q.: The existence of multiple positive solutions for boundary value problems of nonlinear fractional differential equations. Commun. Nonlinear Sci. Numer. Simulat. **16**, 2086–2097 (2011)

Finite-Difference Simulation of Wave Propagation Through Prestressed Elastic Media

Egor Lys[1]([⊠]), Evgeniy Romenski[2], Vladimir Tcheverda[1],
and Mikhail Epov[1]

[1] Trofimuk Institute of Petroleum Geology and Geophysics, Novosibirsk, Russia
lysev@ipgg.sbras.ru
[2] Sobolev Institute of Mathematics, Novosibirsk, Russia

Abstract. The new computational model for the seismic wave propagation is proposed, the governing equations of which are written in terms of velocities, stress tensor and small rotation of element of the medium. The properties of wavefields in the prestressed medium are studied and some examples showing anisotropy of prestressed state are discussed. The staggered grid numerical method is developed for solving the governing equations of the model and numerical examples are presented.

1 Introduction

Analysis of seismoacoustic wavefields is the basic tool for the study of internal structure of Earth and rock masses in the mining technology. The impact of prestressed zones on seismic waves is a poorly studied problem and one can expect that the account of initial stress can have an influence on interpretation of the results of solution of inverse problems and seismic imaging. The basis of the theory of elastic waves in prestressed elastic media goes back to the pioneer work of M.Bio [1]. Its application to seismic problems was not systematic (see, for example, [2,3] and references therein) and there is still an open area for research work.

We propose a new computational model for small amplitude wave propagation in the prestressed medium, the simplified version of which is presented in [4]. The derivation of the model is based on the general theory of finite deformations and as a result, the governing equations in terms of velocities, stress and small rotations are formulated in the form of the first order hyperbolic system.

2 Derivation of Governing Equations

The method of derivation of governing equations for small amplitude wave propagation in the prestressed elastic medium is based on the presented in [5] relationship between stress rate and strain rate in the hypoelastic representation of the hyperelastic model of solid. The governing equations are formulated in Lagrangian coordinates, but the method of derivation requires an introduction of Eulerian coordinates and, in addition, reference unstressed configuration with its own coordinates of unstressed state.

© Springer International Publishing Switzerland 2015
I. Dimov et al. (Eds.): FDM 2014, LNCS 9045, pp. 282–289, 2015.
DOI: 10.1007/978-3-319-20239-6_30

Denote x^i Eulerian coordinates of the particle of the medium and x_0^i corresponding Lagrangian coordinates. Assume that the element of the medium in Lagrangian coordinates containing this particle is prestressed, that is nonzero stress field exists inside this element. Let us introduce coordinates ξ^i of the particle corresponding to the unstressed reference state of the element. Thus, the deformation gradients characterizing deformation from the unstressed reference configuration to the Lagrangian configuration and from the Lagrangian configuration to the current Eulerian configuration $\left(F^0\right)_j^i = \frac{\partial x_0^i}{\partial \xi^j}$, $\left(F\right)_j^i = \frac{\partial x^i}{\partial x_0^j}$ can be introduced. The total deformation from the reference unstressed state to the Eulerian state is characterized by the total deformation gradient $\left(F_{tot}\right)_j^i = \frac{\partial x^i}{\partial \xi^j} = \left(F\right)_\alpha^i \left(F^0\right)_j^\alpha$. For our purpose it is more appropriate to use inverse deformation gradients:

$$\left(f^0\right)_j^i = \frac{\partial \xi^i}{\partial x^j}, \quad \left(f\right)_j^i = \frac{\partial x_0^i}{\partial x^j}, \quad \left(f_{tot}\right)_j^i = \frac{\partial \xi^i}{\partial x^j} = \left(f^0\right)_\alpha^i \left(f\right)_j^\alpha$$

Below the Finger strain tensor is used as a measure of deformation and the total strain from the unstressed state to the current configuration is characterizes by

$$\left(G\right)_{ij} = \left(f_{tot}\right)_i^\alpha \left(f_{tot}\right)_j^\alpha = \left(f\right)_i^\alpha \left(f^0\right)_\alpha^\beta \left(f^0\right)_\gamma^\alpha \left(f\right)_j^\gamma .$$

Thus, $\left(G\right)_{ij} = \left(f\right)_i^\alpha \left(G^0\right)_{\alpha\gamma} \left(f\right)_j^\gamma$, where $\left(G^0\right)_{\alpha\gamma} = \left(f^0\right)_\alpha^\beta \left(f^0\right)_\gamma^\alpha$, is the Finger strain tensor characterizing deformation from the unstressed state to Lagrangian configuration. Further we will use the matrix form of the Finger tensor, which reads as $G = f^T G_0 f$, where the superscript T denotes a matrix transposition.

For the derivation of governing equations we use the so-called hyperelastic model which is based on the fundamental laws of thermodynamics. If the specific elastic energy $E\left(G_{11}, \ldots, G_{33}\right)$ is the known function of the strain tensor, then according to [6], the Cauchy stress tensor in Eulerian configuration is given as

$$s_{ij} = -2\rho \frac{\partial E}{\partial G_{\alpha j}} G_{\alpha i}, \tag{1}$$

where $\rho = \rho_{00}/\det\left(F_{tot}\right)$ is the mass density and ρ_{00} is the density of the medium in the unstressed state. For the isotropic medium the elastic energy depends on invariants of the strain tensor. The density is a function of the Finger tensor

$$\rho = \rho_{00}\sqrt{\det G} = \rho_{00}\sqrt{\det G_0 \det\left(f^T f\right)} = \rho_0 \sqrt{\det\left(f^T f\right)}, \quad \rho_0 = \rho_{00}\sqrt{\det G_0}.$$

The governing equations for the prestressed medium motion consist of the momentum conservation laws and evolution equations for the parameters characterizing deformation. Denoting u^i the velocity vector the momentum equation in Eulerian coordinates can be written in a standard form and reads as

$$\frac{\partial \rho u^i}{\partial t} + \frac{\partial \left(\rho u^i u^k - s^{ik}\right)}{\partial x^k} = 0. \tag{2}$$

What concerns an evolution of the strain parameters, it requires thorough consideration. As a consequence of the definition of deformation gradient $(F)^i_j = \frac{\partial x^i}{\partial x^j_0}$ the following evolution equation in matrix form can be derived:

$$\frac{df}{dt} = -fU, \qquad (3)$$

where $f = \left[(f)^i_j\right] = F^{-1}$, $U = \left[\frac{\partial u^i}{\partial x^j}\right]$ is the velocity gradient and $\frac{d}{dt} = \frac{\partial}{\partial t} + u^\alpha \frac{\partial}{\partial x^\alpha}$ is the material derivative. Note that the Finger tensor G_0, characterizing deformation from the reference unstressed state to Lagrangian configuration, does not change during the motion, i.e. $\frac{dG^0}{dt} = 0$.

As a consequence the evolution equation for the Finger tensor can be derived

$$\frac{dG}{dt} = -GU - U^T G. \qquad (4)$$

With the use of stress-strain relation (1) one can derive the following equation

$$\frac{ds}{dt} = -sU - U^T s + \beta_0 I \, trW + \beta_1 W + \beta_2 G \, trW + \beta_3 I \, tr\,(GW)$$

$$+ \frac{1}{2}\beta_4 \,(GW + WG) + \beta_5 G^2 \, trW + \beta_6 G \, tr\,(GW) + \beta_7 I \, tr\,(G^2 W)$$

$$+ \frac{1}{2}\beta_8 \,(G^2 W + WG^2) + \beta_9 G^2 \, tr\,(GW) + \beta_{10} G \, tr\,(G^2 W) + \beta_{11} G^2 \, tr\,(G^2 W) \qquad (5)$$

Here $W = \frac{1}{2}\,(U + U^T)$ is the strain rate tensor in Eulerian coordinates, coefficients $\beta_0, \beta_1, \ldots, \beta_{11}$ are functions of invariants of G and depend on the choice of energy E. Thus, Eqs. (2) – (5) can be used for the derivation of the small amplitude wave propagation in the prestressed isotropic medium with the arbitrary dependence of the elastic energy on three invariants of the strain tensor.

Assume that the energy is given as a function of the Almansi strain tensor $\varepsilon = [\varepsilon_{ij}] = \frac{1}{2}\,(I - G)$ in the following form

$$E = \frac{\lambda}{2\rho_{00}}(\varepsilon_{11} + \varepsilon_{22} + \varepsilon_{33})^2 + \frac{\mu}{\rho_{00}}\,(\varepsilon_{ij}\varepsilon_{ji})$$

where λ, μ are the Lame parameters. The stress tensor (1) takes a form

$$s = \frac{\rho}{\rho_{00}}\,\left(\lambda \, tr\varepsilon \, I + 2\mu\,\varepsilon - 2\lambda \, tr\varepsilon\,\varepsilon - 4\mu\,\varepsilon^2\right) \qquad (6)$$

Assume that the stress tensor $s = \Sigma + \sigma$ is the sum of the initial stress Σ and its perturbation σ. The initially prestressed state satisfies equilibrium equations in Lagrangian configuration $\frac{\partial \Sigma^{ij}}{\partial x^j_0} = 0$, where Σ^{ij} is connected with the Almansi tensor of the prestressed state by the linearized relation (6) which reduces to Hooke's law $\Sigma = \lambda \, tr\varepsilon_0 \, I + 2\mu\,\varepsilon_0$.

We will derive equations in Lagrangian coordinates but with the use of the Cauchy stress tensor referred to Eulerian coordinates. Introduce the small deformation and small rotation tensors by the following relations:

$$\varepsilon_{ij} = \frac{1}{2}\left(\frac{\partial V^i}{\partial x^j} + \frac{\partial V^j}{\partial x^i}\right), \qquad \omega_{ij} = \frac{1}{2}\left(\frac{\partial V^i}{\partial x^j} - \frac{\partial V^j}{\partial x^i}\right)$$

where V^i is the displacement vector, $V^i = x^i - x_0^i$, so that $f_j^i = \delta_j^i - \frac{\partial V^i}{\partial x^j}$. The evolution equations for ε_{ij} and ω_{ij} read as

$$\frac{\partial \varepsilon_{ij}}{\partial t} = \frac{1}{2} \left(\frac{\partial u^i}{\partial x^j} + \frac{\partial u^j}{\partial x^i} \right), \quad \frac{\partial \omega_{ij}}{\partial t} = \frac{1}{2} \left(\frac{\partial u^i}{\partial x^j} - \frac{\partial u^j}{\partial x^i} \right). \tag{7}$$

Using all above definitions, transforming Eqs. (2), (3), (5)-(7) to Lagrangian coordinates assuming that u^i, ε_{ij}, ω_{ij}, ε_{0ij} are small, and neglecting all terms of order higher than the first one, we obtain the following system:

$$\rho_{00} \left(1 - tr\varepsilon_0 \right) \frac{\partial u^i}{\partial t} = \frac{\partial \sigma^{ij}}{\partial x_0^j} + \left(\varepsilon_{j\alpha} + \omega_{j\alpha} \right) \frac{\partial \Sigma^{i\alpha}}{\partial x_0^j},$$

$$\frac{\partial \sigma}{\partial t} = -\Sigma U_0 - U_0^T \Sigma - tr W_0 \, \Sigma + \lambda \, tr W_0 \, I + 2\mu \, W_0 - \lambda \, tr\varepsilon_0 \, tr W_0 \, I -$$

$$2\mu \, tr\varepsilon_0 \, W_0 - 2\lambda \, tr \left(\varepsilon_0 U_0 \right) I - 2\lambda \, tr W_0 \, \varepsilon_0 - 4\mu \, \varepsilon_0 W_0 - 4\mu \, W_0 \varepsilon_0 \tag{8}$$

$$\frac{\partial \omega_{ij}}{\partial t} = \frac{1}{2} \left(\frac{\partial u^i}{\partial x_0^j} - \frac{\partial u^j}{\partial x_0^i} \right), \quad \frac{\partial \varepsilon_{ij}}{\partial t} = \frac{1}{2} \left(\frac{\partial u^i}{\partial x_0^j} + \frac{\partial u^j}{\partial x_0^i} \right)$$

Here $U_0 = \left[\frac{\partial u^i}{\partial x_0^j} \right]$ is the Lagrangian velocity gradient, $W_0 = \frac{1}{2} \left(U_0 + U_0^T \right)$. The initial Almansi strain tensor can be expressed as $\varepsilon_0 = \frac{1}{2\mu} \left(\Sigma - \frac{\lambda}{3\lambda + 2\mu} I \, tr \Sigma \right)$.

3   Properties of Wavefields in Prestressed Media

It is obvious that system (8) and conventional linear elasticity equations for isotropic media are different. Coefficients of equations (8) depend on the values of initial stress tensor and their spatial derivatives. It turns out that this difference drastically changes the character of elastic waves and leads to their anisotropy and dispersion. To prove this fact one can consider the second order equations system for velocities which can be derived from (8) by the differentiating velocity equations with respect to t and exclusion of stress derivatives with the use of equation for s:

$$\rho_0 \frac{\partial^2 u^i}{\partial t^2} = C_{ijkl} \frac{\partial^2 u^l}{\partial x_0^j \partial x_0^k} + B_{ijk} \frac{\partial u^j}{\partial x_0^k} \tag{9}$$

where $B_{ijk} = \partial \Sigma^{ik} / \partial x_0^j$. Moduli C_{ijkl} depend on the initial stress Σ^{ik}. It is obvious that the prestressed state results in the anisotropy of the medium. Moreover, the term containing first derivatives of the velocities in (9) can result in attenuation and dispersion of the waves.

On Fig. 1 one can see the plane waves velocity distribution for unidirectionally stretched medium with $\Sigma^{11} = \rho V_p^2/50$, $\Sigma^{ij} = 0$ $(ij \neq 11)$ (left) and compressed medium with $\Sigma^{11} = \rho V_p^2/50$, $\Sigma^{ij} = 0$ $(ij \neq 11)$ (right) with parameters $V_p = 3000\,m/s$, $V_s = 2000\,m/s$, $\rho_{00} = 2000\,kg/m^3$. Dots correspond to the wave velocity distribution in the unstressed medium. The anisotropy in longitudinal and shear wave propagation is clearly can be seen in both cases.

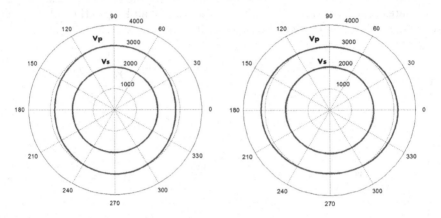

Fig. 1. Velocity distribution for the stretched (left) and compressed (right) media.

4   Finite Difference Staggered Grid Numerical Method

As a numerical tool for solving differential equations (8) the staggered grid finite difference method has been developed, which is similar to proposed in [7] and has the second order accuracy in space and time. Below we denote spacial coordinates as $x_1 = x_0^1$, $x_2 = x_0^2$ and do not distinguish upper and inferior indices. On Figs. 1 and 2 the definition of staggered grid is presented and the finite difference method for (8) reads as

$$\rho_0 \frac{u_i^{n+1} - u_i^n}{\tau} = D_{x_j}\sigma_{ij}^{n+1/2} + \left(\varepsilon_{j\alpha}^{n+1/2} + \omega_{j\alpha}^{n+1/2}\right) D_{x_j}\Sigma_{ij}^{n+1/2}$$

$$\frac{\sigma_{ij}^{n+1/2} - \sigma_{ij}^{n-1/2}}{\tau} = C_{ijkl}D_{x_l}u_k^n$$

$$\frac{\varepsilon_{ij}^{n+1/2} - \varepsilon_{ij}^{n-1/2}}{\tau} = \frac{D_{x_i}u_j^n + D_{x_j}u_i^n}{2}, \qquad \frac{\omega_{ij}^{n+1/2} - \omega_{ij}^{n-1/2}}{\tau} = \frac{D_{x_i}u_j^n - D_{x_j}u_i^n}{2}$$

Here D_{x_i} are the difference approximation of spatial derivatives:

$$D_{x_1}(x_1, x_2)u_i = \frac{1}{h_1}\Big[\frac{u_i(x_1 + h_1\ 2, x_2 + h_2/2) + u_i(x_1 + h_1/2, x_2 - h_2/2)}{2}$$

$$-\frac{u_i(x_1 - h_1/2, x_2 + h_2/2) + u_i(x_1 - h_1/2, x_2 - h_2/2)}{2}\Big] \approx \frac{\partial u_i}{\partial x_1}(x_1, x_2)$$

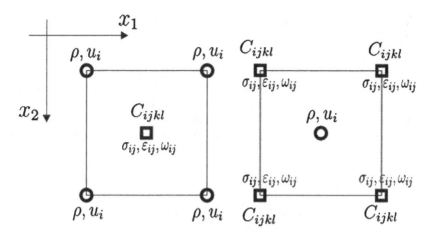

Fig. 2. Staggered grid definition. The density and velocities are related to circles. Stress, strain, rotation tensors and elastic moduli are related to squares.

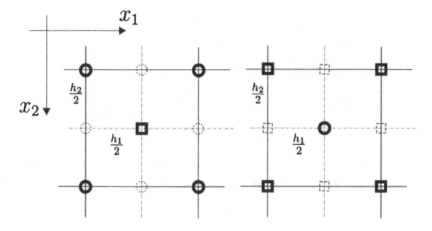

Fig. 3. The structure of finite differences. At the midpoint between circles or squares (solid) a mean value (dashed) of corresponding variables is used.

$$D_{x_2}(x_1, x_2)u_i = \frac{1}{h_2}[\frac{u_i(x_1 - h_1/2, x_2 + h_2/2) + u_i(x_1 + h_1/2, x_2 + h_2/2)}{2}$$

$$-\frac{u_i(x_1 - h_1/2, x_2 - h_2/2) + u_i(x_1 + h_1/2, x_2 - h_2/2)}{2}] \approx \frac{\partial u_i}{\partial x_2}(x_1, x_2)$$

The stability condition for this method in the two-dimensional case is similar to that formulated in [7]: $\tau \leq \frac{h}{\max V_{p\alpha}}$, $h = \sqrt{h_1^2 + h_2^2}$, where $V_{p\alpha}$ is the maximal speed of longidudinal waves (Fig. 1).

5 Wave Propagation in the Unidirectionally Prestressed Media

Consider a numerical test problem aimed to demonstrate an influence of the initial stress on the wave field generated by the Ricker wavelet. The computational domain $(x_1, x_2) \in [0, L] \times [0, L]$ is a square, in which the initial stress is given as $\Sigma^{11} = P(x_2)$, $\Sigma^{ij} = 0$, $(ij \neq 11)$, where $P(x_2)$ is the linear function of x_2: $P(x_2) = \frac{C}{10} \left(1 - \frac{2x_2}{L} \right)$. It is obvious that the above stress field

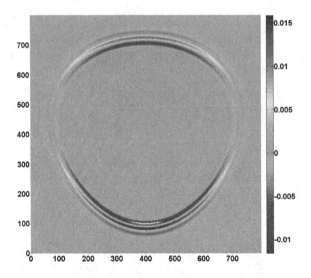

Fig. 4. The snapshot of the vertical stress component. correspond to the unstressed elastic medium. It is clearly seen the dependence of wave velocities on the spacial direction.

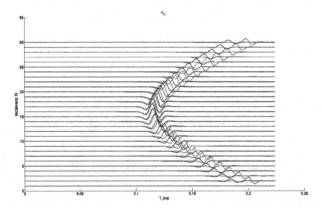

Fig. 5. Seismograms recorded by receivers in the upper region. Red lines correspond to the prestressed medium, black lines correspond to the unstressed medium (Color figure online).

satisfies equilibrium equations. The maximal tensile stress $\Sigma^{11} = C/10$ is on the bottom of the computational domain and the maximal compression with $\Sigma^{11} = -C/10$ is on the top of the domain. The parameters of the medium are $V_p = 3000\,m/s$, $V_s = 2000\,m/s$, $\rho_{00} = 2000\,kg/m^3$, and $C = \rho_{00}V_p^2$. The source of elastic waves with $60\,Hz$ dominant frequency is located in the centre of the domain. On Fig. 4 the snapshot of x_2 stress component σ_{22} is presented. It obvious that the wave velocities decrease towards to the bottom and increase towards to the top of computational domain. This effect is caused by the effect of compression of the upper part of the domain and tension of the lower part. On Fig. 5 the comparison of seismograms recorded by receivers on the top boundary of the domain for unstressed and prestressed medium is presented. One can see that the difference is significant.

6 Conclusions

The method of derivation of the governing equations for the small amplitude wave propagation in the initially prestressed medium is proposed. Governing equation for elastic waves in prestressed medium are derived in the case of quadratic dependence of internal energy on the strain tensor. It is proved that the initial stress can have a significant influence on the character of wave propagation.

Acknowledgements. The financial support of the Russian Foundation for Basic Research (grants 15-05-01310, 13-05-12051) is greatly acknowledged.

References

1. Biot, M.A.: Mechanics of Incremental Deformations. Wiley, New York (1965)
2. Liu, Q.H., Sinha, B.K.: A 3D cylindrical PML/FDTD method for elastic waves in fluid-filled pressurized boreholes in triaxially stressed formations. Geophysics **68**, 1731–1743 (2003)
3. Sharma, M.D.: Wave propagation in a prestressed anisotropic generalized thermoelastic medium. Earth, Planets and Space **62**, 381–390 (2010)
4. Lys, E.V., Romenski, E.I., Cheverda, V.A., Epov, M.I.: Interaction of seismic waves with zones of concentration of initial stresses. Dokl. Earth Sci. **449**, 402–405 (2013)
5. Romenskii, E.I.: Hypoelastic form of equations in nonlinear elastic theory. J. Appl. Mech. Tech. Phys. **15**, 255–259 (1974)
6. Godunov, S.K., Romenskii, E.I.: Elements of Continuum Mechanics and Conservation Laws. Kluwer Academic/Plenum Publishers, New York (2003)
7. Saenger, E.H., Gold, N., Shapiro, S.A.: Modeling the propagation of elastic waves using a modified finite-difference grid. Wave Motion **31**, 77–92 (2000)

Modeling the Wind Influence
on Acoustic-Gravity Propagation Waves
in a Heterogeneous Earth-Atmosphere Model

B. Mikhailenko, A.A. Mikhailov, and G.V. Reshetova[✉]

The Institute of Computational Mathematics and Mathematical Geophysics,
pr. Lavrentieva 6, Novosibirsk 630090, Russia
kgv@nmsf.sscc.ru

Abstract. A numerical-analytical algorithm for modeling of seismic and acoustic-gravity waves propagation is applied to a heterogeneous "Earth-Atmosphere" model. Seismic wave propagation in an elastic half-space is described by a system of first-order dynamic equations of elasticity theory. The propagation of acoustic-gravity waves in the atmosphere is described by the linearized Navier-Stokes equations with the wind. The algorithm is based on the integral Laguerre transform with respect to time, the finite integral Fourier transform with respect to a spatial coordinate combined with a finite difference method for the reduced problem.

Keywords: Seismic waves · Acoustic-gravity waves · Navier-stokes equations · Laguerre transform · Finite difference method

1   Introduction

In the numerical modeling of seismic wave fields in an elastic medium, it is typically assumed that the medium borders on vacuum, and boundary conditions are specified on a free surface. Specifically, at the boundary seismic waves are assumed to be absolutely reflected, and the generation of acoustic-gravity waves by elastic waves in the atmosphere and their interaction at the boundary are ignored.

In the last decade, some theoretical and experimental investigations have shown that there is a striking correlation between waves in the lithosphere and atmosphere. Paper [1] describes the effect of acoustoseismic induction of an acoustic wave produced by a vibrator. Owing to refraction in the atmosphere, the wave excites intensive surface seismic waves at a distance of tens of kilometers from the source. The lithospheric seismic waves produced by earthquakes and explosions generate atmospheric acoustic-gravitational waves of high intensity in the upper layers of the atmosphere and of small density in the ionosphere.

Papers [2,3] deal with to theoretical investigations of wave processes at the boundary between an elastic half-space and an isothermal homogeneous atmosphere. In these papers, the properties of the surface Stoneley-Scholte and modified Lamb waves are studied.

© Springer International Publishing Switzerland 2015
I. Dimov et al. (Eds.): FDM 2014, LNCS 9045, pp. 290–298, 2015.
DOI: 10.1007/978-3-319-20239-6_31

In this paper, an efficient numerical algorithm to simulate and investigate the propagation of seismic and acoustic-gravitational waves in a spatially inhomogeneous "Atmosphere-Earth" model is proposed. A peculiarity of the algorithm is a combination of integral transforms with a finite-difference method. A similar approach to solving the problem for a vertically inhomogeneous model in a cylindrical system of coordinates with no wind in the atmosphere was considered in [4]. In the problem statement, the initial system is written down as a first order hyperbolic system in terms of the velocity vector and stress tensor in a 3D Cartesian system of coordinates. The medium parameters (densities and velocities of longitudinal and transverse waves) are assumed to be functions of only two coordinates, and the medium is assumed to be homogeneous in the third coordinate. This problem statement is called a 2.5D one. The algorithm is based on the integral Laguerre transform with respect to the temporal coordinate. This method can be considered to be an analog to the well-known spectral method based on the Fourier transform, where, instead of the frequency ω, we have a parameter p(the degree of the Laguerre polynomials). The integral Laguerre transform with respect to time (in contrast to the Fourier transform) makes possible to reduce the initial problem to solving a system of equations in which the parameter is present only in the right-hand side of the equations and has a recurrence relation. This method for solving the dynamic problems of elasticity theory was first considered in [5,6] and then developed in problems of viscoelasticity [7,8] and porous media [9]. The above-mentioned papers concern the peculiarities of this method and the advantages of the integral Laguerre transform over the difference methods and the Fourier transform with respect to time.

2   Problem Statement

The system of equations for the propagation of acoustic-gravitational waves in the inhomogeneous non-ionized isothermal atmosphere in the Cartesian system of coordinates (x, y, z) with the wind directed along the horizontal axis x and the vertical stratification along the axis z has the following form:

$$\frac{\partial u_x}{\partial t} + v_x \frac{\partial u_x}{\partial x} = -\frac{1}{\rho_0}\frac{\partial P}{\partial x} - u_z \frac{\partial v_x}{\partial z}, \tag{1}$$

$$\frac{\partial u_y}{\partial t} + v_x \frac{\partial u_y}{\partial x} = -\frac{1}{\rho_0}\frac{\partial P}{\partial y}, \tag{2}$$

$$\frac{\partial u_z}{\partial t} + v_x \frac{\partial u_z}{\partial x} = -\frac{1}{\rho_0}\frac{\partial P}{\partial z} - \frac{\rho g}{\rho_0}, \tag{3}$$

$$\frac{\partial P}{\partial t} + v_x \frac{\partial P}{\partial x} = c_0^2 \left[\frac{\partial \rho}{\partial t} + v_x \frac{\partial \rho}{\partial x} + u_z \frac{\partial \rho_0}{\partial z} \right] - u_z \frac{\partial P_0}{\partial z} \tag{4}$$

$$\frac{\partial \rho}{\partial t} + v_x \frac{\partial \rho}{\partial x} = -\rho_0 \left[\frac{\partial u_x}{\partial x} + \frac{\partial u_y}{\partial y} + \frac{\partial u_z}{\partial z} \right] - u_z \frac{\partial \rho_0}{\partial z} + F(x, y, z, t). \tag{5}$$

Here g is the acceleration of gravity, $\rho_0(z)$ is the reference atmosphere density, $c_0(z)$ is the sound speed, $v_x(z)$ is the wind speed along the axis x, $\vec{u} = (u_x, u_y, u_z)$ is the velocity vector of the air particles displacement, P and ρ are the pressure and density perturbations, respectively, generated by the wave propagating from the source of mass $F(x, y, z, t) = \delta(r - r_0)f(t)$. In this case $f(t)$ is a given time signal in the source. Assume that the axis z is directed upwards. Zero subscripts for the medium physical parameters show their values for the reference atmosphere. The atmospheric pressure P_0 and the density ρ_0 for the reference atmosphere in a homogeneous gravitational field are

$$\frac{\partial P_0}{\partial z} = -\rho_0 g, \quad \rho_0(z) = \rho_1 \exp(-z/H),$$

where H is the height of an isothermal homogeneous atmosphere, and ρ_1 is the density of the atmosphere at the Earth's surface, that is, at $z = 0$.

The seismic waves propagation in an elastic medium is described by a well-known system of first-order equations of elasticity theory as the following relation between the displacement velocity vector components and the stress vector components:

$$\frac{\partial u_i}{\partial t} = \frac{1}{\rho_0} \frac{\partial \sigma_{ik}}{\partial x_k} + F_i f(t), \tag{6}$$

$$\frac{\partial \sigma_{ik}}{\partial t} = \mu \left(\frac{\partial u_k}{\partial x_i} + \frac{\partial u_i}{\partial x_k} \right) + \lambda \delta_{ik} div\, \vec{u}. \tag{7}$$

Here δ_{ij} is the Kronecker symbol, $\lambda(x_1, x_2, x_3)$ and $\mu(x_1, x_2, x_3)$ are the elastic parameters of the medium, $\rho_0(x_1, x_2, x_3)$ is the density, $\vec{u} = (u_1, u_2, u_3)$ is the displacement velocity vector, and σ_{ij} are the stress vector components. The equality $\vec{F}(x, y, z) = F_1 \vec{e}_x + F_2 \vec{e}_y + F_3 \vec{e}_z$ describes the distribution of a source located in space, and $f(t)$ is a given time signal in the source.

The combined system of equations for the propagation of seismic and acoustic-gravitational waves in the Cartesian system of coordinates $(x, y, z) = (x_1, x_2, x_3)$ can be written down as

$$\frac{\partial u_i}{\partial t} = \frac{1}{\rho_0} \frac{\partial \sigma_{ik}}{\partial x_k} + F_i f(t) - K_{atm} \left[v_x \frac{\partial u_i}{\partial x_1} + \frac{\rho g}{\rho_0} e_z - u_z \frac{\partial v_x}{\partial x_3} e_x \right], \tag{8}$$

$$\frac{\partial \sigma_{ik}}{\partial t} = \mu \left(\frac{\partial u_k}{\partial x_i} + \frac{\partial u_i}{\partial x_k} \right) + \lambda \delta_{ik} div\, \vec{u} - \delta_{ik} K_{atm} \left[v_x \frac{\partial \sigma_{ik}}{\partial x_1} + \rho_0 g u_z \right], \tag{9}$$

$$K_{atm} \left[\frac{\partial \rho}{\partial t} + v_x \frac{\partial \rho}{\partial x} \right] = -\rho_0 div\, \vec{u} - u_z \frac{\partial \rho_0}{\partial z}. \tag{10}$$

Here δ_{ij} is the Kronecker symbol, $\rho_0(x, z)$ is the density, $\lambda(x, z)$ and $\mu(x, z)$ are the elastic parameters of the medium, $\vec{u} = (u_1, u_2, u_3)$ is the displacement velocity vector, and σ_{ij} are the stress tensor components. $\vec{F}(x, y, z) = F_1 \vec{e}_x + F_2 \vec{e}_y + F_3 \vec{e}_z$ describes the distribution of a source located in space, and $f(t)$ is a given time signal in the source. The medium is assumed to be homogeneous along the y axis.

System (1)-(5) for the atmosphere is obtained from system (8)-(10) at $\sigma_{11} = \sigma_{22} = \sigma_{33} = -P$, $\mu = 0$, $\lambda = c_0^2 \rho_0$, $\sigma_{12} = \sigma_{13} = \sigma_{23} = 0$, $K_{atm} = 1$. Let us set $K_{atm} = 0$ in system (9)-(10), and obtain the system of equations (6)-(7) for the propagation of seismic waves in an elastic medium.

In our problem, the atmosphere-elastic half-space interface is assumed to be the plane $z = x_3 = 0$. In this case, the condition of contact of the two media at $z = 0$ is written down as

$$u_z|_{z=-0} = u_z|_{z=+0}; \qquad \left.\frac{\partial \sigma_{zz}}{\partial t}\right|_{z=-0} = \left.\left(\frac{\partial \sigma_{zz}}{\partial t} + \rho_0 g u_z\right)\right|_{z=+0};$$

$$\sigma_{xz}|_{z=-0} = \sigma_{yz}|_{z=-0} = 0. \qquad (11)$$

The problem is solved with the following zero initial data:

$$u_i|_{t=0} = \sigma_{ij}|_{t=0} = P|_{t=0} = \rho|_{t=0} = 0, \qquad i = 1,2,3 \quad j = 1,2,3. \qquad (12)$$

All the functions of the wave field components are assumed to be sufficiently smooth so that the transformations presented below are valid.

3   The Solution Algorithm

At the first step, we use the finite cosine-sine Fourier transform with respect to the spatial coordinate y where the medium is assumed to be homogeneous. For each component of the system, we introduce the corresponding transform [10]:

$$\overrightarrow{W}(x,z,n,t) = \int_0^a \overrightarrow{W}(x,y,z,t) \left\{ \begin{matrix} \cos(k_n y) \\ \sin(k_n y) \end{matrix} \right\} d(y), \qquad n = 0,1,2,...,N \qquad (13)$$

with the corresponding inversion formula

$$\overrightarrow{W}(x,y,z,t) = \frac{1}{\pi}\overrightarrow{W}(x,0,z,t) + \frac{2}{\pi}\sum_{n=1}^{N} \overrightarrow{W}(x,n,z,t)\cos(k_n y) \qquad (14)$$

or

$$\overrightarrow{W}(x,y,z,t) = \frac{2}{\pi}\sum_{n=1}^{N} \overrightarrow{W}(x,n,z,t)\sin(k_n y), \qquad (15)$$

where $k_n = \frac{n\pi}{a}$.

At rather a large distance a, consider a wave field up to the time $t < T$, where T is a minimum propagation time of a longitudinal wave to the boundary $r = a$. As a result of this transformation, we obtain $N + 1$ independent 2D unsteady problems.

At the second step, we apply to the thus obtained $N+1$ independent problems the integral Laguerre transform with respect to time

$$\overrightarrow{W}_p(x,n,z) = \int_0^\infty \overrightarrow{W}(x,n,z,t)(ht)^{-\frac{\alpha}{2}} l_p^\alpha(ht) d(ht), \qquad p = 0,1,2,... \qquad (16)$$

with the inversion formula

$$\overrightarrow{W}(x, n, z, t) = (ht)^{\frac{\alpha}{2}} \sum_{p=0}^{\infty} \frac{p!}{(p+\alpha)!} \overrightarrow{W}_p(x, n, z) l_p^{\alpha}(ht), \qquad (17)$$

where $l_p^{\alpha}(ht)$ are the orthogonal Laguerre functions.

The Laguerre functions $l_p^{\alpha}(ht)$ can be expressed in terms of the classical standard Laguerre polynomials $L_p^{\alpha}(ht)$ (see paper [11]). Here we select an integer parameter $\alpha \geq 1$ to satisfy the initial data and introduce the shift parameter $h > 0$. Then we have the following representation:

$$l_p^{\alpha}(ht) = (ht)^{\frac{\alpha}{2}} e^{-\frac{ht}{2}} L_p^{\alpha}(ht).$$

We take the finite cosine-sine Fourier transform with respect to the coordinate x, similar to the previous transform with respect to the coordinate y with the corresponding inversion formulas:

$$\overrightarrow{W}_p(x, n, z_i, p) = \frac{1}{\pi} \overrightarrow{W}_0(n, z_i, p) + \frac{2}{\pi} \sum_{m=1}^{M} \overrightarrow{W}(m, n, z_i, p) \cos(k_m x) \qquad (18)$$

or

$$\overrightarrow{W}(x, n, z_i, p) = \frac{2}{\pi} \sum_{m=1}^{M} \overrightarrow{W}(m, n, z_i, p) \sin(k_m x), \qquad (19)$$

where $k_m = \frac{m\pi}{b}$. It should be noted that in this case the medium is inhomogeneous.

The finite difference approximation for the system of linear algebraic equations with respect to z was applied using the staggered grid method [12] providing second order accuracy approximation. This scheme is used for the FD approximation within the computation domains in the atmosphere and in the elastic half-space, the fitting conditions at the interface being exactly satisfied. As a result of the above transformations, we obtain $N + 1$ systems of linear algebraic equations, where N is the number of harmonics in the Fourier transform with respect to the coordinate y.

The sought for solution vector $\overrightarrow{W}$ is represented as follows:

$$\overrightarrow{W}(p) = (\overrightarrow{V}_0(p), \overrightarrow{V}_1(p), ..., \overrightarrow{V}_K(p))^T,$$

$$\overrightarrow{V}_i = (\bar{\rho}^p(m = 0, ..., M; z_i), \bar{\sigma}_{xx}^p(m = 0, ..., M; z_i), \bar{u}_x^p(m = 0, ..., M; z_i), ...)^T.$$

Then for every n-th harmonic $(n = 0, ..., N)$ the system of linear algebraic equations can be written in the vector form as

$$(A + \frac{h}{2} E) \overrightarrow{W}(p) = \overrightarrow{F}(p - 1). \qquad (20)$$

A sequence of wave field components in the solution vector $\overrightarrow{V}$ is chosen to minimize the number of diagonals in matrix A. The main diagonal of the matrix has the components of this system multiplied by the parameter h(the Laguerre transform parameter). By changing the parameter h, the conditioning of the matrix can be considerably improved. Solving the system of linear algebraic equations (20) determines the spectral values for all the wave field components $\vec{W}(m, n, p)$. Then, using the inversion formulas for the Fourier transform, (14), (15), (18) and (19), and the Laguerre transform, (16), we obtain a solution to the initial problem (8)-(12). In the analytical Fourier and Laguerre transforms, when determining functions by their spectra, inversion formulas in the form of infinite sums are used. A necessary condition in the numerical implementation is to determine the number of terms of the summable series to construct a solution with a given accuracy. For instance, the number of harmonics in the inversion formulas of the Fourier transform (14), (15), (18) and (19) depends on a minimal spatial wavelength in the medium and the size of the spatial calculation domain of the field given by the finite limits of the integral transform. In addition, the convergence rate of a summable series depends on the smoothness of functions of the wave field. The number of the Laguerre harmonics for determining functions by formula (17) depends on a signal given in the source $f(t)$, the parameter h, and the time interval of the wave field. Papers [5–8] consider in detail the way of determining the required number of harmonics and choosing an optimal value of the parameter h.

The iterative conjugate gradient method [13,14] has turned out to be the most efficient for solving the system of linear algebraic equations (20). In this case, the entire matrix need not be stored in computer memory (which is good for large matrices). Another advantage of this method is the fast convergence to the problem solution if the matrix of a system is well-conditioned. Our matrix has this property owing to the parameter h. By specifying the values of h, we can greatly accelerate the convergence of the iterative process. An optimal value of h is chosen by minimizing the number of the Laguerre harmonics in inversion formula (17) and decreasing the number of iterations needed for finding a solution for each of the harmonics.

4   Results

Figures 1 and 2 show the results of numerical calculations of the waves fields propagation in the form of snapshots at a fixed time for the two models of media with wind in the atmosphere and without it. The wind speed was set equal to *50 m/sec* in order to obtain the main physical effects of wave propagation without carrying out calculations at considerable distances.

Figure 1 presents a snapshot of the wave field for $u_x(x, y, z)$ in the plane XZ at the time $t = 50$ sec. This model of the medium consists of a homogeneous elastic layer and an atmospheric layer separated by a plane boundary. The physical characteristics of the layers are as follows:

1. the atmosphere: sound speed $c_p = 340$ m/sec. Density versus coordinate z calculated by the formula $\rho_0(z) = \rho_1 \exp(-z/H)$, where $\rho_1 = 1.225 * 10^{-3}$ g/cm^3, $H = 6700$ m;

2. the elastic layer: longitudinal wave velocity: $c_p = 400$ m/sec, transverse wave velocity: $c_s = 300$ m/sec, density: $\rho_0 = 1.5$ g/cm^3.

A bounded domain, $(x, y, z) = (42\,km,\ 40\,km,\ 39\,km)$, was used for the calculations. A wave field from a point source (a pressure center) located in the elastic medium at a depth of $1/4$ of the length of a longitudinal wave with the coordinates $(x_0, y_0, z_0) = (21\,\text{km},\ 20\,\text{km}, -0.09\,\text{km})$ was simulated. The time signal in the source was given in the form:

$$f(t) = exp\left(-\frac{2\pi f_o(t - t_0)^2}{\gamma^2}\right) sin(2\pi f_0(t - t_0)),\qquad(21)$$

where $\gamma = 4$, $f_0 = 1$ Hz, $t_0 = 1.5$ sec.

The Fig. 1 shows the wave fields for the horizontal component of the displacement velocity u_x in the plane XZ at $y = y_0 = 20$ km: without wind (left), with wind speed in the atmosphere of 50 m/sec (right). The elastic medium-atmosphere interface is shown by the solid line. This Figure demonstrates that in the elastic medium, in addition to the spherical longitudinal wave P and the conic transverse wave S, there also propagates a "non-ray" spherical wave S^*, and then there follows a surface Stoneley-Scholte wave. An acoustic-gravitational wave refracted at the Earth-atmosphere interface propagates in the atmosphere. At the boundary, this wave generates the corresponding longitudinal and transverse waves in the elastic medium.

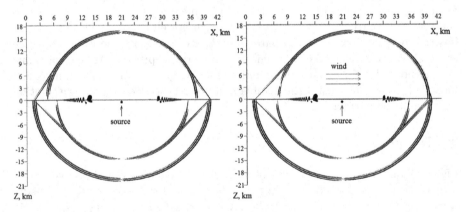

Fig. 1. A snapshot at $t = 50$ sec for the velocity component u_x in the plane (XZ) Without wind (left), with wind (right) (wind speed: 50 m/sec)

Figure 2 present snapshots of the wave field when the seismic waves velocity in the elastic medium is less than the sound speed in the atmosphere. In this model, the physical characteristics of the elastic medium and the atmosphere are as follows:

1. the atmosphere: sound speed $c_p = 340$ m/sec. Density versus coordinate z calculated by the formula $\rho_0(z) = \rho_1 \exp(-z/H)$, where $\rho_1 = 1.225 * 10^{-3}$ g/cm^3, $H = 6700$ m;
2. the elastic layer: longitudinal wave velocity $c_p = 300$ m/sec, transverse wave velocity $c_s = 200$ m/sec, density $\rho_0 = 1.2$ g/cm^3.

A bounded domain, $(x, y, z) = (35\,km, \ 30\,km, \ 30\,km)$, was used for the calculations. A wave field from a point source (a pressure center) located in the elastic medium at a depth of $1/4$ of the length of a longitudinal wave with the coordinates $(x_0, y_0, z_0) = (15\,km, \ 15\,km, \ -0.06\,km)$ was simulated. The time signal in the source was given by formula (21).

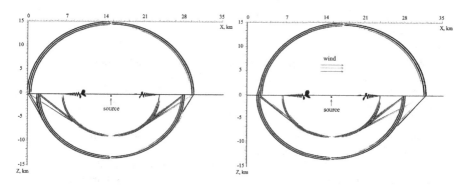

Fig. 2. A snapshot at $t = 45$ sec for the velocity component u_x in the plane (XZ) Without wind (left), with wind (right) (wind speed: 50 m/sec)

The Fig. 2 shows wave fields for the horizontal component of the displacement velocity u_x in the plane XZ at $y = y_0 = 15$ km: without wind (left), with wind speed in the atmosphere of 50 m/sec (right). The elastic medium-atmosphere interface is shown by the solid line. This Figure shows that in the atmosphere, in addition to the conical longitudinal wave P and the conical transverse wave S, there also propagates a "non-ray" spherical wave P^*, and then there follows a surface Stoneley-Scholte wave.

The results of the numerical simulation have revealed some new peculiarities of wave propagation with wind in the atmosphere. Specifically, the influence of the wind on the propagation velocity of the surface Stoneley waves in an elastic medium has been demonstrated. The numerical results have also shown that the velocity of these waves increases downwind and, hence, it decreases upwind by the quantity equal to the wind speed. The same influence of the wind is on a non-ray spherical exchange acoustic-gravitaty wave propagating in the atmosphere from a source located in a solid medium. Another evidence of the wind influence that has been detected is that the surface wave changes in amplitude along its front. This manifests itself as an increase in the amplitude in that part of the wavefront that propagates downwind and a decrease in the wavefront propagating upwind but with conservation of the total wave energy.

5   Conclusions

The above-proposed approach to the statement and solution of the problem makes possible to simulate the effects of the wave field propagation in a unified mathematical Earth-Atmosphere model and the study the exchange waves at their boundary. The numerical simulation of these processes makes possible to investigate the peculiarities of the wind effects on the propagation of acoustic-gravity atmospheric waves and the surface Stoneley waves.

Acknowledgements. This work was supported by the Russian Foundation for Basic Research (project 14-05-00867).

References

1. Alekseev, A.S., Glinsky, B.M., Dryakhlov, S.I., et al.: The effect of acoustic-seismic induction in vibroseismic sounding. Dokl. RAN. **346**(5), 664–667 (1996)
2. Gasilova, L.A., Petukhov, Y.V.: On the theory of surface wave propagation along different interfaces in the atmosphere. Izv. RAN. Fizika atmosfery i okeana **35**(1), 14–23 (1999)
3. Razin, A.V.: Propagation of a spherical acoustic delta wavelet along the gas-solid interface. Izv. RAN. Fizika zemli. **2**, 73–77 (1993)
4. Mikhailenko, B.G., Reshetova, G.V.: Mathematical simulation of propagation of seismic and acoustic-gravitational waves for an inhomogeneous Earth-Atmosphere model. Geologiya i geofizika **47**(5), 547–556 (2006)
5. Mikhailenko, B.G.: Spectral Laguerre method for the approximate solution of time dependent problems. Appl. Math. Lett. **12**, 105–110 (1999)
6. Konyukh, G.V., Mikhailenko, B.G., Mikhailov, A.A.: Application of the integral Laguerre transforms for forward seismic modeling. J. Comput. Acoust. **9**(4), 1523–1541 (2001)
7. Mikhailenko, B.G., Mikhailov, A.A., Reshetova, G.V.: Numerical modeling of transient seismic fields in viscoelastic media based on the Laguerre spectral method. J. Pure Appl. Geophys. **160**, 1207–1224 (2003)
8. Mikhailenko, B.G., Mikhailov, A.A., Reshetova, G.V.: Numerical viscoelastic modeling by the spectral Laguerre method. Geophys. Prospect. **51**, 37–48 (2003)
9. Imomnazarov, K.K., Mikhailov, A.A.: Use of the spectral Laguerre method to solve a linear 2D dynamic problem for porous media. Sib. Zh. Industr. matem. **11**(2), 86–95 (2008)
10. Mikhailov, A.A.: Simulation of seismic fields for 2.5D inhomogeneous viscoelastic media. Proc. of the Intern. Conference "Mathematical Methods in Geophysics". Novosibirsk. Part 1, 146–152 (2003)
11. Suetin, P.K.: Classical orthogonal polynomials, p. 327. Nauka, Moscow (1999)
12. Virieux, J.: P-, SV- wave propagation in heterogeneous media: velocity-stress finite-difference method. Geophysics **51**, 889–901 (1986)
13. Saad, Y., Van der Vorst, H.A.: Iterative solution of linear systems in the 20th century. J. Comput. Appl. Math. **123**, 1–33 (2000)
14. Sonneveld, P.: CGS, a fast Lanczos-type solver for nonsymmetric linear system. SIAM J. Sci. Stat. Comput. **10**, 36–52 (1989)

Numerical Solutions of Fractional Differential Equations by Extrapolation

Kamal Pal[1], Fang Liu[2], and Yubin Yan[1(✉)]

[1] Department of Mathematics, University of Chester, Thornton Science Park,
Chester CH2 4NU, UK
kamal_pal08@yahoo.co.uk, y.yan@chester.ac.uk
[2] Department of Mathematics, Lvliang University,
Lishi, People's Republic of China
1393587028@163.com

Abstract. An extrapolation algorithm is considered for solving linear fractional differential equations in this paper, which is based on the direct discretization of the fractional differential operator. Numerical results show that the approximate solutions of this numerical method has the expected asymptotic expansions.

1 Introduction

We consider the Richardson extrapolation algorithms for solving the following fractional order differential equation

$$_{0}^{C}D_{t}^{\alpha}y(t) = \beta y(t) + f(t), \quad 0 \le t \le 1, \tag{1}$$

$$y(0) = y_0, \tag{2}$$

where $\beta < 0$ and f is a given function on $[0, 1]$.

Extrapolation can be used to accelerate the convergence of a given sequence, [1,2,13]. Its applicability depends on the fact that a given sequence of the approximate solutions of the problem possesses an asymptotic expansion. Diethelm [3] introduced an algorithm for solving the above linear differential equation of fractional order, with $0 < \alpha < 1$, Diethelm and Walz [12] proved that the approximate solution of the numerical algorithm in [3] has an asymptotic expansion. See [5–11] for the numerical methods of the general nonlinear fractional differential equations.

Recently, Yan, Pal and Ford [14] extended the numerical method in [3] and obtained a high order numerical method for solving (1) and (2) and proved that the approximate solution has an asymptotic expansion. In this paper, we give some numerical results to show that the approximate solutions of the proposed numerical methods in this paper have the expected asymptotic expansions.

The paper is organized as follows: in Sect. 2, we introduce the numerical method for solving (1) and (2) and discuss how to approximate the starting values and the starting integrals appeared in the numerical method. In Sect. 3, we give some numerical examples to show that the approximate solutions of the proposed numerical methods in this paper have the expected asymptotic expansions.

© Springer International Publishing Switzerland 2015
I. Dimov et al. (Eds.): FDM 2014, LNCS 9045, pp. 299–306, 2015.
DOI: 10.1007/978-3-319-20239-6_32

2   Higher Order Numerical Method

In this section we will consider a higher order numerical method for solving (1)
and (2). It is well-known that, when $0 < \alpha < 1$, (1) and (2) is equivalent to,
with $0 < \alpha < 1$,

$$
{}_0^R D_t^\alpha [y(t) - y_0] = \beta y(t) + f(t), \quad 0 \le t \le 1, \tag{3}
$$

where ${}_0^R D_t^\alpha y(t)$ denotes the Riemann-Liouville fractional derivative defined by,
with $0 < \alpha < 1$,

$$
{}_0^R D_t^\alpha y(t) = \frac{1}{\Gamma(1-\alpha)} \frac{d}{dt} \int_0^t (t-u)^{-\alpha} y(u) \, d\tau. \tag{4}
$$

By using Hadamard finite-part integral, ${}_0^R D_t^\alpha$ can be written into

$$
{}_0^R D_t^\alpha y(t) = \frac{1}{\Gamma(-\alpha)} \fint_0^t (t-u)^{-1-\alpha} y(u) \, du. \tag{5}
$$

Here the integral $\fint$ denotes a Hadamard finite-part integral [3].

Yan, Pal and Ford [14] extended the numerical method in Diethelm and
Walz [12] and obtained a high order numerical method for solving (1) and (2)
for $0 < \alpha < 1$. Let M be a fixed positive integer and let $0 = t_0 < t_1 < t_2 < \cdots <$
$t_{2j} < t_{2j+1} < \cdots < t_{2M} = 1$ be a partition of $[0,1]$ and h the step size. At the
nodes $t_{2j} = \frac{2j}{2M}$, the Eqs. (1) and (2) satisfy

$$
{}_0^R D_t^\alpha [y(t_{2j}) - y_0] = \beta y(t_{2j}) + f(t_{2j}), \quad j = 1, 2, \ldots, M,
$$

and at the nodes $t_{2j+1} = \frac{2j+1}{2M}$, the Eqs. (1) and (2) satisfy

$$
{}_0^R D_t^\alpha [y(t_{2j+1}) - y_0] = \beta y(t_{2j+1}) + f(t_{2j+1}), \quad j = 0, 1, 2, \ldots, M-1. \tag{6}
$$

Note that

$$
{}_0^R D_t^\alpha y(t_{2j}) = \frac{1}{\Gamma(-\alpha)} \fint_0^{t_{2j}} (t_{2j} - \tau)^{-1-\alpha} y(\tau) \, d\tau = \frac{t_{2j}^{-\alpha}}{\Gamma(-\alpha)} \fint_0^1 w^{-1-\alpha} y(t_{2j} - t_{2j} w) \, dw. \tag{7}
$$

For every j, we denote $g(w) = y(t_{2j} - t_{2j} w)$ and approximate the integral
$\fint_0^1 w^{-1-\alpha} g(w) \, dw$ by $\fint_0^1 w^{-1-\alpha} g_2(w) \, dw$, where $g_2(w)$ is the piecewise quadratic
interpolation polynomials on the nodes $w_l = l/2j$, $l = 0, 1, 2, \ldots, 2j$. More
precisely, we have, for $k = 1, 2, \ldots, j$,

$$
\begin{aligned}
g_2(w) = {}& \frac{(w - w_{2k-1})(w - w_{2k})}{(w_{2k-2} - w_{2k-1})(w_{2k-2} - w_{2k})} g(w_{2k-2}) \\
& + \frac{(w - w_{2k-2})(w - w_{2k})}{(w_{2k-1} - w_{2k-2})(w_{2k-1} - w_{2k})} g(w_{2k-1}) \\
& + \frac{(w - w_{2k-2})(w - w_{2k-1})}{(w_{2k} - w_{2k-2})(w_{2k} - w_{2k-1})} g(w_{2k}), \quad \text{for } w \in [w_{2k-2}, w_{2k}].
\end{aligned}
$$

Thus

$$
\begin{aligned}
{}^{R}_{0}D^{\alpha}_{t}y(t_{2j}) &= \frac{t_{2j}^{-\alpha}}{\Gamma(-\alpha)} \oint_{0}^{1} w^{-1-\alpha}y(t_{2j} - t_{2j}w)\,dw \\
&= \frac{t_{2j}^{-\alpha}}{\Gamma(-\alpha)} \Big(\sum_{k=1}^{j} \oint_{w_{2k-2}}^{w_{2k}} w^{-1-\alpha}g_{2}(w)\,dw + R_{2j}(g) \Big) \\
&= \frac{t_{2j}^{-\alpha}}{\Gamma(-\alpha)} \Big(\sum_{k=0}^{2j} \alpha_{k,2j}y(t_{2j-k}) + R_{2j}(g) \Big)
\end{aligned}
$$

where $R_{2j}(g)$ is the remainder term and $\alpha_{k,2j}, k = 0, 1, 2, \ldots, 2j$ are weights given by

$$
\begin{aligned}
&(-\alpha)(-\alpha + 1)(-\alpha + 2)(2j)^{-\alpha}\alpha_{l,2j} \\
&= \begin{cases}
2^{-\alpha}(\alpha + 2), & \text{for } l = 0, \\
(-\alpha)2^{2-\alpha}, & \text{for } l = 1, \\
(-\alpha)(-2^{-\alpha}\alpha) + \frac{1}{2}F_{0}(2), & \text{for } l = 2, \\
-F_{1}(k), & \text{for } l = 2k - 1, \quad k = 2, 3, \ldots, j, \\
\frac{1}{2}(F_{2}(k) + F_{0}(k + 1)), & \text{for } l = 2k, \quad k = 2, 3, \ldots, j - 1, \\
\frac{1}{2}F_{2}(j), & \text{for } l = 2j.
\end{cases}
\end{aligned}
$$

Here

$$
\begin{aligned}
F_{0}(k) =& (2k - 1)(2k)\Big((2k)^{-\alpha} - (2k - 2)^{-\alpha}\Big)(-\alpha + 1)(-\alpha + 2) \\
& - \Big((2k - 1) + 2k\Big)\Big((2k)^{-\alpha+1} - (2k - 2)^{-\alpha+1}\Big)(-\alpha)(-\alpha + 2) \\
& + \Big((2k)^{-\alpha+2} - (2k - 2)^{-\alpha+2}\Big)(-\alpha)(-\alpha + 1),
\end{aligned}
$$

$$
\begin{aligned}
F_{1}(k) =& (2k - 2)(2k)\Big((2k)^{-\alpha} - (2k - 2)^{-\alpha}\Big)(-\alpha + 1)(-\alpha + 2) \\
& - \Big((2k - 2) + 2k\Big)\Big((2k)^{-\alpha+1} - (2k - 2)^{-\alpha+1}\Big)(-\alpha)(-\alpha + 2) \\
& + \Big((2k)^{-\alpha+2} - (2k - 2)^{-\alpha+2}\Big)(-\alpha)(-\alpha + 1),
\end{aligned}
$$

and

$$
\begin{aligned}
F_{2}(k) =& (2k - 2)(2k - 1)\Big((2k)^{-\alpha} - (2k - 2)^{-\alpha}\Big)(-\alpha + 1)(-\alpha + 2) \\
& - \Big((2k - 2) + (2k - 1)\Big)\Big((2k)^{-\alpha+1} - (2k - 2)^{-\alpha+1}\Big)(-\alpha)(-\alpha + 2) \\
& + \Big((2k)^{-\alpha+2} - (2k - 2)^{-\alpha+2}\Big)(-\alpha)(-\alpha + 1).
\end{aligned}
$$

Hence (3) satisfies for $j = 1, 2, \ldots, M$,

$$
y(t_{2j}) = \frac{1}{\alpha_{0,2j} - t_{2j}^{\alpha}\Gamma(-\alpha)\beta} \Big[t_{2j}\Gamma(-\alpha)f(t_{2j}) - \sum_{k=1}^{2j} \alpha_{k,2j}y(t_{2j-k}) + y_{0}\sum_{k=0}^{2j} \alpha_{k,2j} - R_{2j}(g) \Big]. \tag{8}
$$

At the nodes $t_{2j+1} = \frac{2j+1}{2M}$, $j = 0, 1, 2, \ldots, M - 1$, we have

$$
\begin{aligned}
{}_0^R D_t^\alpha y(t_{2j+1}) &= \frac{1}{\Gamma(-\alpha)} \oint_0^{t_{2j+1}} (t_{2j+1} - \tau)^{-1-\alpha} y(\tau)\, d\tau \\
&= \frac{1}{\Gamma(-\alpha)} \oint_0^{t_1} (t_{2j+1} - \tau)^{-1-\alpha} y(\tau)\, d\tau \\
&\quad + \frac{t_{2j+1}^{-\alpha}}{\Gamma(-\alpha)} \oint_0^{\frac{2j}{2j+1}} w^{-1-\alpha} y(t_{2j+1} - t_{2j+1} w)\, dw.
\end{aligned}
$$

For every j, we denote $g(w) = y(t_{2j+1} - t_{2j+1}w)$ and approximate the integral $\oint_0^{\frac{2j}{2j+1}} w^{-1-\alpha} g(w)\, dw$ by $\oint_0^{\frac{2j}{2j+1}} w^{-1-\alpha} g_2(w)\, dw$, where $g_2(w)$ is the piecewise quadratic interpolation polynomials on the nodes $w_l = \frac{l}{2j+1}$, $l = 0, 1, 2, \ldots, 2j$. We then get

$$
\begin{aligned}
{}_0^R D_t^\alpha y(t_{2j+1}) &= \frac{1}{\Gamma(-\alpha)} \int_0^{t_1} (t_{2j+1} - \tau)^{-1-\alpha} y(\tau)\, d\tau \\
&\quad + \frac{t_{2j+1}^{-\alpha}}{\Gamma(-\alpha)} \left(\sum_{k=1}^{j} \oint_{w_{2k-2}}^{w_{2k}} w^{-1-\alpha} g_2(w)\, dw + R_{2j+1}(g) \right) \\
&= \frac{1}{\Gamma(-\alpha)} \int_0^{t_1} (t_{2j+1} - \tau)^{-1-\alpha} y(\tau)\, d\tau \\
&\quad + \frac{t_{2j+1}^{-\alpha}}{\Gamma(-\alpha)} \left(\sum_{k=0}^{2j} \alpha_{k,2j+1} y(t_{2j+1-k}) + R_{2j+1}(g) \right)
\end{aligned}
$$

where $R_{2j+1}(g)$ is the remainder term and $\alpha_{k,2j+1} = \alpha_{k,2j}$, $k = 0, 1, 2, \ldots, 2j$. Hence

$$
\begin{aligned}
y(t_{2j+1}) &= \frac{1}{\alpha_{0,2j+1} - t_{2j+1}^\alpha \Gamma(-\alpha)\beta} \left[t_{2j+1}^\alpha \Gamma(-\alpha) f(t_{2j+1}) - \sum_{k=1}^{2j} \alpha_{k,2j+1} y(t_{2j+1-k}) \right. \\
&\quad \left. + y_0 \sum_{k=0}^{2j} \alpha_{k,2j+1} - R_{2j+1}(g) - t_{2j+1}^\alpha \int_0^{t_1} (t_{2j+1} - \tau)^{-1-\alpha} y(\tau)\, d\tau \right]. \qquad (9)
\end{aligned}
$$

Here $\alpha_{0,l} - t_l^\alpha \Gamma(-\alpha)\beta < 0$, $l = 2j, 2j + 1$, which follow from $\Gamma(-\alpha) < 0$, $\beta < 0$ and $\alpha_{0,2j+1} = \alpha_{0,2j} < 0$.

Let $y_{2j} \approx y(t_{2j})$ and $y_{2j+1} \approx y(t_{2j+1})$ denote the approximate solutions of $y(t_{2j})$ and $y(t_{2j+1})$, respectively. We define the following numerical methods for solving (1) and (2), with $j = 1, 2, \ldots, M$,

$$
y_{2j} = \frac{1}{\alpha_{0,2j} - t_{2j}^\alpha \Gamma(-\alpha)\beta} \left[t_{2j} \Gamma(-\alpha) f(t_{2j}) - \sum_{k=1}^{2j} \alpha_{k,2j} y_{2j-k} + y_0 \sum_{k=0}^{2j} \alpha_{k,2j} \right], \qquad (10)
$$

and, with $j = 1, 2, \ldots, M - 1$,

$$
y_{2j+1} = \frac{1}{\alpha_{0,2j+1} - t_{2j+1}^{\alpha}\Gamma(-\alpha)\beta}\left[t_{2j+1}^{\alpha}\Gamma(-\alpha)f(t_{2j+1}) - \sum_{k=1}^{2j}\alpha_{k,2j+1}y_{2j+1-k}\right.
$$

$$
\left. + y_0\sum_{k=0}^{2j}\alpha_{k,2j+1} - t_{2j+1}^{\alpha}\int_0^{t_1}(t_{2j+1} - \tau)^{-1-\alpha}y(\tau)\,d\tau\right]. \tag{11}
$$

Yan, Pal and Ford [14] proved the following Theorem.

Theorem 1 (Theorem 2.1 in [14]). *Let $0 < \alpha < 1$ and M be a positive integer. Let $0 = t_0 < t_1 < t_2 < \cdots < t_{2j} < t_{2j+1} < \cdots < t_{2M} = 1$ be a partition of $[0,1]$ and h the step size. Let $y(t_{2j}), y(t_{2j+1}), y_{2j}$ and y_{2j+1} be the exact and the approximate solutions of (8)–(11), respectively. Assume that $y \in C^{m+2}[0,1]$, $m \geq 3$. Further assume that we can approximate the starting value y_1 and the starting integral $\int_0^{t_1}(t_{2j+1}-\tau)^{-1-\alpha}y(\tau)\,d\tau$ in (11) by using some numerical methods and obtain the required accuracy. Then there exist coefficients $c_\mu = c_\mu(\alpha)$ and $c_\mu^* = c_\mu^*(\alpha)$ such that the sequence $\{y_l\}, l = 0, 1, 2, \ldots, 2M$ possesses an asymptotic expansion of the form*

$$
y(t_{2M}) - y_{2M} = \sum_{\mu=3}^{m+1}c_\mu(2M)^{\alpha-\mu} + \sum_{\mu=2}^{\mu^*}c_\mu^*(2M)^{-2\mu} + o((2M)^{\alpha-m-1}), \quad \textit{for } M \to \infty,
$$

that is,

$$
y(t_{2M}) - y_{2M} = \sum_{\mu=3}^{m+1}c_\mu h^{\mu-\alpha} + \sum_{\mu=2}^{\mu^*}c_\mu^* h^{2\mu} + o(h^{m+1-\alpha}), \quad \textit{for } h \to 0,
$$

where μ^* is the integer satisfying $2\mu^* < m+1-\alpha < 2(\mu^*+1)$, and c_μ and c_μ^* are certain coefficients that depend on y.

3   Numerical Simulations

Example 1. Consider the following example in [9], with $0 < \alpha < 1$,

$$
{}_0^C D_t^\alpha y(t) + y(t) = 0, \quad t \in [0,1], \tag{12}
$$

$$
y(0) = 1. \tag{13}
$$

It is well known that the exact solution is

$$
y(t) = E_\alpha(-t^\alpha),
$$

where

$$
E_\alpha(z) = \sum_{k=0}^\infty \frac{z^k}{\Gamma(\alpha k + 1)}
$$

is the Mittag-Leffler function of order α. Here the given function f is smooth and $f = 0$ (Table 1).

Table 1. Errors for Eqs. (12) and (13) with $\alpha = 0.3$, taken at $t = 1$.

Step size	Error of the method	1st extra. error	2nd extra. error
1/10	1.1296e-003		
1/20	2.0412e-004	5.3779e-006	
1/40	3.6454e-005	4.5025e-007	2.7527e-008
1/80	6.4759e-006	3.8539e-008	1.3800e-009
1/160	1.1475e-006	3.3431e-009	6.9354e-011
1/320	2.0310e-007	2.9225e-010	3.5604e-012

Table 2. Orders ("EOC ") for Eqs. (12) and (13) with $\alpha = 0.3$, taken at $t = 1$.

Step size	The method	1st extrapolation	2nd extrapolation
1/10			
1/20	2.46		
1/40	2.49	3.58	
1/80	2.49	3.55	4.32
1/160	2.5	3.53	4.31
1/320	2.5	3.52	4.28

Choose the step size $h = 1/10$. In Table 2, we displayed the errors of the algorithms (10) and (11) at $t = 1$ and of the first two extrapolation steps in the Romberg tableau with $\alpha = 0.3$. We observe that the first column (the errors of the basic algorithm without extrapolation) converges as $h^{3-\alpha}$. The second column (errors using one extrapolation step)converges as $h^{4-\alpha}$, and the last column (two extrapolation steps) converges as h^4. We also consider other values of $\alpha \in (0, 1)$. We observe that when α is close to 1, the convergence seems to be even a bit faster. But when α is close to 0, the convergence is a bit slower than expected which is consistent with the numerical observation in [12] for the lower order method.

Example 2. Consider the following example in [4], with $0 < \alpha < 1$,

$$_0^C D_t^\alpha y(t) + y(t) = t^4 - \frac{1}{2}t^3 - \frac{3}{\Gamma(4-\alpha)}t^{3-\alpha} + \frac{24}{\Gamma(5-\alpha)}t^{4-\alpha}, \quad t \in [0, 1], \quad (14)$$

$$y(0) = 0, \quad (15)$$

whose exact solution is given by $y(t) = t^4 - \frac{1}{2}t^3$.

Choose the step size $h = 1/10$. In Table 3, we displayed the errors of the algorithms (10) and (11) at $t = 1$ and of the first two extrapolation steps in the Romberg tableau with $\alpha = 0.3$. We observe that the first column converges as $h^{3-\alpha}$. The second column converges as $h^{4-\alpha}$ and the last column converges as h^4. We also consider other values of $\alpha \in (0, 1)$. We observe that when α is close

Table 3. Errors for Eqs. (12)–(15) with $\alpha = 0.3$, taken at $t = 1$.

Step size	Error of the method	1st extra. error	2nd extra. error
1/10	1.4571e-004		
1/20	2.3118e-005	8.2097e-007	
1/40	3.6127e-006	6.5021e-008	2.0039e-009
1/80	5.6030e-007	5.1186e-009	1.2514e-010
1/160	8.6565e-008	4.0106e-010	7.8051e-012
1/320	1.3348e-008	3.1315e-011	4.9268e-013

Table 4. Orders ("EOC") for Eqs. (12)–(15) with $\alpha = 0.3$, taken at $t = 1$.

Step size	The method	1st extrapolation	2nd extrapolation
1/10			
1/20	2.66		
1/40	2.68	3.66	
1/80	2.69	3.67	4.00
1/160	2.70	3.67	4.00
1/320	2.70	3.68	3.98

to 1, the convergence seems to be even a bit faster. But when α is close to 0, the convergence is a bit slower than expected which is consistent with the numerical observation in [12] for the lower order method (Table 4).

Acknowledgements. We wish to express our sincere gratitude to Professor Neville. J. Ford for his encouragement, discussions and valuable criticism during the research of this work.

References

1. Brezinski, C.: A general extrapolation algorithm. Numer. Math. **35**, 175–187 (1980)
2. Brezinski, C., Redivo Zaglia, M.: Extrapolation Methods, Theory and Practice. North Holland, Amsterdam (1992)
3. Diethelm, K.: Generalized compound quadrature formulae for finite-part integrals. IMA J. Numer. Anal. **17**, 479–493 (1997)
4. Diethelm, K.: An algorithm for the numerical solution of differential equations of fractional order. Electron. Trans. Numer. Anal. **5**, 1–6 (1997)
5. Diethelm, K.: The Analysis of Fractional Differential Equations: An Application-Oriented Using Differential Operators of Caputo Type. Lecture Notes in Mathematics 2004. Springer, Heidelberg (2010)
6. Diethelm, K.: Monotonicity results for a compound quadrature method for finite-part integrals. J. Inequalities Pure Appl. Math. **5**(2) (2004). (Article 44)

7. Diethelm, K., Ford, N.J.: Analysis of fractional differential equations. J. Math. Anal. Appl. **265**, 229–248 (2002)
8. Diethelm, K., Ford, N.J., Freed, A.D.: Detailed error analysis for a fractional Adams method. Numer. Algorithms **36**, 31–52 (2004)
9. Diethelm, K., Ford, N.J., Freed, A.D.: A predictor-corrector approach for the numerical solution of fractional differential equations. Nonlinear Dynam. **29**, 3–22 (2002)
10. Diethelm, K., Luchko, Y.: Numerical solution of linear multi-term initial value problems of fractional order. J. Comput. Anal. Appl. **6**, 243–263 (2004)
11. Dimitrov, Y.: Numerical approximations for fractional differential equations. J. Frac. Calc. Appl. **5**(3S), 1–45 (2014)
12. Diethelm, K., Walz, G.: Numerical solution of fractional order differential equations by extrapolation. Numer. Algorithms **16**, 231–253 (1997)
13. Walz, G.: Asymptotics and Extrapolation. Akademie-Verlag, Berlin (1996)
14. Yan, Y., Pal, K., Ford, N.J.: Higher order numerical methods for solving fractional differential equations. BIT Numer. Math. **54**, 555–584 (2014). doi:10.1007/s10543-013-0443-3

Finite Difference Method for Two-Sided Space-Fractional Partial Differential Equations

Kamal Pal[1], Fang Liu[2], Yubin Yan[1]([✉]), and Graham Roberts[1]

[1] Department of Mathematics, University of Chester, Thornton Science Park,
Chester CH2 4NU, UK
{y.yan,graham.roberts}@chester.ac.uk, kamal_pal08@yahoo.co.uk
[2] Department of Mathematics, Luliang University,
Lishi, PR China
1393587028l@163.com

Abstract. Finite difference methods for solving two-sided space-fractional partial differential equations are studied. The space-fractional derivatives are the left-handed and right-handed Riemann-Liouville fractional derivatives which are expressed by using Hadamard finite-part integrals. The Hadamard finite-part integrals are approximated by using piecewise quadratic interpolation polynomials and a numerical approximation scheme of the space-fractional derivative with convergence order $O(\Delta x^{3-\alpha})$, $1 < \alpha < 2$ is obtained. A shifted implicit finite difference method is introduced for solving two-sided space-fractional partial differential equation and we prove that the order of convergence of the finite difference method is $O(\Delta t + \Delta x^{\min(3-\alpha,\beta)})$, $1 < \alpha < 2$, $\beta > 0$, where $\Delta t, \Delta x$ denote the time and space step sizes, respectively. Numerical examples are presented and compared with the exact analytical solution for its order of convergence.

1 Introduction

Consider the following two-sided space-fractional partial differential equation, with $1 < \alpha < 2$, $t > 0$,

$$u_t(t,x) = C_+(t,x)\,{}_0^R D_x^\alpha u(t,x), + C_-(t,x)\,{}_x^R D_1^\alpha u(t,x) + f(t,x), \quad 0 < x < 1, \quad (1)$$

$$u(t,0) = \varphi_1(t), \quad u(t,1) = \varphi_2(t), \quad (2)$$

$$u(0,x) = u_0(x), \quad 0 < x < 1. \quad (3)$$

Here the function $f(t,x)$ is a source/sink term. The functions $C_+(t,x) \geq 0$ and $C_-(t,x) \geq 0$ may be interpreted as transport related coefficients. The addition of a classical advective term $-\nu(t,x)\frac{\partial u(t,x)}{\partial x}$ in (1) does not impact the analysis performed in this paper, and has been omitted to simplify the notation [6]. The left-handed fractional derivative ${}_0^R D_x^\alpha f(x)$ and right-handed fractional derivative ${}_x^R D_1^\alpha f(x)$ in (1) are Riemann-Liouville fractional derivatives of order α.

© Springer International Publishing Switzerland 2015
I. Dimov et al. (Eds.): FDM 2014, LNCS 9045, pp. 307–314, 2015.
DOI: 10.1007/978-3-319-20239-6_33

Let us review some numerical methods for solving space-fractional partial differential equation. There are many different numerical methods for solving space-fractional partial differential equations in literature, see [8–20].

In this paper, we will use the idea in Diethelm, [2] to define a finite difference method for solving (1)–(3), see our recent works for this method [3–5,7]. We first express the fractional derivative by using Hadamard finite-part integral, i.e., with $1 < \alpha < 2$,

$$
{}_0^R D_x^\alpha f(x) = \frac{1}{\Gamma(2-\alpha)} \int_0^x (x-\xi)^{1-\alpha} f''(\xi)\, d\xi = \frac{1}{\Gamma(-\alpha)} \fint_0^x (x-\xi)^{-\alpha-1} f(\xi)\, d\xi,
$$

where $\fint_0^x$ denotes the Hadamard finite-part integral [2]. We approximate $f(\xi)$ by using piecewise quadratic interpolation polynomials and obtain an approximation scheme of Riemann-Liouville fractional derivative. Similarly we can approximate the right-handed Riemann-Liouville fractional derivative ${}_x^R D_1^\alpha f(x)$. Based on these approximate schemes, we define a shifted finite difference method for solving (1)–(3). We proved that the convergence order of the numerical method is $O(\Delta t + \Delta x^{\min(3-\alpha,\beta)}), 1 < \alpha < 2, \beta > 0$.

The paper is organized as follows. In Sect. 2, we consider the implicit shifted finite difference method for solving (1)–(3) where the Hadamard integral of the space-fractional derivative is approximated by using piecewise quadratic interpolation polynomials. In Sect. 3, we give two numerical examples. The numerical experiments are consistent with the theoretical results.

2   Numerical Method

In this section, we will introduce a new finite difference method for solving (1)–(3). For simplicity, we assume $C_+(t,x) = C_-(t,x) = 1$ and $\varphi_1(t) = \varphi_2(t) = 0$.

Lemma 1. *Let $1 < \alpha < 2$ and let $M = 2m$ where m is a fixed positive integer. Let $0 = x_0 < x_1 < x_2 < \cdots < x_{2j} < x_{2j+1} < \cdots < x_M = 1$ be a partition of $[0,1]$. Assume that $f(x)$ is a sufficiently smooth function. Then we have, with $j = 1, 2, \ldots, m$,*

$$
{}_0^R D_x^\alpha f(x)\Big|_{x=x_{2j}} = \frac{x_{2j}^{-\alpha}}{\Gamma(-\alpha)}\left(\sum_{l=0}^{2j} \alpha_{l,2j} f(x_{2j-l}) + R_{2j}(f)\right)
$$

$$
= \Delta x^{-\alpha} \sum_{l=0}^{2j} w_{l,2j} f(x_{2j-l}) + \frac{x_{2j}^{-\alpha}}{\Gamma(-\alpha)} R_{2j}(f), \tag{4}
$$

and, with $j = 1, 2, \ldots, m-1$,

$$
{}_0^R D_x^\alpha f(x)\Big|_{x=x_{2j+1}} = \frac{1}{\Gamma(-\alpha)} \int_0^{x_1} (x_{2j+1} - \xi)^{-1-\alpha} f(\xi)\, d\xi
$$

$$
+ \frac{x_{2j+1}^{-\alpha}}{\Gamma(-\alpha)}\left(\sum_{l=0}^{2j} \alpha_{l,2j+1} f(x_{2j+1-l}) + R_{2j+1}(f)\right)
$$

$$= \frac{1}{\Gamma(-\alpha)} \int_0^{x_1} (x_{2j+1} - \xi)^{-1-\alpha} f(\xi)\, d\xi + \Delta x^{-\alpha} \sum_{l=0}^{2j} w_{l,2j+1} f(x_{2j+1-l})$$

$$+ \frac{x_{2j+1}^{-\alpha}}{\Gamma(-\alpha)} R_{2j+1}(f), \tag{5}$$

where

$$(-\alpha)(-\alpha+1)(-\alpha+2)(2j)^{-\alpha}\alpha_{l,2j}$$

$$= \begin{cases}
2^{-\alpha}(\alpha+2), & for\ l = 0, \\
(-\alpha)2^{2-\alpha}, & for\ l = 1, \\
(-\alpha)(-2^{-\alpha}\alpha) + \frac{1}{2}F_0(2), & for\ l = 2, \\
-F_1(k), & for\ l = 2k-1, \quad k = 2,3,\ldots,j, \\
\frac{1}{2}(F_2(k) + F_0(k+1)), & for\ l = 2k, \quad k = 2,3,\ldots,j-1, \\
\frac{1}{2}F_2(j), & for\ l = 2j,
\end{cases}$$

$$F_0(k) = (2k-1)(2k)\big((2k)^{-\alpha} - (2(k-1))^{-\alpha}\big)(-\alpha+1)(-\alpha+2)$$
$$- \big((2k-1) + 2k\big)\big((2k)^{-\alpha+1} - (2(k-1))^{-\alpha+1}\big)(-\alpha)(-\alpha+2)$$
$$+ \big((2k)^{-\alpha+2} - (2(k-1))^{-\alpha+2}\big)(-\alpha)(-\alpha+1), \tag{6}$$

$$F_1(k) = (2k-2)(2k)\big((2k)^{-\alpha} - (2k-2)^{-\alpha}\big)(-\alpha+1)(-\alpha+2)$$
$$- \big((2k-2) + 2k\big)\big((2k)^{-\alpha+1} - (2k-2)^{-\alpha+1}\big)(-\alpha)(-\alpha+2)$$
$$+ \big((2k)^{-\alpha+2} - (2k-2)^{-\alpha+2}\big)(-\alpha)(-\alpha+1), \tag{7}$$

$$F_2(k) = (2k-2)(2k-1)\big((2k)^{-\alpha} - (2k-2)^{-\alpha}\big)(-\alpha+1)(-\alpha+2)$$
$$- \big((2k-2) + (2k-1)\big)\big((2k)^{-\alpha+1} - (2k-2)^{-\alpha+1}\big)(-\alpha)(-\alpha+2)$$
$$+ \big((2k)^{-\alpha+2} - (2k-2)^{-\alpha+2}\big)(-\alpha)(-\alpha+1). \tag{8}$$

Further we have, with $l = 0, 1, 2, \ldots, 2j,$

$$\Gamma(3-\alpha)w_{l,2j} = (-\alpha)(-\alpha+1)(-\alpha+2)(2j)^{-\alpha}\alpha_{l,2j}, \tag{9}$$

and

$$\alpha_{l,2j+1} = \alpha_{l,2j}, \quad w_{l,2j+1} = w_{l,2j}. \tag{10}$$

The remainder term $R_l(f)$ *satisfy, for every* $f \in C^3(0,1)$,

$$|R_l(f)| \leq C\Delta x^{3-\alpha}\|f'''\|_\infty, \ l = 2,3,4,\ldots,M, \ with\ M = 2m. \tag{11}$$

Similarly we can consider the approximation of right-handed fractional derivative ${}_x^R D_1^\alpha f(x)$ at $x = x_l, l = 0, 1, 2, \ldots, 2m - 2$. Using the same argument as for the approximation of $_0^R D_x^\alpha f(x)$ at $x = x_l$, we can show that, with $j = 0, 1, 2, \ldots, \text{m-1}$,

$$_x^R D_1^\alpha f(x)\Big|_{x=x_{2j}} = \Delta x^{-\alpha} \sum_{l=0}^{M-2j} w_{l,M-2j} f(x_{2j+l}) + \frac{x_{2j}^{-\alpha}}{\Gamma(-\alpha)} R_{2j}(f), \qquad (12)$$

and, with $j = 0, 1, 2, \ldots, m - 2$,

$$_x^R D_1^\alpha f(x)\Big|_{x=x_{2j+1}} = \frac{1}{\Gamma(-\alpha)} \int_{x_{M-1}}^{x_M} (\xi - x_{2j+1})^{-1-\alpha} u(\xi, t_{n+1})\, d\xi$$

$$+ \Delta x^{-\alpha} \sum_{l=0}^{M-(2j+1)-1} w_{l,M-(2j+1)} f(x_{2j+1+l}) + \frac{x_{2j+1}^{-\alpha}}{\Gamma(-\alpha)} R_{2j+1}(f). \qquad (13)$$

Discretizing $u_t(t_{n+1}, x_l)$ by using backward Euler method and discretizing $_0^R D_x^\alpha u(t_{n+1}, x_l)$ and $_x^R D_1^\alpha u(t_{n+1}, x_l)$ by using (4)–(5) and (12)–(13), respectively, we get, with $u_j^n = u(t_n, x_j)$, $f_j^n = f(t_n, x_j)$

$$\Delta t^{-1}\left(u_{2j}^{n+1} - u_{2j}^n\right) = \Delta x^{-\alpha}\left(\sum_{k=0}^{2j} w_{k,2j+1} u_{2j+1-k}^{n+1} + \sum_{k=0}^{M-(2j-1)-1} w_{k,M-(2j-1)} u_{2j-1+k}^{n+1}\right)$$

$$+ f_{2j}^{n+1} + S_{2j}^{n+1} + \sigma_{2j}^{n+1} + \tau_{2j}^{n+1}, \; j = 1, 2, \ldots, m-1, \qquad (14)$$

$$\Delta t^{-1}\left(u_{2j+1}^{n+1} - u_{2j+1}^n\right) = \Delta x^{-\alpha}\left(\sum_{k=0}^{2j+2} w_{k,2j+2} u_{2j+2-k}^{n+1} + \sum_{k=0}^{M-2j} w_{k,M-2j} u_{2j+k}^{n+1}\right)$$

$$+ f_{2j+1}^{n+1} + \sigma_{2j+1}^{n+1} + \tau_{2j+1}^{n+1}, \; j = 0, 1, 2, \ldots, m-1, \qquad (15)$$

where the truncation errors $\tau_l^{n+1} = O(\Delta t + \Delta x^{3-\alpha})$, $l = 1, 2, \ldots, \dot{M} - 1$, [2] and

$$S_{2j}^{n+1} = \frac{1}{\Gamma(-\alpha)} \int_0^{x_1} (x_{2j+1} - \xi)^{-1-\alpha} u(\xi, t_{n+1})\, d\xi$$

$$+ \frac{1}{\Gamma(-\alpha)} \int_{x_{M-1}}^{x_M} (\xi - x_{2j+1})^{-1-\alpha} u(\xi, t_{n+1})\, d\xi, \qquad (16)$$

and

$$\sigma_{2j}^{n+1} = -\left(_0^R D_x^\alpha u(t_{n+1}, x_{2j+1}) - _0^R D_x^\alpha u(t_{n+1}, x_{2j})\right)$$

$$- \left(_x^R D_1^\alpha u(t_{n+1}, x_{2j-1}) - _x^R D_1^\alpha u(t_{n+1}, x_{2j})\right),$$

$$\sigma_{2j+1}^{n+1} = -\left(_0^R D_x^\alpha u(t_{n+1}, x_{2j+2}) - _0^R D_x^\alpha u(t_{n+1}, x_{2j+1})\right)$$

$$- \left(_x^R D_1^\alpha u(t_{n+1}, x_{2j}) - _x^R D_1^\alpha u(t_{n+1}, x_{2j+1})\right).$$

Let $U_{2j}^n \approx u(t_n, x_{2j})$ and $U_{2j+1}^n \approx u(t_n, x_{2j+1})$ denote the approximate solutions of $u(t_n, x_{2j})$ and $u(t_n, x_{2j+1})$, respectively. We define the following implicit shifted numerical method for solving (1)–(3).

$$\Delta t^{-1}\big(U_{2j}^{n+1} - U_{2j}^n\big) = \Delta x^{-\alpha}\Big(\sum_{k=0}^{2j} w_{k,2j+1}U_{2j+1-k}^{n+1} + \sum_{k=0}^{M-(2j-1)-1} w_{k,M-(2j-1)}u_{2j-1+k}^{n+1}\Big)$$

$$+ f_{2j}^{n+1} + Q_{2j}^{n+1}, \; j = 1, 2, \ldots, m-1, \tag{17}$$

$$\Delta t^{-1}\big(U_{2j+1}^{n+1} - U_{2j+1}^n\big) = \Delta x^{-\alpha}\Big(\sum_{k=0}^{2j+2} w_{k,2j+2}U_{2j+2-k}^{n+1} + \sum_{k=0}^{M-2j} w_{k,M-2j}U_{2j+k}^{n+1}\Big)$$

$$+ f_{2j+1}^{n+1}, \; j = 0, 1, 2, \ldots, m-1, \tag{18}$$

where Q_{2j}^{n+1} is the approximation of S_{2j}^{n+1}.

Theorem 1. *Let $1 < \alpha < 2$ and let $u(t_{n+1}, x_l)$ and $U_l^{n+1}, l = 1, 2, \ldots, M-1$ be the solutions of (14)–(15) and (17)–(18), respectively. Assume that $u(t,x)$ satisfies the Lipschitz conditions, with some $\beta > 0$,*

$$\left| {}_0^R D_x^\alpha u(t,x) - {}_0^R D_x^\alpha u(t,y) \right| \leq C_\alpha |x-y|^\beta, \tag{19}$$

$$\left| {}_x^R D_1^\alpha u(t,x) - {}_x^R D_1^\alpha u(t,y) \right| \leq C_\alpha |x-y|^\beta. \tag{20}$$

We have

$$\max_{1 \leq l \leq M-1} |u(t_{n+1}, x_l) - U_l^{n+1}| \leq C(\Delta t + \Delta x^{\min(\beta, 3-\alpha)}).$$

3 Numerical Simulations

In this section, we will consider two examples.

Example 1. Consider [1]

$$u_t(t,x) = {}_0^R D_x^\alpha u(t,x) + f(t,x), \quad 0 < x < 1, \; t > 0 \tag{21}$$

$$u(t,0) = 0, \quad u(t,1) = e^{-t}, \tag{22}$$

$$u(0,x) = x^3, \quad 0 < x < 1, \tag{23}$$

where

$$f(t,x) = -e^{-t}x^3 - e^{-t}\frac{\Gamma(4)}{\Gamma(4-\alpha)}x^{3-\alpha}.$$

The exact solution is $u(t,x) = e^{-t}x^3$. In our numerical simulations, we will choose $\alpha = 1.8$ as in [1].

In Table 1, we will compute numerically the convergence orders of the different numerical methods. By Theorem 1, we have

$$\|e(t)\| \leq C(\Delta t + \Delta x^\gamma), \text{ with } \gamma = \min(3-\alpha, \beta),$$

where $\|e(t)\|$ denotes the L_2 norm of the error at time t. We choose $\Delta t = 2^{-10}$ sufficiently small and the different space step size $h_l = \Delta x = 2^{-l}$, $l = 3, 4, 5, 6, 7$.

For the shifted implicit method proposed in Meerschaert and Tadjeran [6], we have,

$$\|e(t)\| \approx C(\Delta t + \Delta x).$$

In Table 1, we choose $\alpha = 1.8$ and list the convergence orders for the Diethelm's method and Meerschaert and Tadjeran's method, respectively. Here we call the shifted implicit method in this paper as Diethelm's method. We observe that the convergence order of the Diethelm's method is $\gamma \approx 1$. This is as expected because of the Lipschitz assumptions for the exact solutions.

Table 1. The convergence orders in Example 1 for $\alpha = 1.8$

Δt	Δx	Conv. order (Diethelm)	Conv. order (Grunwald)
2^{-10}	2^{-3}		
2^{-10}	2^{-4}	0.8439	0.8390
2^{-10}	2^{-5}	0.9372	0.9220
2^{-10}	2^{-6}	0.9818	0.9618
2^{-10}	2^{-7}	1.0030	0.9816

Example 2. Consider [6]

$$u_t(t, x) = c_+(t, x)\,{}^R_0D^\alpha_x u(t, x) + c_-(t, x)\,{}^R_xD^\alpha_1 u(t, x) + f(t, x), \quad 0 < x < 2,\ t > 0 \tag{24}$$

$$u(t, 0) = u(t, 2) = 0, \tag{25}$$

$$u(0, x) = 4x^2(2 - x)^2, \quad 0 < x < 2, \tag{26}$$

where

$$c_+(t, x) = \Gamma(1.2)x^{1.8} \quad \text{and} \quad c_-(t, x) = \Gamma(1.2)(2 - x)^{1.8}$$

$$f(t, x) = -32e^{-t}\left(x^2 + (2 - x)^2 - 2.5(x^3 + (2 - x)^3) + \frac{25}{22}(x^4 + (2 - x)^4)\right).$$

The exact solution is $u(t, x) = 4e^{-t}x^2(2 - x)^2$. In our numerical simulations, we will choose $\alpha = 1.8$ as in [6].

In Table 2, we will compute numerically the convergence orders of the different numerical methods. As in Example 1, we choose $\alpha = 1.8$ and list the convergence orders for the Diethelm's method and Meerschaert and Tadjeran's method, respectively. We observe that the convergence order of the Diethelm's method is $\gamma \approx 1$. This is as expected because of the Lipschitz assumptions for the exact solutions.

Table 2. The convergence orders in Example 2 for $\alpha = 1.8$

Δt	Δx	Conv. order (Diethelm)	Conv. order (Grunwald)
2^{-10}	2^{-3}		
2^{-10}	2^{-4}	0.9982	1.4690
2^{-10}	2^{-5}	1.0340	1.1946
2^{-10}	2^{-6}	1.0701	1.0674
2^{-10}	2^{-7}	1.0250	1.0248

Acknowledgements. We wish to express our sincere gratitude to Professor Neville. J. Ford for his encouragement, discussions and valuable criticism during the research of this work.

References

1. Choi, H.W., Chung, S.K., Lee, Y.J.: Numerical solutions for space fractional dispersion equations with nonlinear source terms. Bull. Korean Math. Soc. **47**, 1225–1234 (2010)
2. Diethelm, K.: An algorithm for the numerical solution of differential equations of fractional order. Electron. Trans. Numer. Anal. **5**, 1–6 (1997)
3. Ford, N.J., Rodrigues, M.M., Xiao, J., Yan, Y.: Numerical analysis of a two-parameter fractional telegraph equation. J. Comput. Appl. Math. **249**, 95–106 (2013)
4. Ford, N.J., Xiao, J., Yan, Y.: A finite element method for time fractional partial differential equations. Fract. Calc. Appl. Anal. **14**, 454–474 (2011)
5. Ford, N.J., Xiao, J., Yan, Y.: Stability of a numerical method for space-time-fractional telegraph equation. Comput. Methods Appl. Math. **12**, 1–16 (2012)
6. Meerschaert, M.M., Tadjeran, C.: Finite difference approximations for two-sided space-fractional partial differential equations. Appl. Numer. Math. **56**, 80–90 (2006)
7. Yan, Y., Pal, K., Ford, N.J.: Higher order numerical methods for solving frational differential equations. BIT Numer. Math. **54**(2), 555–584 (2014). doi:10.1007/s10543-013-0443-3
8. Dimitrov, Y.: Numerical approximations for fractional differential equations. J. Frac. Calc. Appl. **5**(3S), 1–45 (2014)
9. Ervin, V.J., Roop, J.P.: Variational formulation for the stationary fractional advection dispersion equation. Numer. Methods Partial Differ. Equ. **22**, 558–576 (2006)
10. Ervin, V.J., Roop, J.P.: Variational solution of fractional advection dispersion equations on bounded domains in $\mathbb{R}^d$. Numer. Methods Partial Differ. Equ. **23**, 256–281 (2007)
11. Ervin, V.J., Heuer, N., Roop, J.P.: Numerical approximation of a time dependent nonlinear, space-fractional diffusion equation. SIAM J. Numer. Anal. **45**, 572–591 (2007)
12. Li, C.P., Zeng, F.: Finite difference methods for fractional differential equations. Int. J. Bifurc. Chaos **22**, 1230014 (2012)
13. Liu, F., Anh, V., Turner, I.: Numerical solution of space fractional Fokker-Planck equation. J. Comp. Appl. Math. **166**, 209–219 (2004)

14. Meerschaert, M.M., Tadjeran, C.: Finite difference approximations for fractional advection-dispersion flow equations. J. Comput. Appl. Math. **172**, 65–77 (2004)
15. Podlubny, I.: Matrix approach to discrete fractional calculus. Fract. Calc. Appl. Anal. **3**, 359–386 (2000)
16. Roop, J.P.: Computational aspects of FEM approximation of fractional advection dispersion equations on bouinded domains in $\mathbb{R}^2$. J. Comput. Appl. Math. **193**, 243–268 (2006)
17. Shen, S., Liu, F., Anh, V., Turner, I.: The fundamental solution and numerical solution of the Riesz fractional advection-dispersion equation. IMA J. Appl. Math. **73**, 850–872 (2008)
18. Sousa, E.: Finite difference approximations for a fractional advection diffusion problem. J. Comput. Phys. **228**, 4038–4054 (2009)
19. Tadjeran, C., Meerschaert, M.M., Scheffler, H.: A second-order acurate numerical approximation for the fractional diffusion equation. J. Comput. Phys. **213**, 205–213 (2006)
20. Yang, Q., Liu, F., Turner, I.: Numerical methods for fractional partial differential equations with Riesz space fractional derivatives. Appl. Math. Model. **34**, 200–218 (2010)

Spline Collocation for Fractional Integro-Differential Equations

Arvet Pedas$^{(\boxtimes)}$, Enn Tamme, and Mikk Vikerpuur

Institute of Mathematics, University of Tartu, J. Liivi 2, 50409 Tartu, Estonia
{Arvet.Pedas,Enn.Tamme,azzo}@ut.ee

Abstract. We consider a class of boundary value problems for fractional integro-differential equations. Using an integral equation reformulation of the boundary value problem, we first study the regularity of the exact solution. Based on the obtained regularity properties and spline collocation techniques, the numerical solution of the boundary value problem by suitable non-polynomial approximations is discussed. Optimal global convergence estimates are derived and a super-convergence result for a special choice of grid and collocation parameters is given. A numerical illustration is also presented.

1 Introduction

In this paper we consider the numerical solution of a class of boundary value problems for linear fractional integro-differential equations of the form

$$(D_*^\alpha y)(t) + h(t)y(t) + \int_0^t K(t,s)y(s)ds = f(t),\ 0 \le t \le b,\ 0 < \alpha < 1, \quad (1.1)$$

$$\gamma_0 y(0) + \gamma_1 y(b_1) = \gamma,\ 0 < b_1 \le b,\ \gamma_0, \gamma_1, \gamma \in \mathbb{R} := (-\infty, \infty), \quad (1.2)$$

where $D_*^\alpha y$ is the Caputo fractional derivative of y and h, f, K are some given continuous functions: $h, f \in C[0,b]$, $K \in C(\Delta)$, $\Delta = \{(t,s) : 0 \le s \le t \le b\}$. The Caputo differential operator D_*^α of order $\alpha \in (0,1)$ is defined by (see, e.g., [1])

$$(D_*^\alpha y)(t) := (D^\alpha[y - y(0)])(t),\ t > 0.$$

Here $D^\alpha y$ is the Riemann–Liouville fractional derivative of y:

$$(D^\alpha y)(t) := \frac{d}{dt}(J^{1-\alpha} y)(t),\quad 0 < \alpha < 1,\quad t > 0,$$

with J^β, the Riemann–Liouville integral operator, defined by the formula

$$(J^\beta y)(t) := \frac{1}{\Gamma(\beta)} \int_0^t (t-s)^{\beta-1} y(s)\,ds,\quad t > 0,\quad \beta > 0. \quad (1.3)$$

where Γ is the Euler gamma function.

© Springer International Publishing Switzerland 2015
I. Dimov et al. (Eds.): FDM 2014, LNCS 9045, pp. 315–322, 2015.
DOI: 10.1007/978-3-319-20239-6_34

It is well known (see, e.g., [2]) that J^β, $\beta > 0$, is linear, bounded and compact as an operator from $L^\infty(0, b)$ into $C[0, b]$, and we have for any $y \in L^\infty(0, b)$ that (see, e.g. [3])

$$J^\beta y \in C[0, b], \quad (J^\beta y)(0) = 0, \quad \beta > 0, \tag{1.4}$$

$$D^\delta J^\beta y = D_*^\delta J^\beta y = J^{\beta - \delta} y, \quad 0 < \delta \leq \beta. \tag{1.5}$$

Fractional differential equations arise in various areas of science and engineering. In the last few decades theory and numerical analysis of fractional differential equations have received an increasing attention (see, e.g. [1,3–5]). Some recent results about the numerical solution of fractional differential equations can be found in [1,6–12].

In the present paper, the numerical solution of (1.1)-(1.2) by piecewise polynomial collocation techniques is considered. We use an integral equation reformulation of the problem and special non-uniform grids reflecting the possible singular behavior of the exact solution. Our aim is to study the attainable order of the proposed algorithms in a situation where the higher order (usual) derivatives of $h(t)$ and $f(t)$ may be unbounded at $t = 0$. Our approach is based on some ideas and results of [10]. In particular, the case where (1.1)-(1.2) is an initial value problem ($\gamma_1 = 0$) or a terminal (boundary) value problem ($\gamma_0 = 0$, see [7,8]) is under consideration.

2   Smoothness of the Solution

In order to characterize the behavior of higher order derivatives of a solution of equation (1.1), we introduce a weighted space of smooth functions $C^{q,\nu}(0, b]$ (cf., e.g., [2,13]). For given $q \in \mathbb{N}$ and $-\infty < \nu < 1$, by $C^{q,\nu}(0, b]$ we denote the set of continuous functions $y : [0, b] \to \mathbb{R}$ which are q times continuously differentiable in $(0, b]$ and such that for all $t \in (0, b]$ and $i = 1, \ldots, q$ the following estimates hold:

$$\left| y^{(i)}(t) \right| \leq c \begin{cases} 1 & \text{if } i < 1 - \nu, \\ 1 + |\log t| & \text{if } i = 1 - \nu, \\ t^{1-\nu-i} & \text{if } i > 1 - \nu. \end{cases}$$

Here $c = c(y)$ is a positive constant. Clearly,

$$C^q[0, b] \subset C^{q,\nu}(0, b] \subset C^{m,\mu}(0, b] \subset C[0, b], \quad q \geq m \geq 1, \quad \nu \leq \mu < 1. \tag{2.1}$$

Note that a function of the form $y(t) = g_1(t) t^\mu + g_2(t)$ is included in $C^{q,\nu}(0, b]$ if $\mu \geq 1 - \nu > 0$ and $g_j \in C^q[0, b]$, $j = 1, 2$.

In what follows we use an integral equation reformulation of (1.1)-(1.2). Let $y \in C[0, b]$ be such that $D_*^\alpha y \in C[0, b]$. Introduce a new unknown function $z := D_*^\alpha y$. Then (see [1,3])

$$y(t) = (J^\alpha z)(t) + c, \tag{2.2}$$

where c is an arbitrary constant. The function (2.2) satisfies the boundary conditions (1.2) if and only if (see (1.4))

$$c(\gamma_0 + \gamma_1) = \gamma - \gamma_1 (J^\alpha z)(b_1).$$

In the sequel we assume that $\gamma_0 + \gamma_1 \neq 0$. Therefore

$$c = \frac{\gamma}{\gamma_0 + \gamma_1} - \frac{\gamma_1}{\gamma_0 + \gamma_1}(J^\alpha z)(b_1). \tag{2.3}$$

Thus, the function (2.2) satisfies the conditions (1.2) if and only if

$$y(t) = (J^\alpha z)(t) + \frac{\gamma}{\gamma_0 + \gamma_1} - \frac{\gamma_1}{\gamma_0 + \gamma_1}(J^\alpha z)(b_1), \quad 0 \leq t \leq b. \tag{2.4}$$

Substituting (2.4) into (1.1) and using (1.5), we obtain for z an operator equation of the form

$$z = Tz + g, \tag{2.5}$$

with an operator T, defined by formula

$$(Tz)(t) = -\frac{h(t)}{\Gamma(\alpha)} \int_0^t (t-s)^{\alpha-1} z(s)ds - \frac{1}{\Gamma(\alpha)} \int_0^t K(t,s) \int_0^s (s-\tau)^{\alpha-1} z(\tau)d\tau ds$$
$$+ \frac{\gamma_1}{\Gamma(\alpha)(\gamma_0 + \gamma_1)} \left[h(t) + \int_0^t K(t,s)ds \right] \int_0^{b_1} (b_1-s)^{\alpha-1} z(s)ds \tag{2.6}$$

and

$$g(t) = f(t) - \frac{\gamma}{\gamma_0 + \gamma_1} \left(h(t) + \int_0^t K(t,s)ds \right), \quad 0 \leq t \leq b. \tag{2.7}$$

We observe that equation (2.5) is a linear weakly singular Fredholm integral equation of the second kind with respect to z.

The existence and regularity of a solution to (1.1)-(1.2) is described by the following lemma which can be proved similarly to Theorem 2.1 in [10].

Lemma 1. *Assume that $K \in C^q(\Delta)$ and $h, f \in C^{q,\mu}(0,b]$, where $q \in \mathbb{N}$ and $-\infty < \mu < 1$. Moreover, assume that $\gamma_0 + \gamma_1 \neq 0$ and the boundary value problem (1.1)-(1.2) with $f = 0$ and $\gamma = 0$ has in $C[0,b]$ only the trivial solution $y = 0$.*

Then problem (1.1)-(1.2) possesses a unique solution $y \in C[0,b]$ such that $D_*^\alpha y \in C^{q,\nu}(0,b]$, where $0 < \alpha < 1$ and

$$\nu := \max\{\mu, 1 - \alpha\}.$$

3   Numerical Method

Let $N \in \mathbb{N}$ and let $\Pi_N := \{t_0, \ldots, t_N\}$ be a partition (a graded grid) of the interval $[0,b]$ with the grid points

$$t_j := b \left(\frac{j}{N} \right)^r, \quad j = 0, 1, \ldots, N, \tag{3.1}$$

where the grading exponent $r \in \mathbb{R}$, $r \geq 1$. If $r = 1$, then the grid points (3.1) are distributed uniformly; for $r > 1$ the points (3.1) are more densely clustered near the left endpoint of the interval $[0, b]$.

For given integer $k \geq 0$ by $S_k^{(-1)}(\Pi_N)$ is denoted the standard space of piecewise polynomial functions :

$$S_k^{(-1)}(\Pi_N) := \left\{ v : v\big|_{(t_{j-1},t_j)} \in \pi_k, \ j = 1, \ldots, N \right\}.$$

Here $v\big|_{(t_{j-1},t_j)}$ is the restriction of $v : [0, b] \to \mathbb{R}$ onto the subinterval (t_{j-1}, t_j) $\subset [0, b]$ and π_k denotes the set of polynomials of degree not exceeding k. Note that the elements of $S_k^{(-1)}(\Pi_N)$ may have jump discontinuities at the interior points $t_1, \ldots, t_{N-1}$ of the grid Π_N.

In every interval $[t_{j-1}, t_j]$, $j = 1, \ldots, N$, we define $m \in \mathbb{N}$ collocation points $t_{j1}, \ldots, t_{jm}$ by formula

$$t_{jk} := t_{j-1} + \eta_k(t_j - t_{j-1}), \quad k = 1, \ldots, m, \ j = 1, \ldots, N, \tag{3.2}$$

where $\eta_1 \ldots, \eta_m$ are some fixed (collocation) parameters which do not depend on j and N and satisfy

$$0 \leq \eta_1 < \eta_2 < \ldots < \eta_m \leq 1. \tag{3.3}$$

We look for an approximate solution y_N to (1.1)-(1.2) in the form (cf. (2.4))

$$y_N(t) = (J^\alpha z_N)(t) + \frac{\gamma}{\gamma_0 + \gamma_1} - \frac{\gamma_1}{\gamma_0 + \gamma_1}(J^\alpha z_N)(b_1), \quad 0 \leq t \leq b, \tag{3.4}$$

where $z_N \in S_{m-1}^{(-1)}(\Pi_N)$ $(m, N \in \mathbb{N})$ is determined by the following collocation conditions:

$$z_N(t_{jk}) = (T z_N)(t_{jk}) + g(t_{jk}), \quad k = 1, \ldots, m, \ j = 1, \ldots, N. \tag{3.5}$$

Here T, g and t_{jk} are defined by (2.6), (2.7) and (3.2), respectively. If $\eta_1 = 0$, then by $z_N(t_{j1})$ we denote the right limit $\lim_{t \to t_{j-1}, t > t_{j-1}} z_N(t)$. If $\eta_m = 1$, then $z_N(t_{jm})$ denotes the left limit $\lim_{t \to t_j, t < t_j} z_N(t)$. Conditions (3.5) have an operator equation representation

$$z_N = \mathcal{P}_N T z_N + \mathcal{P}_N g \tag{3.6}$$

with an interpolation operator $\mathcal{P}_N = \mathcal{P}_{N,m} : C[0, T] \to S_{m-1}^{(-1)}(\Pi_N)$ defined for any $v \in C[0, b]$ by the following conditions:

$$\mathcal{P}_N v \in S_{m-1}^{(-1)}(\Pi_N), \ (\mathcal{P}_N v)(t_{jk}) = v(t_{jk}), \ k = 1, \ldots, m, \ j = 1, \ldots, N. \tag{3.7}$$

The collocation conditions (3.5) form a system of equations whose exact form is determined by the choice of a basis in $S_{m-1}^{(-1)}(\Pi_N)$. If $\eta_1 > 0$ or $\eta_m < 1$ then we can use the Lagrange fundamental polynomial representation:

$$z_N(t) = \sum_{\lambda=1}^{N} \sum_{\mu=1}^{m} c_{\lambda\mu} \varphi_{\lambda\mu}(t), \quad t \in [0, b], \tag{3.8}$$

where $\varphi_{\lambda\mu}(t) := 0$ for $t \notin [t_{\lambda-1}, t_\lambda]$ and

$$\varphi_{\lambda\mu}(t) := \prod_{i=1, i\neq\mu}^{m} \frac{t - t_{\lambda i}}{t_{\lambda\mu} - t_{\lambda i}} \quad \text{for} \quad t \in [t_{\lambda-1}, t_\lambda], \ \mu = 1, \ldots, m, \ \lambda = 1, \ldots, N.$$

Then $z_N \in S_{m-1}^{(-1)}(\Pi_N)$ and $z_N(t_{jk}) = c_{jk}$, $k = 1, \ldots, m$, $j = 1, \ldots, N$. Searching the solution of (3.5) in the form (3.8), we obtain a system of linear algebraic equations with respect to the coefficients $c_{jk} = z_N(t_{jk})$:

$$c_{jk} = \sum_{\lambda=1}^{N} \sum_{\mu=1}^{m} (T\varphi_{\lambda\mu})(t_{jk})c_{\lambda\mu} + g(t_{jk}), \quad k = 1, \ldots, m, \ j = 1, \ldots, N. \quad (3.9)$$

Note that this algorithm can be used also in the case if in (3.3) $\eta_1 = 0$ and $\eta_m = 1$. In this case we have $t_{jm} = t_{j+1,1} = t_j$, $c_{jm} = c_{j+1,1} = z_N(t_j)$ ($j = 1, \ldots, N - 1$), and hence in the system (3.9) there are $(m-1)N + 1$ equations and unknowns.

4   Convergence Estimates

In this section we formulate two theorems about the convergence and convergence order of the proposed algorithms.

Theorem 1. *Let $m \in \mathbb{N}$ and assume that the collocation points (3.2) with grid points (3.1) and arbitrary parameters $\eta_1, \ldots, \eta_m$ satisfying (3.3) are used. Assume that $h, f \in C[0, b]$ and $K \in C(\Delta)$. Moreover, assume that $\gamma_0 + \gamma_1 \neq 0$ and the problem (1.1)-(1.2) with $f = 0$ and $\gamma = 0$ has in $C[0, b]$ only the trivial solution $y = 0$.*

Then (1.1)-(1.2) has a unique solution $y \in C[0, b]$ such that $D_*^\alpha y \in C[0, b]$. Moreover, there exists an integer N_0 such that for all $N \geq N_0$ equation (3.6) possesses a unique solution $z_N \in S_{m-1}^{(-1)}(\Pi_N)$ and

$$\|y - y_N\|_\infty \to 0 \quad \text{as} \quad N \to \infty \quad (4.1)$$

where y_N is defined by (3.4).

If, in addition, $K \in C^m(\Delta)$ and $h, f \in C^{m,\mu}(0, b]$ with $-\infty < \mu < 1$, then for all $N \geq N_0$ and $r \geq 1$ (given by (3.1)) the following error estimate holds:

$$\|y - y_N\|_\infty \leq c \begin{cases} N^{-r(1-\nu)} & \text{for} \quad 1 \leq r < \frac{m}{1-\nu}, \\ N^{-m} & \text{for} \quad r \geq \frac{m}{1-\nu}. \end{cases} \quad (4.2)$$

Here c is a constant which is independent of N, $\nu = \max\{\mu, 1 - \alpha\}$, $0 < \alpha < 1$ and

$$\|v\|_\infty := \sup_{0 < t < b} |v(t)|, \quad v \in L^\infty(0, b).$$

It follows from Theorem 1 that in the case of sufficiently smooth h, f and K, using sufficiently large values of the grid parameter r, for method (3.4), (3.6) by every choice of collocation parameters $0 \leq \eta_1 < \cdots < \eta_m \leq 1$ a convergence of order $O(N^{-m})$ can be expected. The following result shows that by a careful choice of parameters $\eta_1, \ldots, \eta_m$ it is possible to establish a faster convergence of this method.

Theorem 2. *Let the following conditions be fulfilled:*
(i) $\mathcal{P}_N = \mathcal{P}_{N,m}$ $(N, m \in \mathbb{N})$ *is defined by* (3.7) *where the interpolation nodes* (3.2) *with grid points* (3.1) *and parameters* (3.3) *are used;*
(ii) *the assumptions of Lemma 1 hold with* $q := m + 1$;
(iii) *the quadrature approximation*

$$\int_0^1 F(x)\, dx \approx \sum_{k=1}^m w_k\, F(\eta_k), \tag{4.3}$$

with the knots $\{\eta_k\}$ *satisfying* (3.3) *and appropriate weights* $\{w_k\}$ *is exact for all polynomials of degree* m.

   Then (1.1)-(1.2) *has a unique solution* $y \in C[0, b]$ *such that* $D_*^\alpha y \in C^{q,\nu}(0, b]$. *There exists an integer* N_0 *such that, for* $N \geq N_0$, *equation* (3.6) *possesses a unique solution* $z_N \in S_{m-1}^{(-1)}(\Pi_N)$, *determining by* (3.4) *a unique approximation* y_N *to* y, *the solution of* (1.1)-(1.2), *and the following error estimate holds:*

$$\|y - y_N\|_\infty \leq c \begin{cases} N^{-r(1+\alpha-\nu)} & for \quad 1 \leq r < \frac{m+\alpha}{1+\alpha-\nu}, \\ N^{-m-\alpha} & for \quad r \geq \frac{m+\alpha}{1+\alpha-\nu}. \end{cases} \tag{4.4}$$

Here $0 < \alpha < 1$, $\nu = \max\{\mu, 1 - \alpha\}$, $r \in [1, \infty)$ *is the grading exponent of the grid (see* (3.1)) *and* c *is a positive constant not depending on* N.

   The proofs of Theorems 1 and 2 are based on Lemma 1 and are similar to the corresponding proofs of Theorems 4.1 and 4.2 in [10].

5 Numerical Illustration

We consider the following boundary value problem:

$$(D_*^{\frac{1}{2}} y)(t) + h(t)y(t) + \int_0^t K(t, s)y(s)ds = f(t), \quad y(0) + y(1) = 2, \quad 0 \leq t \leq 2, \tag{5.1}$$

where $K(t, s) := 1$ for $0 \leq s \leq t \leq 2$ and

$$h(t) := t^{\frac{1}{2}}, \quad f(t) := \frac{5\Gamma(\frac{7}{4})}{2\Gamma(\frac{9}{4})} t^{\frac{1}{4}} + 2t^{\frac{5}{4}} + \frac{8}{7}t^{\frac{7}{4}}, \quad 0 \leq t \leq 2.$$

This is a special problem of (1.1)-(1.2) with $\alpha = \frac{1}{2}, b = 2, b_1 = 1, \gamma_0 = \gamma_1 = 1$ and $\gamma = 2$. Clearly, $h, f \in C^{q,\mu}(0, 2]$ with $\mu = \frac{3}{4}$ and arbitrary $q \in \mathbb{N}$. To solve (5.1) by (3.4)-(3.6) we set $z := D_*^{\frac{1}{2}} y$. For z we have equation (2.5) with T and g given by (2.6) and (2.7), respectively. Approximations $z_N \in S_{m-1}^{(-1)}(\Pi_N)$ for $m = 2$ and $N \in \mathbb{N}$ to the solution z of equation (2.5) on the interval $[0,2]$ are found by (3.5) using $m = 2$ and (3.2) with $\eta_1 = (3 - \sqrt{3})/6$, $\eta_2 = 1 - \eta_1$, the knots of the Gaussian quadrature formula (4.3). Actually, $z_N(t_{jk}) = c_{jk}$ ($k = 1, 2$, $j = 1, \ldots, N$) and $z_N(t)$ for $t \in [0, 2]$ are determined by (3.9) and (3.8), respectively. After that the approximate solution y_N for the boundary value problem (5.1) has been found by formula (3.4).

Table 1. Numerical results for the problem (5.1).

N	$r = 1$		$r = 2$		$r = 10/3$		$r = 5$	
	ε_N	ϱ_N	ε_N	ϱ_N	ε_N	ϱ_N	ε_N	ϱ_N
16	$1.81 \cdot 10^{-2}$	1.64	$2.28 \cdot 10^{-3}$	3.04	$4.99 \cdot 10^{-4}$	6.54	$8.75 \cdot 10^{-4}$	5.45
32	$1.08 \cdot 10^{-2}$	1.67	$8.23 \cdot 10^{-4}$	2.77	$7.14 \cdot 10^{-5}$	5.57	$1.50 \cdot 10^{-4}$	5.82
64	$6.45 \cdot 10^{-3}$	1.68	$2.85 \cdot 10^{-4}$	2.89	$1.23 \cdot 10^{-5}$	5.84	$2.53 \cdot 10^{-5}$	5.95
128	$3.85 \cdot 10^{-3}$	1.68	$1.01 \cdot 10^{-4}$	2.81	$1.97 \cdot 10^{-6}$	6.22	$4.31 \cdot 10^{-6}$	5.87
256	$2.29 \cdot 10^{-3}$	1.68	$3.57 \cdot 10^{-5}$	2.85	$2.83 \cdot 10^{-7}$	6.97	$7.51 \cdot 10^{-7}$	5.73
512	$1.36 \cdot 10^{-3}$	1.68	$1.26 \cdot 10^{-5}$	2.83	$6.53 \cdot 10^{-8}$	4.34	$1.35 \cdot 10^{-7}$	5.55
1024	$8.09 \cdot 10^{-4}$	1.68	$4.45 \cdot 10^{-6}$	2.83	$1.54 \cdot 10^{-8}$	4.24	$2.37 \cdot 10^{-8}$	5.71
2048	$4.81 \cdot 10^{-4}$	1.68	$1.58 \cdot 10^{-6}$	2.83	$2.51 \cdot 10^{-9}$	6.13	$4.23 \cdot 10^{-9}$	5.61
		1.68		2.83		5.66		5.66

In Table 1 some results of numerical experiments for different values of the parameters N and r are presented. The errors ε_N in Table 1 are calculated as follows:

$$\varepsilon_N := \max_{j=1,\ldots,N} \max_{k=0,\ldots,10} |y(\tau_{jk}) - y_N(\tau_{jk})|, \tag{5.2}$$

where $\tau_{jk} := t_{j-1} + k(t_j - t_{j-1})/10$, $k = 0, \ldots, 10$, $j = 1, \ldots, N$ (the grid points t_j and collocation points t_{jk} are determined by (3.1) and (3.2), respectively). In (5.2) we have taken into account that the exact solution of (5.1) is

$$y(t) = 2 t^{\frac{3}{4}}, \quad t \in [0, 2],$$

and thus

$$z(t) = (D_*^{\frac{1}{2}} y)(t) = \frac{5\Gamma(\frac{7}{4})}{2\Gamma(\frac{9}{4})} t^{\frac{1}{4}}, \quad t \in [0, 2].$$

The ratios

$$\varrho_N := \frac{\varepsilon_{N/2}}{\varepsilon_N},$$

characterizing the observed convergence rate, are also presented.

Since $\alpha = \frac{1}{2}$, $\mu = \frac{3}{4}$ and $\nu = \max\{\mu, 1 - \alpha\} = \frac{3}{4}$ we obtain from Theorem 2 (see (4.4)) that, for sufficiently large N,

$$\varepsilon_N \leq c \begin{cases} N^{-0.75r} & \text{if } 1 \leq r < \frac{10}{3}, \\ N^{-2.5} & \text{if } r \geq \frac{10}{3}. \end{cases} \qquad (5.3)$$

Due to (5.3) the ratios ϱ_N for $r = 1$, $r = 2$ and $r \geq \frac{10}{3}$ ought to be approximatively $2^{0.75} \approx 1.68$, $2^{1.5} \approx 2.83$ and $2^{2.5} \approx 5.66$, respectively. These values are given in the last row of Table 1. As we can see from Table 1 the estimate (4.4) expresses well enough the actual rate of convergence of y_N to y.

Acknowledgements. This work was supported by Estonian Science Foundation Grant No. 9104 and Estonian Institutional Research Project IUT 20-57.

References

1. Diethelm, K.: The Analysis of Fractional Differential Equations. Lecture Notes in Mathematics, vol. 2004. Springer, Berlin (2010)
2. Brunner, H., Pedas, A., Vainikko, G.: Piecewise polynomial collocation methods for linear Volterra integro-differential equations with weakly singular kernels. SIAM J. Numer. Anal. **39**, 957–982 (2001)
3. Kilbas, A.A., Srivastava, H.M., Trujillo, J.J.: Theory and Applications of Fractional Differential Equations. North Holland Mathematics Studies, vol. 204. Elsevier, Amsterdam (2006)
4. Podlubny, I.: Fractional Differential Equations. Academic Press, San Diego (1999)
5. Agarwal, R.P., Benchohra, M., Hamani, S.: A survey of existence results for boundary value problems of nonlinear fractional differential equations and inclusions. Acta. Appl. Math. **109**, 973–1033 (2010)
6. Pedas, A., Tamme, E.: Spline collocation methods for linear multi-term fractional differential equations. J. Comput. Appl. Math. **236**, 167–176 (2011)
7. Ford, N.J., Morgado, M.L.: Fractional boundary value problems: analysis and numerical methods. Fract. Calc. Appl. Anal. **14**, 554–567 (2011)
8. Ford, N.J., Morgado, M.L., Rebelo, M.: High order numerical methods for fractional terminal value problems. Comput. Methods Appl. Math. **14**, 55–70 (2014)
9. Doha, E.H., Bhrawy, A.H., Ezz-Eldien, S.S.: A Chebyshev spectral method based on operational matrix for initial and boundary value problems of fractional order. Comput. Math. Appl. **62**, 2364–2373 (2011)
10. Pedas, A., Tamme, E.: Piecewise polynomial collocation for linear boundary value problems of fractional differential equations. J. Comput. Appl. Math. **236**, 3349–3359 (2012)
11. Pedas, A., Tamme, E.: Numerical solution of nonlinear fractional differential equations by spline collocation methods. J. Comput. Appl. Math. **255**, 216–230 (2014)
12. Ma, X., Huang, C.: Spectral collocation method for linear fractional integro-differential equations. Appl. Math. Model. **38**, 1434–1448 (2014)
13. Vainikko, G.: Multidimensional Weakly Singular Integral Equations. Lecture Notes in Mathematics, vol. 1549. Springer, Berlin (1993)

Rational Spectral Collocation Method for Pricing American Vanilla and Butterfly Spread Options

Edson Pindza, Kailash C. Patidar$^{(\boxtimes)}$, and Edgard Ngounda

Department of Mathematics and Applied Mathematics,
University of the Western Cape, Private Bag X17, Bellville 7535, South Africa
kpatidar@uwc.ac.za

Abstract. We present a rational spectral collocation method for pricing American vanilla and butterfly spread options. Due to the early exercise possibilities, free boundary conditions are associated with both of these PDEs. The problem is first reformulated as a variational inequality. Then, by adding a penalty term, the resulting variational inequality is transformed into a nonlinear advection-diffusion-reaction equation on fixed boundaries. This nonlinear PDE is discretised in asset (space) direction by means of rational interpolation using suitable barycentric weights and transformed Chebyshev points. This gives a system of stiff nonlinear ODEs which is then integrated using an implicit fourth-order Lobatto time-integration method. We carried out extensive comparisons with other results obtained by using some existing methods found in literature and observed that our approach is very competitive.

Keywords: American options · Butterfly spread options · Rational spectral collocation method

1 Introduction

Among a huge variety of financial derivative securities traded in exchange markets such as call or put on dividend paying stocks, foreign currency options, callable bonds and others, American options are the most traded ones [8]. The fact that the American options give their owner the right (but not the obligation) to exercise or exit the contract early makes these options the most attractive to investors as compared to other derivatives of similar types. However they are more difficult to price than the European options which can only be exercised at expiration. In this paper, we consider a financial market model $\mathcal{M} = \left(\Omega, \mathcal{F}, \mathbb{P}, (\mathcal{F}_\tau)_{\tau \geq 0}, (S_\tau)_{\tau \geq 0} \right)$ where Ω is the set of all possible outcomes of the experiment known as the sample space, $\mathcal{F}$ is the set of all events, i.e., permissible combinations of outcomes, $\mathbb{P}$ is a map $\mathcal{F} \longrightarrow [0, 1]$ which assigns a probability to each event, $\mathcal{F}_\tau$ is a natural filtration and S_τ a risky underlying asset price process. The triplet $(\Omega, \mathcal{F}, \mathbb{P})$ is defined as a probability space.

© Springer International Publishing Switzerland 2015
I. Dimov et al. (Eds.): FDM 2014, LNCS 9045, pp. 323–331, 2015.
DOI: 10.1007/978-3-319-20239-6_35

Let B_τ be a $\mathbb{P}$-Brownian motion, $\sigma > 0$ the volatility of the underlying asset, $r > 0$ a risk-free interest rate and $\delta \geq 0$ a continuous dividend yield. Without loss of generality, we assume that both the risk-free interest rate and the dividend yield are constants. Then under the equivalent martingale measure $\mathbb{Q}$, the dynamics of the Black-Scholes model satisfies the stochastic differential equation

$$dS_\tau = (r - \delta)S_\tau d\tau + \sigma S_\tau dB_\tau. \tag{1}$$

Using Ito's formula and appropriate boundary and the final conditions along with $t = T - \tau$, we obtain the following free boundary-values problem for American put options

$$\left.\begin{array}{l} V_t = L_{BS}V, \ \ S_f(t) \leq S < \infty, \ 0 \leq t \leq T, \\[2mm] V(S,0) = \max(E - S, 0), \\[2mm] \lim_{S \to \infty} V(S,t) = 0, \\[2mm] V(S_f(t),t) = E - S_f(t), \\[2mm] \dfrac{\partial V}{\partial S}(S_f(t),t) = -1, \end{array}\right\} \tag{2}$$

where T the expiry time, and

$$L_{BS} \equiv \frac{1}{2}\sigma^2 S^2 \frac{\partial^2}{\partial S^2} + (r - \delta)S\frac{\partial}{\partial S} - r.$$

The problem (2) can be formulated as a linear complementarity problem [9]:

$$\left.\begin{array}{l} V_t - L_{BS}V \geq 0, \\[2mm] V(S,t) - V^*(S) \geq 0, \\[2mm] (V_t - L_{BS}V)(V(S,t) - V^*(S)) = 0, \\[2mm] V(S,0) = \max(E - S, 0), \\[2mm] \lim_{S \to \infty} V(S,t) = 0, \end{array}\right\} \tag{3}$$

where V is the value of the option and V^* denotes its exercise value. In next section, we proposed a spectral discretisation based on a penalty approach to solve the American vanilla and butterfly spread option problems.

The rest of this paper is organised as follows. In Sect. 2, we describe the spacial approximations using rational spectral collocation method. Discretisation of associated nonlinear PDEs is discussed in Sect. 3. Finally, in Sect. 4, we present and discuss some comparative numerical results.

2   Spectral Collocation Approximation

We first transform the variational inequality (3) into a nonlinear partial differential equation on a fixed domain by adding a penalty term. This gives

$$
\left.
\begin{aligned}
&\frac{\partial V_\epsilon}{\partial \tau} = L_{BS}V_\epsilon + \frac{1}{\epsilon}[V_\epsilon(S,t) - V_\epsilon^*(S)]^+, \quad S_m \le S \le S_M, \ 0 \le t \le T, \\[2mm]
&V_\epsilon(S,0) = V_\epsilon^*(S) = \max(E - S, 0), \\[2mm]
&(V_t - L_{BS}V)(V(S,t) - V^*(S)) = 0, \\[2mm]
&V_\epsilon(S_m,t) = E, \ V_\epsilon(S_M,t) = 0,
\end{aligned}
\right\}
\tag{4}
$$

where $0 < \epsilon \ll 1$ is the penalty constant and $[V_\epsilon(S,t) - V_\epsilon^*(S)]^+ = \max(V_\epsilon(S,t) - V_\epsilon^*(S), 0)$ is the penalty term. To the best of our knowledge, most of the schemes used to value American options using the penalty term are finite difference, finite element and finite volume methods. However, our method in this paper is based on a rational spectral collocation (RSC) method which is described as follows. In RSC method (in barycentric form) we approximate the unknown solution $u(x,t)$ at the nodes x_j as

$$
\widetilde{u}(x) = \sum_{j=0}^{N} \widetilde{u}(x_j) L_j^{(\omega)}(x) \in \mathcal{R}_N^{(\omega)},
\tag{5}
$$

where $x_0, x_1 \ldots x_N$ are the collocation points and the set $\omega = [\omega_0, \ldots, \omega_N]^T$, $\omega_j \ne 0$ consists of the barycentric weights $\omega_j = 1/\prod_{j \ne i}(x_j - x_i)$, to which we associate the linear space, denoted by $\mathcal{R}_N^{(\omega)}$ and spanned by the function

$$
L_j^{(\omega)}(x) = \frac{\omega_j}{x - x_j} \Big/ \sum_{k=0}^{N} \frac{\omega_k}{x - x_k}, \quad j = 0, \ldots, N,
\tag{6}
$$

which satisfies the Langrange property $L_j^{(\omega)}(x_i) = \delta_{ij}$.

The next step is to construct conformal mappings to obtain high resolution of non-smooth initial conditions. We may note that in the most common pseudo-spectral Chebyshev method, the interpolation points in the interval $[-1,1]$ are the Chebyshev-Gauss-Lobatto (CGL) collocation points $y_j = \cos(j\pi/N)$ for $j = 0, \ldots, N$. The CGL points are clustered near the boundaries of $[-1,1]$, which in present case need to be accumulated in the vicinity of the region where the solution changes rapidly. To do so, we use the conformal map g derived in [1]:

$$
x = g(y) = \beta + \frac{1}{\alpha} \tan[\lambda(y - \mu)], \quad \lambda = \frac{\gamma + \delta}{2}, \quad \mu = \frac{\gamma - \delta}{\gamma + \delta},
\tag{7}
$$

where $\gamma = \tan^{-1}[\alpha(1 + \beta)]$, $\delta = \tan^{-1}[\alpha(1 - \beta)]$. Here, β and α determine the location and the magnitude of the region of rapid change(s), respectively.

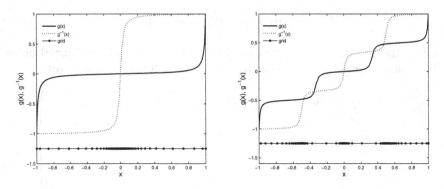

Fig. 1. Function $g(x)$ and its inverse $g^{-1}(x)$ for one region of rapid changes (left figure) and for three regions of rapid changes (right figure).

The conformal map g is constructed from its inverse as $y = g^{-1}(x)$. Figure 1 (left) represents the function g and its inverse g^{-1} for $\alpha = 20$ and $\beta = 0.0$. In the case of multiple regions of singularity, it is possible to combine single point of singularity maps in order to accommodate more points around these regions, see Fig. 1 (right). We address such problems with a single conformal map involving two parameters α_k and β_k. The construction of these maps with R singularities is done through the inverse map g^{-1} as

$$y(x) = g^{-1}(x) = \mu + \frac{1}{\lambda} \sum_{k=1}^{R} \tan^{-1}[\alpha_k(x - \beta_k)], \tag{8}$$

where λ and μ satisfy $g^{-1}(-1) = -1$ and $g^{-1}(1) = 1$. Finally, we compute $g(y) = x$ pointwise by solving the nonlinear equation

$$\sum_{k=1}^{R} \tan^{-1}[\alpha_k(x - \beta_k)] = \lambda(y - \mu). \tag{9}$$

3   Discretisation of the Nonlinear PDE

To discretise the PDE in asset direction by means of rational Chebyshev collocation method, let $x = g(y_j)$ be the transformed Chebyshev points. Then we transform $x \in [-1, 1]$ into $S \in [S_m, S_M]$ that better suits the option problems at hand as $x = (2S - (S_M - S_m))/(S_M + S_m)$. Now writing $V_\epsilon(S, t) = u(x, t)$, the PDE (4) together with its initial and boundary conditions yields

$$
\left.
\begin{aligned}
u_t &= \tfrac{1}{2}\sigma^2 S^2 \left(\frac{2}{S_M - S_m}\right)^2 u_{xx} + (r - \delta)S\left(\frac{2}{S_M - S_m}\right) u_x + ru + \tfrac{1}{\epsilon}\left[u - u^0\right]^+, \\
u(x, 0) &= \max(E - S, 0), \quad -1 \le x \le 1, \quad S_m \le S \le S_M, \\
u(-1, t) &= E - S_m, \quad u(1, t) = 0, \quad 0 \le t \le T.
\end{aligned}
\right\}
$$
$$\tag{10}$$

The rational collocation method can be obtained by replacing the solution u in (10) with

$$\tilde{u}(x) = \sum_{j=0}^{N} \tilde{u}_j L_j^{(\omega)}(x), \tag{11}$$

and collocating at $N-1$ points $x_1, \ldots, x_{N-1}$. This yields

$$\left.\begin{aligned}
\tilde{u}_{i,t} &= p(x_i) \sum_{j=0}^{N} \tilde{u}_j L_j^{(\omega)''}(x_i) + q(x_i) \sum_{j=0}^{N} \tilde{u}_j L_j^{(\omega)'}(x_i) + r\tilde{u}_i + f(\tilde{u}_i), \\
\tilde{u}_0 &= \tilde{u}(-1,t), \quad \tilde{u}_N = \tilde{u}(1,t),
\end{aligned}\right\} \tag{12}$$

where

$$p(x_i) = \frac{1}{2}\sigma^2 S_i^2 \left(\frac{2}{S_M - S_m}\right)^2, \quad q(x_i) = (r-\delta)S_i\left(\frac{2}{S_M - S_m}\right), \quad f(\tilde{u}_i) = \frac{1}{\epsilon}\left[\tilde{u}_i - \tilde{u}_i^0\right]^+.$$

Above expression (12) can be written as

$$\frac{d\tilde{u}}{dt} = \left(PD^{(2)} + QD^{(1)} - rI\right)\tilde{u} + g, \tag{13}$$

where

$$\tilde{u} = [\tilde{u}_1, \tilde{u}_2 \ldots, \tilde{u}_{N-1}]^T,$$

$$D^{(1)} = \left(D_{ij}^{(1)}\right), \quad D_{ij}^{(1)} = L_j^{(\omega)'}(x_i); \quad i, j = 1, 2, \ldots, N-1,$$

$$D^{(2)} = \left(D_{ij}^{(2)}\right), \quad D_{ij}^{(2)} = L_j^{(\omega)''}(x_i); \quad i, j = 1, 2, \ldots, N-1, \tag{14}$$

$$P = \mathrm{diag}(p(x_i)), \quad Q = \mathrm{diag}(q(x_i)); \quad i = 1, 2, \ldots, N-1,$$

$$g = \left[f(u_i) + \left(p(x_i)D_{i0}^{(2)} + q(x_i)D_{i0}^{(1)} + rI_{i0}\right)\tilde{u}_0\right.$$

$$\left. + \left(p(x_i)D_{iN}^{(2)} + q(x_i)D_{iN}^{(1)} + rI_{iN}\right)\tilde{u}_N\right]^T,$$

with $i = 1, 2, \ldots, N-1$ and I is a $(N+1) \times (N+1)$ identity matrix. Above system (13) is then solved using a fourth-order Lobatto method ([6]).

4   Numerical Results and Discussions

We present numerical results of our rational collocation method to price American vanilla put and butterfly spread options for the realistic parameters. In all numerical experiments, unless otherwise indicated, we choose the grid stretching parameter as $\alpha = 20$ and truncate the computational domain in such a way that $S_m = 0$ and $S_M = 200$.

4.1   Results for American Put Options

In order to illustrate the utility of our RSC method, two experiments are performed. As a benchmark, we take the value of the options obtained by using the penalty method in [5] with 50000 time and space grid points. The first experiment checks the behaviour of the method with respect to the penalty parameter ϵ. Table 1 shows the results of American put options obtained by using the set of parameters $r = 0.05$, $\sigma = 0.2$, $\delta = 0$, $E = 100$, $T = 0.5$. We vary the value of the penalty parameter ϵ, and for each fixed number of spacial nodes we then compute the error of the American put options, the ratio between two consecutive errors and evaluate the number of time steps of the adaptive Lobatto time integrator (with both absolute and relative error tolerances as 10^{-5}). We observe that the error is nearly independent of the spacial mesh points and is of order $\mathcal{O}(\epsilon)$. We obtain very satisfactory results for a small number of grid points. For instance, for $\epsilon = 10^{-5}$ and $N = 50$ an error of order 10^{-4} is obtained. Similar results were obtained in [4,7] but they required to use a larger number of mesh points.

Table 1. Results of the rational spectral collocation method with respect to the penalty term for valuing American put options.

ϵ	Nodes	Error	Ratio	Time steps
10^{-3}	50	1.61E-3	7.1200	64
10^{-4}	50	5.88E-4	2.7395	190
10^{-5}	50	4.79E-4	1.2285	627
10^{-3}	100	1.20E-3	9.5250	100
10^{-4}	100	1.36E-4	8.8438	209
10^{-5}	100	3.09E-5	4.4014	706
10^{-3}	200	1.19E-3	9.6912	102
10^{-4}	200	1.46E-4	8.1322	215
10^{-5}	200	4.28E-5	3.4092	713

Now we investigate the spacial convergence of the rational spectral collocation (RSC) method and compare with the convergence of some existing methods which we recompute in order to use the same set of parameters. To ensure that errors in numerical results are dominated by spacial rather than temporal errors we impose (here as well as for the case of butterfly spread options) the absolute and relative error tolerances of the implicit fourth-order Lobatto method and the penalty term to be 10^{-8}. From Table 2, we observe that our RSC has higher order of convergence compared to the existing methods. Our method produces a satisfactory error of order 10^{-3} with only 20 grid points, whereas other methods require almost 16 times more points, to obtain a similar error. In Table 2 as well as in Tables 3-5, the columns with headings as Method A, Method B and

Table 2. Comparison of the rational spectral collocation (RSC) method with other existing methods for valuing American put options.

N	Method A V(E)	Error	Method B V(E)	Error	Method C V(E)	Error	RSC V(E)	Error
20	2.953828	1.70E-0	4.215878	4.39E-1	4.222097	4.34E-1	4.651765	3.92E-3
40	4.221527	4.34E-1	4.550680	1.05E-1	4.555841	9.98E-2	4.654778	9.05E-4
80	4.555846	9.98E-2	4.627469	2.82E-2	4.630338	2.53E-2	4.655634	4.92E-5
160	4.630491	2.52E-2	4.647740	7.91E-3	4.649266	6.38E-3	4.655679	4.80E-6
320	4.649285	6.36E-3	4.653227	2.42E-3	4.654034	1.61E-3	4.655684	4.96E-7

Method C represent the results obtained by our computation for the penalty method of Forsyth and Vetzal [5], the Crank-Nicolson method based on the Brennan and Schwartz approach [2] and the projected successive overrelaxation (PSOR) method of Cryer [3], respectively.

4.2 Results for American Butterfly Spread Options

A more challenging problem is the valuation of the American butterfly options. A butterfly option has the payoff

$$V(S,0) = \max(S-E_1,0) - 2\max(S-E_2,0) + \max(S-E_3,0), \quad E_2 = (E_1+E_3)/2, \tag{15}$$

and the boundary conditions

$$V(S,t) = 0 \ \text{ as } \ S \to 0; \ \ V(S,t) = 0 \text{ as } S \to \infty. \tag{16}$$

Like the previous case, we compute the benchmark solution in a similar manner. The grid stretching parameter is taken to be $\alpha = 50$. Other parameters used in this case are $r = 0.05$, $\sigma = 0.2$, $\delta = 0$, $E_1 = 90$, $E_2 = 100$, $E_3 = 110$, $T = 0.5$. Rapid convergence of the RSC is observed while valuing these options at strike prices E_1, E_2 and E_3 as illustrated in Tables 3, 4 and 5, respectively. Our RSC approach shows very accurate results obtained with few grid points. For example, at the strike price E_2 an accuracy of order 10^{-2} is attained with only 20 points in the case of RSC method which cannot be achieved in the case of the penalty method of Forsyth [5] with even 320 points.

We have also experimented (but not reported in this paper due to space limitations) that the numerical values of the Greeks for these options are free of spurious oscillations around the strike price(s) that are typically present in numerical solution of the American style options, in particular while computing Γ, due to the discontinuity of the second order derivative in the optimal exercise price.

Currently, we are investigating our proposed approach to solve other types of financial options, in particular, multi-asset American options and American options under jump diffusion processes.

Table 3. Comparison of the convergence of the rational spectral collocation (RSC) method with other existing methods for valuing American butterfly spread options at strike price E_1.

	Method A		Method B		Method C		RSC	
N	$V(E_1)$	Error	$V(E_1)$	Error	$V(E_1)$	Error	$V(E_1)$	Error
20	3.850580	1.40E-0	4.644988	6.08E-1	4.800868	4.52E-1	5.212192	4.09E-2
40	4.219791	1.03E-0	5.000685	2.52E-1	5.161263	9.19E-2	5.255192	2.05E-3
80	4.743739	5.09E-1	5.077096	1.76E-1	5.233733	1.94E-2	5.253491	3.55E-4
160	5.166896	8.62E-2	5.105809	1.47E-1	5.255290	2.15E-3	5.253196	5.49E-5
320	5.169988	8.32E-2	5.134745	1.18E-1	5.251017	2.12E-3	5.253147	4.90E-6

Table 4. Comparison of the convergence of the rational spectral collocation (RSC) method with other existing methods for valuing American butterfly spread options at strike price E_2.

	Method A		Method B		Method C		RSC	
N	$V(E_2)$	Error	$V(E_2)$	Error	$V(E_2)$	Error	$V(E_2)$	Error
20	3.977636	6.01E-0	9.302053	6.89E-1	9.871566	1.19E-1	9.955600	3.55E-2
40	8.653698	1.34E-0	9.512456	4.78E-1	9.913464	7.76E-2	9.994326	3.26E-3
80	8.679782	1.31E-0	9.822351	1.69E-1	9.957679	3.34E-2	9.991326	2.59E-4
160	9.478241	5.13E-1	9.875372	1.16E-1	9.976355	1.47E-2	9.991026	4.06E-5
320	9.648192	3.43E-1	9.967679	2.34E-2	9.997890	6.82E-3	9.991062	4.64E-6

Table 5. Comparison of the convergence of the rational spectral collocation (RSC) method with other existing methods for valuing American butterfly spread options at strike price E_3.

	Method A		Method B		Method C		RSC	
N	$V(E_3)$	Error	$V(E_3)$	Error	$V(E_3)$	Error	$V(E_3)$	Error
20	1.758169	3.14E-0	4.454015	4.45E-1	4.600968	2.98E-1	4.811230	8.73E-2
40	4.535713	3.63E-1	4.686623	2.12E-1	4.832751	6.58E-2	4.897053	1.51E-3
80	4.722869	1.76E-1	4.745747	1.52E-1	4.882757	1.58E-2	4.898053	5.05E-4
160	4.715394	1.83E-1	4.771779	1.27E-1	4.890226	8.33E-3	4.898503	5.51E-5
320	4.848402	5.02E-2	4.806425	9.21E-2	4.893375	5.18E-3	4.898553	5.10E-6

Acknowledgments. E. Pindza and E. Ngounda acknowledge the financial support from the Agence Nationale des Bourses du Gabon. Patidar's research was also supported by the South African National Research Foundation.

References

1. Bayliss, A., Turkel, E.: Mappings and accuracy for Chebyshev pseudo-spectral approximations. J. Comput. Phy. **101**, 349–359 (1992)
2. Brennan, M.J., Schwartz, E.S.: The valuation of American put options. J. Finance **32**(2), 449–462 (1977)
3. Cryer, C.W.: The solution of a quadratic programme using systematic overrelaxation. SIAM J. Cont. Opt. **9**, 385–392 (1971)
4. De Frutos, J.: Implicit-explicit Runge-Kutta methods for financial derivatives pricinf models. Eur. J. Operat. Res. **171**, 991–1004 (2006)
5. Forsyth, P.A., Vetzal, K.R.: Quadratic convergence of a penalty method for valuating American options. SIAM J. Sci. Comput. **23**(6), 1823–1835 (2002)
6. Hairer, E., Wanner, G.: Solving Ordinary Differential Equations II: Stiff and Differential-Algebraic Problems. Springer, Berlin (1996)
7. Khaliq, A.Q.M., Voss, D.A., Kazmi, S.H.K.: A linearly implicit predicto-corrector scheme for pricing American options using a penalty methods approach. J. Bank. Fin. **30**, 489–502 (2006)
8. Wilmott, P., Dewynne, J., Howison, S.: Option Pricing: Mathematical Models and Computation. Cambridge University Press, New York (1995)
9. Zhu, Y.L., Wu, X., Chern, I.L.: Derivatives Securities and Difference Methods. Springer, New York (2004)

Riemann Problem for First-Order Partial Equations Without the Convexity of a State Functions

Mahir Rasulov[1][✉] and S. Ozgur Ulas[2]

[1] Department of Mathematics and Computing, Beykent University,
Sisli-Ayazaga Campus, 34396 Istanbul, Turkey
mresulov@beykent.edu.tr
[2] Institute of Sciences, Beykent University, Taksim, Istanbul, Turkey
ozgur_ulas78@hotmail.com

Abstract. In this work, the exact solution of the Riemann problem for first-order nonlinear partial equation with non-convex state function in $Q_T = \{(x,t) | x \in I = (-\infty, \infty), \ t \in [0,T)\} \subset R^2$ is found. Here $F \in C^2(Q_T)$ and $F''(u)$ change their signs, that is $F(u)$ has convex and concave parts. In particular, the state function $F(u) = -\cos u$ on $\left[\frac{\pi}{2}, \frac{3\pi}{2}\right]$ and $\left[\frac{\pi}{2}, \frac{5\pi}{2}\right]$ is discussed. For this, when it is necessary, the auxiliary problem which is equivalent to the main problem is introduced. The solution of the proposed problem permits constructing the weak solution of the main problem that conserves the entropy condition. In some cases, depending on the nature of the investigated problem a convex or a concave hull is constructed. Thus, the exact solutions are found by using these functions.

Keywords: First order nonlinear partial differential equations · Riemann problem · Characteristics · Weak solution · Shock wave · Convex and concave hull

1 Introduction

Many solutions of the problems of mechanics, particularly of gas dynamics are reduced to investigate the discontinuous solution of the nonlinear equations of the first-order hyperbolic type. It is known that properties of the discontinuous solutions of the nonlinear equations have some properties which are absent in discontinuous solutions of the linear equations. Relevant results in the theory of nonlinear equation of hyperbolic type have been obtained by O.A. Oleinik, A.N. Tikhonov, A.A. Samarskii, P. Lax and I.M. Gelfand. The case when $F'''(u) \leq 0$ or $F''(u) \geq 0$ have been investigate in [2,3,5–9] in detail.

2 The Problem with a Convex State Function

As usual, let R^2 be an Euclidean space in the plane (x,t). Here x and t are spatial and time variables, respectively. Let us denote $Q_T = \{(x,t) | x \in I = (-\infty, \infty), \ t \in [0,T)\} \subset R^2$ and in Q_T we consider the following problem

© Springer International Publishing Switzerland 2015
I. Dimov et al. (Eds.): FDM 2014, LNCS 9045, pp. 332–339, 2015.
DOI: 10.1007/978-3-319-20239-6_36

$$\frac{\partial u}{\partial t} + \frac{\partial F(u)}{\partial x} = 0, \tag{1}$$

$$u(x,0) = \begin{cases} u_1, & x < 0 \\ u_2, & x > 0. \end{cases} \tag{2}$$

Here $F \in C^2(Q_T)$, is given function and F'' changes its sign, i.e. $F(u)$ has convex and concave parts. The problem (1), (2) particularly, describes the mathematical models of the process of displacement of oil by water in porous medium and traffic flow on highway, [1–4, 10, 11].

In this paper the problem (1), (2) is investigated for the case $F(u) = \cos u$ if $u \in [\frac{\pi}{2}, \frac{3\pi}{2}]$ and $u \in [\frac{\pi}{2}, \frac{5\pi}{2}]$ respectively and u_1, u_2 are constants.

As seen from an admissibility condition of jump, when $F''(u) < 0$ and $u_1 < u_2$ jump occurs immediately beginning from the origin of coordinates in the solution.

Firstly we will study the case $u_1 = \frac{3\pi}{2}$, $u_2 = \frac{\pi}{2}$. At this stage, due to the entropy condition which is not fulfilled we can't extract the solution of the problem (1), (2) in shock wave form.

According to the nature of the characteristics method we must seam the three lines $u = u_1$, $u = u_2$ and $u = G\left(\frac{x}{t}\right)$ together retaining of continuity of the solution. Here G is an inverse function of the $\sin u$ on the $[\frac{\pi}{2}, \frac{3\pi}{2}]$. Therefore for the solution we get

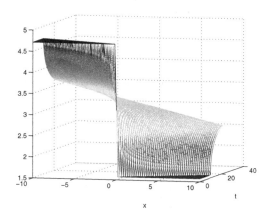

Fig. 1. The graph of the $u(x,t)$ when $u_1 = \frac{3\pi}{2}$, $u_2 = \frac{\pi}{2}$,

$$u(x,t) = \begin{cases} \frac{3\pi}{2}, & \frac{x}{t} < -1 \\ \pi - \arcsin\left(\frac{x}{t}\right), & -1 < \frac{x}{t} < 1 \\ \frac{\pi}{2}, & \frac{x}{t} > 1 \end{cases} \tag{3}$$

The graph of this solution is given in Fig. 1.

It is easy to show that the function given by (3) is the generalized solution of the problem (1), (2). For this aim we must prove that the following integral equality

$$\int\int_D \{\varphi_t(x,t)u(x,t) - \varphi_x \cos u(x,t)\}dxdt + \int_{-a}^a u_0(x,t)\varphi(x,0)dx = 0$$

is valid for any compactly supported continuously differentiable test functions $\varphi(x,t)$ with respect to both variables x, t vanishing outside of some bounded domain D such that $D = \{(-a,a) \times ([0,T)\} \subset Q_T$.

Taking account of (3) from the last equality can be rewritten as

$$\int_0^T \int_{-a}^a \{u(x,t)\varphi_t(x,t) - \cos u(x,t)\varphi_x\} dxdt + \int_{-a}^a u_0(x,t)\varphi(x,0)dx$$

$$= \int_0^T \int_{-a}^{-t} \frac{3\pi}{2}\varphi_t(x,t) dxdt + \int_0^T \int_{-t}^t \left(\pi - \arcsin\left(\frac{x}{t}\right)\right)\varphi_t(x,t) dxdt$$

$$- \int_0^T \int_{-t}^t \cos\left(\pi - \arcsin\left(\frac{x}{t}\right)\right) \varphi_x dxdt$$

$$+ \int_0^T \int_t^a \frac{\pi}{2}\varphi_t(x,t) dxdt + \int_{-a}^0 \frac{3\pi}{2}\varphi(x,0) dx + \int_0^a \frac{\pi}{2}\varphi(x,0) dx.$$

By interchanging the order of integration we get

$$-\frac{3\pi}{2} \int_{-a}^{-T} \varphi(x,0) dx + \frac{3\pi}{2} \int_{-T}^0 \varphi(x,-x) dx$$

$$-\frac{3\pi}{2} \int_{-T}^0 \varphi(x,0) dx - \int_{-T}^0 (\pi - \arcsin(-1)) \varphi(x,-x) dx$$

$$-\int_{-T}^0 \int_{-x}^T \varphi(x,t) \frac{x}{t^2\sqrt{1 - \left(\frac{x}{t}\right)^2}} dtdx - \int_0^T (\pi - \arcsin(1)) \varphi(x,x) dx$$

$$-\int_0^T \int_x^T \varphi(x,t) \frac{x}{t^2\sqrt{1 - \left(\frac{x}{t}\right)^2}} dtdx$$

$$\int_0^T \int_{-t}^t \varphi(x,t) \frac{x}{t^2\sqrt{1 - \left(\frac{x}{t}\right)^2}} dxdt + \frac{\pi}{2} \int_{-a}^T \varphi(x,x) dx - \frac{\pi}{2} \int_{-T}^0 \varphi(x,0) dx$$

$$-\frac{\pi}{2}\int_T^a \varphi\,(x,0)\,dx + \frac{3\pi}{2}\int_{-a}^{-T}\varphi\,(x,0)\,dx + \frac{3\pi}{2}\int_{-T}^0 \varphi\,(x,0)\,dx = 0.$$

Now we consider the case $u_1 = \frac{\pi}{2}$, $u_2 = \frac{3\pi}{2}$. Since the entropy condition takes place in this case the shock arises in solution. According to [1,10] the following auxiliary problem is introduced.

$$\frac{\partial w}{\partial t} - \cos\left(\frac{\partial w}{\partial x}\right) = 0. \tag{4}$$

The initial condition for (4) is

$$w\,(x,0) = w_0(x). \tag{5}$$

Here the function $w_0(x)$ is any continuously differentiable solution of the equation

$$\frac{dw_0(x)}{dx} = u_0(x).$$

The problem (4), (5) is called auxiliary problem. It is easy to prove that

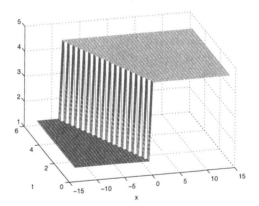

Fig. 2. The graph of the $u(x,t)$ when $u_1 = \frac{\pi}{2}$, $u_2 = \frac{3\pi}{2}$,

$$u(x,t) = \frac{\partial w(x,t)}{\partial x}. \tag{6}$$

The solution of this auxiliary problem (4), (5) is, [1,10]

$$w(x,t) = \begin{cases} w_-, & \xi < 0 \\ w_+, & \xi > 0. \end{cases}$$

Here

$$w_- = (\cos u_1 - u_1 \sin u_1)\, t + \frac{\pi}{2} x, \quad w_+ = (\cos u_2 - u_2 \sin u_2)\, t + \frac{3\pi}{2} x$$

and $\xi = x - t \sin u$. From $w_- = w_+$ we obtain $w = \frac{x}{t} = -2$. Therefore for the solution of the auxiliary problem we have

$$w(x,t) = \begin{cases} w_-, & \frac{x}{t} < -2, \\ w_+, & \frac{x}{t} > -2. \end{cases} \tag{7}$$

With regard to (6) the physical genuine solution of the main problem (1), (2) is found. The graph of this solution is given in Fig. 2. As it is seen from Fig. 2 the shock available in the wave's initial profile is moved with speed w to left side of coordinate axis.

3   The Problem Without Convexity of State Function

In this section we will study the problem (1), (2) when $u \in \left[\frac{\pi}{2}, \frac{5\pi}{2}\right]$.

Case 1. Firstly the (1) is investigated with

$$u(x,0) = \begin{cases} \frac{5\pi}{2}, & x < 0, \\ \frac{\pi}{2}, & x > 0 \end{cases}$$

initial condition.

As in [3,4,10] we use the convex hull of the function $F(u) = -\cos u$ on interval $\left[\frac{\pi}{2}, \frac{5\pi}{2}\right]$ graph which is shown in Fig. 3.

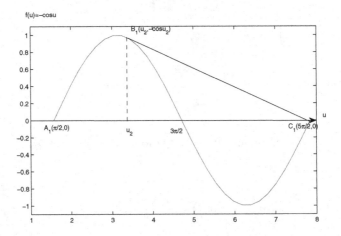

Fig. 3. Convex hull of $F(u) = -\cos u$ on the interval $\left[\frac{\pi}{2}, \frac{5\pi}{2}\right]$

As seen from Fig. 3 that the convex hull of the function $F(u) = -\cos u$ consists of the part of the graph lying between points $A_1\left(\frac{\pi}{2}, 0\right)$ and $B_1(u_2, -\cos u_2)$, and of the straight line B_1C_1. In order to obtain the value of u_2 we will use tangent line leaving from point $C_1\left(\frac{5\pi}{2}, 0\right)$ to graph of $F(u) = -\cos u$

$$g(u_2) \equiv u_2 - \frac{5\pi}{2} + \cot u_2 = 0.$$

Since $g(\pi)g\left(\frac{3\pi}{2}\right) < 0$ this equation has one root on the interval $\left[\pi, \frac{3\pi}{2}\right]$. Using the Newton iteration method the mentioned root is found by

$$u_{n+1} = u_n + \frac{g(u_n)}{g'(u_n)}, \quad (n = 0, 1, 2, \dots.)$$

equation as $u_1 = 3.36$, here $g(u) = -u + \frac{5\pi}{2} - \cot u$ and $g'(u) = -1 + \frac{1}{\sin^2 u} = \cot^2 u$.

Therefore the generalized solution of the problem is

$$u(x, t) = \begin{cases} \frac{5\pi}{2}, & x < Kt \\ \pi - \arcsin \frac{x}{t}, & Kt < x < t \\ \frac{\pi}{2}, & t < x, \end{cases} \tag{8}$$

here $F'(u) = \sin 3.36 = -0.2167 = K$. The graph of this solution is shown in Fig. 4.

Case 2. In this section we will investigate the equation (1) with

$$u(x, 0) = \begin{cases} \frac{\pi}{2}, & x < 0 \\ \frac{5\pi}{2}, & x > 0 \end{cases}$$

initial condition. For this we use concave hull of the function $F(u) = -\cos u$ on the interval $\left[\frac{\pi}{2}, \frac{5\pi}{2}\right]$, (Fig. 5).

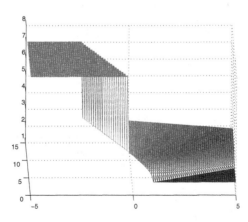

Fig. 4. The graph of the solution $u(x, t)$ when $u_1 = \frac{5\pi}{2}$, $u_2 = \frac{\pi}{2}$.

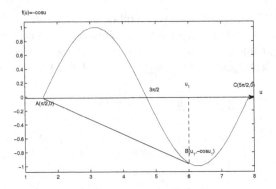

Fig. 5. Concave hull of $F(u) = -\cos u$ on the interval $\left[\frac{\pi}{2}, \frac{5\pi}{2}\right]$

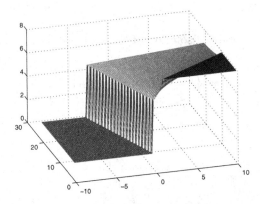

Fig. 6. Concave hull of $F(u) = -\cos u$ on the interval $\left[\frac{\pi}{2}, \frac{5\pi}{2}\right]$

The concave hull of the function $F(u) = -\cos u$ consists of the straight line AB and of the part of the graph $F(u) = -\cos u$ lying between points $B(u_1, -\cos u_1)$ and $C\left(\frac{5\pi}{2}, 0\right)$.

Since the graph of the convex and concave hulls of function $-\cos u$ is symmetrical on the interval $\left[\frac{\pi}{2}, \frac{5\pi}{2}\right]$ the value of u_1 is found from equation $u_1 - \frac{3\pi}{2} = \frac{3\pi}{2} - u_2$ as $u_1 \cong 1.93\pi$.

In this case obtained root is relatively symmetrical to point $\left(\frac{3\pi}{2}, 0\right)$. Then $\sin u_2 \cong \sin 1.93\pi = -0.217 = K$ and the tangent line leaving from point $\left(\frac{5\pi}{2}, 0\right)$ is $x = f'(u)|_{u=\frac{5\pi}{2}} . t = \sin \frac{5\pi}{2} \; t = t$. Similarly as shown above the shock occurs at both lines $\xi = x - Kt$ and $\xi = x - t$. In addition to this, this shock takes place from $\frac{\pi}{2}$ to u_2. Finally the solution is present as

$$u(x,t) = \begin{cases} \frac{\pi}{2}, & x < Kt, \\ 2\pi + \arcsin \frac{x}{t}, & Kt < x < t, \\ \frac{5\pi}{2}, & t < x \end{cases}$$

and graph of which is given in Fig. 6.

4   Conclusion

1. The exact solutions of the Riemann problem for first order partial equation with state function $F(u) = -\cos u$ on the intervals $\left[\frac{\pi}{2}, \frac{3\pi}{2}\right]$ and $\left[\frac{\pi}{2}, \frac{5\pi}{2}\right]$ are found.

2. In order to obtain the location of the shock point the special auxiliary problem having some advantages over main problem is introduced.

3. It is proved that the obtained solutions are weak solutions of the investigated problem.

References

1. Abasov, M.T., Rasulov, M.A., Ibrahimov, T.M., Ragimova, T.A.: On a method of solving the cauchy problem for a first order nonlinear equation of hyperbolic type with a smooth initial condition. Soviet Math. Dok. **43**(1), 150–153 (1991)

2. Gelfand, I.M.: Some problems in the theory of quasi linear equations. Amer. Math. Soc. Trans. **29**, 295–381 (1963)

3. Godlewski, E., Raviart, P.A.: Hyperbolic Systems of Conservation Laws. Ellipses, Paris (1991)

4. Goritski, A.Y., Kruzhkov, S.N., Chechkin, G.A.: Quasi Linear First Order Partial Differential Equations. Moscow State University Press, Moscow (1997)

5. Kruzhkov, S.N.: First order quasi linear equation in several independent variables. Math. USSR Sb. **10**(2), 217–243 (1970)

6. Lax, P.D.: Weak solutions of nonlinear hyperbolic equations and their numerical computations. Comm. Pure App. Math. **7**, 159–193 (1954)

7. Lax, P.: The Formation and Decay of Shock Waves. Amer. Math. Monthly **79**, 227–241 (1972)

8. Lax, P.D.: Development of singularities of solutions of nonlinear hyperbolic partial differential equations. J. Math. Phys. **5**(5), 611–613 (1964)

9. Oleinik, O.A.: Discontinuous solutions of nonlinear differential equations. Uspekhi Mat. Nauk. **12**, 3–73 (1957)

10. Rasulov, M.A., Süreksiz Fonksiyonlar Sınıfında Korunum Kuralları, Seçkin Yayınevi, Istanbul, 2011

11. Whitham, G.B.: Linear and Nonlinear Waves. Wiley, New York (1974)

Numerical Modeling of a Block Medium as an Orthotropic Cosserat Continuum

Oxana V. Sadovskaya$^{(\boxtimes)}$, Vladimir M. Sadovskii, and Mariya A. Pokhabova

Institute of Computational Modeling SB RAS, Akademgorodok 50/44,
660036 Krasnoyarsk, Russia
{o_sadov,sadov}@icm.krasn.ru, pahomova_mariya_@mail.ru

Abstract. Based on the equations of the dynamics of a piecewise-homogeneous elastic material, parallel computational algorithms are developed to simulate the process of stress and strain wave propagation in a medium consisting of a large number of blocks, interacting through compliant interlayers. For the description of waves in a block medium with thick interlayers the orthotropic couple-stress continuum theory, taking into account the symmetry of elastic properties relative to the coordinate planes, can be applied. By comparing the elastic wave velocities in the framework of piecewise-homogeneous model and continuum model, the simple method is obtained to estimate mechanical parameters of an orthotropic Cosserat continuum, modeling a block medium. In two-dimensional formulation of the orthotropic model, the computational algorithm and the program system are worked out for the analysis of propagation of elastic waves. The comparison showed good qualitative agreement between the results of computations of waves, caused by localized impulses, by the model of a block medium with compliant interlayers and the model of an orthotropic Cosserat continuum.

Keywords: Dynamics · Elasticity · Block medium · Cosserat continuum · Shock-capturing method · Parallel computational algorithm

1 Introduction

The model of a block medium with compliant interlayers can be used to estimate the deformation and strength characteristics of a masonry. Wave processes in soils and rocks with layered and block structure can be analyzed with the help of this model. Various approximations of the model are applied to the analysis of such processes, in particular, the approximation of absolutely rigid blocks and elastic interlayers [1]. In a more simple variant, a block medium is replaced by a discrete lattice, in which the dimensions of blocks do not matter [2]. In this paper, we consider the case of elastic blocks and interlayers. On the basis of developed algorithms, in the plane formulation computations for a large number of blocks were performed, showing anisotropy of a block medium even at sufficiently thin interlayers. Obtained results serve as justification of a hypothesis, that the processes of wave propagation in a multiblock medium

© Springer International Publishing Switzerland 2015
I. Dimov et al. (Eds.): FDM 2014, LNCS 9045, pp. 340–347, 2015.
DOI: 10.1007/978-3-319-20239-6_37

can be approximately described by the equations of an orthotropic Cosserat continuum. These equations take into account the asymmetry of stress tensor, associated with the rotational motion of blocks, and the couple stresses, caused by the curvature of a regular block structure due to the inhomogeneity of the field of rotations.

2   A Block Medium

Let's consider the plane strain state of a block medium, consisting of rectangular elastic blocks with sides h_1, h_2, parallel to the axes x_1, x_2 of a Cartesian coordinate system, and interlayers with thicknesses δ_1, δ_2. Blocks are numbered by pairs of indices k_1, k_2, taking the values from 1 to N_1 and from 1 to N_2, respectively. The motion of each block is described by the system of equations of a homogeneous isotropic elastic medium:

$$
\begin{aligned}
&\rho\,\dot{v}_1 = \sigma_{11,1} + \sigma_{12,2}, \quad \rho\,\dot{v}_2 = \sigma_{12,1} + \sigma_{22,2}, \quad \dot{\sigma}_{12} = \rho\,c_2^2\,(v_{2,1} + v_{1,2}), \\
&\dot{\sigma}_{11} = \rho\,c_1^2\,(v_{1,1} + v_{2,2}) - 2\,\rho\,c_2^2\,v_{2,2}, \quad \dot{\sigma}_{22} = \rho\,c_1^2\,(v_{1,1} + v_{2,2}) - 2\,\rho\,c_2^2\,v_{1,1}.
\end{aligned}
\tag{1}
$$

Here ρ is the density, c_1 and c_2 are the velocities of longitudinal and transverse elastic waves, dot over the symbol and indices after a comma denote partial derivatives with respect to time and spatial variables. The system is written relative to the projections of the velocity vector v_k and the components of the stress tensor σ_{jk}.

Elastic interlayer between neighboring blocks in the horizontal direction with the indices (k_1, k_2) and $(k_1 + 1, k_2)$ is described by the ordinary differential equations, taking into account its mass and the longitudinal and transverse stiffnesses:

$$
\begin{aligned}
&\rho'\frac{\dot{v}_1^+ + \dot{v}_1^-}{2} = \frac{\sigma_{11}^+ - \sigma_{11}^-}{\delta_1}, \quad \frac{\dot{\sigma}_{11}^+ + \dot{\sigma}_{11}^-}{2} = \rho'c_1'^2\frac{v_1^+ - v_1^-}{\delta_1}, \\
&\rho'\frac{\dot{v}_2^+ + \dot{v}_2^-}{2} = \frac{\sigma_{12}^+ - \sigma_{12}^-}{\delta_1}, \quad \frac{\dot{\sigma}_{12}^+ + \dot{\sigma}_{12}^-}{2} = \rho'c_2'^2\frac{v_2^+ - v_2^-}{\delta_1},
\end{aligned}
\tag{2}
$$

where ρ', c_1' and c_2' are the density and the velocities of longitudinal and transverse waves in the interlayer. The interlayer between blocks in the vertical direction with the indices (k_1, k_2) and $(k_1, k_2 + 1)$ is described by the equations:

$$
\begin{aligned}
&\rho'\frac{\dot{v}_2^+ + \dot{v}_2^-}{2} = \frac{\sigma_{22}^+ - \sigma_{22}^-}{\delta_2}, \quad \frac{\dot{\sigma}_{22}^+ + \dot{\sigma}_{22}^-}{2} = \rho'c_1'^2\frac{v_2^+ - v_2^-}{\delta_2}, \\
&\rho'\frac{\dot{v}_1^+ + \dot{v}_1^-}{2} = \frac{\sigma_{12}^+ - \sigma_{12}^-}{\delta_2}, \quad \frac{\dot{\sigma}_{12}^+ + \dot{\sigma}_{12}^-}{2} = \rho'c_2'^2\frac{v_1^+ - v_1^-}{\delta_2}.
\end{aligned}
\tag{3}
$$

The quantities with superscripts $\pm$ are related to the boundaries of interacting blocks. It can be shown that these equations are thermodynamically consistent with the system of Eqs. (1), i.e. for a regular block structure the integral energy conservation law is fulfilled, in which kinetic and potential energies are the sums of kinetic and potential energies of blocks and interlayers separately.

For numerical solution of the system of Eqs. (1)–(3) under given initial data and boundary conditions, the parallel computational algorithm is worked out [3]. In this algorithm the method of two-cyclic splitting with respect to the spatial variables is realized. This method preserves the second order of approximation, if second-order schemes are applied for the solution of one-dimensional systems. One-dimensional systems in blocks are solved on the basis of the Godunov gap decay scheme with an uniform grid and a maximum permissible time step according to the Courant–Friedrichs–Lewy condition. A piecewise-linear ENO–reconstruction is used to improve the accuracy of solution. The Eqs. (2) and (3), playing the role of internal boundary conditions at the stages of two-cyclic splitting, are solved by means of the nondissipative finite-difference scheme, constructed by the Ivanov method.

Computer programs are implemented in Fortran using the MPI (Message Passing Interface) technology. The parallelization of computations is performed by means of the domain decomposition: each processor of a cluster makes computations for a chain of blocks in the direction of the x_1 axis with data exchange between neighboring processors in the boundary meshes of the upper and lower boundaries of the chain of blocks.

Created programs are used to solve a series of problems on the propagation of elastic waves, caused by the action of short-term and long-term concentrated loads on a block medium. In Figs. 1 and 2 one can see the results of computations for the problem, in which the boundary effects on a block medium are absent and the initial data for velocities correspond to the rotation of central block around the center of mass with given angular velocity ω_0:

$$v_1^{k_1 k_2} = -\omega_0(x_2 - h_2/2), \quad v_2^{k_1 k_2} = \omega_0(x_1 - h_1/2). \tag{4}$$

Initial stresses are equal to zero. Initial velocities in all blocks, except the central one, are zero, too. Blocks are square with the side of 0.1 m and interlayers have the same thickness in both directions. A block rock mass consists of 200×200 blocks, each of which has a grid with 10×10 nodes. The parameters of materials are taken close to the parameters of a masonry: $\rho = 3700$, $\rho' = 1200\,\mathrm{kg/m^3}$,

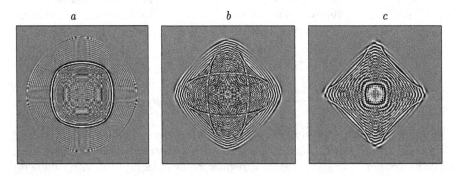

a   *b*   *c*

Fig. 1. Instant rotation of the central block. Level curves of the angular velocity $\bar{\omega}$ depending on the thickness of interlayers: (*a*) $\delta = 0.1$ mm, (*b*) $\delta = 1$ mm, (*c*) $\delta = 5$ mm

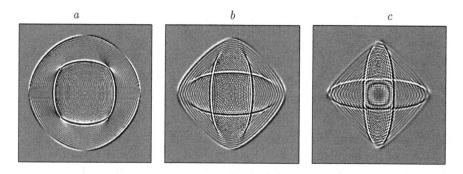

Fig. 2. Instant rotation of the central block. Level curves of the tangential stress $\bar{\sigma}_{21}$ depending on the thickness of interlayers: (a) $\delta = 0.1$ mm, (b) $\delta = 1$ mm, (c) $\delta = 5$ mm

$c_1 = 3500$, $c_2 = 2100$, $c'_1 = 1500$, $c'_2 = 360$ m/s. Computations were performed for interlayers of different thickness $\delta = 0.1$, 1 and 5 mm on time intervals of varying duration $t = 2.4$, 2.5 and 3.1 ms, during which the head fronts of waves pass approximately equal distances.

For visualization of numerical solution the angular velocities of blocks and the tangential stresses, averaged over the parallel lateral sides, were calculated. Level curves of these values are represented in Figs. 1 and 2, respectively. A comparison of results for different thicknesses of interlayers shows that the wave fields may differ essentially. In the case of thin interlayers the wave fronts are almost circular, their velocities are close to the velocities of longitudinal and transverse waves in blocks (Figs. 1a and 2a). The form of fronts becomes elliptical, if interlayers become thicker (Figs. 1b, 2b and Figs. 1c, 2c).

Computations show that the velocities of waves in the direction of semiaxes of ellipses are much lower than the average velocities of longitudinal and transverse waves in a block medium along the coordinate directions. This applies especially to the transverse waves. The block-averaged amplitudes of the head longitudinal and transverse waves are small in comparison with the amplitudes of waves with elliptical fronts, which are caused by oscillatory motions of the blocks as rigid bodies. These are so-called pendulum waves. Velocities of pendulum waves slow down, if interlayers become thicker. The characteristic oscillations, caused by the rotational motion of blocks, appear behind the fronts of these waves.

3   An Orthotropic Couple-Stress Medium

For averaged description of the deformation processes in a multiblock medium with interlayers it is possible to use the model of the Cosserat continuum in the approximation of an orthotropic material with planes of elastic symmetry, parallel to the coordinate planes. In the plane strain state the complete system of equations has the form:

$$\rho_0 \, \dot{v}_1 = \sigma_{11,1} + \sigma_{12,2}, \quad \rho_0 \, \dot{v}_2 = \sigma_{21,1} + \sigma_{22,2},$$
$$j_0 \, \dot{\omega}_3 = \mu_{31,1} + \mu_{32,2} + \sigma_{21} - \sigma_{12},$$
$$\dot{\sigma}_{11} = a_1 v_{1,1} + b_1 v_{2,2}, \quad \dot{\sigma}_{22} = a_1 v_{2,2} + b_1 v_{1,1},$$
$$\dot{\sigma}_{33} = b_1 (v_{1,1} + v_{2,2}), \tag{5}$$
$$\dot{\sigma}_{21} = a_2(v_{2,1} - \omega_3) + b_2(v_{1,2} + \omega_3),$$
$$\dot{\sigma}_{12} = a_2(v_{1,2} + \omega_3) + b_2(v_{2,1} - \omega_3),$$
$$\dot{\mu}_{31} = \alpha_2 \, \omega_{3,1}, \quad \dot{\mu}_{32} = \alpha_2 \, \omega_{3,2}.$$

Here v_1 and v_2 are the projections of the linear velocity vector, ω_3 is the nonzero projection of the angular velocity vector, σ_{ij} and μ_{ij} are the components of the asymmetric tensors of stresses and couple stresses, respectively. The symbol j_0 denotes the inertial characteristic of a material, which is equal to the product of the moment of inertia of a block and the number of blocks in a unit volume. The quantities a_1, a_2, b_1, b_2, α_2 are the elasticity parameters of a material.

The system (5) can be represented in the matrix form

$$A\frac{\partial U}{\partial t} = B^1 \frac{\partial U}{\partial x_1} + B^2 \frac{\partial U}{\partial x_2} + Q\,U \tag{6}$$

relative to the vector–function $U = (v_1, v_2, \omega_3, \sigma_{11}, \sigma_{22}, \sigma_{33}, \sigma_{21}, \sigma_{12}, \mu_{31}, \mu_{32})$ with constant matrices–coefficients A, B^1, B^2 and Q. Matrices are written by Eqs. (5). The characteristic equation $\det(n_1 B^1 + n_2 B^2 - c\,A) = 0$ has six nontrivial roots $c = \pm c_1''$, $\pm c_2''$ and $\pm c_3''$, that define three velocities of weak shock waves, propagating in the direction of the unit vector (n_1, n_2), – velocities of longitudinal waves, transverse waves and waves of rotational motion:

$$c_1'' = \sqrt{\frac{\lambda_1}{\rho_0}}, \quad c_2'' = \sqrt{\frac{\lambda_2}{\rho_0}}, \quad c_3'' = \sqrt{\frac{\alpha_2}{j_0}}, \tag{7}$$

where $\lambda_{1;2} = \dfrac{a_1 + a_2}{2} \pm \sqrt{\left(\dfrac{a_1 - a_2}{2}\right)^2 (n_1^2 - n_2^2)^2 + (b_1 + b_2)^2 \, n_1^2 \, n_2^2}$, and the triple zero root, corresponding to the contact discontinuities.

The state of a simple shear in the plane $x_1 x_2$ with constant shear rate $\dot{\chi}$, where $v_1 = 0$, $v_2 = \dot{\chi} x_1$, and ω_3 depends only on time, is described by the equation

$$j_0 \, \ddot{\omega}_3 = -(a_2 - b_2)(\omega_3 - \dot{\chi}). \tag{8}$$

The general solution of this equation: $\omega_3 = \dot{\chi} + C_1 \cos 2\pi\nu_0 t + C_2 \sin 2\pi\nu_0 t$ (C_1 and C_2 are the constants of integration) indicates to the oscillatory nature of rotation of the blocks under shear. The oscillation frequency ν_0, calculated by the formula: $\nu_0 = \sqrt{(a_2 - b_2)/j_0}$, is a phenomenological parameter of a material. On the basis of exact solutions and using numerical computations, in [4,5] it was shown that for an isotropic couple-stress medium this frequency is the resonance frequency: by means of applying a periodic external load with such frequency it is possible to excite the resonance of rotational motion of particles in a couple-stress medium.

The problem of determining the coefficients of the system of Eq. (5) by given parameters of materials and sizes of blocks and interlayers belongs to the class of inverse problems, and it has no trivial solution, although some of these coefficients one can find by simple ways. In particular, the coefficients a_1 and a_2 are calculated by the formulas $a_i = \rho_0 c_i''^2$ via the average density of a block medium and the velocities of longitudinal and transverse pendulum waves in the direction of the coordinate axes, received with the help of numerical results in the framework of the model of a block medium.

For a medium with square blocks, elongated in the direction of the x_3 axis, and interlayers of uniform thickness in both directions:

$$\rho_0 = \rho' + (\rho - \rho') \frac{h^2}{(h+\delta)^2}, \quad j_0 = \rho \frac{h^4}{6(h+\delta)^2}. \tag{9}$$

The coefficient b_1 is determined by a_1 for a given Poisson's ratio $\eta = b_1/(a_1+b_1)$, it is equal to 0.3 in computations. The coefficient $\alpha_2 = j_0 c_3''^2$ is calculated via the velocity of waves of rotational motion for blocks, which is approximately equal to $0.95 c_2''$.

Sum of the coefficients b_1 and b_2, characterizing the anisotropy of a medium, is estimated by the curvature of the wave fronts on the basis of formulas (7). Such estimate is possible, when the sign of the sum $b_1 + b_2$ is known in advance, because the velocities c_1'' and c_2'' are independent of this sign, and the permissible range of the coefficients $a_1 + a_2 > b_1 + b_2 > -a_1/2 + a_2$ (where the conditions on the strong convexity of a quadratic potential of stresses are fulfilled) includes both positive and negative values. Figure 3 shows typical velocity hodographs of longitudinal and transverse waves, which are calculated for $\delta = 0.1, 1$ and $5\,\mathrm{mm}$ by means of mechanical parameters of a block medium. Curves 1 correspond to $b_1 + b_2 = 0$, curves 3 – to $b_1 + b_2 = a_1 - a_2$ (the case of an isotropic medium), curves 5 correspond to the limiting value $b_1 + b_2 = a_1 + a_2$, curves 2 and 4 – to the intermediate values $b_1 + b_2 = (a_1 - a_2)/2$ and $b_1 + b_2 = a_1$. Analysis of the wave fields in Figs. 1 and 2 shows that for the parameters of a block medium,

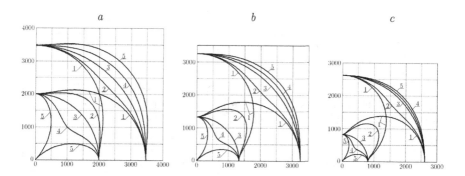

Fig. 3. Fronts of longitudinal and transverse waves in an orthotropic medium for different values of the parameter $b_1 + b_2$: (a) $\delta = 0.1$ mm, (b) $\delta = 1$ mm, (c) $\delta = 5$ mm

being under consideration, the hodographs of velocities lie between the curves 2 and 3, approaching the curve 3 if the thickness of interlayers tends to zero.

Selected from these considerations coefficients of the system of equations of the dynamics of an orthotropic couple-stress medium for materials of blocks and interlayers, whose parameters were used in solving problems for a block medium, are represented in the Table 1.

Table 1. Mechanical parameters of the couple-stress medium

δ, mm	ρ_0, kg/m^3	J_0, kg/m	a_1, GPa	b_1, GPa	a_2, GPa	b_2, GPa	α_2, MN	c_1'', m/s	c_2'', m/s	c_3'', m/s	ν_0 KHz
0.1	3690	6.15	44.5	19.1	14.8	6.77	22.2	3470	2000	1900	5.74
1.0	3650	6.05	38.6	16.5	6.46	4.05	9.60	3250	1330	1260	3.17
5.0	3470	5.59	23.8	10.2	2.33	0.53	3.39	2620	820	780	2.85

Numerical solution of boundary-value problems for the system of Eq. (5) is carried out by means of the procedure of two-cyclic splitting with respect to spatial variables. Parallel version of the algorithm is implemented on multiprocessor computers of the cluster architecture. Program codes are written in Fortran using the basic procedures of the MPI library. A detailed description of the numerical algorithm and the structure of computer programs by example of equations of an isotropic Cosserat continuum one can find in [5,6].

For comparison with computations based on the model of a block medium with compliant interlayers, results of which are shown in Figs. 1 and 2, the same problem was solved numerically in the framework of the theory of orthotropic couple-stress continuum. More coarser finite-difference grid of 1000×1000 square meshes with respect to spatial variables was used. Computations were performed on 100 processors of the MVS-100k cluster of the Joint Supercomputer Center of RAS (Moscow). As compared with the problem for a block medium, such problem for an orthotropic continuum is about eight times less computationally

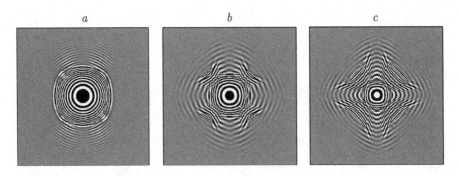

a   *b*   *c*

Fig. 4. Instant rotation of the central particle of an orthotropic Cosserat continuum. Level curves of the angular velocity ω_3: (*a*) $\delta = 0.1$ mm, (*b*) $\delta = 1$ mm, (*c*) $\delta = 5$ mm

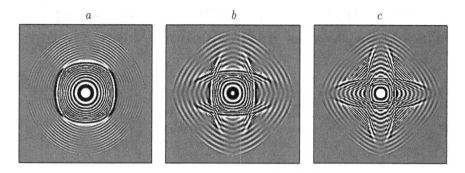

Fig. 5. Instant rotation of the central particle of an orthotropic Cosserat continuum. Level curves of the tangential stress σ_{21}: (a) $\delta = 0.1$ mm, (b) $\delta = 1$ mm, (c) $\delta = 5$ mm

intensive because of reduction of the dimension of a spatial grid and increase of the time step, maximum permissible by stability condition. Results of computations for the problem with free boundaries of a body and with initial localized rotation of a central mesh of finite-difference grid under $t = 0$ are represented in Figs. 4 and 5. Obtained wave patterns of angular velocity ω_3 and tangential stress σ_{21} are qualitatively similar to wave patterns, shown in Figs. 1 and 2.

Acknowledgements. This work was supported by the Russian Foundation for Basic Research (grant no. 14-01-00130) and the Complex Fundamental Research Program no. 18 of the Presidium of RAS.

References

1. Ayzenberg-Stepanenko, M.V., Slepyan, L.I.: Resonant-frequency primitive waveforms and star waves in lattices. J. Sound Vib. **313**(3–5), 812–821 (2008)
2. Alessandrini, B., Raganelli, V.: The propagation of elastic waves in a discrete medium. Eur. J. Mech., Ser. A: Solids **8**(2), 129–160 (1989)
3. Sadovskii, V.M., Sadovskaya, O.V., Pokhabova, M.A.: Modeling of elastic waves in a block medium based on equations of the Cosserat continuum. Comput. Continuum Mech. **7**(1), 52–60 (2014). [in Russian]
4. Sadovskaya, O.V., Sadovskii, V.M.: Analysis of rotational motion of material microstructure particles by equations of the Cosserat elasticity theory. Acoust. Phys. **56**(6), 942–950 (2010)
5. Sadovskaya, O.V., Sadovskii, V.M.: Mathematical Modeling in Mechanics of Granular Materials. Advanced Structured Materials, vol. 21. Springer, Heidelberg (2012)
6. Sadovskii, V.M., Sadovskaya, O.V., Varygina, M.P.: Numerical solution of dynamic problems in couple-stressed continuum on multiprocessor computer systems. Int. J. Numer. Anal. Modeling, Ser. B **2**(2–3), 215–230 (2011)

Computational Algorithm for Modeling Viscoelastic Waves in the Kelvin–Voigt Medium

Vladimir M. Sadovskii[✉]

Institute of Computational Modeling SB RAS, Akademgorodok 50/44,
660036 Krasnoyarsk, Russia
sadov@icm.krasn.ru

Abstract. Numerical algorithm for solving dynamic problems of the theory of viscoelastic medium of Kelvin–Voigt is worked out on the basis of Ivanov's method of constructing finite difference schemes with prescribed dissipative properties. In one-dimensional problem the results of computations are compared with the exact solution, describing the propagation of plane monochromatic waves. When solving two-dimensional problems, the total approximation method based on the splitting of the system with respect to the spatial variables is applied. The algorithm is tested on solving the problem of traveling surface waves. For illustration of the method, the numerical solution of Lamb's problem about instantaneous action of concentrated force on the boundary of a half-plane is represented in viscoelastic formulation.

Keywords: Viscoelasticity · Wave motion · Dissipative difference scheme · Computational algorithm

1 Introduction

Mathematical model of Kelvin–Voigt is one of two basic models of the mechanics of viscoelastic materials. Alternative Maxwell's model is used to describe sols – elastic fine particles suspended in a viscous liquid. Deformation of these materials consists of elastic and viscous parts. Under the influence of prolonged constant stresses the phenomenon of creep can be observed – irreversible strains increase monotonically and infinitely in time. The system of dynamic equations of the Maxwell viscoelastic medium refers to the hyperbolic type. This determines the choice of numerical methods for solving wave problems based on it.

The Kelvin–Voigt model describes gels – porous elastic skeletons, filled by viscous fluid. In this case stresses (but not strains) are the sum of elastic and viscous components. Deformation of these materials under constant load is limited – it grows only up to a certain limit. This is an argument in favor of the applicability of the Kelvin–Voigt equations to analysis of the stress-strain state of geomaterials (soils, grounds and fractured rocks).

The system of equations of the dynamics of the Kelvin–Voigt medium is not a hyperbolic system, so the choice of reliable numerical methods is not trivial.

© Springer International Publishing Switzerland 2015
I. Dimov et al. (Eds.): FDM 2014, LNCS 9045, pp. 348–355, 2015.
DOI: 10.1007/978-3-319-20239-6_38

When solving the problems of wave propagation, the methods based on integral transforms on time are effective [1]. Slow wave motions of a viscoelastic medium are analyzed using the finite element approximation in combination with the discontinuous Galerkin method [2,3], whose indisputable advantage is the ability to consider the domains of complex shape.

The method for constructing dissipative difference schemes with a predetermined artificial energy dissipation for the solution of boundary value problems of the linear dynamic elasticity was proposed by Ivanov [4]. This method proved to be effective for a wide range of problems, including boundary value problems in the theory of plates and shells. In the present paper, the Ivanov method is applied to the model of the Kelvin–Voigt viscoelastic medium.

2   One-Dimensional Model

The system of equations of one-dimensional motions of the Kelvin–Voigt viscoelastic medium for longitudinal waves can be written in the following dimensionless form:

$$\frac{\partial v}{\partial t} = \frac{\partial \sigma}{\partial x}, \qquad \frac{\partial s}{\partial t} = \frac{\partial v}{\partial x}, \qquad \sigma = s + \eta \frac{\partial v}{\partial x}. \tag{1}$$

Here v is the velocity of particles, σ and s are the full stress and its elastic part, η is the dimensionless parameter of viscosity.

Under numerical solution of boundary value problems for the system of Eq. (1), according to the Ivanov method, the extended system is considered:

$$\frac{\partial v}{\partial t} = \frac{\partial \sigma'}{\partial x}, \qquad \frac{\partial s}{\partial t} = \frac{\partial v'}{\partial x}, \qquad \sigma = s + \eta \frac{\partial v'}{\partial x}, \tag{2}$$

where v' and σ' are the auxiliary functions, which are not equal to v and σ in general case. Multiplying the first equation of (2) by v, the second one – by s, and summing the results, one can obtain the equation

$$\frac{1}{2}\frac{\partial}{\partial t}(v^2 + s^2) = v\frac{\partial \sigma'}{\partial x} + s\frac{\partial v'}{\partial x} = (v - v')\frac{\partial \sigma'}{\partial x} + (\sigma - \sigma')\frac{\partial v'}{\partial x} + \frac{\partial(v'\sigma')}{\partial x} - \eta\left(\frac{\partial v'}{\partial x}\right)^2,$$

which is reduced to the equation of the energy change for (1), if $v = v'$ and $\sigma = \sigma'$. Closing equations to the system (2) are written as

$$\begin{pmatrix} v - v' \\ \sigma - \sigma' \end{pmatrix} = -D\frac{\partial}{\partial x}\begin{pmatrix} \sigma' \\ v' \end{pmatrix}, \tag{3}$$

where D is the given positively semidefinite matrix of the dimension 2×2.

For the system (2), (3) the energy equation holds. Additional dissipative term appears in this equation as compared with the similar equation for (1). This term is a non-negative quadratic form with respect to spatial derivatives of σ' and v', which coefficients are the elements of matrix D. Using this equation, it is easy to obtain a priori estimates, which allow us to prove the uniqueness and continuous dependence on the initial data of the solutions of boundary value problems with dissipative boundary conditions for the extended system.

In fact, when constructing the difference scheme, these calculations are repeated at a discrete level. Let τ and h be the steps of a uniform grid in time and in spatial variable. The discrete analog of (2) can be written as a system of difference equations:

$$\frac{v^j - v_j}{\tau} = \frac{\sigma_{j+1/2} - \sigma_{j-1/2}}{h}, \qquad \frac{s^j - s_j}{\tau} = \frac{v_{j+1/2} - v_{j-1/2}}{h},$$

$$\sigma_j = \frac{s^j + s_j}{2} + \eta \frac{v_{j+1/2} - v_{j-1/2}}{h}. \tag{4}$$

Here the quantities with integer indices $j = 1, ..., n$ approximate the basic functions at lower and upper sides of the space-time grid mesh, and the quantities with half-integer indices approximate the auxiliary functions at lateral sides of the mesh.

Discrete analog of the equation of energy is obtained by multiplying the first and second equations of (4) by $(v^j + v_j)/2$ and $(s^j + s_j)/2$, respectively. So, the equations approximating (3) have the next form:

$$\begin{pmatrix} v^j + v_j - v_{j+1/2} - v_{j-1/2} \\ 2\,\sigma_j - \sigma_{j+1/2} - \sigma_{j-1/2} \end{pmatrix} = -\frac{2}{h} \, D \begin{pmatrix} \sigma_{j+1/2} - \sigma_{j-1/2} \\ v_{j+1/2} - v_{j-1/2} \end{pmatrix}. \tag{5}$$

Using the method of a priori estimates, one can obtain the rigorous proof of stepwise stability for difference scheme in the mean-square norm, if dissipative boundary conditions are formulated at the boundary of computational domain in terms of v, σ, which guarantee the fulfillment of inequalities

$$\left(\Delta v \, \Delta\sigma\right)_{1/2} \leq 0, \qquad \left(\Delta v \, \Delta\sigma\right)_{n+1/2} \geq 0 \tag{6}$$

for the differences $\Delta v = \hat{v} - v$, $\Delta\sigma = \hat{\sigma} - \sigma$ of arbitrary pairs of functions v, σ and $\hat{v}$, $\hat{\sigma}$, satisfying these conditions.

The scheme (4), (5) approximates the system (1) only if the coefficients of matrix D are small: $D = O(h)$. For simplicity, we choose as D a diagonal matrix of the special form:

$$D = \begin{pmatrix} \alpha & 0 \\ 0 & \beta \end{pmatrix}, \qquad \alpha = \beta = \frac{h - \tau}{2} \geq 0.$$

Ivanov showed [4], that for the hyperbolic system of linear elasticity this choice corresponds to the Godunov gap decay scheme. In this case, in view of (4), the system of equations (5) is transformed to the next one:

$$\frac{v_{j+1/2} + v_{j-1/2}}{2} - \frac{\sigma_{j+1/2} - \sigma_{j-1/2}}{2} = v_j,$$

$$\frac{\sigma_{j+1/2} + \sigma_{j-1/2}}{2} - \left(1 + \frac{2\eta}{h}\right)\frac{v_{j+1/2} - v_{j-1/2}}{2} = s_j. \tag{7}$$

To construct efficient algorithm of numerical implementation of this scheme, we choose resolving equations for the system (7) in the following form:

$$v_{j+1/2} - \sigma_{j+1/2} = v_j - X_j, \quad v_{j-1/2} + \sigma_{j-1/2} = v_j + Y_j.$$

Unknown values X_j and Y_j can be determined from a comparison of these equations with the equations (7)

$$X_j = Y_j = s_j + \eta \, \frac{v_{j+1/2} - v_{j-1/2}}{h}.$$

The resolving equations are a generalization of the relationships on characteristics of the equations of a non-viscous elastic medium. For the interior cells of computational domain, they lead to the following system of difference equations:

$$v_{j-1/2} = \frac{v_j + v_{j-1}}{2} + \frac{s_j - s_{j-1}}{2} + \eta \, \frac{v_{j+1/2} - 2\,v_{j-1/2} + v_{j-3/2}}{2\,h},$$

$$\sigma_{j-1/2} = \frac{s_j + s_{j-1}}{2} + \frac{v_j - v_{j-1}}{2} + \eta \, \frac{v_{j+1/2} - v_{j-3/2}}{2\,h}. \tag{8}$$

The first equation of (8) allows to calculate the values $v_{j-1/2}$ using the tridiagonal matrix algorithm. To do this, the boundary conditions are set, the main of which are the kinematic conditions for velocities $v_{1/2} = v_{1/2}^0$, $v_{n+1/2} = v_{n+1/2}^0$ and the dynamic conditions for stresses $\sigma_{1/2} = \sigma_{1/2}^0$, $\sigma_{n+1/2} = \sigma_{n+1/2}^0$, satisfying the inequalities (6). Also there are allowed the mixed boundary conditions of various types on different boundaries. Under numerical implementation the dynamic conditions are rewritten in terms of velocities using the resolving equations.

Note that the initial data of the problem are also formulated in terms of velocities and total stresses. The initial values of s_j, which are necessary to organize the computational process, are determined by virtue of the third equation of (1). With the help of the second equation of (8) and the boundary conditions the values $\sigma_{j-1/2}$ for all $j = 1, ..., n+1$ are calculated after realization of the tridiagonal matrix algorithm. This is a predictor step of the difference scheme. A corrector step is performed by the formulas (4).

3 Governing Equations in Plane Case

Two-dimensional equations are written in terms of projections of the velocity vector v_j on the axes of a Cartesian coordinate system and of the components of symmetric tensors of total stresses σ_{jk} and elastic stresses s_{jk}. The closed system of equations is reduced to the matrix form:

$$A\frac{\partial U}{\partial t} = B^1 \frac{\partial V}{\partial x_1} + B^2 \frac{\partial V}{\partial x_2}, \qquad V = U + C^1 \frac{\partial V}{\partial x_1} + C^2 \frac{\partial V}{\partial x_2}, \tag{9}$$

$$U = \begin{Vmatrix} v_1 \\ v_2 \\ s_{11} \\ s_{22} \\ s_{12} \end{Vmatrix}, \quad V = \begin{Vmatrix} v_1 \\ v_2 \\ \sigma_{11} \\ \sigma_{22} \\ \sigma_{12} \end{Vmatrix}, \quad A = \begin{Vmatrix} \rho & 0 & 0 & 0 & 0 \\ 0 & \rho & 0 & 0 & 0 \\ 0 & 0 & a_1 & a_2 & 0 \\ 0 & 0 & a_2 & a_1 & 0 \\ 0 & 0 & 0 & 0 & a_3 \end{Vmatrix}, \quad B^1 = \begin{Vmatrix} 0 & 0 & 1 & 0 & 0 \\ 0 & 0 & 0 & 0 & 1 \\ 1 & 0 & 0 & 0 & 0 \\ 0 & 0 & 0 & 0 & 0 \\ 0 & 1 & 0 & 0 & 0 \end{Vmatrix},$$

$$B^2 = \left\|\begin{matrix} 0\,0\,0\,0\,1 \\ 0\,0\,0\,1\,0 \\ 0\,0\,0\,0\,0 \\ 0\,1\,0\,0\,0 \\ 1\,0\,0\,0\,0 \end{matrix}\right\|, \quad C^1 = \left\|\begin{matrix} 0 & 0 & 0\,0\,0 \\ 0 & 0 & 0\,0\,0 \\ \alpha_1 & 0 & 0\,0\,0 \\ \alpha_2 & 0 & 0\,0\,0 \\ 0 & \alpha_3 & 0\,0\,0 \end{matrix}\right\|, \quad C^2 = \left\|\begin{matrix} 0 & 0 & 0\,0\,0 \\ 0 & 0 & 0\,0\,0 \\ 0 & \alpha_2 & 0\,0\,0 \\ 0 & \alpha_1 & 0\,0\,0 \\ \alpha_3 & 0 & 0\,0\,0 \end{matrix}\right\|.$$

Here ρ is the density, the coefficients

$$a_1 = \frac{3\kappa + 4\mu}{4\mu(3\kappa + \mu)}, \quad a_2 = -\frac{3\kappa - 2\mu}{4\mu(3\kappa + \mu)}, \quad a_3 = \frac{1}{\mu}, \quad \alpha_1 = \frac{4\eta}{3}, \quad \alpha_2 = -\frac{2\eta}{3}, \quad \alpha_3 = \eta$$

are expressed in terms of the phenomenological parameters of a medium – the isothermic bulk modulus κ, the shear modulus μ and the coefficient of viscosity η. Matrices A, B^1 and B^2 are symmetric. Matrix A is positive definite under natural constraints on the parameters: $\rho > 0$, $\kappa > 0$, $\mu > 0$ and $\eta \geq 0$. For the system (9) the energy equation

$$\frac{1}{2}\frac{\partial}{\partial t} U A U = \frac{1}{2}\frac{\partial}{\partial x_1} V B^1 V + \frac{1}{2}\frac{\partial}{\partial x_2} V B^2 V - \left\|\frac{\partial V}{\partial x_1}\ \frac{\partial V}{\partial x_2}\right\| D^0 \left\|\begin{matrix} \dfrac{\partial V}{\partial x_1} \\[2pt] \dfrac{\partial V}{\partial x_2} \end{matrix}\right\|$$

holds, where D^0 is the symmetric and positively semidefinite matrix which is the product of blocky matrices (asterisk denotes the transposition):

$$D^0 = \left\|\begin{matrix} C^1 & 0 \\ 0 & C^2 \end{matrix}\right\|^{*} \left\|\begin{matrix} B^1 & B^2 \\ B^1 & B^2 \end{matrix}\right\|.$$

The nonzero coefficients of this matrix are:

$$D^0_{11} = D^0_{77} = \alpha_1, \quad D^0_{17} = D^0_{71} = \alpha_2, \quad D^0_{22} = D^0_{66} = D^0_{26} = D^0_{62} = \alpha_3.$$

Sequential construction of the difference scheme with specified dissipative properties for the solution of two-dimensional system by analogy with one-dimensional one leads to a special class of implicit schemes, for numerical implementation of which it is difficult to find effective algorithms. Therefore numerical solution of boundary value problems for the system (9) was carried out by the procedure of splitting with respect to spatial variables. Along with classical version of the splitting method (method of weak approximation), two-cyclic version of splitting was used, in which at each time interval $(t, t + \Delta t)$ the next series of one-dimensional problems is solved:

$$A\frac{\partial U^{(1)}}{\partial t} = B^1 \frac{\partial V^{(1)}}{\partial x_1}, \qquad V^{(1)} = U^{(1)} + C^1\frac{\partial V^{(1)}}{\partial x_1} + C^2\frac{\partial V^{(0)}}{\partial x_2},$$

$$A\frac{\partial U^{(2)}}{\partial t} = B^2 \frac{\partial V^{(2)}}{\partial x_2}, \qquad V^{(2)} = U^{(2)} + C^1\frac{\partial V^{(1)}}{\partial x_1} + C^2\frac{\partial V^{(2)}}{\partial x_2},$$

$$A\frac{\partial U^{(3)}}{\partial t} = B^2 \frac{\partial V^{(3)}}{\partial x_2}, \qquad V^{(3)} = U^{(3)} + C^1\frac{\partial V^{(2)}}{\partial x_1} + C^2\frac{\partial V^{(3)}}{\partial x_2},$$

$$A\frac{\partial U^{(4)}}{\partial t} = B^1 \frac{\partial V^{(4)}}{\partial x_1}, \qquad V^{(4)} = U^{(4)} + C^1 \frac{\partial V^{(4)}}{\partial x_1} + C^2 \frac{\partial V^{(3)}}{\partial x_2}.$$

The vector–function $V^{(0)}$ is taken from the previous time step, and at $t = 0$ it is taken from the initial data of the problem. The initial data for systems of equations at splitting stages and the solution, related to the new time layer, are determined by the formulas:

$$U^{(1)}(t) = U(t), \qquad U^{(2)}(t) = U^{(1)}(t + \Delta t/2), \qquad U^{(3)}(t + \Delta t/2) = U^{(2)}(t + \Delta t/2),$$

$$U^{(4)}(t + \Delta t/2) = U^{(3)}(t + \Delta t), \qquad U(t + \Delta t) = U^{(4)}(t + \Delta t).$$

A large series of methodological computations showed, that the obtained difference scheme is stable under the fulfillment of conditions of stability for one-dimensional schemes at the stages of splitting, or under a little more strict conditions, where an additional energy dissipation suppresses the influence of the terms with derivatives in transverse direction in the formulas for recalculation of vector V via U, taken from previous stages.

4   Numerical Results

When testing the algorithm and the program, the exact solution of the form $V = \bar{V} \exp(i\omega t - \lambda x_1 - i l x_2)$ is used, which describes the monochromatic wave, damped in depth, traveling with the constant velocity $c = \omega/l$ along the axis x_2 on the boundary of viscoelastic half-plane $x_1 \geq 0$. The solution changes qualitatively depending on that, how this velocity correlates to the velocities of longitudinal and transverse elastic waves $c_1 = \sqrt{(\kappa + 4\mu/3)/\rho}$ and $c_2 = \sqrt{\mu/\rho}$. In supersonic regimes the degree of damping of the wave is essentially less than in corresponding subsonic regimes.

Numerical solution of the problem is constructed on the basis of two-dimensional system of dimensionless equations, obtained with the help of formulas:

$$x'_j = x_j/a, \quad t' = c_p t/a, \quad v'_j = v_j/c_p, \quad \sigma'_{jk} = \sigma_{jk}/(\rho c_1^2), \quad s'_{jk} = s_{jk}/(\rho c_1^2),$$

where a is the spatial scale of the problem, the ratio of elastic longitudinal and transverse waves is $c_1 = 2 c_2$, the dimensionless viscosity $\eta/(\rho a c_1) = 0.001$. In Fig. 1$a$ the surface of normal stress σ_{11} is shown in the case of supersonic longitudinal wave moving with the velocity $c = 1.5 c_1$. In Fig. 1b the case of subsonic wave with the velocity $c = c_1/2$ is represented. A posteriori analysis of the error of numerical solution for different ratios of grid steps in time and spatial variables showed, that the scheme corresponds to the first order of approximation, however, the procedure of reconstruction of the solution at the stages of splitting leads to a significant increase in accuracy.

.

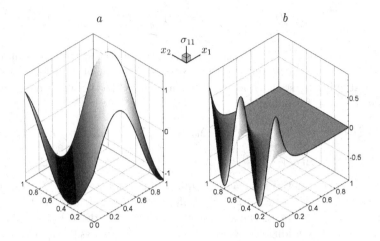

Fig. 1. The surface of normal stress σ_{11} in the problem of traveling wave: (a) supersonic wave with $c = 1.5\,c_1$, (b) subsonic wave with $c = c_1/2$

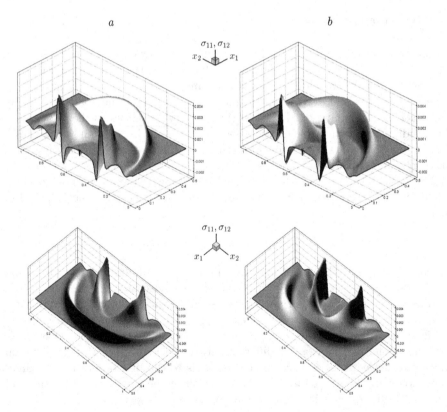

Fig. 2. The surface of normal stress σ_{11} (a) and the surface of tangential stress σ_{12} (b) in the Lamb problem

Figure 2 shows the surfaces of normal and tangential stresses in the Lamb problem of instantaneous action of a concentrated load, normal to the boundary of a viscoelastic half-plane. Computations were performed for the same values of parameters. The time in the figures corresponds to the time of arrival of a head longitudinal wave at the boundary of computational domain. Obtained spatial patterns allow to estimate visually the ratio of amplitudes of longitudinal, transverse, conical and surface waves, which take the more flat and more smooth form with the increasing of viscosity.

Note in conclusion, that the objective point of this research is to develop a reliable numerical algorithm for computation of the granular flows with stagnant zones in moving stream based on the original mathematical model, proposed in [5–7]. The finite difference scheme, constructed by the Ivanov method, has an important advantage because of it is implicit in the predictor step and explicit in the corrector step. So, it allows reasonably apply the nonlinear procedures of the solution correction, such as the Wilkins correction of stresses, when calculating the problems taking into account plastic deformations and different resistance of a material to tension and compression.

Acknowledgements. This work was supported by the Complex Fundamental Research Program no. 18 "Algorithms and Software for Computational Systems of Superhigh Productivity" of the Presidium of RAS and the Russian Foundation for Basic Research (grant no. 14–01–00130).

References

1. Kosloff, D., Baysal, E.: Forward modeling by a fourier method. Geophys. **47**(10), 1402–1412 (1982)
2. Carcione, J.M., Poletto, F., Gei, D.: 3-D wave simulation in anelastic media using the Kelvin-Voigt constitutive equation. J. Comput. Phys. **196**(1), 282–297 (2004)
3. Bécache, E., Ezziani, A., Joly, P.: A mixed finite element approach for viscoelastic wave propagation. Comput. Geosci. **8**(3), 255–299 (2005)
4. Ivanov, G.V., Volchkov, Y.M., Bogulskii, I.O., Anisimov, S.A., Kurguzov, V.D.: Numerical Solution of Dynamic Elastic-Plastic Problems of Deformable Solids. Sib. Univ. Izd, Novosibirsk (2002). (in Russian)
5. Sadovskaya, O.V., Sadovskii, V.M.: The theory of finite strains of a granular material. J. Appl. Math. Mech. **71**(1), 93–110 (2007)
6. Sadovskaya, O., Sadovskii, V.: Mathematical Modeling in Mechanics of Granular Materials. Advanced Structured Materials, vol. 21. Springer, Heidelberg (2012)
7. Krasnenko, A.N., Sadovskaya, O.V.: Mathematical modeling of shear flows of a granular medium with stagnant zones. In: Proceedings of Conference Fundamental and Applied Problems of Mechanics and Control Processes, pp. 90–96. Dal'nauka, Vladivostok (2011) (in Russian)

Investigating the Dynamics of Traffic Flow on a Highway in a Class of Discontinuous Functions

Bahaddin Sinsoysal[1][(✉)], Hakan Bal[2], and E. Ilhan Sahin[3]

[1] Department of Mathematics and Computing, Beykent University, Istanbul, Turkey
{bsinsoysal,hbal}@beykent.edu.tr
[2] Department of International Trade, Beykent University, Istanbul, Turkey
[3] Satellite Communication and Remote Sensing, ITU, Informatics Institute, Ayazaga Campus, 34396 Istanbul, Turkey
shnethem@gmail.com

Abstract. This work is devoted to finding a solution of Riemann problem for the first order nonlinear partial equation which describes the traffic flow on highway. When $\rho_\ell > \rho_r$, the solution is presented as a piecewise continuous function, where ρ_ℓ and ρ_r are the densities of cars on the left and right side of the intersection respectively. On the contrary case, a shock of which the location is unknown beforehand arises in the solution. In this case, a special auxiliary problem is introduced, the solution of which makes it possible to write the exact solution showing the locations of shock. For the realization of the proposed method, the parameters of the flow are also found.

Keywords: Traffic flow equation · Shock · Characteristics · Weak solution

1 Introduction

As known, to study the flow of vehicles on a highway is one of today's most contemporary problems. To solve the problem of traffic congestion, it is not enough to intervene at the level of instrumentality. It is essential to the creation of mathematical models to solve such problems. First and considerable researches in this field have been worked through in [1,4–6].

Let $\rho(x,t)$ and $q(x,t)$ denote vehicle density per unit length and number of vehicles passing through any highway section per unit time, respectively. In this instance, $\int_a^b \rho(x,t_1)dx$ and $\int_a^b \rho(x,t_2)dx$ refer to the numbers of vehicles on any $[a,b]$ section of the highway at times $t = t_1$ and $t = t_2$. Since the integrals $\int_{t_1}^{t_2} q(a,s)ds$ and $\int_{t_1}^{t_2} q(b,s)ds$ denote the number of vehicles entering into the highway at the point a and leaving at the point b at the time period $\Delta t = t_2 - t_1$, respectively, the following balance equation is valid

$$\int_a^b \rho(x,t_2)dx - \int_a^b \rho(x,t_1)dx = \int_{t_1}^{t_2} q(a,s)ds - \int_{t_1}^{t_2} q(b,s)ds. \qquad (1)$$

© Springer International Publishing Switzerland 2015
I. Dimov et al. (Eds.): FDM 2014, LNCS 9045, pp. 356–363, 2015.
DOI: 10.1007/978-3-319-20239-6_39

Applying the mean value theorem to Eq. (1), we get

$$\frac{\partial \rho(x,t)}{\partial t} + \frac{\partial q(x,t)}{\partial x} = 0. \tag{2}$$

To investigate the flow dynamics, it is necessary to know the functional relation of the function q with the local density ρ in Eq. (2). In the process of derivation of the Eq. (2), the physical assumptions below are made

1. The vehicle flux on the highway is sufficiently dense.
2. No vehicles enter into and leave the highway in the region where the vehicles density is high.
3. Driving reflexes are not considered.

2   Finding the Flow Parameters

First, let us formulate the speed of vehicles on the highway. However, the creation of this formula can be obtained based on theoretical and experimental data. To express the speed of a vehicle these formulas can be considered as the first approach.

In the considering interval $[a, b]$, let us suppose that the vehicles are aligned bumper to bumper. In addition, according to the kinematic theory, the flow speed is as follows

$$v = \frac{q}{\rho}. \tag{3}$$

The connection between the speed of vehicle and density is found as

$$v(\rho) = v_{\max} - \frac{v_{\max}\rho}{\rho_{\max}} \tag{4}$$

from the equation of the line passing through the points $A(0, v_{\max})$ and $B(\rho_{\max}, 0)$. Taking Eq. (3) into account we get

$$q = Q(\rho) = \rho v(\rho) = v_{\max}\left(\rho - \frac{\rho^2}{\rho_{\max}}\right). \tag{5}$$

The actual observed number of vehicles on the highway indicate that it is

$$\rho_{max} = 225\frac{vehicle}{mil}, \quad \rho_j = 80\frac{vehicle}{mil}, \quad q_{max} = 1500\frac{vehicle}{mil}$$

in a single-lane roads. These numbers for a single-lane highway are accepted as a first approximation. The number of vehicles in a highway can be expressed as the product of these numbers with the number of lanes. According to the observations on the highway, the maximum flow value in low speeds are $v = \frac{q_{max}}{\rho_{max}} = 20\frac{mil}{hour}$, [6].

Dispersion speed of the wave is

$$c(\rho) = Q'(\rho) = v(\rho) + \rho v'(\rho). \tag{6}$$

Since $v'(\rho) < 0$, the speed $c(\rho)$ is less than the speed of the moving vehicles. Since the wave propagates in the opposite direction of the traffic flow, it informs the drivers that there is a problem ahead. The speed $c(\rho)$ is equal to the slope of curves $Q(\rho)$, thus the wave propagates forward if $\rho < \rho_j$, and backwards if $\rho > \rho_j$. If $\rho = \rho_j$, in other words $\rho = \rho_j$ takes the maximum value, the wave remains stationary relative to the road.

3 Simulation Model and Its Solution

To analyze the dynamics of traffic flow, we will investigate the equation

$$\frac{\partial \rho(x,t)}{\partial t} + \frac{\partial Q(x,t)}{\partial x} = 0 \tag{7}$$

with following initial condition

$$\rho(x,t) = \begin{cases} \rho_\ell, & x < 0, \\ \rho_r, & x > 0 \end{cases} \tag{8}$$

where the numbers ρ_ℓ and ρ_r denote the values of the density ρ at the behind and in front of the jump. Let us assume $\rho_\ell > \rho_r$, firstly.

Using the method of characteristics for the solution of the problem (7) and (8) we get

$$\rho(x,t) = \begin{cases} \rho_\ell, & \xi < 0, \\ \rho_r, & \xi > 0. \end{cases} \tag{9}$$

Here, ξ is the equation of characteristics, and can be written as

$$\xi = x - Q'(\rho)t. \tag{10}$$

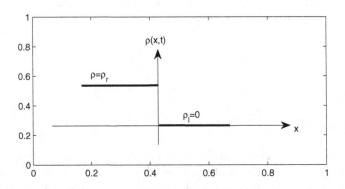

Fig. 1. Initial condition when $\rho_\ell > \rho_r$

As known from the general theory, when $\rho_\ell > \rho_r$ multi-valuable state does not occur in the solution, Fig. 1. Since $\rho_\ell > \rho_r$ and $\rho_r \leq \rho \leq \rho_\ell$, all automodel

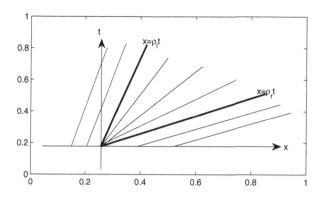

Fig. 2. Characteristics of the Equation (7)

solutions of the Eq. (7) remaining between the lines $x = \rho_r t$ and $x = \rho_\ell t$ are lines crossing the origin, that is, $\xi = 0$, Fig. 2

$$x = \rho_\ell t, \quad x = \rho_r t.$$

To express a physically meaningful solution of the problem (7) and (8), It is necessary to combine the line

$$x = v_{\max}\left(1 - \frac{2}{\rho_{\max}}\rho\right)t \tag{11}$$

with the lines $\rho = \rho_r$ and $\rho = \rho_\ell$ so that the initial condition is satisfied and the solution is continuous. Therefore we get

1. if $\xi < 0$, then $\rho = \rho_\ell$ and

$$\frac{x}{t} < v_{\max}\left(1 - \frac{2}{\rho_{\max}}\rho_\ell\right),$$

2. if $\xi > 0$, then $\rho = \rho_r$ and

$$\frac{x}{t} > v_{\max}\left(1 - \frac{2}{\rho_{\max}}\rho_r\right).$$

3. When $\xi = 0$, using the Eq. (11) we have

$$\rho = -\frac{\rho_{\max}}{v_{\max}}\frac{x}{2t} + \frac{\rho_{\max}}{2}.$$

Taking these expressions into account, we get the solution of the problem as follows

$$\rho(x,t) = \begin{cases} \rho_\ell, & \frac{x}{t} < v_{\max}\left(1 - \frac{2}{\rho_{\max}}\rho_\ell\right), \\[2mm] -\frac{\rho_{\max}}{v_{\max}}\frac{x}{2t} + \frac{\rho_{\max}}{2}, & v_{\max}\left(1 - \frac{2}{\rho_{\max}}\rho_\ell\right) < \frac{x}{t} < v_{\max}\left(1 - \frac{2}{\rho_{\max}}\rho_r\right), \\[2mm] \rho_r, & \frac{x}{t} > v_{\max}\left(1 - \frac{2}{\rho_{\max}}\rho_r\right). \end{cases} \tag{12}$$

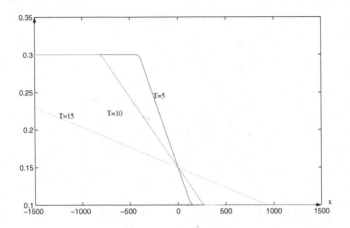

Fig. 3. Graphs of the solution (12)

The graph of the solution defined by the expression (12) is shown in Fig. 3.

Now, we will investigate the case where the initial distribution of vehicles are as shown in Fig. 4, where $\rho_\ell < \rho_r$. In this case, instead of solving the problem (7) and (8), we need to solve the following problem as in [2, 3]

$$\frac{\partial u(x,t)}{\partial t} + Q\left(\frac{\partial u(x,t)}{\partial x}\right) = 0 \tag{13}$$

$$u(x,0) = \begin{cases} \rho_\ell x, & x < 0, \\ \rho_r x, & x > 0. \end{cases} \tag{14}$$

The problem (13) and (14) is called the auxiliary problem. The solution becomes

$$u(x,t) = \begin{cases} u_-, & \xi < 0, \\ u_+, & \xi > 0. \end{cases} \tag{15}$$

The problem (13) and (14) was examined in [2, 3], and it was proven that the expression $\rho(x,t) = \frac{\partial u(x,t)}{\partial x}$ is the solution of the problem (7) and (8). Here,

$$u_- = -\frac{v_{\max}}{\rho_{\max}}\rho_\ell^2 t + \rho_\ell\left[x - v_{\max}\left(1 - \frac{2\rho_\ell}{\rho_{\max}}\right)\right]t \tag{16}$$

and

$$u_+ = -\frac{v_{\max}}{\rho_{\max}}\rho_r^2 t + \rho_r\left[x - v_{\max}\left(1 - \frac{2\rho_r}{\rho_{\max}}\right)\right]t. \tag{17}$$

The characteristics of the problem corresponding to this situation are shown in Fig. 4. According to the general theory, in this case a jump occurs in the solution. Location of the jump is found from the equation as follows

$$\frac{x}{t} = v_{\max} - \frac{v_{\max}}{\rho_{\max}}(\rho_\ell + \rho_r) \equiv U \tag{18}$$

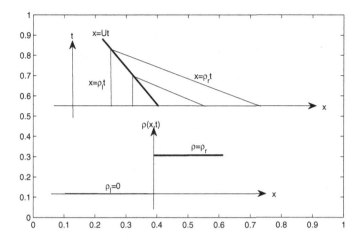

Fig. 4. Initial distribution of vehicles and characteristics when $\rho_\ell < \rho_r$

using the equality $u_- = u_+$. Taking the expression (18) into account, the solution of the problem (7) and (8) is written as

$$\rho(x,t) = \begin{cases} \rho_\ell, & \frac{x}{t} < U \\ \\ \rho_r, & \frac{x}{t} > U. \end{cases} \tag{19}$$

The graphs of the solution in this case are shown in Fig. 5.

The traffic lights can be regulated using this results. To this end, it is sufficient to establish the graphs of the family of characteristics on the (x, t) plane. They are constant density lines such that, their $c(\rho)$ slopes determine the constant values of $\rho(x, t)$ taken on these lines. Let us apply the above theory to the green light problem on the traffic. We let $\rho_\ell = \rho_{max}$ and $\rho_r = 0$. This case physically refers to the traffic stop as the traffic light turns red at the point $x = 0$. As shown, the vehicles density has a jump at the point $x = 0$. The distribution of the vehicles (initial state) is shown in Fig. 4.

As the traffic light turns green from red, we are required to determine the dynamic distribution of vehicles. According to the general theory,

$$\frac{dx}{dt} = c(\rho) = Q'(\rho) = v_{max}\left(1 - \frac{2}{\rho_{max}}\rho\right)$$

and remains constant on characteristics. The slope of a characteristic that intersects the x axis at any point $x = x_0 > 0$ in the positive direction is given as

$$\frac{dx}{dt} = c(\rho) = Q'(\rho) = v_{max}\left(1 - \frac{2\rho}{\rho_{max}}\right) = v_{max}.$$

The equation of the family of characteristics is given as $x = v_{max}t + x_0$. On the other hand, the slope of the characteristics that intersects the x axis at any point in the negative x axis direction is

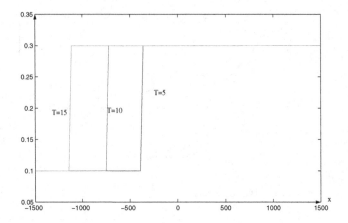

Fig. 5. Graphs of the solution (19)

$$\frac{dx}{dt} = c(\rho) = Q'(\rho) = v_{max}\left(1 - \frac{2\rho}{\rho_{max}}\right) = -v_{max}.$$

It is obvious that, the equation of the family of these characteristics is written as $x = -v_{max}t + x_0$.

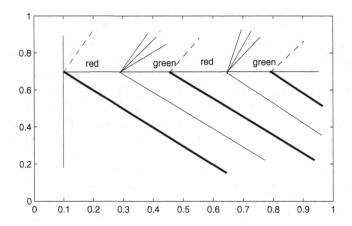

Fig. 6. Family of the characteristics for determining traffic lights

The characteristics settled in the domain $-v_{max}t < x < v_{max}t$ should have a common point, so $\frac{dx}{dt} = \frac{x}{t}$. In this case, the equation of the family of characteristics is as shown in Fig. 6.

The traffic flow dynamic that flows slowly across the traffic lights can be seen in Fig. 6.

4   Conclusion

The advantages of the proposed model are listed below:
1. The vehicle density at any point on the highway is specified.
2. Average vehicle speed is determined to estimate the maximum flow of vehicles on the highway.
3. The traffic lights are managed to avoid possible traffic congestion.

References

1. Lighthill, M.J., Whitham, G.B.: On kinematic waves, i. flood movement in long rivers; ii. theory of traffic flow on long crowded roads. Proc. Royal Soc. London Ser. A **229**, 281–345 (1955)
2. Rasulov, M.: Sureksiz Fonksiyonlar Sinifinda Korunum Kurallari. Seckin Yayinevi, Istanbul (2011)
3. Rasulov, M., Sinsoysal, B., Hayta, S.: Numerical simulation of initial and initial-boundary value problems for traffic flow in a class of discontinuous functions. WSEAS Trans. Math. **5**(12), 1339–1342 (2006)
4. Richards, P.I.: Shock waves on the highway. Oper. Res. **4**, 42–51 (1956)
5. Siebel, F., Mauser, W.: On the fundamental diagram of traffic flow. SIAM J. Appl. Math. **66**(4), 1150–1162 (2006)
6. Whitham, G.B.: Linear and Nonlinear Waves. Wiley, New York (1974)

Numerical Simulation of Thermoelasticity Problems on High Performance Computing Systems

Petr V. Sivtsev[1]([✉]), Petr N. Vabishchevich[2], and Maria V. Vasilyeva[1]

[1] North-Eastern Federal University, 58, Belinskogo, 677000 Yakutsk, Russia
sivkapetr@mail.ru
[2] Nuclear Safety Institute, 52, B. Tulskaya, 115191 Moscow, Russia

Abstract. In this work we consider the coupled linear system of equations for temperature and displacements which describes the thermoelastic behaviour of the body. For numerical solution we approximate our system using finite element method. As model problem for simulation we consider the thermomechanical state of the ceramic substrates with metallization, which are used for the manufacturing of light-emitting diode modules. The results of numerical simulation of the 3D problem in the complex geometric area are presented.

1 Introduction

Many applied problems of mathematical modeling are connected with the calculation of the stress-strain state of solids. In many cases, the deformation is caused by thermal expansion. The thermoelasticity models are used for their research.

Basic mathematical models include heat conduction equation and Lame thermoelasticity equation for displacements [1–4]. The fundamental point is that the system is tied up, the equation for displacement comprises volumetric force proportional to the temperature gradient and the temperature equation includes a term that describes the compressibility of the medium.

In this work we consider the coupled linear system of equations for temperature and displacements which describes the thermoelastic behavior of the body. For numerical solution we approximate our system using finite element method [5–10].

As model problem we consider simulation of the thermomechanical state of the ceramic substrates with metallization, which are used for the manufacturing of light-emitting diode (LED) modules. The results of numerical simulation of the 3D problem in the complex geometric area are presented. Calculations are performed using the North-Eastern Federal University computational cluster *Arian Kuzmin*.

2 Problem Statement

Under mechanical and thermal effects in an elastic body displacement $\boldsymbol{u}$, strain $\boldsymbol{\varepsilon}$ and stress $\boldsymbol{\sigma}$ occur in an elastic body. Let T be the constant absolute temperature

I. Dimov et al. (Eds.): FDM 2014, LNCS 9045, pp. 364–370, 2015.
DOI: 10.1007/978-3-319-20239-6_40

at which body is in initial state of equilibrium, and θ be temperature increment. External forces that impact the body are treated as mechanical effects, whereas for the thermal influences one realizes heat exchange processes between the body surface and environment, and release or absorption of heat by the sources inside the body.

Mathematical model of thermoelastic state is defined by coupled system of equations for displacement $\boldsymbol{u}$ and temperature increment θ in domain Ω [1–4]:

$$-\operatorname{div}(k \operatorname{grad} \theta) = f. \tag{1}$$

$$-\mu \Delta \boldsymbol{u} - (\lambda + \mu) \operatorname{grad} \operatorname{div} \boldsymbol{u} + \alpha \operatorname{grad} \theta = 0, \tag{2}$$

Here μ, λ are Lame constants, k is heat conduction coefficient, c is strain-free volumetric heat capacity, $\alpha = \alpha_T(3\lambda + 2\mu)$, where α_T is linear thermal expansion coefficient, $\boldsymbol{\varepsilon}$ is strain tensor:

$$\varepsilon = \frac{1}{2}(\boldsymbol{\nabla u} + (\boldsymbol{\nabla u})^T),$$

and $\boldsymbol{\sigma}$ is stress tensor:

$$\boldsymbol{\sigma} = \lambda \boldsymbol{\nabla u I} + 2\mu\varepsilon.$$

Here $\boldsymbol{I}$ defines unit tensor.

Also Eqs. (1) and (2) are supplemented with appropriate boundary conditions:

$$\boldsymbol{\sigma n} = 0, \quad \boldsymbol{x} \in \Gamma_N^u, \quad \boldsymbol{u} = \boldsymbol{u}_0, \quad \boldsymbol{x} \in \Gamma_D^u,$$

$$-k\frac{\partial \theta}{\partial n} = 0, \quad \boldsymbol{x} \in \Gamma_N^T, \quad \theta = \theta_0, \quad \boldsymbol{x} \in \Gamma_D^T,$$

where $\partial \Omega = \Gamma_D^u + \Gamma_N^u = \Gamma_D^\theta + \Gamma_R^\theta$.

3   Approximation by Space

For numerical solution, we rewrite Eqs. (1) and (2) in weak form, using integration by parts to eliminate second derivatives [5–10].

Let $H = L_2(\Omega)$ be the Hilbert space for temperature increment with following scalar product and norm:

$$(u, v) = \int_\Omega u(\boldsymbol{x})\, v(\boldsymbol{x})\, dx, \quad ||u|| = (u, u)^{1/2},$$

and $\boldsymbol{H} = (L_2(\Omega))^d$ be space for displacement, where $\Omega \in \mathbb{R}^d$, $d = 2, 3$.

Then letting test functions q and $\boldsymbol{v}$ vanish on the appropriate Dirichlet boundaries Γ_D^θ and Γ_D^u, respectively, where solutions are known, we receive following variational problem: find $\theta \in V_\theta$ and $\boldsymbol{u} \in \boldsymbol{V_u}$ such that

$$\int_\Omega (k \operatorname{grad} \theta, \operatorname{grad} q) dx + \int_\Omega f\, q\, dx \quad \forall q \in \hat{V}_\theta = 0, \tag{3}$$

$$\int_\Omega \boldsymbol{\sigma(u)}\,\boldsymbol{\varepsilon(v)}dx + \int_\Omega \alpha(\operatorname{grad}\theta, \boldsymbol{v})dx = 0 \quad \forall \boldsymbol{v} \in \hat{\boldsymbol{V}}_u, \tag{4}$$

where test spaces $\hat{V}_\theta$ and $\hat{\boldsymbol{V}}_u$ are defined by

$$\hat{V}_\theta = \{q \in H^1(\Omega) : q(\boldsymbol{x}) = 0, \quad \boldsymbol{x} \in \Gamma_D^\theta\},$$

$$\hat{\boldsymbol{V}}_u = \{\boldsymbol{v} \in H^d(\Omega) : \boldsymbol{v}(\boldsymbol{x}) = 0, \quad \boldsymbol{x} \in \Gamma_D^u\},$$

and the trial spaces V_θ and $\boldsymbol{V}_u$ are shifted from test spaces by the Dirichlet boundary conditions:

$$\hat{V}_\theta = \{q \in H^1(\Omega) : q(\boldsymbol{x}) = \theta_0, \quad \boldsymbol{x} \in \Gamma_D^\theta\},$$

$$\hat{\boldsymbol{V}}_u = \{\boldsymbol{v} \in H^d(\Omega) : \boldsymbol{v}(\boldsymbol{x}) = \boldsymbol{u}_0, \quad \boldsymbol{x} \in \Gamma_D^u\}.$$

Further, we define the following bilinear and linear forms on the defined spaces

$$b(\theta, q) = \int_\Omega (k \operatorname{grad}\theta, \operatorname{grad}q)dx, \quad l(q) = (f, q) = \int_\Omega f\,q\,dx,$$

$$a(\boldsymbol{u}, \boldsymbol{v}) = \int_\Omega \boldsymbol{\sigma(u)}\,\boldsymbol{\varepsilon(v)}dx, \quad g(\theta, \boldsymbol{v}) = \int_\Omega \alpha(\operatorname{grad}\theta, \boldsymbol{v})dx.$$

Then problem becomes: find $\theta \in V_\theta$ and $\boldsymbol{u} \in \boldsymbol{V}_u$ that satisfy the following relations

$$b(\theta, q) + l(q) = 0 \quad \forall q \in \hat{V}_\theta, \tag{5}$$

$$a(\boldsymbol{u}, \boldsymbol{v}) + g(\theta, \boldsymbol{v}) = 0 \quad \forall \boldsymbol{v} \in \hat{\boldsymbol{V}}_u. \tag{6}$$

Note that these parts of problem are solved successively. First, we find distribution of temperature field from (5). And then we use it for calculation of displacement in (6).

4   Numerical Results

The object of research is ceramic substrates with metallization, which are used for the manufacturing of LED modules. During the creation process these substrates are subjected to significant heating, thereby an elastically-stressed state occurs, which leads to cracking of the substrate in some cases.

One of the ways of improvement is the minimization of the elastic stresses in the ceramic substrate with metallization. To find solution of this problem we need calculate elastic stress state of ceramic substrates under the influence of thermal stress raises.

The substrate has the length of 130 mm, the width of 72 mm and the thickness of 0.635 mm and 0.03 mm for ceramic and metal layers, respectively. On both sides it has technological holes of 1 mm and 1.5 mm diameter. Also the ceramic side has deepening of 0.2 mm with width of 0.1 mm. The full geometry of the object is shown in Fig. 1.

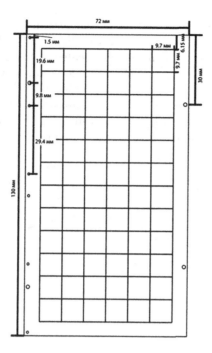

Fig. 1. Ceramic substrate geometry

To verify the model with the experimental data we use the temperature distribution along the middle line of the substrate. The boundary conditions for heating and cooling the substrate are modeled by appropriate Robin boundary conditions corresponding to convection with ambient air and metal rails. In this case the heat flux is modeled as convection with strongly heated air. These boundary conditions can be represented as following equations:

$$k\frac{\partial\theta}{\partial n} = \beta_i(\theta - \theta_i), \quad x \in \Gamma_i, \quad i = 1, 2, 3, 4, \tag{7}$$

where k is coefficient of thermal conductivity, β_i is heat transfer coefficient with air when $i = 1, 4$ and with metal when $i = 2, 3$, θ_i is difference between

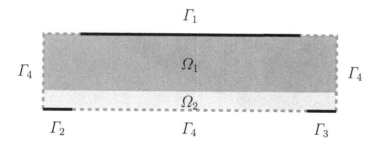

Fig. 2. Boundaries and domains in a slice

temperature of i-th boundary and initial temperature of substrate. Respective boundaries of heat flux Γ_1, convection with rails Γ_2, Γ_3 and air Γ_4, also ceramic layer Ω_1 and metal layer Ω_2 domains represented in Fig. 2.

For the following values: $k = 20\text{W}/(\text{m·K})$ for ceramic and $k = 400\text{W}/(\text{m·K})$ for metallization, $\beta_{1,4} = 5\text{W}/(\text{m}^2\text{·K})$, $\beta_{2,3} = 400\text{W}/(\text{m}^2\text{·K})$, $\theta_1 = 270C^0$, $\theta_2 = 90C^0$, $\theta_3 = 75C^0$ and $\theta_4 = 95C^0$, temperature distribution along the midline was obtained, which agrees with field experiments. In Fig. 3 a comparison of the temperature along the midline for the model and experiment is shown.

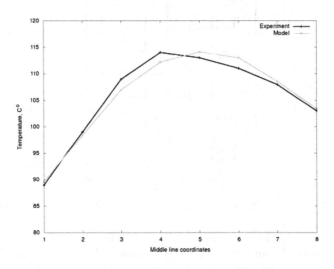

Fig. 3. Temperature distribution along the middle line (experiment, model)

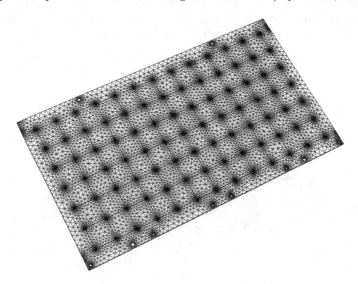

Fig. 4. Computational domain

In this work the simulation is performed using the first-degree polynomial approximation for temperature and first-degree for displacement. To solve the arising system of linear equations a standard direct method of ILU-factorization is used. A collection of free software FEniCS [11] is used for the numerical solution, and open-source application Paraview is used for visualization of the results.

For numerical modeling of thermoelasticity problem for ceramic substrate with metallization three grids containing about 250 000, 450 000 and 1 000 000 cells are used. These grids were made in Netgen mesh generator program. As for example, the finest grid with more than million cells is shown in Fig. 4. As results of numerical computation temperature distribution across the substrate (Fig. 5) and von Mizes stress distribution in technological hole (Fig. 6) are presented.

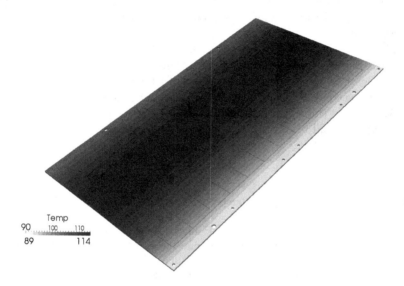

Fig. 5. Temperature distribution

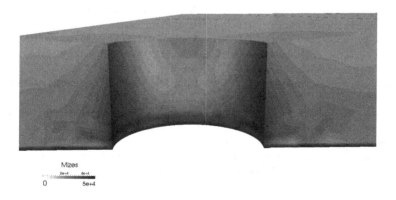

Fig. 6. Mizes stress distribution in technological hole

Table 1. Dependence of computation time (in seconds) from number of processes for different grids

Grid size	Number of processes					
	1	2	4	8	16	32
250 000	119	87	53	44	49	53
450 000	199	133	87	70	78	86
1 000 000	433	347	206	181	161	160

To illustrate the effectiveness of parallelization on a cluster, a series of computational experiments on three grids with different amounts of running processes are made. The results of gained dependence of computation time from number of processes are given in Table 1.

Table 1 shows the effectiveness of parallelization on different amount of running processes. Note that the effectiveness is evident for all presented grids. Moreover, for each grid we have optimal number of running processes. For instance, the first and second grids have the fastest computation when 8 processes are used. And for the third grid it is optimal to use 16 processes, as further growth of the computational resources does not gain any acceleration.

References

1. Biot, M.: Thermoelasticity and irreversible thermodynamics. J. Appl. Phys. **27**(3), 240–253 (1956)
2. Lubliner, J.: Plasticity theory. Dover Publications, New York (2008)
3. Simo, J.C., Hughes, T.J.R.: Computational Inelasticity. Interdisciplinary Applied Mathematics, vol. 7. Springer, New York (1998)
4. Nowacki, W.: Dynamic Problems of Thermoelasticity, 1st edn. Springer, Heidelberg (1975)
5. Samarskii, A., Nikolaev, E.: Methods of Solving Grid Equations (1978)
6. Gaspar, F., Lisbona, F., Vabishchevich, P.: A finite difference analysis of Biot's consolidation model. Appl. Numer. Math. **44**(4), 487–506 (2003)
7. Lisbona, F., Vabishchevich, P.: Operator-splitting schemes for solving unsteady elasticity problems. Comput. Methods Appl. Math. **1**(2), 188–198 (2001)
8. Afanas'eva, N., Vabishchevich, P., Vasil'eva, M.: Unconditionally stable schemes for convection-diffusion problems. Russian Math. **57**(3), 1–11 (2013)
9. Vabishchevich, P., Vasil'eva, M.: Explicit-implicit schemes for convection-diffusion-reaction problems. Numer. Anal. Appl. **5**(4), 297–306 (2012)
10. Kolesov, A., Vabishchevich, P., Vasil'eva, M.: Splitting schemes for poroelasticity and thermoelasticity problems. Comput. Math. Appl. **67**(12), 2185–2198 (2014)
11. Logg, A., Mardal, K., Wells, G. (eds.): Automated Solution of Differential Equations by the Finite Element Method. Lecture Notes in Computational Science and Engineering, vol. 84. Springer, Heidelberg (2012)

Multifrontal Hierarchically Solver for 3D Discretized Elliptic Equations

Sergey Solovyev[✉]

Institute of Petroleum Geology and Geophysics SB RAS,
3, Akademika Koptyuga Prosp., Novosibirsk 630090, Russia
`solovevsa@ipgg.sbras.ru`

Abstract. This paper presents a fast direct solver for 3D discretized
linear systems using the supernodal multifrontal method together with
low-rank approximations. For linear systems arising from certain partial differential equations (PDEs) such as elliptic equations, during the
Gaussian elimination of the matrices with Nested Dissection ordering, the
fill-in of L and U factors loses its sparsity and contains dense blocks with
low-rank property. Off-diagonal blocks can be efficiently approximated
with low-rank matrices; diagonal blocks approximated with semiseparable structures called hierarchically semiseparable (HSS) representations.
Matrix operations in the multifrontal method are performed in low-rank
arithmetic. We present efficient way to organize the HSS structured operations along the elimination. To compress dense blocks into low-rank or
HSS structures, we use effective cross approximation (CA) approach.
We also use idea of adaptive balancing between robust arithmetic for
computing the small dense blocks and low-rank matrix operations for
handling with compressed ones while performing the Gaussian elimination. This new proposed solver can be essentially parallelized both on
architecture with shared and distributed memory and can be used as
effective preconditioner. To check efficient of our solver we compare it
with Intel MKL PARDISO - the high performance direct solver. Memory and performance tests demonstrate up to 3 times performance and
memory gain for the 3D problems with more than 10^6 unknowns. Therefore, proposed multifrontal HSS solver can solve large problems, which
cannot be resolved by direct solvers because of large memory consump-
tions.

1 Introduction

Solving systems of linear algebraic equations (SLAE) is one of the fundamental
important problems in computational mathematic. Usually these SLAEs arise
from discretization of partial differential equations (PDEs). The common property of them is the high sparsity. Lets we have a system

$$Ax = b, \tag{1}$$

where $A \in R^{N \times N}$ is the result of descritization a PDF. High sparsity means that
number of nonzero elements is about $O(N)$. There are two basic ways to solve

© Springer International Publishing Switzerland 2015
I. Dimov et al. (Eds.): FDM 2014, LNCS 9045, pp. 371–378, 2015.
DOI: 10.1007/978-3-319-20239-6_41

such systems: iterative methods and direct ones. Iterative methods are based on sequential multiplying matrix A on some vector and on the using precondition technique. These methods are very efficient in memory usage. But they may diverge or convergence can be very slow if preconditioner is not so good.

Directs methods are reliable and efficient for many right hand sides, but very expensive because of matrix loss of sparsity during factorization process. Redusing fill-in of $L(U)$ factors can be done by using reordering technique: permute columns and rows of initial matrix A before factorization process. Nested Dissection (ND) approach is one of the effective well known reordering algorithms [1]. In this algorithm, the matrix A is associated with mesh which is recursively divided by small sets of mesh points (separators). However, some preconditioned iterative methods (such as *Multigrid* ones) are faster for many 3D problems.

Nowadays new popular approaches of additional decreasing memory of direct solvers are based on *low-rank* properties. During the factorization process of matrix A, raised from some PDEs (such as elliptic equations), certain off-diagonal blocks of Shur complements and $L(U)$ factors are dense and can be well approximated by low-rank blocks [2]. The diagonal blocks also have dense structure and can be effective compressed into HSS format. This format and HSS matrix arithmetic was proposed by Hackbusch [3]. Modern multifrontal HSS algorithms of solving sparse SLAEs are based on ND technique and low-rank/HSS approximation [5–8]. They are efficient and compatible with iterative solvers for 3D problems. For example, on 3D parallelepipedal mesh with n nodes in each direction and by using 7-point discretization stencil the number of operations P to compute $L(U)$ factors (without reordering) is $O(n^7)$ and number of non-zero elements Q is $O(n^5)$. Reordering technique reduces these numbers down to $P = O(n^4)$ and $Q = O(n^3)$.

In this paper we propose algorithm based on ND and low-rank/HSS techniques. High quality is based on the proposed features, i.e.:

- Using adopt low-rank approximation while Gauss elimination process;
- Applying CA algorithm to compress dense blocks of Shur-complement ;
- Using memory saving algorithm of factorization diagonal blocks, which are written in HSS format [9];
- Improve accuracy of solution by using iterative refinement technique.

2   Description of the Algorithm

There are three stages of the algorithm: reordering, factorization and solving. To simplify the understanding let consider the matrix A as the result of finite difference approximation the Laplace equation on the parallelepepidal mesh $n_1 \times n_2 \times n_3$ with 7-point stencil. So the pattern of A is 7-diagonal (Fig. 1, left); L-factor is banded with width $n_1 n_2$ (Fig. 1, right).

Reordering technique is based on row and column permutation of the initial matrix A. The details of ND reordering are described in [1]. As result of reordering we have matrices $\hat{A} = PAP^t$ and $\hat{L}\hat{L}^t = \hat{A}$ with some complicated

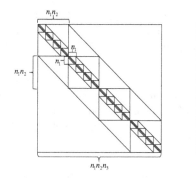
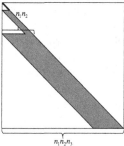

Fig. 1. Pattern matrix A for the 7-point stencil in 3D domain (left) and for it L-factor (right).

structure, but the number of non-zero elements in $\hat{L}$ less than in L (Fig. 3, left). So, the solution the system (1) equivalent the solution of the next one:

$$
\begin{cases}
\hat{b} = Pb \\
\hat{A} = PAP^t \\
\hat{A}\hat{x} = \hat{b} \\
x = P^t\hat{x}
\end{cases}
\tag{2}
$$

Let me note that permutation matrix P is the permutation vector with size N.

The second stage is the **factorization** of $\hat{A}$. It based on multi-level algorithm which are discribed in details in papers [6,7]. To improve the performance, the columns of Schur complement and L-factors are incorporated into panels. During Gauss elimination process these panels become wider, denser and can be effectively approximated by low-rank technique.

Definition 1. *Matrix $F \in R^{m \times n}$ can be low-rank approximated if there are two matrices $U \in R^{m \times k}$ and $V \in R^{n \times k}$ which satisfied the condition:*

$$
\frac{\|UV^t - F\|}{\|F\|} < \varepsilon,
\tag{3}
$$

for some ε. In practice, k is very small. The $rank_\varepsilon$ of a matrix F is the minimum k, such that (3) is true.

One of the important issue of Gauss elimination process is making decision "Should be the block of Shur complement approximated or not?". Well-known way is using *switching level* parameter [6,7]. In our approach we use more flexible criteria when the *compress coefficient* is estimated for each block.

Definition 2. *The compressing coefficient is $CC = k(m + n)/mn$, where the $(m + n)k$ elements need to store U and V; mn to store F.*

If $CC \geq 1$, so dense block F does not change; if $CC < 1$ then low-rank approximation is using instead of F and matrix-vector multiplications are performed in low-rank arithmetic. To compress F, the CA approach is using. It is faster than well-known truncated SVD and rank revealing QR factorization [4].

The diagonal blocks cannot be approximated by low-rank matrices because they are well-conditioned. The technique of HSS approximation is very efficient for such blocks. The main idea of HSS approximation of some dense matrix F is representative it in 2×2 blocks, where off-diagonal blocks can be low-rank approximated (Fig. 2, left).

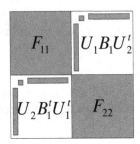 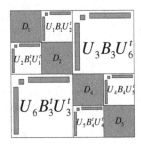

Fig. 2. HSS format of matrix F (2-level – left image and 3-level – right one).

The diagonal blocks can be divided on two diagonal and two off-diagonal sub-blocks, which can be low-rank approximated as well (Fig. 2, right). This process is performed recursively while compression coefficient less than 1.

Low-rank/HSS compression is performed on the fly during factorization process. I.e. to handle with the panel A_i of matrix $\hat{A}$ Shur complement F_i is computed, compressed and factorized either in dense or low-rank/HSS arithmetic. As result, the pattern of L-factor of matrix $\hat{A}$ has the structure which is presented on Fig. 3 (right). In fact it is not LL^t decomposition of matrix $\hat{A}$, but can be considered as decomposition $\tilde{L}\tilde{L}^t$ of some matrix $\tilde{A}$.

The **solving** stage consist of *Forward* step (solve $\tilde{L}y = b$) and *Backward* step (solve $\tilde{L}^t x = y$). Before *Forward* step, right hand side (rhs) should be permuted by reordering matrix $P : b' = Pb$. Solution vector should be reordered come back after *Backward* step : $x = P^t x'$.

Matrix $\tilde{L}$ has block structure, so to inverse $\tilde{L}$ the matrix-vector operation are performed. If block has low-rank/HSS structure, so its arithmetic is used; if block is dense, so standard dense operations are performed.

To improve quality of solution the *iterative refinement* process is performed:

$$
\begin{aligned}
&x_0 = \tilde{A}^{-1}b \\
&r_0 = b - \hat{A}x_0 \\
&\text{while } \frac{||r_i||}{||b||} < \varepsilon \text{ do} \\
&\quad r_i = b - \hat{A}x_i \\
&\quad \delta x_i = \tilde{A}^{-1}r_i \\
&\quad x_{i+1} = x_i + \delta x_i
\end{aligned}
$$

Let me note, that this process can be diverge because of poor quality of low-rank/HSS approximation.

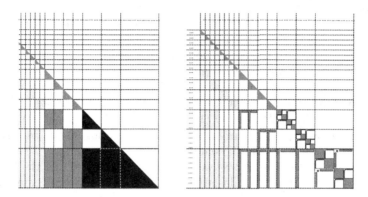

Fig. 3. Pattern matrix $\hat{L}$ after ND reordering (left) and $\tilde{L}$ after HSS/Low-rank approximation (right).

The current version of programm has been implemented and tested for the acoustic problem in 3D parallelepipedal computational domain. The Helmholtz equation is approximated on the parallelepipedal grid by finite difference approach. The second order approximation on 7-point stencil is used. On the all external faces we set Perfect Matching Layer (PML) boundary condition, so our matrix in SLAE is complex symmetric (non-Hermitian).

3 Numerical Experiments

We performed testing to check the correctness, performance and memory consumption of program implementation for proposed algorithm. In the **first** example we show the correlation between the relative residual $\|f - Ax\|/\|f\|$ with internal low-rank accuracy ϵ_{in}. The **second** and **third** tests show the memory consumption and performance dependence on low-rank accuracy ε_{in}. At the **fourth** example we investigate the behavior of memory and performance consumption while increasing the size of problem. In each tests we compare proposed HSS solver with high performance Intel MKL PARDISO direct solver which is optimized on multi-core systems with shared memory. All computational were performed on server Intel(R) Xeon(R) CPU E5-2690 v2 3.0 GHz, (Sandy Bridge) with 512 GB RAM. To make clear experiments we try to avoid impact of OMP parallelization of all MKL functions by switching off threading (set OMP_NUM_TREADS=1).

The first three test were computed in homogenius media on $141 \times 141 \times 141$ mesh with the size domain $1000\,m \times 1000\,m \times 1000\,m$. The wave velocity is $1500\,m/s$, frequency $\nu = 12\,Hz$. Th PML layer has ten points in each direction. Internal accuracy is varied from 10^{-4} to 10^{-15}.

The **first** example demonstrates the strong dependence of residual $\|f - Ax\|/\|f\|$ on the ε_{in} (Fig. 4, bold line). If this accuracy increased up to 10^{-15}, the residual of HSS archives the similar quality like PARDISO solver (Fig. 4, thin line).

The second test shows that the memory gain against PARDISO is increasing linearly (in log scale) while decreasing the low-rank accuracy (Fig. 5). It achives up to 2.5 for $\varepsilon_{in} = 10^{-4}$. Let me note that number of non-zero elements in L-factor of PARDISO for this task (number of unknowns is $2.8 * 10^6 \sim 141 \times 141 \times 141$) is $3.5 * 10^9$ elements.

In performance tests we switch on the iterative refinement step to achieve the high residual like in PARDISO (10^{-14}). The number of right hand sides is increased up to 500. Performance behavior of factorization step is decreasing as well as memory behavior (Fig. 6, bold dashed line).

Factorization time is up to 3.5 times faster than PARDISO for the low accuracy ($\varepsilon_{in} = 10^{-4}$, Fig. 6, thin dashed line). However, the total timing for the

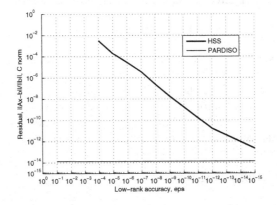

Fig. 4. Correlation between relative residual and internal accuracy in log scale.

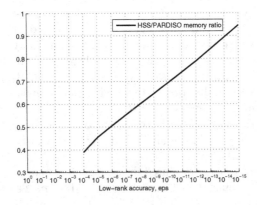

Fig. 5. Correlation between memory storage and internal accuracy.

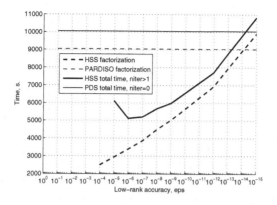

Fig. 6. Performance of factorization step and total time with iterative refinement.

Table 1. Trend of increasing memory and factorization time on cube domain

	Memory to store L-factors	Factorization time
PARDISO	$O(n^{4.2})$	$O(n^{6.2})$
HSS	$O(n^{3.3})$	$O(n^{4.2})$

same ε_{in} is large than for bigger ones ($\varepsilon_{in} = 10^{-5} \ldots 10^{-6}$) because of increasing number of iterations for the lowest accuracy.

To investigate the trend of time and memory increasing while increasing size domain, we perform computing on mesh $n \times n \times n$ with different $n = 101 \ldots 301$. The order of increasing time and memory both for PARDISO and HSS solvers is presented in the Table 1.

As result, the memory behavior is better than in PARDISO (about 1 order), the performance is better as well (about 2 order).

4 Conclusions

Algorithm of solving 3D discretized elliptic equations based on Nested Dissection reordering and low-rank/HSS techniques has been developed and tested.

High performance of proposed algorithm is based on the using nowadays features: adopt low-rank approximation, applying Cross Approximation algorithm, using memory saving HSS-factorization algorithm and improve accuracy of solution by using iterative refinement technique.

Validation tests show high quality of program implementation. Memory and performance tests demonstrate up to 3 times performance and memory gain for the 3D problems with more than 10^6 unknowns in compare with PARDISO Intel MKL.

Acknowledgements. The research described was partially supported by RFBR grants 14-01-31340, 14-05-31222, 14-05-00049.

References

1. George, A.: Nested dissection of a regular finite elementmesh. SIAM J. Numer. Anal. **10**(2), 345–363 (1973)
2. Chandrasekaran, S., Dewilde, P., Gu, M., Somasunderam, N.: On the numerical rank of the off-diagonal blocks of schur complements of discretized elliptic PDEs. Siam J. Matrix Anal. Appl. **31**(5), 2261–2290 (2010)
3. Hackbusch, W.: A sparse matrix arithmetic based on H-matrices. part I: Introduction to H-matrices. Computing **62**(2), 89–108 (1999)
4. Rjasanow, S.: Adaptive cross approximation of dense matrices. In: IABEM 2002, International Association for Boundary Element Methods, UT Austin, 28–30 May 2002
5. Le Borne, S., Grasedyck, L., Kriemann, R.: Domain-decomposition based H-LU preconditioners. In: Widlund, O.B., Keyes, D.E. (eds.) Domain Decomposition Methods in Science and Engineering XVI. LNCSE, vol. 55, pp. 667–674. Springer, Heidelberg (2007)
6. Xia, J.: Robust and efficient multifrontal solver for large discretized PDEs. In: Berry, M.W., Gallivan, K.A., Gallopoulos, E., Grama, A., Phillippe, B., Saad, Y., Saied, F. (eds.) High-Performance Scientific Computing, pp. 199–217. Springer, London (2012)
7. Wang, S., de Hoo, M.V., Xia, J., Li, X.S.: Massively parallel structured multifrontal solver for time-harmonic elastic waves in 3-D anisotropic media. Proc. Proj. Rev. Geo-Math. Imag. Group (Purdue Univ., West Lafayette IN) **1**, 175–192 (2011)
8. Xia, J.: A robust inner-outer HSS preconditioner. Numer. Linear Algebra Appl. **00**, 1 (2011)
9. Chandrasekaran, S., Gu, M., Li, X.S., Xia, J.: Some fast algorithms for hierarchically semiseparable matrices. Numer. Linear Algebra Appl. **17**, 953–976 (2010)

Solving the 3D Elasticity Problems by Rare Mesh FEM Scheme

S.V. Spirin[✉], Dmitry T. Chekmarev, and A.V. Zhidkov

Lobachevsky State University of Nizhni Novgorod, Nizhny Novgorod, Russia
4ekm@mm.unn.ru

Abstract. The article describes the rare mesh scheme based on finite element method, describes the methods of constructing such schemes are described arrangements of nodes, describes methods of calculation tasks based on rare mesh schemes, the problem of static, Numerical solutions of different tasks based on rare mesh scheme circuit compares the results with the known systems.

1 Introduction

Principles of construction and use of openwork patterns FEM presented in [1,2]. The basis of such schemes is an rare mesh of finite elements. At the same time for an rare mesh hexagonal cell is broken down into five tetrahedra (Fig. 1), left central tetrahedron remaining tetrahedra removed. The resulting rare mesh compared with the traditional grid consisting of tetrahedra solid way to fill the entire volume is five times smaller elements and almost half the nodes (Figs. 2, 3 and 4).

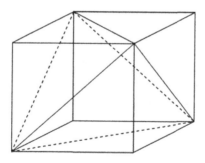

Fig. 1. Rare mesh element

2 Solving Static Problems Based Rare Scheme

Numerical implementation of solutions of static problems in elastic formulation is based on the finite element approximation of the variational equation (principle of virtual displacements)

$$\int_V \sigma_{ij} \delta\varepsilon_{ij}\, dV = \int_V \rho F_i \delta u_i dV + \int_{S_p} P_i \delta u_i dS, \qquad (1)$$

I. Dimov et al. (Eds.): FDM 2014, LNCS 9045, pp. 379–384, 2015.
DOI: 10.1007/978-3-319-20239-6_42

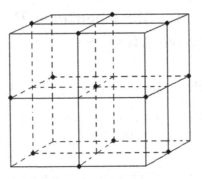

Fig. 2. Rare pattern grid

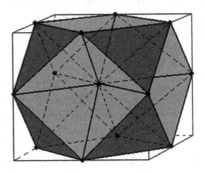

Fig. 3. Elements rare mesh filling pattern

where V – the amount of elastic isotropic body, on some part of the boundary surface is configured for load; - components of the stress tensor; σ_{ij} – components of the linear strain tensor; u_i – components of the displacement vector; F_i and P_i - components, respectively, mass and surface loads ρ – density. Displacement, rotation angles are considered small deformation

$$\varepsilon_{ij} = \frac{1}{2}\left(u_{i,j} + u_{j,i}\right), \tag{2}$$

deformation associated with the stresses by Hooke's law

$$\sigma_{ij} = \lambda\delta_{ij}\varepsilon_{kk} + 2\mu\varepsilon_{ij}, \tag{3}$$

where λ and μ – Lame parameters.

Discusses the implementation options rare FEM scheme based on linear quadrangular element. Distribution of displacements in the elements taken linear strain and stress are considered permanent. Bulk and surface forces acting on the tetrahedra removed and their surfaces are distributed among nodes of elements involved in the calculation. On uniform grids this delicate scheme has second-order approximation.

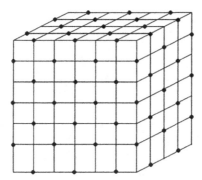

Fig. 4. The regular grid nodes involved in the calculations

As is customary in the FEM, will not use the tensor, and matrix- vector form of the relations. Vector finite element nodal displacements denote

$$(\mathbf{u})^T = \left(u_1^1, u_1^2, u_1^3, u_2^1, u_2^2, u_2^3, u_3^1, u_3^2, u_3^3, u_4^1, u_4^2, u_4^3\right), \tag{4}$$

component vectors of linear strain and stress tensor (by symmetry we consider only six components), respectively

$$(\boldsymbol{\varepsilon})^T = (\varepsilon_{11}, \varepsilon_{22}, \varepsilon_{33}, \gamma_{12}, \gamma_{23}, \gamma_{31}), \tag{5}$$

$$(\boldsymbol{\sigma})^T = (\sigma_{11}, \sigma_{22}, \sigma_{33}, \sigma_{12}, \sigma_{23}, \sigma_{31}), \tag{6}$$

matrix of the elastic constants of an isotropic material

$$(\mathbf{C}) = \left(\begin{array}{ccc|ccc}
\lambda + 2\mu & \lambda & \lambda & & & \\
\lambda & \lambda + 2\mu & \lambda & & 0 & \\
\lambda & \lambda & \lambda + 2\mu & & & \\
\hline
 & & & \mu & 0 & 0 \\
 & 0 & & 0 & \mu & 0 \\
 & & & 0 & 0 & \mu
\end{array}\right) \tag{7}$$

matrix differential operator

$$(\mathbf{B}) = \left(\begin{array}{cccccccccccc}
\beta_1^1 & 0 & 0 & \beta_2^1 & 0 & 0 & \beta_3^1 & 0 & 0 & \beta_4^1 & 0 & 0 \\
0 & \beta_1^2 & 0 & 0 & \beta_2^2 & 0 & 0 & \beta_3^2 & 0 & 0 & \beta_4^2 & 0 \\
0 & 0 & \beta_1^3 & 0 & 0 & \beta_2^3 & 0 & 0 & \beta_3^3 & 0 & 0 & \beta_4^3 \\
\beta_1^2 & \beta_1^1 & 0 & \beta_2^2 & \beta_2^1 & 0 & \beta_3^2 & \beta_3^1 & 0 & \beta_4^2 & \beta_4^1 & 0 \\
0 & \beta_1^3 & \beta_1^2 & 0 & \beta_2^3 & \beta_2^2 & 0 & \beta_3^3 & \beta_3^2 & 0 & \beta_4^3 & \beta_4^2 \\
\beta_1^3 & 0 & \beta_1^1 & \beta_2^3 & 0 & \beta_2^1 & \beta_3^3 & 0 & \beta_3^1 & \beta_4^3 & 0 & \beta_4^1
\end{array}\right) \tag{8}$$

where

$$\beta_s^m = \frac{1}{D}\begin{vmatrix} x_p^n - x_4^n & x_p^k - x_4^k \\ x_q^n - x_4^n & x_q^k - x_4^k \end{vmatrix}$$

$\beta_4^m = -\beta_1^m + \beta_2^m - \beta_3^m$, $m = 1,2,3$, $s = 1,2,3$, index sequence mnk and spq form a cyclic permutation of the sequence numbers 123,

$$D = \begin{vmatrix} 1 & x_1^1 & x_1^2 & x_1^3 \\ 1 & x_2^1 & x_2^2 & x_2^3 \\ 1 & x_3^1 & x_3^2 & x_3^3 \\ 1 & x_4^1 & x_4^2 & x_4^3 \end{vmatrix}$$

(x_i^1, x_i^2, x_1^3) – coordinates of the i node of the final ($i = 1,2,3,4$).

In matrix form the Cauchy (2), Hooke's law (3) and the total potential energy of a single element can be written as

$$(\varepsilon) = (\mathbf{B})(\mathbf{u}), \tag{9}$$

$$(\sigma) = (\mathbf{C})(\varepsilon) = (\mathbf{C})(\mathbf{B})(\mathbf{u}), \tag{10}$$

$$\Pi = \frac{1}{2}(\mathbf{u})^T(\mathbf{K})(\mathbf{u}) - (\mathbf{u})^T(\mathbf{q}), \tag{11}$$

where $(\mathbf{K})$ – stiffness matrix element

$$(\mathbf{K}) = \int_{V_i} (\mathbf{B})^T(\mathbf{C})(\mathbf{B})dV,$$

$(\mathbf{q})$ – vector of nodal forces statically equivalent to the current element of the distributed mass and surface forces.

Stationarity condition (1) energy functional leads to a system of linear algebraic equations of equilibrium of the body, which is modified by the boundary conditions on the movement. Solving the resulting system with respect to displacement or direct iterative method defined nodal displacements whole computational domain and the formulas (9) and (10) components are calculated strain and stress tensors.

3   Solution of the Model Problem

We consider the problem of determining the elastic contact condition of a thick-walled tube of finite length, compressing absolutely rigid plates parallel to the axis of the tube. Inner radius of the tube - 5 cm, external radius of the tube - 10 cm, length of the pipe - 60 cm, the offset of each rigid plate is squeezed at - 0.5 mm, the modulus of elasticity of the material - 200 GPa, Poisson's ratio - 0.3. Because of the symmetry of the problem the design scheme is one-eighth of the pipe with symmetry conditions on the respective planes (Fig. 5).

Contact problem solution implemented by planting nodes finite element mesh beyond the plane of the junction, on the plane and iterative refinement to the equilibrium conditions.

Figure 6 shows the convergence of the solution to the radial displacement along the generator inside of the tube beneath the contact zone, with nested grids. Opening 10 contains a grid of elements through the thickness of the pipe

Fig. 5. Calculation scheme

elements 16 in the circumferential direction of elements 60 and along the pipe. In two subsequent calculations the number of elements in each direction, respectively, increased in two and four -fold compared to the initial mesh.

The solution obtained using the openwork scheme compared with the solution of a similar formulation obtained in the system ANSYS. The comparison results in the radial displacement along the generatrix of the inner surface of the tube, beneath the contact zone, for one embodiment, the calculations are shown in Fig. 7. As can be seen, solutions give a good qualitative and quantitative agreement with the exception of a small (1 item) zone of the edge effect near the free end of the tube.

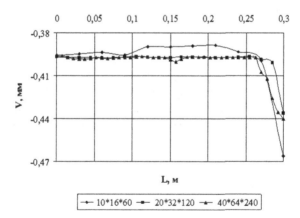

Fig. 6. Convergence solutions

The obtained results of model problems demonstrate the feasibility and effectiveness of the use of openwork schemes for static elasticity problems.

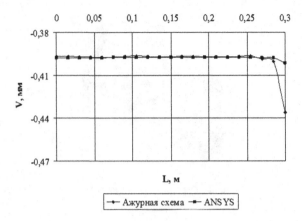

Fig. 7. Comparison of solutions

Conclusion

From the examples and descriptions of the method of construction and calculations based on rare mesh scheme based on finite element method can be
concluded that the using of this scheme provides significant gains over time, and
the use of super rare mesh scheme winning time increases even almost doubled.
At the same time a significant loss of accuracy compared to other schemes not
observed. Thus the using of rare mesh circuit on models with a very large number
of constituent elements leads to a significant gain in time.

References

1. Chekmarev, D.T.: Fishnet finite element method. In: Applied Problems of Strength
   and Ductility, Issue 55, pp. 157–159. UNN, Nizhny Novgorod (1997)
2. Chekmarev, D.T.: Numerical schemes for the finite element method "fishnet" grids.
   In: Problems of Atomic Science and Technology, Ser. Mathematical modeling of
   physical processes, Issue 2, pp. 49–54 (2009)

Computational Algorithm for Identification of the Right-Hand Side of the Parabolic Equation

Petr N. Vabishchevich[1,2], Maria V. Vasilyeva[1,2(✉)], and Vasily I. Vasilyev[1,2]

[1] Nuclear Safety Institute, 52, B. Tulskaya, 115191 Moscow, Russia
[2] North-Eastern Federal University, 58, Belinskogo, 677000 Yakutsk, Russia
vasilyevadotmdotv@gmail.com

Abstract. Among inverse problems for PDEs we distinguish coefficient inverse problems, which are associated with the identification of the right-hand side of an equation using some additional information. When considering time-dependent problems, the identification of the right-hand side dependences on space and on time is usually separated into individual problems. We have linear inverse problems; this situation essentially simplify their study. This work deals with the problem of determining in a multidimensional parabolic equation the right-hand side that depends on time only. To solve numerically a inverse problem we use standard finite difference approximations in space. The computational algorithm is based on a special decomposition, where the transition to a new time level is implemented via solving two standard elliptic problems.

1 Introduction

In the theory and practice of inverse problems for partial differential equations (PDEs), much attention is paid to the problem of the identification of coefficients from some additional information [1,2]. Particular attention should be given to inverse problems for PDEs [3,4]. In this case, a theoretical study includes the fundamental questions of uniqueness of the solution and its stability from the viewpoint of the theory of differential equations [4,5]. Many inverse problems are formulated as non-classical problems for PDEs. To solve approximately these problems, emphasis is on the development of stable computational algorithms that take into account peculiarities of inverse problems [6,7].

Much attention is paid to the problem of determining the right-hand side, lower and leading coefficients of a parabolic equation of second order, where, in particular, the right-hand side and the coefficients depends on time only. An additional condition is most often formulated as a specification of the solution at an interior point or as the average value that results from integration over the whole domain. The existence and uniqueness of the solution to such an inverse problem and well-posedness of this problem in various functional classes are examined, for example, in the book [5].

Numerical methods for solving problems of the identification of the right-hand side, lower and leading coefficients for parabolic equations are considered in many works. In view of the practical use, we highlight separately the studies

© Springer International Publishing Switzerland 2015
I. Dimov et al. (Eds.): FDM 2014, LNCS 9045, pp. 385–392, 2015.
DOI: 10.1007/978-3-319-20239-6_43

dealing with numerical solving inverse problems for multidimensional parabolic equations [7,8]. Approximation in space is performed using the standard finite differences or finite elements. In this paper, for a multidimensional parabolic equation, we consider the problem of determining the right-hand side of an equation that depends on time only. Non-classical problems at every time level are solved on the basis of a special decomposition into two standard elliptic problems.

2   Problem Formulation

For simplicity, we restrict ourselves to a 2D problem in a rectangle. Let $x = (x_1, x_2)$ and

$$\Omega = \{x \mid x = (x_1, x_2), \quad 0 < x_\alpha < l_\alpha, \quad \alpha = 1, 2\}.$$

The direct problem is formulated as follows. We search $u(x, t), 0 \leq t \leq T, T > 0$ such that it is the solution of the parabolic equation of second order:

$$\frac{\partial u}{\partial t} - \operatorname{div}(k(x)\operatorname{grad} u) = p(t)\psi(x, t), \quad x \in \Omega, \quad 0 < t \leq T. \tag{1}$$

The boundary and initial conditions are also specified:

$$k(x)\frac{\partial u}{\partial n} = 0, \quad x \in \partial\Omega, \quad 0 < t \leq T, \tag{2}$$

$$u(x, 0) = u_0(x), \quad x \in \Omega, \tag{3}$$

where n is the normal to Ω. The formulation (1)–(3) presents the direct problem, where the right-hand side, coefficients of the equation as well as the boundary and initial conditions are given.

Let us consider the inverse problem, where in Eq. (1), the coefficient $p(t)$ is unknown. An additional condition is often formulated as

$$\int_\Omega u(x, t)\omega(x)dx = \varphi(t), \quad 0 < t \leq T, \tag{4}$$

where $\omega(x)$ is a weight function. In particular, choosing $\omega(x) = \delta(x - x^*)$ $(x^* \in \Omega)$, where $\delta(x)$ is the Dirac δ-function, from (4), we get

$$u(x^*, t) = \varphi(t), \quad 0 < t \leq T. \tag{5}$$

We assume that the above inverse problem of finding a pair of $u(x, t)$, $p(t)$ from Eqs. (1)–(3) and additional conditions (4) or (5) is well-posed. The corresponding conditions for existence and uniqueness of the solution are available in the above-mentioned works. In the present work, we consider only numerical techniques for solving these inverse problems omitting theoretical issues of the convergence of an approximate solution to the exact one.

3   Semi-discrete Problem

To solve numerically the parabolic problem, we introduce the uniform grid in the domain Ω:

$$\omega = \left\{ x \mid x = (x_1, x_2), \quad x_\alpha = \left(i_\alpha + \frac{1}{2} \right) h_\alpha, \quad i_\alpha = 0, 1, ..., N_\alpha, \right.$$

$$\left. (N_\alpha + 1) h_\alpha = l_\alpha, \quad \alpha = 1, 2 \right\}.$$

For grid functions, we define the Hilbert space $H = L_2(\omega)$, where the scalar product and norm are given as follows:

$$(y, w) \equiv \sum_{x \in \omega} y(x) w(x) h_1 h_2, \quad \|y\| \equiv (y, y)^{1/2}.$$

The difference operator for the diffusion transport D has the following additive representation:

$$D = \sum_{\alpha=1}^{2} D_\alpha, \quad \alpha = 1, 2, \quad x \in \omega, \tag{6}$$

where D_α, $\alpha = 1, 2$ are associated with the corresponding differential operator in one spatial direction.

For all nodes except ajoining the boundary, and for sufficiently smooth diffusion coefficients $k(x)$, the grid operator D_1 can be written as:

$$D_1 y = - \frac{1}{h_1^2} k(x_1 + 0.5h_1, h_2)(y(x_1 + h_1, h_2) - y(x))$$

$$+ \frac{1}{h_1^2} k(x_1 - 0.5h_1, h_2)(y(x) - y(x_1 - h_1, h_2)),$$

$$x \in \omega, \quad x_1 \neq 0.5h_1, \quad x_1 \neq l_1 - 0.5h_1.$$

At the nodes ajoining the boundary, approximation should take into account the boundary condition (2):

$$D_1 y = - \frac{1}{h_1^2} k(x_1 + 0.5h_1, h_2)(y(x_1 + h_1, h_2) - y(x)),$$

$$x \in \omega, \quad x_1 = 0.5h_1,$$

$$D_1 y = \frac{1}{h_1^2} k(x_1 - 0.5h_1, h_2)(y(x) - y(x_1 - h_1, h_2)),$$

$$x \in \omega, \quad x_1 = l_1 - 0.5h_1.$$

The grid operator D_2 is constructed in a similarly way. Direct calculations yield (see, e.g., [9,10]):

$$D_\alpha = D_\alpha^* \geq 0, \quad \alpha = 1, 2.$$

This grid operator of diffusion approximates the corresponding differential operator with an accuracy of $O\left(|h|^2\right)$. As in the differential case, the difference operator of diffusive transport (6) is self-adjoint and positive definite in H:

$$D = D^* \geq 0. \tag{7}$$

In view of (7), we can obtain the corresponding a priori estimates for the solution of the boundary value problem (1)–(3) in H that ensure the stability of the solution with respect to the initial data and the right-hand side.

After discretization in space, from the problem (1)–(3), we arrive at the Cauchy problem for the semi-discrete equation:

$$\frac{dy}{dt} + Dy = p(t)\psi(t), \quad 0 < t \leq T, \tag{8}$$

$$y(0) = u_0. \tag{9}$$

For Cauchy problem (8), (9) we have

$$\|y(t)\| \leq \|u_0\| + \int_0^t p(\theta)\|\psi(\theta)\|d\theta. \tag{10}$$

The a priori estimate (10) holds in the Banach space of grid functions $L_\infty(\omega)$, where

$$\| \cdot \| = \| \cdot \|_\infty, \quad \|y\|_\infty \equiv \max_{x \in \omega} |y|.$$

This fact can be established on the basis of the maximum principle for grid functions and the relevant comparison theorems [9] taking into account the diagonal dominance of the matrix (operator) D.

4   Time-Stepping Techniques

Let us define a uniform grid in time $t^n = n\tau$, $n = 0, 1, ..., N$, $\tau N = T$ and denote $y^n = y(t^n)$, $t^n = n\tau$. We start with discretization in time for the numerically solving direct problem (8), (9). To solve numerically boundary value problems for transient diffusion Eq. (1) we use unconditionally stable implicit scheme

$$\frac{y^{n+1} - y^n}{\tau} + Dy^{n+1} = p^{n+1}\psi^{n+1}, \quad n = 0, 1, ..., N - 1. \tag{11}$$

The initial condition (9) yields

$$y^0 = u_0. \tag{12}$$

The difference solution of the problem (11), (12) satisfies the following level-wise estimate in $L_\infty(\omega)$:

$$\|y^{n+1}\| \leq \|y^n\| + \tau p^{n+1}\|\psi^{n+1}\|, \quad n = 0, 1, ..., N - 1. \tag{13}$$

The estimate (13) is a discrete analog of the estimate (11) for the solution of the problem (8)–(10). To prove (13), we cam apply the maximum principle for grid functions [9]. The second possibility to check the a priori estimate (13) is associated with the use of the concept of the logarithmic norm.

5   Algorithm for Solving the Inverse Problem

For the fully discretized (both in space and in time) direct problem (1)–(3), we can solve the inverse problem of the identification of the right-hand side $p(t)$. We restrict ourselves to the case, where an additional information on the solution is defined (see (5)) at some interior node $\boldsymbol{x}^* \in \omega$ of the grid:

$$y^{n+1}(\boldsymbol{x}^*) = \varphi^{n+1}, \quad n = 0, 1, ..., N - 1. \tag{14}$$

For the approximate solution of the problem (11), (12), (14) at the new time level y^{n+1}, we introduce the following decomposition (see, e.,g., [7,8]):

$$y^{n+1}(\boldsymbol{x}) = v^{n+1}(\boldsymbol{x}) + p^{n+1}w^{n+1}(\boldsymbol{x}), \quad \boldsymbol{x} \in \omega. \tag{15}$$

To find $v^{n+1}(\boldsymbol{x})$, we employ the equation

$$\frac{v^{n+1} - y^n}{\tau} + Dv^{n+1} = 0, \quad n = 0, 1, ..., N - 1. \tag{16}$$

The function $w^{n+1}(\boldsymbol{x})$ is determined from

$$\frac{1}{\tau}w^{n+1} + Dw^{n+1} = \psi^{n+1}, \quad n = 0, 1, ..., N - 1. \tag{17}$$

Using the decomposition (15)–(17), Eq. (11) holds automatically for any p^{n+1}.

To evaluate p^{n+1}, we apply the condition (14). The substitution of (15) into (14) yields

$$p^{n+1} = \frac{1}{w^{n+1}(\boldsymbol{x}^*)}(\varphi^{n+1} - v^{n+1}(\boldsymbol{x}^*)). \tag{18}$$

The fundamental point of applicability of this algorithm is associated with the condition $w^{n+1}(\boldsymbol{x}^*) \neq 0$. The auxiliary function $w^{n+1}(\boldsymbol{x}^*)$ is determined from the grid elliptic Eq. (17). The property of having fixed sign for $w^{n+1}(\boldsymbol{x}^*)$ is followed, in particular, from

$$\psi^{n+1}(\boldsymbol{x}) \geq 0, \quad \boldsymbol{x} \in \omega, \quad \|\psi^{n+1}\| > 0, \quad n = 0, 1, ..., N - 1.$$

Such constraints on the solution can be provided by the corresponding restrictions on the input data of the inverse problem.

6   Numerical Examples

To demonstrate possibilities of the above schemes for solving the right-hand side
identification problem for the parabolic equation, we consider a model problem.
In the examples below, we consider the problem in the unit square ($l_1 = l_2 = 1$).
Suppose

$$k(x) = 1, \quad \psi(x,t) = x_1 x_2, \quad u_0(x) = 1, \quad x \in \Omega.$$

The problem is considered on the grid $N_1 = N_2 = 51$, the observation point is
located at the square centre ($x^* = (0.5, 0.5)$). The coefficient $p(t)$ is taken in the
form

$$p(t) = \begin{cases} 1000t, & 0 < t \leq 0.5T, \\ 0, & 0.5T < t \leq T. \end{cases}$$

The solution of the direct problem (1)–(3) at the observation point is depicted
in Fig. 1. The solution at the final time moment ($T = 0.1$) is presented in Fig. 2.

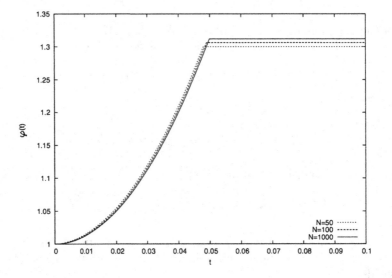

Fig. 1. The solution of the direct problem at the point of observation

The results of solving the inverse problem with various grids in time are shown
in Fig. 3. To study the influence of parameters of the computational algorithm,
we need to use the same input data. In our case, as the input data we use
the numerical solution of the direct problem obtained using a very fine grid in
time. The solution of the direct problem obtained with $N = 1000$ is employed
as the input data (the function $\varphi(t)$ in the condition (5)). It is easy to see that
the approximate solution of the inverse problem converges with decreasing the
time step.

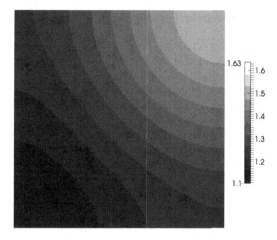

Fig. 2. The solution of the direct problem at $t = T$

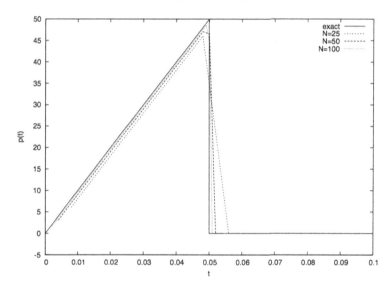

Fig. 3. The solution of the inverse problem

Acknowledgements. This work was supported by RFBR (project 14-01-00785).

References

1. Alifanov, O.M.: Inverse Heat Transfer Problems. Springer, New York (2011)
2. Aster, R.C., Borchers, B., Thurber, C.H.: Parameter Estimation and Inverse Problems. Elsevier Science, Burlington (2011)
3. Lavrent'ev, M.M., Romanov, V.G., Shishatskii, S.P.: Ill-posed Problems of Mathematical Physics and Analysis. American Mathematical Society, Providence (1986)

4. Isakov, V.: Inverse Problems for Partial Differential Equations. Springer, New York (1998)
5. Prilepko, A.I., Orlovsky, D.G., Vasin, I.A.: Methods for Solving Inverse Problems in Mathematical Physics. Marcel Dekker Inc., New York (2000)
6. Vogel, C.R.: Computational Methods for Inverse Problems. Society for Industrial and Applied Mathematics, Philadelphia (2002)
7. Samarskii, A.A., Vabishchevich, P.N.: Numerical Methods for Solving Inverse Problems of Mathematical Physics. De Gruyter, Berlin (2007)
8. Borukhov, V.T., Vabishchevich, P.N.: Numerical solution of the inverse problem of reconstructing a distributed right-hand side of a parabolic equation. Comput. Phys. Commun. **126**(1), 32–36 (2000)
9. Samarskii, A.A.: The Theory of Difference Schemes. Marcel Dekker, New York (2001)
10. Samarskii, A.A., Nikolaev, E.S.: Numerical Methods for Grid Equations, vol. I, II. Birkhauser Verlag, Basel (1989)

Simulation and Numerical Investigation of Temperature Fields in an Open Geothermal System

N.A. Vaganova[1]([✉]) and M.Yu. Filimonov[1,2]([✉])

[1] Institute of Mathematics and Mechanics of Ural Branch of Russian
Academy of Sciences, Ekaterinburg, Russia
{vna,fmy}@imm.uran.ru
[2] Ural Federal University, Ekaterinburg, Russia

Abstract. An open geothermal system consisting of injection and productive wells is considered. Hot water from production well is used and became cooler, and injection well returns the cold water into the aquifer. To simulate this open geothermal system a three–dimensional nonstationar mathematical model of the geothermal system is developed taking into account the most important physical and technical parameters of the wells to describe processes of heat transfer and thermal water filtration in a aquifer. Results of numerical calculations, which, in particular, are used to determine an optimal parameters for a geothermal system in North Caucasus, are presented. For example, a distance in the productive layer between the point of hot water inflow and of cold water injection point is considered.

1 Introduction

A geothermal system is a system used for heating that utilizes the earth as a heat source. Geothermal systems simply take advantage of the relatively constant temperature within the earth. For example, in North Caucasus the earth's temperature ranges 90–$102°C$ at a depth of $900\,\mathrm{m}$ throughout the year.

Geothermal systems have the potential for significant savings in energy costs and for reducing of oil and natural gas consumption. However, some of these systems also have the potential for adverse environmental effects if installed or operated improperly or if they use inappropriate materials. It is also important to ensure a long-term service and appropriate heat efficiency of these installations.

Let consider an open geothermal system consisting of injection and productive wells. Hot water from production well is used and became cooler, and injection well returns the cold water into the aquifer. This cold water is filtered in porous soil towards the inflow of hot water of the production well. It is required to describe propagation of the cold front in the productive layer of water, depending on the different thermal soil parameters and initial data defined filtration rate in the productive layer, and to answer the question about the time of the system effective operation (operation of a geothermal system is stopped when the front of the cold water will reach the inflow of the production well).

© Springer International Publishing Switzerland 2015
I. Dimov et al. (Eds.): FDM 2014, LNCS 9045, pp. 393–399, 2015.
DOI: 10.1007/978-3-319-20239-6_44

In simulations of underground flow, Darcy's law and law of mass conservation (continuity equation) are used [1]. A convection-diffusion equation with dominant diffusion is considered. A system of equations for temperature in aquifer is solved using a finite difference method based on an approach of works of A.A.Samarskii and P.N.Vabishevich [2].

Let note that the model and the numerical algorithms are convenient (with some adaptations) to simulate different problems of heat and mass transfer, for example, to find thermal fields of underground pipelines [3,4] and the problems related to phase transitions in the soil around engineering constructions [5–7] in permafrost.

2   Mathematical Model and a Method of Solving the Problem

Let consider a mathematical model of a Underground Water Source (Geothermal Open Loop) Heat Pump System. A geothermal open loop (GOL) consists of two wells: an injection well and a production well (Fig. 1), which are inserted into as aquifer Ω as the heat source and sink. Water is taken from the aquifer by a productive well (Ω_1), circulated to the individual pump, cooled, and returned via an injection well (Ω_2). Let injection well has a cold water with temperature $T_1(t)$, production well has a hot water with temperature $T_2(t)$.

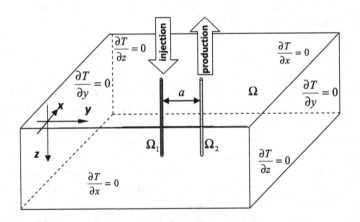

Fig. 1. A model of Geothermal Open Loop.

Let $T(t, x, y, z)$ be temperature in the aquifer, $p = p(t, x, y, z)$ — pressure field. Thermal exchange is described by equation

$$\frac{\partial T}{\partial t} + b \left(\frac{\partial T}{\partial x} u + \frac{\partial T}{\partial y} v + \frac{\partial T}{\partial z} w \right) = \lambda_0 \Delta T, \qquad (1)$$

here $b = \dfrac{\sigma \rho c_f}{\rho_0 c_0 (1 - \sigma) + \rho c_f \sigma}$, $\lambda_0 = \dfrac{\kappa_0}{\rho_0 c_0 (1 - \sigma) + \rho c_f \sigma}$, ρ_0 and ρ_f are density of aquifer soil and of water, c_0 and c_f are specific heats of aquifer soil and of water, κ_0 is thermal conductivity coefficient of soil, σ is porosity, (u, v, w) is vector of velocity of water filtration in the soil.

This equation is necessary to be considered with the following system, describing the water filtration:

$$
\begin{aligned}
\frac{\partial u}{\partial t} &= -\frac{1}{\rho} \frac{\partial p}{\partial x} - \frac{g \sigma u}{k}, \\
\frac{\partial v}{\partial t} &= -\frac{1}{\rho} \frac{\partial p}{\partial y} - \frac{g \sigma v}{k}, \\
\frac{\partial w}{\partial t} &= -\frac{1}{\rho} \frac{\partial p}{\partial z} - \frac{g \sigma w}{k} - g, \\
\frac{\partial u}{\partial x} &+ \frac{\partial v}{\partial y} + \frac{\partial w}{\partial z} = 0.
\end{aligned}
\tag{2}
$$

In simulations described by Eq. (1) and (2) it is necessary to estimate a distance a between the injection and productive wells due to the temperature in the productive well be appropriate for the considered system.

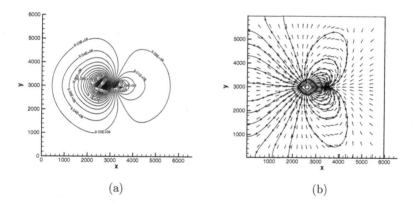

(a)   (b)

Fig. 2. (a) — density, (b) — velocity.

After transformation of system (2) we get Laplace equation for pressure

$$
\frac{\partial^2 p}{\partial x^2} + \frac{\partial^2 p}{\partial y^2} + \frac{\partial^2 p}{\partial z^2} = 0
\tag{3}
$$

with corresponding boundary conditions. Pressure may be computed by method Fig. 2 shows pressure distribution in an aquifer.

Let consider an exact solutions, which satisfy equations of water filtration in the aquifer. For the surfaces Ω_1 and Ω_2 let conside the given pressure

$$
P(t, x, y, z)\Big|_{\Omega_1} = P_1 - \rho g z,
\tag{4}
$$

and

$$P(t, x, y, z)\Big|_{\Omega_2} = P_2 - \rho g z. \tag{5}$$

Let find the solution in the form

$$p = b_1(t)x - \rho g z + b_0(t), \tag{6}$$

where $b_1(t)$ and $b_2(t)$ are unkown functions. Taking into consideration (4) and (5) this function has the form

$$p = -\frac{P_2 - P_1}{a}x - \rho g z + P_1. \tag{7}$$

Then, for the velocity u we get an ordinary differential equation

$$\frac{du}{dt} = -\frac{1}{\rho}\frac{P_1 - P_2}{a} - \frac{g\sigma u}{k}, \tag{8}$$

with the zero boundary condition $u(0) = 0$. Then the solution of Eq. (8) has the form

$$u = \frac{k(P_1 - P_2)}{\rho g \sigma a}\left(1 - e^{-\frac{g\sigma t}{k}}\right). \tag{9}$$

Thus, a partial solution of system (2) has a solution (9) and $v = w = 0$. We have to note, that when the time tends to infinity, the solution tends to the stationary

$$u^* = \frac{k(P_1 - P_2)}{\rho g \sigma a}, \tag{10}$$

Naturally, the resulting partial solution does not satisfy all the boundary and initial conditions of the problem, but suggests that over time the problem under consideration is a stationary regime. So in the model we can use a steady-state flow to describe convective transport terms in Eq. (1).

After finding the pressure field, a vector of velocity of filtered water is determined in the aquifer.

On the base of ideas in [2] a finite difference method is used with splitting by the spatial variables in three-dimensional domain to solve the problem (1)–(5). We construct an orthogonal grid, uniform, or condensing near the ground surface or to the surfaces of Ω_1 and Ω_2. The original equation for each spatial direction is approximated by an implicit central-difference scheme and a three-point sweep method to solve a system of linear differential algebraic equations is used.

3   Numerical Results

Let a computational domain be a parallelepiped 6000 m·6000 m·50 m size (Fig. 1). The choice of such large computational domain is related with decreasing the influence of boundary conditions. Mesh size is 201·201·51 = 2060451 nodes. Injection well is in point (2600 m, 3000 m), productive — in (3400 m, 3000 m). The distance between the wells is 800 m. Soil thermal parameters correspond to Hankal geothermal fields in the North Caucasus. The initial temperature of water in the aquifer at a depth of 950–1000 m is 95°C, temperature of the injected water is 55°C.

Computations are carried out with 1 day time step for 50 years and for various pressures difference.

We present the results of calculations for the differential pressure between production and injection wells 420000Pa. Figure 2 shows a typical pressure (a) and velocity (b) distribution in the aquifer.

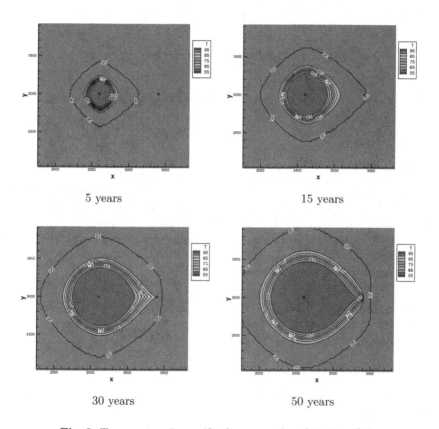

5 years   15 years

30 years   50 years

Fig. 3. Temperature in aquifer for years of exploitation, °C.

The transition of filtered water allow to compute temperature in the aquifer and to determine how the a cold injection well influences to the productive well. Figure 3 shows a temperature distribution in (x, y)-plane after 5, 15, 30, and 50 years of exploitation of the system.

Figure 4 is an average temperature in productive well during the time.

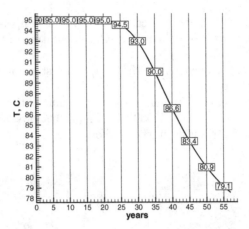

Fig. 4. Temperature in productive well during the time of exploitation, °C.

4 Conclusion

Computations alows to choose an optimal parameters of an open geothermal systems, in particular to determine an appropriate distance between injection and production wells depending on the operating conditions of the geothermal system. Taking into consideration a geothermal gradient allows to increase time of the system operation. Note that many researchers do not consider this factor in the simulation, although the results obtained, for example, in [8] are in qualitative agreement with our results.

Acknowledgement. Supported by Russian Foundation for Basic Research 14–01–00155 and by Program of UD RAS 15–16–1–10.

References

1. Polubarinova-Kochina, P.Ya.: Theory of Ground Water Movement. Princeton University Press, Princeton (1962)
2. Samarsky, A.A., Vabishchevich, P.N.: Computational Heat Transfer. The Finite Difference Methodology, vol. 2. Wiley, New York (1995)
3. Bashurov, V.V., Vaganova, N.A., Filimonov, M.Y.: Numerical simulation of thermal conductivity processes with fluid filtration in soil. J. Comput. Technol. **16**(4), 3–18 (2011). (in Russian)
4. Vaganova, N.: Mathematical model of testing of pipeline integrity by thermal fields. In: AIP Conference Proceedings, vol. 1631, pp. 37–41 (2014)
5. Filimonov, M.Yu., Vaganova, N.A.: Simulation of thermal fields in the permafrost with seasonal cooling devices. In: Proceedings of the ASME. 45158. Pipelining in Northern and Offshore Environments; Strain-Based Design; Risk and Reliability; Standards and Regulations, vol. 4, pp. 133–141 (2012)

6. Filimonov, M.Yu., Vaganova, N.A.: Simulation of thermal stabilization of soil around various technical systems operating in permafrost. Appl. Math. Sci. **7**(144), 7151–7160 (2013)
7. Vaganova, N.A., Filimonov, M.Yu.: Simulation of engineering systems in permafrost. Vestnik Novosibirskogo Gosudarstvennogo Universiteta. Seriya Matematika, Mekhanika, Informatika **13**(4), 37–42 (2013). (in Russian)
8. Le Brun, M., Hamm, V., Lopez, V., Ungemach, P., Antics, M., Ausseur, J.Y., Cordier, E., Giuglaris, E., Goblet, P., Lalos, P.: Hydraulic and thermal impact modelling at the scale of the geothermal heating doublet in the Paris basin, France. In: Proceedings of the Thirty-Sixth Workshop on Geothermal Reservoir Engineering Stanford University, Stanford, California, 31 January–2 February, SGP-TR-191 (2011)

Modeling of Annual Heat and Moisture Diffusion in a Multilayer Wall

G.P. Vasilyev[1], V.A. Lichman[1], and N.V. Peskov[1,2]($\boxtimes$)

[1] Insolar Group of Companies, Moscow, Russia
[2] Faculty of Computational Mathematics and Cybernetics, Lomonosov Moscow
State University, Moscow 119991, Russia
peskov@cs.msu.ru

Abstract. We present an 1D numerical model of heat, steam, and water
transfer across a wall consisting of several layers of different materials.
The model is the system of coupled diffusion equations for wall temperature; vapor pressure, and water concentration in material pores, with
account of vapor condensation and water evaporation. The system of
nonlinear PDEs is solved numerically using the finite difference method.
The main objective of modeling is simulation of long-term behavior of
building wall moisture distribution under influence of seasonal variations
in atmospheric air temperature and humidity.

1 Introduction

Simultaneous heat, steam, and moisture transfer with condensation and evaporation in porous materials is of practical importance in applications in civil
engineering. The transport of water vapor across building walls and its possible
condensation increases the thermal conductivity of the porous materials and may
cause structural damage. Complexity of the phenomenon and the large variety
of conditions under which it can proceed cause the development of new models
and approaches to the problem in a lot of research works. In particular, we can
refer the works [1–9].

The main goal of this paper is not the development of original model of the
phenomenon, but the presentation of a sufficiently precious and stable numerical
algorithm for solution of the model of coupled heat, steam, and moisture diffusion, invented by Fokin K.F. in his book [10]. The Fokin model is widely used in
civil engineering practice in Russia for estimation of long term thermal and moisture behavior in building envelopes [11]. The model is phenomenological model
based on simple, intuitive physical principles. The diffusion flux of each component is proportional to gradient of corresponding variable: temperature, partial
pressure of steam, and water concentration. All parameters of the model can be
measured experimentally. Evolution equation for each component is the diffusion
equation. Mutual coupling between equations is performed by the dependence
of diffusion coefficients on moisture, and trough the source terms related to the
phase transitions. Main difficult in numerical solution of the model is connected
with variability in time of regions of vapor condensation.

I. Dimov et al. (Eds.): FDM 2014, LNCS 9045, pp. 400–407, 2015.
DOI: 10.1007/978-3-319-20239-6_45

2   Mathematical Formulation of the Model

Model parameters. For definiteness, in this paper we will consider three-ply wall. The thickness of wall layers is denoted as d_1, d_2, and d_3 [m] respectively. Let the first layer is the external (outward building) and the third layer is the internal (inward building) layer of the wall. We will consider the fluxes of heat and moisture in the direction traversal to wall only, – the direction along x axis. The point $L_0 = 0$ is the outer edge of external layer, the points $L_1 = d_1$ and $L_2 = L_1 + d_2$ are the interface points between layers, and the point $L_3 = L_2 + d_3$ is the outer edge of internal layer.

The material of each layer is characterized by following parameters: density – ρ [kg/m^3]; heat capacity – c [kJ/(kg·K)]; coefficients of thermal – λ [kJ/(h·m·K)]; vapor – μ [g/(h·m·Pa)]; and hydraulic – β [g/(h·m·%)], conductivity. The parameters ρ, c, and μ are supposed to be constant in considered range of temperature T [K] and moisture volumetric concentration ω [%], while other parameters can depend on ω. For given material the dependencies $\lambda(\omega)$ and $\beta(\omega)$ used in calculations are polynomial interpolations of experimental measurements. Under equilibrium conditions at constant temperature the relation between air humidity φ and moisture ω is determined from experimentally measured sorption isotherm $\omega = o(\varphi)$.

The external layer of the wall contacts with atmospheric air, and the internal layer contacts with building interior air. The air temperature $T_{ex}(t)$, $T_{in}(t)$, and humidity $\varphi_{ex}(t)$, $\varphi_{in}(t)$ are given function of time t [h]. Heat exchange between wall and air is determined by coefficients α_{ex} and α_{in} [kJ/(h·m^2·K)]. Steam exchange at the wall borders is determined by coefficients γ_{ex} and γ_{in} [g/(h·m^2·Pa)].

Heat conductivity. The transversely heat transfer in each wall layer is described by the heat conduction equation for the material temperature $T(t, x)$:

$$c\rho \frac{\partial T(t, x)}{\partial t} = \frac{\partial}{\partial x}\left(\lambda(\omega)\frac{\partial T(t, x)}{\partial x}\right) + Q(T, \omega). \tag{1}$$

The source term Q takes into consideration the latent heat of vapor condensation and water evaporation (the water-ice transitions are not included in the model).

At the wall borders and layer interfaces there are imposed boundary and conjugation conditions:

$$-\lambda(\omega)\frac{\partial T(t, x)}{\partial x}\bigg|_{x=L_0} = \alpha_{ex}\left(T_{ex}(t) - T(t, x)\right)\bigg|_{x=L_0}; \tag{2}$$

$$T(t, L_i - 0) = T(t, L_i + 0),$$
$$\lambda(\omega)\frac{\partial T(t, x)}{\partial x}\bigg|_{x=L_i-0} = \lambda(\omega)\frac{\partial T(t, x)}{\partial x}\bigg|_{x=L_i+0}, \quad i = 1, 2; \tag{3}$$

$$-\lambda(\omega)\frac{\partial T(t,x)}{\partial x}\bigg|_{x=L_3} = \alpha_{\text{in}}\left(T(t,x) - T_{\text{in}}(t)\right)\bigg|_{x=L_3}. \tag{4}$$

Vapor and liquid water conductivity. It is supposed that the water in wall material can be in three forms: water vapor (steam), and mobile liquid water in material pores, and immobile absorbed water rigidly connected with material skeleton. The concentration of absorbed water ω depends upon local air humidity in pores $\varphi(t,x)$ and assumed to be equal to equilibrium concentration $o(\varphi)$. If $\varphi < 1$, the material contains only steam and absorbed water, the mass of absorbed water in unit volume of material is equal to $0.01\omega\rho$.

The air humidity φ is defined as ratio $\varphi = e/E(T)$, where e [Pa] is partial pressure of water vapor, and $E(T)$ is pressure of saturated vapor at air temperature T. For calculation of E we use the approximation formula:

$$E(T) = \begin{cases} 4.688(1.486 + T/100)^{12.3}, & T < 0, \\ 288.58(1.098 + T/100)^{8.02}, & T \geq 0. \end{cases} \tag{5}$$

The function $\varphi(t,x)$ is supposed to be continuous function of both variables. So, at every t one can define the subset $V_t \subseteq (L_0, L_3)$, $V_t = \{x : \varphi(t,x) < 1\}$, and the complementary subset $W_t = (L_0, L_1) \setminus V_t$.

In subset V_t the moisture moves in the form of steam only, and its motion is governed by the vapor conduction equation

$$\xi(\omega)\rho\frac{\partial e(t,x)}{\partial t} = \mu\frac{\partial^2 e(t,x)}{\partial x^2}. \tag{6}$$

(Thermo-diffusion of steam, that is the diffusion induced by temperature gradient, does not included in the model.) Here the parameter ξ defines the 'vapor capacity' of material and can be estimated by the equation

$$\xi(\omega) = \frac{do(\varphi)}{d\varphi}. \tag{7}$$

As noted above, Eq. (6) is defined in the subset V_t. If the point L_0 and/or L_3 are the boundary points of V_t, then the boundary conditions of convective exchange of steam between air in material pores and surrounding air are imposed similar to (2), (4) with replaced T by e, λ by μ, and α by γ. If any of interface points L_1, L_2 belongs to V_t, then the conjugation condition assuming continuity of vapor pressure and flux is imposed similar to (3).

In subset W_t the material pores contain liquid water together with water vapor, and it is supposed that between water and vapor there keeps up the dynamic equilibrium. That is the partial pressure of vapor in W_t equals to pressure of saturated vapor, $e(t,x) \equiv E(T(t,x))$ for all $x \in W_t$. To estimate the volumetric concentration of liquid water w [%] we propose the following formula

$$w(t,x) = \omega(t,x) - o(1), \tag{8}$$

where $o(1)$ is maximal concentration of absorbed water, corresponding to $\varphi = 1$.

Diffusive motion of liquid water is described by equation

$$10\rho\frac{\partial w(t,x)}{\partial t} = \frac{\partial}{\partial x}\left(\beta(\omega)\frac{\partial w(t,x)}{\partial x}\right) + \mu\frac{\partial^2 E(T)}{\partial x^2}. \tag{9}$$

Here numerical coefficient '10' appears due to different dimensions of parameters (kg in ρ, g in μ and β, and % in w). Second term in the right hand side corresponds to vapor condensation or water evaporation depending on its sign. To avoid extra model complication the condition of water impermeability at border points of W_t is accepted

$$\left.\frac{\partial w}{\partial x}\right|_{\partial W_t} = 0. \tag{10}$$

If any of points L_1, L_2 is inner point of W_t, then there is imposed the condition of continuity of water concentration and water flux, similar to (3). In the points, separating subsets V_t and W_t, we suppose continuity of pressure $e(t,x)$.

Latent heat of phase transition. The term $\nu = \mu(\partial^2 E(T)/\partial x^2)$ [g/(h·m^3)] in Eq. (9) represents the rate of change of water concentration due to water evaporation or vapor condensation, that is in result of phase transition. The phase transitions are accompanied by release or absorption of latent heat, depending on the sign of ν. The latent heat of phase transition is accounted in Eq. (1) via the source term Q, numerical value of which is calculated with the formula

$$Q = \begin{cases} q_L\nu, & x \in W_t, \\ 0, & x \in V_t, \end{cases} \tag{11}$$

where q_L is the specific heat of water vaporization ($= 2.26\,\text{kJ/g}$).

3   Numerical Scheme

The system of Eqs. (1), (6), and (9), with all additional conditions listed above is solved numerically with the help of the finite differences method. To approximate the space derivatives with finite differences we define the uniform grid of N nodes x_k on the interval (L_0, L_3): $x_k = L_0 + (k - 0.5)h$, $k = 1, 2, \ldots, N$, $h = (L_3 - L_0)/N$. The time derivative is approximated on the time grid $t_0 = 0, t_1, t_2, \ldots, t_n, \ldots$, with variable time step $\tau_n = t_{n+1} - t_n$, $n \geq 0$. Each function $f(t,x)$ is replaced its grid approximation $f_k^n = f(t_n, x_k)$. Below, for brevity, we will drop the upper index denoting f_k^n as f_k, and f_k^{n+1} as $\hat{f}_k$.

Finite difference equations. To derive the finite difference equations we use implicit scheme and heat and mass balance method. The finite-difference counterparts of Eqs. (1), (6), and (9) are the following

$$hc_k\rho_k\frac{\hat{T}_k - T_k}{\tau} = \hat{\Lambda}_k^-(\hat{T}_{k-1} - \hat{T}_k) - \hat{\Lambda}_k^+(\hat{T}_k - \hat{T}_{k+1}) + \hat{Q}_k, \quad k = 1, \ldots, N. \tag{12}$$

$$h\hat{\xi}_k \rho_k \frac{\hat{e}_k - e_k}{\tau} = M_k^-(\hat{e}_{k-1} - \hat{e}_k) - M_k^+(\hat{e}_k - \hat{e}_{k+1}), \ x_k \in V_n. \tag{13}$$

$$10h\rho_k \frac{\hat{w}_k - w_k}{\tau} = \hat{B}_k^-(\hat{w}_{k-1} - \hat{w}_k) - \hat{B}_k^+(\hat{w}_k - \hat{w}_{k+1}) + \hat{\nu}_k, \ x_k \in W_n. \tag{14}$$

The coefficients Λ are calculated as follows

$$\Lambda_1^- = \frac{2\alpha_{\text{ex}}\lambda_1}{\alpha_{\text{ex}}h + 2\lambda_1}, \ \Lambda_k^- = \frac{2\lambda_{k-1}\lambda_k}{h(\lambda_{k-1} + \lambda_k)}, \ k = 2, \ldots, N;$$

$$\Lambda_k^+ = \frac{2\lambda_k\lambda_{k+1}}{h(\lambda_k + \lambda_{k+1})}, \ k = 1, \ldots, N-1, \ \Lambda_N^+ = \frac{2\lambda_N\alpha_{\text{in}}}{2\lambda_N + \alpha_{\text{in}}h}. \tag{15}$$

The coefficients M and B are calculated similar to Λ. The term ν is approximated by second difference

$$\nu_k = M_k^-(E_{k-1} - E_k) - M_k^+(E_k - E_{k+1}), \tag{16}$$

where $E_k = E(T_k)$, if $x \in W_n$; $E_k = e_k$, if $x \in V_n$; $E_0 = e_{\text{ex}}(t_n)$, $E_{N+1} = e_{\text{in}}(t_n)$.

Assuming that $T_0 = T_{\text{ex}}(t_n)$, $T_{N+1} = T_{\text{in}}(t_n)$; $e_0 = e_{\text{ex}}(t_n)$, $e_{N+1} = e_{\text{in}}(t_n)$, one obtains the closed system of $2N$ algebraic equations with $2N$ unknowns. First N unknowns are the values of temperature $\hat{T}_k$ at grid nodes x_k in time moment t_{n+1}. Other N unknowns are the values of steam partial pressure $\hat{e}_k$ at nodes $x_k \in V_n$ or the values of water concentration $\hat{w}_k$ at nodes $x_k \in W_n$.

Iterative solution. The system of finite-difference Eqs. (12), (13), (14) is the non-linear system because of the scheme is implicit and the coefficients λ and β depend on the system solution. To solve the non-linear system we use an iterative procedure.

Using known solution at time t_n we calculate the coefficients Λ and B (coefficients M are constant) and substitute them in system (12), (13), (14). Solving the resulting linear system we obtain first approximation to solution at moment t_{n+1}, $\{T_k^{(1)}, e_k^{(1)}, w_k^{(1)}\}$. Then the first approximation is used in the same way to obtain second approximation $\{T_k^{(2)}, e_k^{(2)}, w_k^{(2)}\}$, and so on. The iterations are ended when the relative difference between two successive approximations becomes sufficiently small.

Designation of subset W_t. At initial time moment t_0 all nodes x_k, where initial moisture concentration ω_k is not less than maximal adsorption concentration, $\omega_k \geq o(1)$, are included in subset W_0. Further, the domain W is re-designated after each successful time step. The nodes x_k from V_n, where the vapor pressure $e_k \geq E_k$ pass from V_n in W_{n+1}, while the nodes x_k from W_n, where $w_k < 0$, pass from W_n in V_{n+1}. There are no restrictions on shape and size of subset V (and, therefore, of W).

Choice the value of time step τ. To avoid possible numerical instability time step τ is limited by some empirical maximal value $\tau_{\max}$. Initial time step is assigned less than $\tau_{\max}$ in several orders of magnitude. Time step with assigned value of τ is considered as successful, if (i) the number of iterations in solution of system (12), (13), (14) is not exceed assigned maximal number of iterations, and (ii) relative difference between solution at current time step and solution at previous time step is sufficiently small. If current time step is not successful, the value of τ is diminished by definite factor, and step is repeated until the time step becomes successful. If the number of successive successful steps exceeds assigned number, the value of τ increases by definite factor.

4   Numerical Example

As an example we consider a three-ply wall. The external layer of wall is a thin stucco, and other two layers of equal thickness are concrete and mineral wool as heat insulator. The main goal of our example is to demonstrate the difference (well-known in construction practice) in heat and moisture behavior in two walls consisting of these layers. First wall is 'correct' wall with layers order: stucco - heat insulator - concrete; and second wall is 'incorrect' with layers order: stucco - concrete - heat insulator.

In this paper we use some 'typical' values of material parameters, which can be found in numerous literature. The accepted values of constant material parameters are presented in Table 1 (dimensions were specified in text above). Two additional parameters k and λ_0 are used for calculation of material coefficient of thermal conductivity λ, $\lambda = \lambda_0 + k\omega$.

Table 1. Material parameters.

Material	d	ρ	c	μ	k	λ_0	$o(1)$
Concrete	0.2	2500	0.84	3×10^{-5}	0.1	1.45	3.997
Mineral wool	0.2	125	0.84	3×10^{-4}	0.001	0.06	1.875
Stucco	0.02	1800	0.84	9×10^{-5}	0.07	0.7	1.184

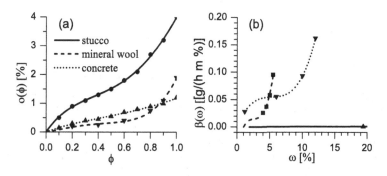

Fig. 1. (a) – sorption isotherm $o(\varphi)$, (b) – coefficient of water conductivity $\beta(\omega)$.

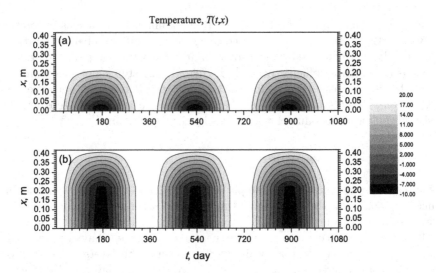

Fig. 2. Space-time plot of temperature in 'correct' (a), and 'incorrect' (b) wall.

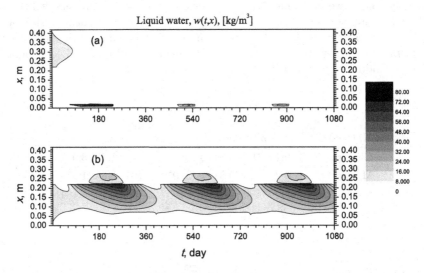

Fig. 3. Space-time plot of water concentration in 'correct' (a), and 'incorrect' (b) wall.

The graphs of the sorption isotherms and the coefficient β in dependence on ω are shown in Fig. 1, where the experimental measurements, depicted by markers, are connected via polynomial interpolations. In the building interior the constant air temperature and humidity are supposed, $T_{in} = 20\,°C$, $\varphi_{in} = 0.6$. The temperature and humidity of external air simulate their seasonal variations for Moscow region. The heat exchange coefficients $\alpha_{in} = 31.4$, $\alpha_{ex} = 85.8$, and the vapor exchange coefficients $\gamma_{in} = 0.075$, $\gamma_{ex} = 0.750$.

The initial time moment corresponds to the middle of July with average daily air temperature of 19.3 °C. The initial temperature distribution in wall is linear between 19.3 °C at outer and 20 °C at inner wall surface. The initial moisture concentration is assigned for every wall layer. In our example it was $0.9 \times o(1) = 3.597$ for stucco, $1.5 \times o(1) = 1.775$ for concrete, and $1.5 \times o(1) = 2.813$ for mineral wool. That is, the wall is waterlogged and have to dry.

Figure 2 shows, that (as it should be expected) the fall of temperature in winter time occurs mostly in the heat insulator layer. Hence, during winter the concrete temperature is high in 'correct' wall, and low in 'incorrect' wall. The vapor condensation proceeds in that regions, where the vapor pressure e becomes higher than the pressure of saturated vapor E. Such a condition is mainly created at low temperatures. Therefore, there are much more possibilities for condensation in 'incorrect' wall than in 'correct' wall. This is confirmed by Fig. 3.

Acknowledgments. This work was financially supported by the Ministry of Science and Education of Russian Federation, contract RFMEFI57614X0034.

References

1. Kohonen, R.: Transient analysis of the thermal and moisture physical behaviors of building constructions. Build. Environ. **19**, 1–11 (1984)
2. Ogniewicz, Y., Tien, C.E.: Analysis of condensation in porous insulation. Int. J. Heat Mass Transf. **24**, 421–429 (1986)
3. Motakef, S., El-Masri, M.A.: Simultaneous heat and mass transfer with phase change in a porous slab. Int. J. Heat Mass Transf. **29**, 1503–1512 (1986)
4. Shapiro, A.P., Motakef, S.: Unsteady heat and mass transfer with phase change in a porous slab: analytical solutions and experimental results. Int. J. Heat Mass Transf. **33**, 163–173 (1990)
5. Künzel, H.M., Kiessl, K.: Calculation of heat and moisture transfer in exposed building components. Int. J. Heat Mass Transf. **40**, 159–617 (1997)
6. Häupl, P., Grunewald, J., Fechner, H., Stopp, H.: Coupled heat air and moisture transfer in building structures. Int. J. Heat Mass Transf. **40**, 1633–1642 (1997)
7. Budaiwi, I., El-Diasty, R., Abdou, A.: Modeling of moisture and thermal transient behaviour of multilayer non-cavity walls. Build. Environ. **34**, 537–551 (1999)
8. Steeman, M., Janssens, A., Steeman, H.J., Van Belleghem, M., De Paepe, M.: On coupling 1D non-isothermal heat and mass transfer in porous materials with a multizone building energy simulation model. Build. Environ. **45**, 865–877 (2010)
9. Konga, F., Zhangb, Q.: Effect of heat and mass coupled transfer combined with freezing process on building exterior envelope. Energy Build. **62**, 486–495 (2013)
10. Fokin, K.F.: Constructional Thermo-technique of the Building. Moscow, Stroiizdat (1973). (in Russian)
11. Vasilyev, G.P.: What can prevent us from making Moscow an energy-efficient city? Therm. Eng. **58**, 682–690 (2011)

Asymptotic-Numerical Method for Moving Fronts in Two-Dimensional R-D-A Problems

Vladimir Volkov, Nikolay Nefedov$^{(\boxtimes)}$, and Eugene Antipov

Department of Mathematics, Faculty of Physics,
Lomonosov Moscow State University, 119991 Moscow, Russia
volkovvt@mail.ru, nefedov@phys.msu.ru

Abstract. A singularly perturbed initial-boundary value problem for a parabolic equation known in applications as the reaction-diffusion equation is considered. An asymptotic expansion of the solution with moving front is constructed. Using the asymptotic method of differential inequalities we prove the existence and estimate the asymptotic expansion for such solutions. The method is based on well-known comparison theorems and formal asymptotics for the construction of upper and lower solutions in singularly perturbed problems with internal and boundary layers.

Keywords: Singularly perturbed parabolic problems · Reaction-diffusion equation · Internal layers · Fronts · Asymptotic methods · Differential inequalities

1 Statement of the Problem

The purpose of the presented paper is to develop an effective numerical-asymptotic approach to study solutions with internal transition layers – moving fronts – in a mathematical model of reaction-diffusion type in the case of two spatial dimensions. We demonstrate our method for the following problem.

Consider the equation

$$\varepsilon^2 \Delta u - \varepsilon \frac{\partial u}{\partial t} = f(u, x, y, \varepsilon), \qquad y \in (0, a), \quad x \in (-\infty, +\infty), \quad t > 0 \qquad (1)$$

with the boundary and initial conditions

$$\left. \frac{\partial u}{\partial y} \right|_{y=0;a} = 0, \quad u(x, y, t, \varepsilon) = u(x+L, y, t, \varepsilon), \quad u(x, y, t, \varepsilon)|_{t=0} = u^0(x, y). \quad (2)$$

In the Eq. (1), $\varepsilon > 0$ is small parameter, which is usually a consequence of the parameters of the physical problem. It should be noted that the appearance of the small parameter before the spatial derivatives is determined by the characteristics of the physical system, while the small parameter before the time derivative determines only the scale of the time, convenient for the further consideration. Functions $u^0(x, y)$ and $f(u, x, y, \varepsilon)$ are assumed to be sufficiently smooth and L- periodic in the variable x.

© Springer International Publishing Switzerland 2015
I. Dimov et al. (Eds.): FDM 2014, LNCS 9045, pp. 408–416, 2015.
DOI: 10.1007/978-3-319-20239-6_46

Stationary solutions of problem (1)–(2) with internal and boundary layers have been thoroughly investigated (see [1] and the references therein). The generation of an internal layer from smooth initial functions has also been studied (see [2,3]). Main purpose of this paper is to study the solution of moving front type and to obtain equations for effective description of its dynamics. We also prove the existence of a solution of such type and construct its asymptotics. The results below extend [4], where the case of one spatial dimension was considered, and the ideas in [5] are used for the proof of the existence of front type solutions.

Suppose the following conditions are satisfied.

(A₁). (a) *The function* $f(u, x, y, \varepsilon)$ *is such that the reduced equation* $f(u, x, y, 0) = 0$ *has exactly three roots* $u = \varphi^{(\pm)}(x, y)$, $u = \varphi^{(0)}(x, y)$. (b) *Assume that* $\varphi^{(-)}(x, y) < \varphi^{(0)}(x, y) < \varphi^{(+)}(x, y)$ *for all* $(x, y) \in \bar{D} = (-\infty, +\infty) \times [0, a]$ *and* $f_u(\varphi^{(\pm)}(x, y), x, y, 0) > 0$, $f_u(\varphi^{(0)}(x, y), x, y, 0) < 0$.

It is known from [2,3] that under condition (A₁) and some quite general conditions for the initial function $u^0(x, y)$ at time of order $t_B(\varepsilon) = B\varepsilon |\ln \varepsilon|$ the solution of problem (1)–(2) quickly generates a thin internal transition layer between the two levels $\varphi^{(-)}(x, y)$ and $\varphi^{(+)}(x, y)$ located in the neighborhood of some curve $C_0^0 : y = h^0(x)$.

(A₂). *Assume that the initial function* $u^0(x, y)$ *has the form of a transition layer:* $u^0(x, y) = \varphi^{(-)}(x, y) + O(\varepsilon)$ *for* $(x, y) \in D_0^{(-)}$, $u^0(x, y) = \varphi^{(+)}(x, y) + O(\varepsilon)$ *for* $(x, y) \in D_0^{(+)}$ *excluding a small neighborhood of the curve* $C_0^0 : y = h^0(x)$.

Our further purpose is to study the front type solution of (1)–(2) and describe its dynamics.

Let us consider the following problem, where $f(u, x, y, \varepsilon)$ satisfies the condition (A₁) and x, y are parameters:

$$\frac{\partial^2 p}{\partial \xi^2} + W \frac{\partial p}{\partial \xi} = f(p, x, y, 0); \quad p(x, 0) = \varphi^{(0)}(x, y), \ p(x, \pm\infty) = \varphi^{(\pm)}(x, y) \quad (3)$$

This problem is well known (see, for example, [6]), and for every x, y there exists a unique pair $(W(x, y), p(\xi; x, y))$ that satisfies problem (3) and the following estimates are valid (C and σ are positive constants)

$$\left| p(x, \xi) - \varphi^{(\pm)}(x, y) \right| \leq C e^{\sigma |\xi|} \quad \text{for} \quad \xi \to \pm\infty.$$

(A₃). *There exists a solution* $h(x, t)$ *of the Cauchy problem*

$$\frac{h_t}{\sqrt{1 + h_x^2}} = W(x, h(x, t)), \quad h(x, 0) = h^0(x), \ h(x, t) = h(x + L, t), \ x \in (-\infty; +\infty).$$

Using this solution for fixed t we define the curve $C(t) \div \{y = h(x, t)\} \in \bar{D}$ if $t \in [0; T]$.

2   Description of the Moving Front

We define the location of the internal layer at fixed t by curve $C_\lambda(t) \div \{y = h^*(x, t, \varepsilon)\}$, which is the intersection of the solution $u(x, y, t, \varepsilon)$ and root $u = \varphi^{(0)}(x, y)$. An asymptotic approximation of $C_\lambda(t)$ will be constructed below. We denote by $D^{(+)}$ and $D^{(-)}$ the domains located at two sides of curve $C_\lambda(t)$.

   2a. Formal asymptotic procedure.

   To construct the formal asymptotics of the solution (1)–(2) we consider:

$$\varepsilon^2 \Delta u - \varepsilon \frac{\partial u}{\partial t} - f(u, x, y, \varepsilon) = 0, \qquad (x, y) \in D^{(\pm)}, \quad t > 0,$$

$$u(x, y, t, \varepsilon) = u(x + L, y, t, \varepsilon), \quad u(x, y, 0, \varepsilon) = u^0(x, y, \varepsilon), \quad (x, y) \in D^{(\pm)}$$

$$u(x, h^*(x, t, \varepsilon), t, \varepsilon) = \varphi^{(0)}(x, h^*(x, t, \varepsilon)), \quad \left. \frac{\partial u}{\partial y} \right|_{y=0} = 0 \qquad (4)$$

and

$$u(x, h^*(x, t, \varepsilon), t, \varepsilon) = \varphi^{(0)}(x, h^*(x, t, \varepsilon)), \quad \left. \frac{\partial u}{\partial y} \right|_{y=a} = 0 \qquad (5)$$

   To find the location of the internal transition layer $C_\lambda(t)$ we introduce local coordinates (r, l) in a neighborhood of some curve $C_0(t) : \{x = l, \ y = h(l, t)\}$, where r is the distance from $C_0(t)$ along the normal to this curve, with the sign "+" in the domain $D^{(+)}$ and with "−" in $D^{(-)}$, l is the coordinate of the point on the curve $C_0(t)$ from which this normal is going. We have

$$x = l + r \cdot n_1(l, t), \quad y = h(l, t) + r \cdot n_2(l, t), \qquad (6)$$

where $n_1(l, t) = \frac{-h_l}{\sqrt{1+h_l^2}}$, $n_2(l, t) = \frac{1}{\sqrt{1+h_l^2}}$ are the components of the unit normal vector to $C_0(t)$ at the point $(l, h(l, t))$. Note, that in these coordinates the curve $C_0(t)$ is determined by $r = 0$. Further we will show, how to find $C_0(t)$.

   Using these local coordinates we define the unknown curve $C_\lambda(t)$ in the form of a power series in ε:

$$r = \lambda^*(l, t, \varepsilon) = \varepsilon \cdot \lambda_1(l, t) + \varepsilon^2 \cdot \lambda_2(l, t) + ..., \qquad (7)$$

The asymptotics of (4), (5) can be constructed in the form including regular and boundary functions

$$U^{(\pm)}(x, y, t, \varepsilon) = \bar{u}^{(\pm)}(x, y, \varepsilon) + P^{(\pm)}(\rho^{(\pm)}, x, \varepsilon) + Q^{(\pm)}(\xi, l, \varepsilon), \qquad (8)$$

where $\xi = \frac{r - \lambda^*(l, t, \varepsilon)}{\varepsilon}$, $\rho^{(+)} = \frac{a-y}{\varepsilon}$, $\rho^{(-)} = \frac{y}{\varepsilon}$, and the functions $\bar{u}^{(\pm)}(x, y, \varepsilon)$, $P^{(\pm)}(\rho^{(\pm)}, x, \varepsilon)$, $Q^{(\pm)}(\xi, l, \varepsilon)$ are power series in ε, which can be find by the standard method of boundary functions [1]. The functions $Q^{(\pm)}(\xi, l, \varepsilon)$ describe the internal transition layer (moving front) near the curve $C_\lambda(t)$, therefore they depend on the variable t by means of ξ. The functions $P^{(\pm)}(\rho^{(\pm)}, x, \varepsilon)$ describe the solution near the boundaries $y = 0$, $y = a$. The regular series $\bar{u}^{(\pm)}(x, y)$ in the domains $D^{(-)}$ and $D^{(+)}$, and also the boundary series $P^{(\pm)}\left(\rho^{(\pm)}, x, \varepsilon\right)$ near the boundaries of D are determined by the standard scheme [1]. Note that the boundary series $P^{(\pm)}\left(\rho^{(\pm)}, x, \varepsilon\right)$ are significant only in a small area near $y = 0$ and $y = a$, and rapidly exponentially decrease and do not influence the behavior of the internal transition layer. By this reason we concentrate only on the describing of the internal layer $Q^{(\pm)}(\xi, l, \varepsilon)$.

To define the terms of (7) and (8) we must write the asymptotic expansions for the solutions of each problems (4) and (5) according standard scheme [1]. Terms of series (7), (8) will be defined in this process from the conditions of continuous matching for the functions $U^{(-)}(x, y, t, \varepsilon)$, $U^{(+)}(x, y, t, \varepsilon)$ - asymptotic expansions in domains $D^{(\pm)}$ - and their normal derivatives on the curve $C_\lambda(t)$ (C^1 - matching conditions):

$$\text{(a)} \quad U^{(-)} = U^{(+)}, \qquad \text{(b)} \quad \varepsilon\frac{\partial U^{(-)}}{\partial n} = \varepsilon\frac{\partial U^{(+)}}{\partial n} \quad \text{on} \quad C_\lambda(t) \tag{9}$$

Conditions (9) must be carried out consistently for zero and all higher degrees of ε.

We briefly describe some details of this asymptotic procedure. Using the local coordinates (6) and introducing the stretched variable $\xi = \frac{r - \lambda^*(l, t, \varepsilon)}{\varepsilon}$, we have for the parabolic operator $\hat{L}u \equiv \varepsilon^2 \Delta u - \varepsilon\frac{\partial u}{\partial t}$ in the form

$$\hat{L}u = \frac{\partial^2 u}{\partial \xi^2} + V_n\frac{\partial u}{\partial \xi} - \varepsilon\left[k\frac{\partial u}{\partial \xi} + \frac{\partial u}{\partial t}\right] + O\left(\varepsilon^2\right), \tag{10}$$

where $V_n = V_0 + \lambda_t^*$ is the normal speed of the point on the curve $C_\lambda(t)$ and $V_0 \equiv \frac{h_t}{\sqrt{1+h_l^2}}$; $k = k(l)$ is the local curvature of $C_\lambda(t)$, $\lambda^*(l, t, \varepsilon)$ is defined in (7).

We represent $f(u, x, y, \varepsilon)$ in the form $f(u, x, y, \varepsilon) = \bar{f}^{(\pm)}(x, y, \varepsilon) + Q^{(\pm)}f(\xi, l, \varepsilon)$, where the functions $\bar{f}^{(\pm)}(x, y, \varepsilon) = f(\bar{u}^{(\pm)}(x, y), x, y, \varepsilon)$ and

$$Q^{(\pm)}f(\xi, l, \varepsilon) = f(\bar{u}^{(\pm)}(x, y) + Q^{(\pm)}(\xi, l, \varepsilon), x, y, \varepsilon) - \bar{f}^{(\pm)}(x, y, \varepsilon)$$

are power series in ε, and the indices $(\pm)$ correspond to the domains $D^{(\pm)}$. Substituting these functions and the operator $\varepsilon^2 \Delta u - \varepsilon\frac{\partial u}{\partial t}$ in the form (10) into (4), (5) and equating the terms depending on (x, y) and (ξ, l) separately, we obtain the relations to determine the coefficients of the asymptotic expansions:

$$\varepsilon^2 \Delta\bar{u}^{(\pm)} - \bar{f}^{(\pm)}(u, x, y, \varepsilon) = 0, \tag{11}$$

$$\left(\frac{\partial^2}{\partial \xi^2} + V_n\frac{\partial}{\partial \xi} - \varepsilon\left(\frac{\partial}{\partial t} + k\frac{\partial}{\partial \xi}\right) + O\left(\varepsilon^2\right)\right)Q^{(\pm)} = Q^{(\pm)}f(\xi, l, t, \varepsilon). \tag{12}$$

2b. Zero order functions (moving front).

At zero order we have for regular part $f(u^{(\pm)}, x, y, 0) = 0$. Thus according to condition (A_1) we can take $u^{(\pm)}(x, y) = \varphi^{(\pm)}(x, y)$.

The functions $Q_0^{(\pm)}(\xi, l)$ satisfy the following problem:

$$\left(\frac{\partial^2}{\partial \xi^2} + V_0 \frac{\partial}{\partial \xi}\right) Q_0^{(\pm)}(\xi, l) = f\left(\varphi^{(\pm)}(x, y) + Q_0^{(\pm)}(\xi, l), x, y, 0\right), \qquad (13)$$

$$Q_0^{(\pm)}(0, l) = \varphi^{(0)}(x, y) - \varphi^{(\pm)}(x, y) \text{ for } (x, y) \in C_\lambda(t); \quad Q_0^{(\pm)}(\pm\infty, l) = 0.$$

We define the continuous function $\tilde{u}(\xi) = \varphi^{(\pm)}(x, y) + Q_0^{(\pm)}(\xi, l)$ for $(x, y) \in C_\lambda t)$, and rewrite (13) as

$$\frac{\partial^2 \tilde{u}}{\partial \xi^2} + V_0 \frac{\partial \tilde{u}}{\partial \xi} = f(\tilde{u}, x, y, 0); \quad \tilde{u}(\pm\infty) = \varphi^{(\pm)}(x, y), \quad \tilde{u}(0) = \varphi^{(0)}(x, y) \quad (14)$$

If the function $f(u, x, y, \varepsilon)$ satisfies the condition (A_1), thus problem (14) has the unique solution $\tilde{u}(\xi)$, and the estimate $\left|\tilde{u}(\xi) - \varphi^{(\pm)}(x, y)\right| \leq M e^{\sigma|\xi|}$ for $\xi \to \pm\infty$ is valid, where M and σ are positive constants [6]. Note, that condition (9a) is fulfilled by the definition of $\tilde{u}(\xi)$. If we suppose $V_0 = W(x, y)$ (see condition (A_3)) and define the curve $C_0(t)$ according to condition (A_3), then the $C^{(1)}$ - matching conditions (9b) in zero order will be fulfilled also. So, the location of the moving front in zero order approximation is the curve $C_0(t)$, which satisfies the following Cauchy problem:

$$\frac{h_t}{\sqrt{1 + h_x^2}} = W(x, h), \quad h(x, 0) = h^0(x), \quad h(x, t) = h(x + L, t) \qquad (15)$$

According to condition (A_3), there exists such $T > 0$, that the solution $h(x, t)$ of (15) defines the curve $C_0(t) \div \{y = h(x, t)\} \in \bar{D}$ for $t \in [0, T]$.

2c. First order asymptotics.

Separating terms with ε^1 in (11), we obtain for the regular functions $\bar{u}_1^{(\pm)}$ the equation $f_u(\varphi^{(\pm)}(x, y), x, y, 0) \cdot \bar{u}^{(\pm)} + f_\varepsilon(\varphi^{(\pm)}(x, y), x, y, 0) = 0$, which has a unique solution (see condition (A_1)).

For the transition layer functions $Q_1^{(\pm)}$ we get the linear differential equations

$$\left(\frac{\partial^2}{\partial \xi^2} + V_0 \frac{\partial}{\partial \xi} - f_u(\tilde{u}(\xi), x, y, 0)\right) Q_1^{(\pm)} = q_1(\xi, l, t) \equiv \frac{\partial Q_0^{(\pm)}(\xi, l)}{\partial t} + \qquad (16)$$

$$+ (k - (\lambda_1)_t) \frac{\partial Q_0^{(\pm)}(\xi, l)}{\partial \xi} + \tilde{f}_\varepsilon + \left[\tilde{f}_r + \tilde{f}_u \cdot \frac{\partial \varphi^{(\pm)}}{\partial r}\right](\lambda_1 + \xi) + \left(\tilde{f}_u - \bar{f}_u\right) \bar{u}_1^{(\pm)}$$

with the boundary conditions $Q_1^{(\pm)}(\pm\infty, l) = 0$,

$$Q_1^{(\pm)}(0, l) = \left[-u_1^{(\pm)}(x, y) + \lambda_1(l, t)\left(\frac{\partial \varphi^{(0)}}{\partial n} - \frac{\partial \varphi^{(\pm)}}{\partial n}\right)\right]\Bigg|_{(x,y) \in C_0(t)}. \qquad (17)$$

In (16), (17) $\lambda_1(l, t)$ is the unknown first term of (7), $k = k(l)$ is the local curvature of $C_0(t)$; $\tilde{g}$ means the function depending on $(\tilde{u}(\xi), (x, y) \in C_0(t), t)$ and $\bar{g}$ - the function depending on $(\varphi^{(\pm)}, (x, y) \in C_0(t), t)$.

If we mark $\Phi(\xi, t) = \frac{\partial \tilde{u}}{\partial \xi}$, the solution of (16) with boundary conditions from (17) can be written explicitly (function $q_1(\xi, l, t)$ defined in (16)):

$$Q_1^{(\pm)}(\xi, l) = \frac{\Phi(\xi, t)}{\Phi(0, t)} Q_1^{(\pm)}(0, l) - \Phi(\xi, t) \int_0^\xi \frac{e^{-V_0 \eta}}{\Phi^2(\eta, t)} \int_\eta^{\pm\infty} \Phi(\tau, t) e^{V_0 \tau} q_1(\tau, l, t) d\tau d\eta$$

(18)

Using the $C^{(1)}$ - matching condition (9) for $Q_1^{(\pm)}(\xi, l)$ we get

$$\frac{\partial Q_1^{(+)}(\xi, l)}{\partial \xi} - \frac{\partial Q_1^{(-)}(\xi, l)}{\partial \xi}\bigg|_{\xi=0} = \frac{\partial \varphi^{(-)}(x, y)}{\partial n} - \frac{\partial \varphi^{(+)}(x, y)}{\partial n}\bigg|_{(x,y)\in C_0(t)}$$

(19)

Substituting the derivatives $\frac{\partial Q_1^{(\pm)}(\xi, l)}{\partial \xi}\bigg|_{\xi=0}$, calculated from (18) into (19), we obtain the linear Cauchy problem for $\lambda_1(l, t)$:

$$\frac{d\lambda_1(l, t)}{dt} - k(l) = B(l, t) \cdot \lambda_1(l, t) + R(l, t), \quad \lambda_1(l, 0) = 0,$$

(20)

where $B(l, t)$ and $R(l, t)$ are known function, which do not depend of $\lambda_1(l, t)$ and of its derivatives, $k(l)$ is the local curvature of the curve $C_0(t)$.

Continuing this procedure for higher orders terms in ε and we get linear problems for all $Q_i^{(\pm)}(\xi, l)$, $i = 2, 3, \ldots$ and also linear Cauchy problems of type (20) for $\lambda_i(l, t)$, $i = 2, 3, \ldots$

As a result, we obtain a nonlinear equation, which determines the location of the moving front at zero order approximation, and linear equations for higher order terms. Note that now we can estimate the location of the moving front and adequately describe the front dynamics not from the original system (1)–(2), but from problem (15) in zero order approximation in ε and from the problems of type (20) at higher order approximations in ε. We present a comparison of asymptotic and numerical results in Sect. 5.

3   Existence of Solution and the Main Theorem

The proof for the existence of a solution to (1)–(2) is based on the asymptotic method of differential inequalities similarly to the case of one spatial dimension (see [4]) with slight changes. Let define $D_n^{(+)}$ and $D_n^{(-)}$ the domains located at two sides of curve $C_n(t)$, where

$$\Lambda_n(l, t) = \sum_{i=1}^{n+1} \varepsilon^i \lambda_i(l, t), \quad \xi_n = \frac{r - \Lambda_n(l, t)}{\varepsilon}, \quad C_n(t) : r = \Lambda_n(l, t)$$

(21)

$$U_n(x, y, t, \varepsilon) = \begin{cases} \sum_{i=0}^n \varepsilon^i \left(\bar{u}_n^{(+)}(x, y) + Q_n^{(+)}(\xi_n, l, t) \right), & (x, y) \in D_n^{(+)} \\ \sum_{i=0}^n \varepsilon^i \left(\bar{u}_n^{(-)}(x, y) + Q_n^{(-)}(\xi_n, l, t) \right), & (x, y) \in D_n^{(-)} \end{cases}$$

(22)

We define upper and lower solution $\alpha(x, y, t, \varepsilon)$, $\beta(x, y, t, \varepsilon)$ as follows:

(1) $\alpha(x, y, t, \varepsilon) \leq \beta(x, y, t, \varepsilon)$, $\alpha, \beta(x, y, t, \varepsilon) = \alpha, \beta(x + L, y, t, \varepsilon)$

(2α) $\varepsilon^2 \Delta\alpha - \varepsilon\dfrac{\partial\alpha}{\partial t} - f(\alpha, x, y, \varepsilon) \geq 0$, $(x, y) \in \bar{D}$, $t \in (0, T]$, $\varepsilon \in (0, \varepsilon_0]$

(2β) $\varepsilon^2 \Delta\beta - \varepsilon\dfrac{\partial\beta}{\partial t} - f(\beta, x, y, \varepsilon) \leq 0$, $(x, y) \in \bar{D}$, $t \in (0, T]$, $\varepsilon \in (0, \varepsilon_0]$

(3) $\left.\dfrac{\partial\alpha}{\partial y}\right|_{y=0} \geq 0$, $\left.\dfrac{\partial\alpha}{\partial y}\right|_{y=a} \leq 0$; $\left.\dfrac{\partial\beta}{\partial y}\right|_{y=0} \leq 0$, $\left.\dfrac{\partial\beta}{\partial y}\right|_{y=a} \geq 0$. Suppose also

that initial function satisfies $\alpha(x, y, 0, \varepsilon) \leq u^0(x, y, \varepsilon) \leq \beta(x, y, 0, \varepsilon)$.

Now we can formulate the main result in the following theorem.

Theorem 1. *Under the conditions* $(A_1) - (A_3)$ *for sufficiently smooth initial function and sufficiently small ε there exists the solution $u(x, y, t, \varepsilon)$ of the problem (1)–(2) and satisfies*

1. $\alpha(x, y, t, \varepsilon) \leq u(x, y, t, \varepsilon) \leq \beta(x, y, t, \varepsilon)$,
2. $u(x, y, t, \varepsilon) = U_n(x, y, t, \varepsilon) + O(\varepsilon^{n+1})$ *for* $(x, y) \in \bar{D}$, $t \in (0, T]$, $\varepsilon \in (0, \varepsilon_0]$.

Main ideas, how to prove this theorem you can see in [4]. We construct upper and lower solutions by modification of (22)–(21), verify inequalities (1)–(3) from the definitions of $\alpha(x, y, t, \varepsilon)$, $\beta(x, y, t, \varepsilon)$ and control proper sign of the jump of first normal derivative at the curve $\bar{C}(t) = C_n(t) - \varepsilon^{n+1}\delta(t)$. Required calculations can be done in the same way as in [4].

4   Examples

In this section we present an example, for which we can calculate some parameters of the front in zero order (e.g., normal speed) explicitly. Consider the problem

$$\varepsilon^2 \Delta u - \varepsilon\frac{\partial u}{\partial t} = \left(u - \varphi^0(x, y)\right) \cdot \left(u^2 - 1\right), \quad y \in (0, 1), \ x \in (-\infty, +\infty), \ t > 0$$

$$u_y|_{y=0, y=1} = 0, \quad u(x, y, t, \varepsilon) = u(x + 1, y, t, \varepsilon), \quad u(x, y, t, \varepsilon)|_{t=0} = u^0(x, y, \varepsilon),$$

where $-1 < \varphi^0(x, y) < 1$ and $u^0(x, y, \varepsilon)$ satisfies condition (A_2). Note, that in this case $\varphi^{(-)}(x, y) = -1$ and $\varphi^{(+)}(x, y) = 1$, and if we mark $\tilde{u}'_\xi = z$, we can write the problem (15) for the zero order function $\tilde{u}(\xi) = \varphi^{(\pm)}(x, y) + Q_0^{(\pm)}(\xi, l)$ in the form

$$z\frac{dz}{d\tilde{u}} + V_0 z = (\tilde{u} - \varphi^0(x, y)) \cdot (\tilde{u}^2 - 1), \qquad \tilde{u}(\pm\infty) = \pm 1. \tag{23}$$

Solution of (23) exists, if there exists a separatrix going from the saddle point $(0; -1)$ to the saddle point $(0; +1)$. If we find this separatrix in the form $z = A \cdot (\tilde{u}^2 - 1)$, $A < 0$, we obtain $A = \frac{-1}{\sqrt{2}}$ and $A \cdot V_0 = -\varphi^0(x, y)$ so $V_0 = \varphi^0(x, y) \cdot \sqrt{2}$, and Eq. (15) for the moving front at zero order of ε takes the form

$$\frac{h_t}{\sqrt{1 + h_x^2}} = \varphi^0(x, h) \cdot \sqrt{2}, \quad h(x, 0) = h^0(x), \quad h(x, t) = h(x + 1, t).$$

5   Numerical Experiment

The asymptotic approximation will be compared with the results of numerical solution of the problem (1)–(2). For this purpose we use a finite-difference scheme for problem (1)–(2) and for the Eq. (15). Calculations are done in D, representing a rectangle with the sides $L = 1$, $a = 1$, for $\varepsilon = 0.01$ and the function $f(u, x, y, \varepsilon) = (u^2 - 1) \cdot (u - \varphi^0(x, y))$ for some cases of $\varphi^0(x, y)$ and the initial curve $y = h^0(x)$. Results are represented in Figs. 1, 2. Figure 1 shows sequent positions of the front (zero order asymptotics and the numerical solution of full problem (1)–(2)) at different times for $\varphi^0(x, y) = 0.15 \cos 4\pi x$, $h^0(x) = 0.5 - 0.15 \sin 2\pi x$. Figure 2 shows sequent positions of the front at different times for $\varphi^0(x, y) = 0.15 \cos 4\pi x$, $h^0(x) = 0.5 - 0.15 \sin 2\pi x$.

The analysis of the numerical calculations showed a good correspondence between the above asymptotic descriptions of the front behavior by (15) and numerical calculations for problem (1)–(2). Thus, the asymptotic approach allows fully to describe the dynamics and the shape of the moving front, its width and the time process of its formation, which is important for the effective estimate of various parameters of the physical system. In addition, the combination of asymptotic and numerical methods gives the possibility to speed up the process of constructing approximate solutions with a suitable accuracy. As a result, we have more efficient numerical calculations.

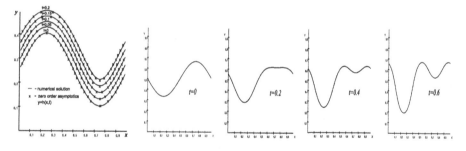

Fig. 1.   **Fig. 2.**

Acknowledgements. This work is supported by RFBR, pr. N 13-01-00200.

References

1. Vasilieva, A.B., Butuzov, V.F., Nefedov, N.N.: Contrast structures in singularly perturbed problems. J. Fund. Prikl. Math. **4**(3), 799–851 (1998)
2. Volkov, V.T., Grachev, N.E., Nefedov, N.N., Nikolaev, A.N.: On the formation of sharp transition layers in two-dimensional reaction-diffusion models. J. Comp. Math. Math. Phys. **47**(8), 1301–1309 (2007)
3. Butuzov, V.F., Nefedov, N.N., Schneider, K.R.: On generation and propagation of sharp transition layers in parabolic problems. Vestnik MGU **3**(1), 9–13 (2005)

4. Bozhevolnov, Y.V., Nefedov, N.N.: Front motion in parabolic reaction-diffusion problem. J. Comp. Math. Math. Phys. **50**(2), 264–273 (2010)
5. Nefedov, N.N.: The method of diff. inequalities for some classes of nonlinear singul. perturbed problems. J. Diff. Uravn. **31**(7), 1142–1149 (1995)
6. Fife, P.C., Hsiao, L.: The generation and propagation of internal layers. Nonlinear Anal. Theory Meth. Appl. **12**(1), 19–41 (1998)
7. Volkov, V., Nefedov, N.: Asymptotic-numerical investigation of generation and motion of fronts in phase transition models. In: Dimov, I., Faragó, I., Vulkov, L. (eds.) NAA 2012. LNCS, vol. 8236, pp. 524–531. Springer, Heidelberg (2013)

A New Approach to Constructing Splitting Schemes in Mixed FEM for Heat Transfer: A Priori Estimates

Kirill Voronin$^{(\boxtimes)}$ and Yuri Laevsky

Institute of Computational Mathematics and Mathematical Geophysics SB RAS,
Prospect Akademika Lavrentjeva, 6, 630090 Novosibirsk, Russia
{kvoronin,laev}@labchem.sscc.ru
http://www.sscc.ru/

Abstract. A priori estimates for a new approach to constructing vector splitting schemes in mixed FEM for heat transfer problems are presented. Heat transfer problem is considered in the mixed weak formulation approximated by Raviart-Thomas finite elements of lowest order on rectangular meshes. The main idea of the considered approach is to develop splitting schemes for the heat flux using well-known splitting scheme for the scalar function of flux divergence. Based on flux decomposition into discrete divergence-free and potential (orthogonal) components, a priori estimates for 2D and 3D vector splitting schemes are presented. Special attention is given to the additional smoothness requirements imposed on the initial heat flux. The role of these requirements is illustrated by several numerical examples.

Introduction

In this paper a priori estimates for splitting schemes in mixed FEM for heat transfer are presented. Heat transfer problem is considered in the mixed weak formulation or, in other words, as a system of first order ODE's written in terms "temperature - heat flux". For space approximation mixed finite element method [2] with Raviart-Thomas finite elements of lowest order on rectangular meshes is implemented. For time discretization a new approach [1] is used for constructing vector splitting schemes for the heat flux [3,4]. The main idea is to use well-known schemes [5] for the scalar function of flux divergence. Actually, any splitting scheme for the heat flux can be obtained using the approach developed (but sometimes one has to introduce additional fractional step variables).

It turns out that due to the connection between schemes for the flux and schemes for the flux divergence, stability and accuracy for the scalar schemes for flux divergence can be used to obtain corresponding results for splitting schemes for the heat flux. First of all, since temperature within considered framework depends only on the flux divergence, this approach almost immediately gives the order of convergence and stability of temperature whenever we know these

This work was financially supported by RFBR grant No. 13-01-00019.

I. Dimov et al. (Eds.): FDM 2014, LNCS 9045, pp. 417–425, 2015.
DOI: 10.1007/978-3-319-20239-6_47

results for the underlying scheme for flux divergence. Second, it can be proved that order of accuracy of a splitting scheme for the flux is the same as order of accuracy of the underlying scheme for the flux divergence in non-commutative case. The main problem is to get global stability results for the flux.

The crucial idea of getting a priori estimates is to decompose the heat flux into two components - discrete divergence-free and discrete-potential components. We present examples in 2D and 3D how this idea can be applied to the stability analysis. The obtained a priori estimates impose additional smoothness requirements on the initial heat flux which is computed through the initial temperature. As it is shown by numerical examples these conditions are not satisfied for some problems and this leads to conditional convergence issues. Therefore, smoothing procedures should be applied either to the initial temperature or to the initial heat flux.

The paper is organized as follows: in Sect. 1 statement of the problem is given together with space approximation details. In Sect. 2 main idea of the proposed approach and examples of splitting schemes for the heat flux are presented. Next, in Sect. 3 a priori estimates for the proposed schemes in 2D and 3D are given, as well as numerical examples which show that the additional smoothness requirements should be imposed in order to guarantee convergence. In the end we give a conclusion of the paper.

1   Problem Statement

Consider the following system of first order differential equations written in terms "temperature - heat flux" which describes heat transfer process for $t \in [0, t_f]$ and $\mathbf{x} \in \Omega \subset R^n, n = 2, 3$:

$$\begin{cases} c_p \rho \frac{\partial T}{\partial t} + \nabla^T \mathbf{w} = f \\ \mathbf{w} = -\lambda \nabla T \end{cases} \mathbf{x} \in \Omega, t \in [0, t_f].$$

Here T and $\mathbf{w}$ are the unknown functions of temperature and heat flux, c_p, ρ and λ are coefficients of heat capacity, density and heat conductivity respectively; righthand side f corresponds to the heat sources in Ω. On the boundary of Ω Dirichlet and Neumann boundary conditions are imposed, and for $t = 0$ initial condition $T = T^0(\mathbf{x})$ is also given, hence completing the statement of initial boundary value problem. One can notice that the first equation is the energy conservation law, the second one is constitutive equation (Fourier law) and gives the connection between temperature and heat flux.

Standard technique allows one to obtain a mixed weak formulation of the problem:

$$\begin{cases} \int_\Omega c_p \rho \frac{\partial T}{\partial t} \chi + \int_\Omega div \, \mathbf{w} \chi = \int_\Omega f \chi, \forall \chi \in L_2(\Omega) \\ \int_\Omega \frac{1}{\lambda} \mathbf{w} \cdot \mathbf{u} = \int_\Omega T \nabla \mathbf{u} - \int_{\partial\Omega} T \mathbf{u} \cdot \mathbf{n}, \forall \mathbf{u} \in \mathbf{H}_{div}(\Omega) \end{cases},$$

where unknown functions T and $\mathbf{w}$ are searched for in $C^1(0, t_f; L_2(\Omega))$ and $C(0, t_f; \mathbf{H}_{div})$ respectively.

Now, consider a rectangular grid covering the domain Ω. Space approximation of the problem is implemented by mixed finite element method based on

Raviart-Thomas finite elements of lowest order (for $\mathbf{H}_{div}$) and piecewise constant elements for scalar functions (for L_2). One can proceed to the following system of first-order ODE's:

$$\begin{cases} M\frac{dT_h}{dt} + B^T \mathbf{w}_h = f_h \\ A\mathbf{w}_h = BT_h + g_h \end{cases},$$

where M is a diagonal mass matrix for temperature, $\mathbf{A}$ - tridiagonal mass matrix for heat flux (symmetric and positive-definite), $\mathbf{B}$ and $\mathbf{B}^T$ are discrete operators of gradient and divergence respectively; g_h stands for inhomogeneous boundary conditions (omitted further), f_h approximates the differential righthand side f. Now it is straightforward to write the implicit α-weighted scheme in the form:

$$\begin{cases} (\mathbf{A} + \alpha\tau\mathbf{G})\frac{\mathbf{w}^{n+1}-\mathbf{w}^n}{\tau} + \mathbf{G}\mathbf{w}^n = BM^{-1}(\alpha F^{n+1} + (1-\alpha)F^n) \\ M\frac{T^{n+1}-T^n}{\tau} + \mathbf{B}^T(\alpha\mathbf{w}^{n+1} + (1-\alpha)\mathbf{w}^n) = \alpha F^{n+1} + (1-\alpha)F^n \end{cases}$$

where $\mathbf{G} = BM^{-1}\mathbf{B}^T$ approximates the second space derivatives. To obtain the first equation for the heat flux only one should "differentiate" on the mesh level the Fourier law and rewrite then the first equation. Besides, initial heat flux is obtained by solving $\mathbf{A}\mathbf{w}^0 = BT^0$. One can notice that temperature is computed at each time step through the flux and in this sense can be called secondary variable in the considered framework.

Now the question is how to factor the operator $\mathbf{G}$. The possible ways are the following:

1. Explicit factorization of the operator, e.g. alternating-triangular or SOR-type factorizations (for vector equation for the heat flux [3]).
2. Uzawa-type algorithms (Arbogast et al. [7]).
3. A new approach [1,4] based on splitting schemes for the scalar function of flux divergence.

Remark 1. Introducing new notations $\mathbf{B} = \mathbf{A}^{-1/2}BM^{-1/2}$, $\mathbf{w} = \mathbf{A}^{-1/2}\mathbf{w}$ and $T = M^{1/2}T$ we can simplify all equations avoiding explicit usage of $\mathbf{A}$ and M. For example, $\mathbf{A} + \alpha\tau\mathbf{G} = \mathbf{A} + \alpha\tau BM^{-1}\mathbf{B}^T$ becomes $\mathbf{E} + \alpha\tau\mathbf{B}\mathbf{B}^T$ in new notations.

2   Constructing Splitting Schemes for the Heat Flux: Examples

As already mentioned, the main idea of the proposed approach is to use well-known scalar splitting schemes [5] for the flux divergence. Consider an example of splitting scheme for the heat flux in 2D case (without righthand side):

$$\frac{w_y^{n+1/2} - w_y^n}{0.5\tau} + B_y\mathbf{B}^T\mathbf{w}^n = 0$$

$$\frac{w_x^{n+1/2} - w_x^n}{0.5\tau} + B_x\mathbf{B}^T\mathbf{w}^{n+1/2} = 0$$

$$\frac{w_x^{n+1} - w_x^{n+1/2}}{0.5\tau} + B_x \mathbf{B}^T \mathbf{w}^{n+1/2} = 0$$

$$\frac{w_y^{n+1} - w_y^{n+1/2}}{0.5\tau} + B_y \mathbf{B}^T \mathbf{w}^{n+1} = 0$$

$$\frac{T^{n+1} - T^n}{\tau} + \mathbf{B}^T \frac{\mathbf{w}^{n+1} + \mathbf{w}^n}{2} = 0$$

Here $\mathbf{B}^T \mathbf{w} = B_x^T w_x + B_y^T w_y$.

Remark 2. 1. Actually, $w_x^{n+1/2} = \frac{w_x^n + w_x^{n+1}}{2}$ and third equation can be simplified. 2. To implement this scheme one has to invert only one-dimensional operators with tridiagonal matrices along the mesh lines.

Now, if one introduces flux divergence $\xi = \mathbf{B}^T \mathbf{w}$ and applies operators B_x^T and B_y^T to equation pairs 1-2 and 3-4 and sums them up, one can get the following scheme for flux divergence $\xi = \mathbf{B}^T \mathbf{w}$:

$$\frac{\xi^{n+1/2} - \xi^n}{0.5\tau} + \Lambda_x \xi^{n+1/2} + \Lambda_y \xi^n = 0$$

$$\frac{\xi^{n+1} - \xi^{n+1/2}}{0.5\tau} + \Lambda_x \xi^{n+1/2} + \Lambda_y \xi^{n+1} = 0$$

with operators $\Lambda_x = B_x^T B_x$ and $\Lambda_y = B_y^T B_y$ approximating second space derivatives. It is easy to recognize the classical alternating-direction scheme for flux divergence ξ. One can also notice that equation for the temperature now takes the form of

$$\frac{T^{n+1} - T^n}{\tau} + \frac{\xi^{n+1} + \xi^n}{2} = 0$$

and properties of the scheme for temperature are completely defined by the properties of the scheme for flux divergence. Obviously, accuracy of temperature computation is the same as for flux divergence and stability of the scheme for flux divergence in any norm $\| \xi \|_*$ is equivalent to stability of heat flux in the seminorm $\| \mathbf{B}^T \mathbf{w} \|_*$. A bit more complicated but simple assertion is also valid - the order of accuracy for heat flux is the same as order of accuracy for flux divergence in non-commutative case $\Lambda_x \Lambda_y \neq \Lambda_y \Lambda_x$. This fact can be illustrated by taking locally one-dimensional scheme for flux divergence as an example. The order of accuracy for heat flux will be 1, while for the underlying scheme it equals 2 in commutative case.

Remark 3. The **main point** in the proposed approach is that we can take **any** splitting scheme for flux divergence and derive from it the corresponding splitting scheme for heat flux. In the next section we'll give the key idea of how the connection between two schemes can be exploited for stability analysis.

In 3D case we can take, for example, scheme of Douglas and Gunn [6] of second order as the underlying scheme for flux divergence. The corresponding vector splitting schemes for heat flux can be written in the following form:

$$\frac{w_y^{n+1/2} - w_y^n}{\tau} + B_y B_x^T w_x^n + B_y B_y^T w_x^n + B_y B_z^T w_z^n = 0$$

$$\frac{w_z^{n+1/2} - w_z^n}{\tau} + B_z B_x^T w_x^n + B_z B_y^T w_x^n + B_z B_z^T w_z^n = 0$$

$$\frac{w_x^{n+1} - w_x^n}{\tau} + B_x B_x^T \frac{w_x^{n+1} + w_x^n}{2} + B_x B_y^T \frac{w_y^{n+1/2} + w_y^n}{2} + + B_x B_z^T \frac{w_z^{n+1/2} + w_z^n}{2} = 0$$

$$\frac{w_y^{n+1} - w_y^{n+1/2}}{\tau} + B_y B_x^T \frac{w_x^{n+1} - w_x^n}{2} + B_y B_y^T \frac{w_y^{n+1} - w_y^n}{2} + + B_y B_z^T \frac{w_z^{n+1/2} - w_z^n}{2} = 0$$

$$\frac{w_z^{n+1} - w_z^{n+1/2}}{\tau} + B_z B_x^T \frac{w_x^{n+1} - w_x^n}{2} + B_z B_y^T \frac{w_y^{n+1} - w_y^n}{2} + B_z B_z^T \frac{w_z^{n+1} - w_z^n}{2} = 0$$

$$\frac{T^{n+1} - T^n}{\tau} + \mathbf{B}^T \frac{\mathbf{w}^{n+1} + \mathbf{w}^n}{2} = 0$$

Analysis of this scheme can be performed the same way as for the scheme based on alternating-direction scheme for flux divergence.

Remark 4. Uzawa-type schemes in 2D and 3D can be obtained using the developed approach by taking predictor-corrector scheme based on fully implicit locally one-dimensional scheme as a predictor part.

3   A Priori Estimates and Smoothness Requirements

3.1   A Priori Estimates

Again, let us restrict ourselves to the presented example of splitting schemes in 2D. An important point is that after fractional steps elimination in the schemes for heat flux and flux divergence the connection between schemes is still valid, i.e. applying divergence operator $\mathbf{B}^T$ to the scheme for the flux

$$(\mathbf{E} + \frac{\tau^2}{4} \begin{pmatrix} B_x \Lambda_y \\ 0 \end{pmatrix} \mathbf{B}^T) \frac{\mathbf{w}^{n+1} - \mathbf{w}^n}{\tau} + \mathbf{B}\mathbf{B}^T \frac{\mathbf{w}^{n+1} + \mathbf{w}^n}{2} = 0$$

one can get the exactly the ADI scheme (with fractional steps eliminated) for flux divergence:

$$(E + \frac{\tau^2}{4} \Lambda_x \Lambda_y) \frac{\xi^{n+1} - \xi^n}{\tau} + (\Lambda_x + \Lambda_y) \frac{\xi^{n+1} + \xi^n}{2} = 0$$

The next step is to represent heat flux $\mathbf{w}^n$ as a sum of two components: $\mathbf{w}^n = \mathbf{w}^{n,0} + \mathbf{w}^{n,1}$, where $\mathbf{w}^{n,0} \in Ker\mathbf{B}^T$ (divergence kernel $\mathbf{B}^T \mathbf{w}^{n,0} = 0$) and $\mathbf{w}^{n,1} \in Im\mathbf{B}$ (orthogonal complement to the kernel = range of the gradient operator $\mathbf{B}$). Then one can obtain the following equations in the subspaces:

$$\frac{\mathbf{w}^{n+1,1} - \mathbf{w}^{n,1}}{\tau} + \mathbf{B}\mathbf{B}^T \frac{\mathbf{w}^{n+1,1} + \mathbf{w}^{n,1}}{2} + (\mathbf{E} - \mathbf{Pr}_{Ker\mathbf{B}^T}) \frac{\tau^2}{4} \begin{pmatrix} B_x \Lambda_y \\ 0 \end{pmatrix} \mathbf{B}^T) \frac{\mathbf{w}^{n+1,1} - \mathbf{w}^{n,1}}{\tau}$$
$$= 0, \ in \ Im\mathbf{B}$$

$$\frac{\mathbf{w}^{n+1,0} - \mathbf{w}^{n,0}}{\tau} + \frac{\tau^2}{4} \mathbf{Pr}_{Ker\mathbf{B}^T} \begin{pmatrix} B_x \Lambda_y \\ 0 \end{pmatrix} \mathbf{B}^T) \frac{\mathbf{w}^{n+1,1} - \mathbf{w}^{n,1}}{\tau} = 0, \quad in \ Ker\mathbf{B}^T$$

Since heat flux should satisfy Fourier law, the component $\mathbf{w}^{\cdot,0}$ is completely parasitic. As it follows from equations above it depends on the component from $\mathrm{Im}\mathbf{B}$ and therefore, all we need is to get stability estimates for the "regular" component in the corresponding norms. First, introduce the discrete $\mathbf{H}_{div}$ norm in the following way: $\| \mathbf{w} \|_{\mathbf{H}} = \| \mathbf{w} \|_2 + \| \mathbf{B}^T\mathbf{w} \|_2$. Here $\| \cdot \|_2$ is the usual mesh L_2-norm. Finally, after some tedious computations, one can get the following a priori estimates for the presented 2D scheme:

Theorem 1 (commutative case, $\Lambda_x\Lambda_y = \Lambda_y\Lambda_x$). The following estimate is valid for splitting scheme for flux (based on alternating-direction scheme for flux divergence) in commutative case with $C \neq C(\tau, h)$:

$$\| \mathbf{w}^n \|_{\mathbf{H}} \leq C \| \mathbf{w}^0 \|_{\widetilde{\mathbf{H}}} \; \forall \mathbf{w}^0 \in \widetilde{\mathbf{H}}$$

where $\| \mathbf{w} \|_{\widetilde{\mathbf{H}}} = \| \mathbf{w} \|_{\mathbf{H}} + \tau^2 \| \Lambda_y\mathbf{B}^T\mathbf{w} \|_{\Lambda_x+\Lambda_y}$.

Theorem 2 (non-commutative case, $\Lambda_x\Lambda_y \neq \Lambda_y\Lambda_x$). The following estimate is valid for splitting scheme for flux (based on alternating-direction scheme for flux divergence) in non-commutative case with $C \neq C(\tau, h)$:

$$\| \mathbf{w}^n \|_{\mathbf{H}} \leq C(1 + \frac{\tau}{h}) \| \mathbf{w}^0 \|_{\widetilde{\mathbf{H}}} \; \forall \mathbf{w}^0 \in \widetilde{\mathbf{H}}$$

where $\| \mathbf{w} \|_{\widetilde{\mathbf{H}}} = \| \mathbf{w} \|_{\mathbf{H}} + \| \Lambda_y\mathbf{B}^T\mathbf{w} \|_{\Lambda_x+\Lambda_y} + \| (E + \frac{\tau}{2}\Lambda_y)\mathbf{B}^T\mathbf{w}^0 \|_2$.

Remark 5. For splitting schemes in 3D case similar a priori estimates can be proved much along the lines it is done above.

As one can see the a priori estimates impose additional smoothness requirements (related to 4th derivatives) on the initial heat flux which is computed from solving $\mathbf{A}\mathbf{w}^0 = \mathbf{B}T^0$. The question arises whether this initial flux fulfill the requirements in various situations (non-uniform meshes, non-constant coefficients, different boundary conditions, etc.). Let us consider some numerical examples illustrating the problem. In model problems the domain $\Omega = [0,1]^2$ with Dirichlet boundary conditions for y-axis and Neumann boundary conditions for x-axis.

3.2 Boundary Conditions

Consider two test solutions: $T_1(t, x, y) = \cos 2\pi x \cdot \sin 2\pi y \cdot e^{-t}$ and $T_2(t, x, y) = \cos 2\pi x \cdot y \sin 2\pi y \cdot e^{-t}$ and uniform mesh, $\frac{\tau}{h} = const$. Then we have the following results in C and L_2-norm of the heat flux (Table 1):

The reason for the lack of convergence for test 2 (order 0 in C-norm and $1/2$ in L_2-norm) is due to the bad approximation of Dirichlet boundary condition and is in agreement with a priori estimates. On the Fig. 1 $\mathbf{B}^T\mathbf{w}^0$ is shown for test 1(blue line) and test 2 (red line) along one mesh line in y-direction. Obviously, for test 2 derivatives of flux divergence contain negative powers of h. One of possible ways to avoid this problem is to use integral (Sobolev) averaging for initial temperature instead of using point-wise data.

Remark 6. If one takes the lumped mass matrix $\mathbf{A}$, integral averaging doesn't help.

Table 1. Comparison of test1 and test 2, Dirichlet+Neumann cnd's.

h	test 1		test 2	
	$\| w - w_{exact} \|_C$	$\| w - w_{exact} \|_{L_2}$	$\| w - w_{exact} \|_C$	$\| w - w_{exact} \|_{L_2}$
1/32	3.5e-1	2.5e-1	**3.1e+2**	**6.4e+2**
1/64	5.9e-2	4.1e-2	**3.1e+2**	**4.6e+1**
1/128	1.3e-2	9.1e-3	**4.1e+2**	**4.3e+1**

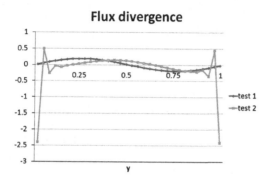

Fig. 1. Heat flux divergence $\mathbf{B}^T \mathbf{w}^0$ along a mesh line in y-direction: test 1 (blue), test 2 (red) (Color figure online).

3.3 Non-Uniform Mesh

Now consider the same solution $T_1(t, x, y) = \cos 2\pi x \cdot \sin 2\pi y \cdot e^{-t}$ (test 1 above) with nonuniform mesh covering the computational domain. As in the previous case, we are focusing only on the flux, since temperature always converges with second order in both C and L_2 norms. The nonuniform mesh is constructed in the following way: $h_x = const, h_y = \begin{cases} h_1, \, if \, y \leq 0.5 \\ h_2, \, else \end{cases}$

Table 2. Convergence issues for nonuniform mesh experiments for test 1.

h		$\tau/h^2 = const$		$\tau/h = const$	
h_1	h_2	$\| w - w_{exact} \|_C$	$\| w - w_{exact} \|_{L_2}$	$\| w - w_{exact} \|_C$	$\| w - w_{exact} \|_{L_2}$
1/30	1/36	1.8e+2	3.4e+1	1.8e+2	3.4e+1
1/60	1/72	6.0e+1	8.1e+0	1.8e+2	2.4e+1
1/120	1/144	2.0e+1	1.9e+0	2.4e+2	2.3e+1

For $\tau/h = const$ there is no convergence in C-norm and convergence of order $1/2$ in L_2-norm. It can be shown explicitly (at least for the lumped mass matrix $\mathbf{A}$) that in this case there is a term with $\frac{\tau^2}{h^2}$ at the line, where space step changes, in the heat flux error. And, again, one can avoid the arising approximation problem by using smoothing procedure for either initial flux or initial temperature. In the next table results for using the same test and nonuniform mesh are shown but, instead of computing initial flux by solving the equation, exact heat flux at time moment $t = 0$ is used (Table 2).

Table 3. Nonuniform mesh experiments for test 2 with exact initial flux.

h		$\| T - T_{exact} \|_C$	$\| T - T_{exact} \|_{L_2}$	$\| \mathbf{w} - \mathbf{w}_{exact} \|_C$	$\| \mathbf{w} - \mathbf{w}_{exact} \|_{L_2}$
1/30	1/36	3.4e-2	1.7e-2	8.5e-1	2.1e-1
1/60	1/72	7.4e-3	3.8e-3	2.1e-1	4.3e-2
1/120	1/144	1.8e-3	9.1e-4	6.4e-2	9.6e-3

As one can notice, accuracy for the heat flux is significantly better and the order of convergence is about 2. Therefore, the main conclusion is that in order to avoid lack of convergence one has to provide initial flux which is smooth enough. Using exact initial flux do not improves temperature significantly, which also indicates that approximation problems appear in the kernel of discrete divergence operator $\mathbf{B}^T$ (Table 3).

4   Conclusion

In the present paper a priori estimates for a new approach to constructing splitting schemes in mixed FEM for heat transfer problem are presented. The main idea of the approach is to use well-known scalar splitting schemes for heat flux divergence. Splitting schemes for flux obtained using the developed approach inherit certain properties related to accuracy and stability from the underlying scalar splitting schemes. For the example of vector splitting scheme in 2D case a priori estimates are presented. It is shown by numerical examples that poor approximation and lack of convergence for heat flux can occur in special cases (nonuniform mesh, nonconstant coefficients, etc.) if the initial heat flux is not smooth enough (while temperature is always computed well). Smoothing procedures (e.g., integral averaging) seem to be a promising way to overcome the arising difficulties.

References

1. Voronin, K.V., Laevsky Y.M.: An approach to the construction of flow splitting schemes in the mixed finite element method. Math. Mod. and Comp. Simul. (to appear)
2. Brezzi, F., Fortin, M.: Mixed and Hybrid Finite Element Methods. Springer, New York (1991)
3. Voronin, K.V., Laevsky, Y.M.: On splitting schemes in the mixed finite element method. Numer. Anal. Appl. 5(2), 150–155 (2010)

4. Voronin, K.V., Laevsky, Y.M.: Splitting schemes in the mixed finite-element method for the solution of heat transfer problems. Math. Mod. Comp. Simul. **5**(2), 167–174 (2012)
5. Yanenko, N.N.: The Method of Fractional Steps: Solution of Problems of Mathematical Physics in Several Variables. Springer, New York (1971)
6. Douglas, J., Gunn, J.E.: A general formulation of alternating direction methods. Numerische Mathematik **6**, 428–453 (1964)
7. Arbogast, T., Huang, C.-S., Yang, S.-M.: Improved accuracy for alternating-direction methods for parabolic equations based on regular and mixed finite elements. Math. Mod. Meth. Appl. Sci. **17**(8), 1279–1305 (2007)

The Analysis of Lagrange Interpolation for Functions with a Boundary Layer Component

Alexander Zadorin$^{(\boxtimes)}$

Sobolev Mathematics Institute SB RAS, Pevtsova 13, Omsk 644043, Russia
`zadorin@ofim.oscsbras.ru`

Abstract. Interpolation formulas for the functions of one variable with a boundary layer component are investigated. An interpolated function corresponds to a solution of a singular perturbed problem. An application of Lagrange interpolation on a uniform mesh leads to significant errors. Two approaches for a interpolation of a function with a boundary layer component are considered: a fitting of the interpolation formula to a boundary layer component and the application of Lagrange interpolation on Shishkin mesh. Numerical results are discussed.

Keywords: Function · Boundary layer · Lagrange interpolation · Shishkin mesh · Nonpolynomial interpolation

1 Introduction

We investigate the interpolation problem for a function with large gradients in a boundary layer. We suppose that the interpolated function corresponds to a solution of a singular perturbed boundary value problem:

$$\varepsilon u''(x) + a(x)u'(x) - b(x)u(x) = f(x), \ x \in (0,1], \ u(0) = A, \ u(1) = B, \quad (1)$$

where

$$a(x) \geq \alpha > 0, b(x) \geq 0, \ \varepsilon \in (0,1],$$

functions a, b, f are smooth enough. The derivatives of the function $u(x)$ are not ε-uniformly bounded [1], therefore an application of Lagrange interpolation formulas [2] can lead to significant errors [3–5].

Let us a function $u(x)$ be given at nodes of a mesh Ω :

$$\Omega = \{x_n : x_n = x_{n-1} + h_n, \ x_0 = 0, \ x_N = 1, \ n = 1, 2, \ldots, N\},$$

$u_n = u(x_n), \ n = 0, 1, 2, \ldots, N$. We investigate interpolation formulas with m interpolation nodes.

We consider two approaches to increase the accuracy of Lagrange interpolation: the fitting of the formula to the boundary layer component and using the mesh, which is dense in the boundary layer.

Through the paper C and C_j denote generic positive constants independent of ε and mesh size.

© Springer International Publishing Switzerland 2015
I. Dimov et al. (Eds.): FDM 2014, LNCS 9045, pp. 426–432, 2015.
DOI: 10.1007/978-3-319-20239-6_48

2   The Fitting of the Interpolation Formula to a Boundary Layer Component

Let us a function $u(x)$ be smooth enough with the following representation:

$$u(x) = p(x) + \gamma \Phi(x), \quad x \in [0,1] \tag{2}$$

Here $\Phi(x)$ is known function with large derivatives, the derivatives $p^{(j)}(x)$ are bounded up to some order, the constant γ is not given.

The solution of the problem (1) has the representation (2). If we take $\Phi(x) = \exp(-a(0)\varepsilon^{-1}x)$, $\gamma = -\varepsilon u'(0)/a(0)$, then for some constant C_1 the estimate $|p'(x)| \le C_1$, $x \in [0,1]$ is correct [1]. The derivatives $\Phi^{(j)}(x), j \ge 1$ are not ε-uniformly bounded.

We investigate the problem of the interpolation on subintervals with m nodes of the mesh Ω. We suppose that $N/(2(m-1))$ is integer. Let us

$$[0,1] = \bigcup_{k=0,\, m-1}^{N-m+1} [x_k, x_{k+m-1}].$$

We use Lagrange polynomial $L_{k,m}(u,x)$ with m interpolation nodes of the interval $[x_k, x_{k+m-1}]$:

$$L_{k,m}(u,x) = \sum_{n=k}^{k+m-1} u_n \prod_{\substack{j=k \\ j \ne n}}^{k+m-1} \frac{x - x_j}{x_n - x_j}. \tag{3}$$

According to [4], the application of the polynomial $L_{k,m}(u,x)$ can lead to the interpolation errors of the order $O(1)$ if the function $u(x)$ is the solution of the problem (1).

In [6] the following modification of Lagrange interpolation was proposed:

$$L_{\Phi,k,m}(u,x) = L_{k,m-1}(u,x) + \frac{[x_k, x_{k+1}, \ldots, x_{k+m-1}]u}{[x_k, x_{k+1}, \ldots, x_{k+m-1}]\Phi} \Big[\Phi(x) - L_{k,m-1}(\Phi,x)\Big],$$

$$x \in [x_k, x_{k+m-1}] \tag{4}$$

where $[x_k, x_{k+1}, \ldots, x_{k+m-1}]u$ is the divided difference for the function $u(x)$ (e.g., see [2]). The interpolation formula (4) is exact on a boundary layer component $\Phi(x)$. We will notice that the mesh Ω may be uniform.

According to [2], for some $s \in (x_k, x_{k+m-1})$

$$[x_k, x_{k+1}, \ldots, x_{k+m-1}]\Phi = \Phi^{(k-1)}(s)/(k-1)!.$$

Therefore, the expression (4) is correct if $\Phi^{(k-1)}(x) \ne 0$, $x \in (x_k, x_{k+m-1})$.

According to [2],

$$L_{k,m}(u,x) = L_{k,m-1}(u,x) + [x_k, x_{k+1}, \ldots, x_{k+m-1}]u \prod_{j=k}^{k+m-2} (x - x_j).$$

Therefore, the interpolant (4) can be written in a form:

$$L_{\Phi,k,m}(u,x) = L_{k,m}(u,x) + \frac{[x_k, x_{k+1}, \ldots, x_{k+m-1}]u}{[x_k, x_{k+1}, \ldots, x_{k+m-1}]\Phi}\Big[\Phi(x) - L_{k,m}(\Phi,x)\Big],$$

where $x \in [x_k, x_{k+m-1}]$ and $L_{k,m}(u,x)$ corresponds to (3).

3   Lagrange Interpolation on Shishkin Mesh

Now we investigate the accuracy of Lagrange interpolation on Shishkin mesh for a function with the exponential boundary layer component. We suppose that a function $u(x)$ can be represented in a form:

$$u(x) = q(x) + \Theta(x), \quad x \in [0,1] \tag{5}$$

where for some constant C_1

$$|q^{(j)}(x)| \le C_1, \quad |\Theta^{(j)}(x)| \le \frac{C_1}{\varepsilon^j}e^{-\alpha x/\varepsilon}, \quad 0 \le j \le m, \tag{6}$$

where m is given integer, functions $q(x)$ and $\Theta(x)$ are not known in the explicit form. According to [7,8], the solution of the problem (1) can be represented in the form (5) with conditions (6).

Let us Ω be Shishkin mesh [7] with steps:

$$h_n = \begin{cases} \dfrac{2\sigma}{N}, & 1 \le n \le \dfrac{N}{2} \\[2mm] \dfrac{2(1-\sigma)}{N}, & \dfrac{N}{2} < n \le N \end{cases}, \qquad \sigma = \min\Big\{\frac{1}{2}, \frac{m\varepsilon}{\alpha}\ln N\Big\}. \tag{7}$$

Lemma 1. *Suppose that a function $u(x)$ can be represented in the form (5) with conditions (6) and the mesh Ω corresponds to (7). Then for some constant C the following estimates on each interval $[x_k, x_{k+m-1}]$*

$$\Big|u(x) - L_{k,m}(u,x)\Big| \le \begin{cases} C\dfrac{\ln^m N}{N^m}, & \varepsilon < \dfrac{\alpha}{2m\ln N}, \; x_{k+m-1} \le \sigma, \\[3mm] \dfrac{C}{N^m}, & \varepsilon < \dfrac{\alpha}{2m\ln N}, \; x_k \ge \sigma, \\[3mm] \dfrac{C}{N^m}\min\Big\{\dfrac{1}{\varepsilon^m}, \ln^m N\Big\}, & \varepsilon \ge \dfrac{\alpha}{2m\ln N} \end{cases} \tag{8}$$

are satisfied.

Proof. We define h as a uniform step of the interval $[x_k, x_{k+m-1}]$. The following estimate is known [2]:

$$|u(x) - L_{k,m}(u,x)| \le \max_{x \in [x_k, x_{k+m-1}]}|u^{(m)}(x)|\frac{h^m}{4m}. \tag{9}$$

We take into account the inequality $h < 2/N$ and from (6), (9) obtain

$$|q(x) - L_{k,m}(q,x)| \le \frac{C_1 2^m}{4mN^m}. \tag{10}$$

Now we estimate the interpolation error for the function $\Theta(x)$.

Let us $\sigma < 1/2$.

Consider the case $x_{k+m-1} \le \sigma$. We use relations (6), (7) and obtain

$$h = \frac{2m\varepsilon}{\alpha} \frac{\ln N}{N}, \quad |\Theta^{(m)}(x)| \le \frac{C_1}{\varepsilon^m}.$$

Now we use the inequality (9) in the case of the function $\Theta(x)$ and obtain

$$|\Theta(x) - L_{k,m}(\Theta,x)| \le \frac{C_1(2m)^m}{4m\alpha^m} \frac{\ln^m N}{N^m}. \tag{11}$$

Consider the case $x_k \ge \sigma$. We use (6), (7) and obtain

$$|\Theta(x)| \le \frac{C_1}{N^m}.$$

Therefore,

$$|\Theta(x) - L_{k,m}(\Theta,x)| \le |\Theta(x)| + |L_{k,m}(\Theta,x)| \le \frac{C_1}{N^m}(1 + \lambda_{k,m}),$$

were $\lambda_{k,m}$ is Lebeg's constant [2] for the polynomial $L_{k,m}(\Theta,x)$. The mesh of the interval $[x_k, x_{k+m-1}]$ is uniform and according to [9] $\lambda_{k,m} \le 2^{m-1}$. Then

$$|\Theta(x) - L_{k,m}(\Theta,x)| \le \frac{C_1}{N^m}(1 + 2^{m-1}). \tag{12}$$

Now we consider the last case $\sigma = 1/2$, when the mesh Ω is uniform. We use relations (6), (7), (9) and obtain

$$|\Theta(x) - L_{k,m}(\Theta,x)| \le \frac{C_1}{4m\varepsilon^m N^m}, \quad \varepsilon \ge \frac{\alpha}{2m\ln N}.$$

Therefore,

$$|\Theta(x) - L_{k,m}(\Theta,x)| \le \frac{C_1}{4mN^m} \min\left\{\frac{1}{\varepsilon^m}, \left(\frac{2m\ln N}{\alpha}\right)^m\right\}. \tag{13}$$

Using estimates (10)–(13), we obtain (8) for some constant C. The lemma is proved. $\square$

4   Numerical Results

Now we consider the function

$$u(x) = \cos\frac{\pi x}{2} + e^{-\varepsilon^{-1}(x+x^2/2)}, \quad x \in [0,1], \quad \varepsilon > 0,$$

which can be considered as the particular solution of the problem (1). We investigate numerically the interpolation formulas with four nodes on each interval $[x_k, x_{k+3}]$, $k = 0, 3, 6, \ldots, N - 3$. We test piecewise-cubic interpolation and piecewise-fitted interpolation (4).

We define the interpolation error of Lagrange polynomial as

$$\Delta_{N,\varepsilon} = \max_{k=0,3,\ldots,N-3} \max_{j=1,2,3} |L_{k,4}(u, \tilde{x}_{k+j}) - u(\tilde{x}_{k+j})|, \tag{14}$$

where $\tilde{x}_{k+j} = (x_{k+j} + x_{k+j-1})/2$. Similarly as (14) we define the interpolation error of the interpolant (4). Then we define the computed order of the accuracy:

$$CR_{N,\varepsilon} = \log_2 \frac{\Delta_{N,\varepsilon}}{\Delta_{2N,\varepsilon}}.$$

In tables $e \pm m$ means $10^{\pm m}$.

Table 1 presents $\Delta_{N,\varepsilon}$ and $CR_{N,\varepsilon}$ of the piecewise-cubic interpolation in the case of a uniform mesh Ω. If $\varepsilon \approx 1$ the piecewise-cubic interpolation is of the fourth order of the accuracy. For small values of ε the interpolation error is $O(1)$.

Table 1. The errors and the accuracy orders of the piecewise-cubic interpolation on the uniform meshes

ε	N					
	24	48	96	192	384	768
1	$4.43e-7$	$2.89e-8$	$1.84e-9$	$1.16e-10$	$7.31e-12$	$4.58e-13$
	3.94	3.97	3.98	3.99	3.99	3.98
10^{-1}	$4.04e-4$	$2.85e-5$	$1.88e-6$	$1.21e-7$	$7.64e-9$	$4.80e-10$
	3.82	3.92	3.96	3.98	3.99	3.99
10^{-2}	$2.03e-1$	$7.14e-2$	$1.28e-2$	$1.44e-3$	$1.23e-4$	$8.99e-6$
	1.51	2.48	3.15	3.55	3.77	3.88
10^{-3}	$3.12e-1$	$3.12e-1$	$3.07e-1$	$2.44e-1$	$1.08e-1$	$2.41e-2$
	0.00	0.02	0.33	1.17	2.16	2.94
10^{-4}	$3.12e-1$	$3.12e-1$	$3.12e-1$	$3.12e-1$	$3.12e-1$	$3.11e-1$
	0.00	0.00	0.00	0.00	0.00	0.18
10^{-5}	$3.12e-1$	$3.12e-1$	$3.12e-1$	$3.12e-1$	$3.12e-1$	$3.12e-1$
	0.00	0.00	0.00	0.00	0.00	0.00

Table 2 presents $\Delta_{N,\varepsilon}$ and $CR_{N,\varepsilon}$ of the piecewise-fitted interpolation (4) with $m = 4$ in the case of a uniform mesh Ω. Similarly as in the previous case, for $\varepsilon \approx 1$ the piecewise-fitted interpolation is of the fourth order of the accuracy. For small values of ε the piecewise-fitted interpolation is of the third order of the accuracy.

Table 2. The errors and the accuracy orders of the piecewise-fitted interpolation (4) with $m = 4$ on the uniform meshes

ε	N					
	24	48	96	192	384	768
1	$1.20e-5$	$7.55e-7$	$4.71e-8$	$2.94e-9$	$1.84e-10$	$1.15e-11$
	3.99	4.00	4.00	4.00	4.00	4.00
10^{-1}	$4.12e-5$	$2.50e-6$	$1.52e-7$	$9.44e-9$	$5.87e-10$	$3.66e-11$
	4.04	4.04	4.01	4.01	4.00	4.01
10^{-2}	$4.68e-4$	$2.99e-5$	$1.70e-6$	$9.81e-8$	$5.86e-9$	$3.57e-10$
	3.97	4.14	4.12	4.07	4.04	4.01
10^{-3}	$6.89e-4$	$8.72e-5$	$1.08e-5$	$1.08e-6$	$7.46e-8$	$4.28e-9$
	2.98	3.01	3.32	3.86	4.12	4.12
10^{-4}	$6.89e-4$	$8.72e-5$	$1.09e-5$	$1.37e-6$	$1.71e-7$	$2.13e-8$
	2.98	3.00	2.99	3.00	3.01	3.17
10^{-5}	$6.89e-4$	$8.72e-5$	$1.09e-5$	$1.37e-6$	$1.71e-7$	$2.14e-8$
	2.98	3.00	2.99	3.00	3.01	3.00

Table 3. The errors and the accuracy orders of the piecewise-cubic interpolation on Shishkin meshes

ε	N					
	24	48	96	192	384	768
1	$4.43e-7$	$2.89e-8$	$1.84e-9$	$1.16e-10$	$7.31e-12$	$4.58e-13$
	3.94	3.97	3.98	3.99	3.99	3.98
10^{-1}	$4.04e-4$	$2.85e-5$	$1.88e-6$	$1.21e-7$	$7.64e-9$	$4.80e-10$
	3.82	3.92	3.96	3.98	3.99	3.99
10^{-2}	$1.34e-2$	$2.94e-3$	$4.84e-4$	$6.46e-5$	$7.44e-6$	$7.73e-7$
	2.19	2.60	2.90	3.11	3.26	3.37
10^{-3}	$1.37e-2$	$3.03e-3$	$5.03e-4$	$6.76e-5$	$7.82e-6$	$8.14e-7$
	2.17	2.59	2.89	3.11	3.26	3.36
10^{-4}	$1.37e-2$	$3.00e-3$	$5.05e-4$	$6.79e-5$	$7.86e-6$	$8.20e-7$
	2.17	2.58	2.89	3.11	3.26	3.36
10^{-5}	$1.37e-2$	$3.04e-3$	$5.05e-4$	$6.79e-5$	$7.86e-6$	$8.19e-7$
	2.17	2.58	2.89	3.11	3.26	3.36
M_N	2.86	3.05	3.18	3.28	3.36	3.43

Table 3 presents $\Delta_{N,\varepsilon}$ and $CR_{N,\varepsilon}$ of the piecewise-cubic interpolation on Shishkin mesh (7). In the bottom of the Table 3 the theoretical order of the accuracy is presented as

$$M_N = \log_2 \left(\frac{\ln^4 N}{N^4} : \frac{\ln^4(2N)}{(2N)^4} \right).$$

According to the Table 3, $CR_{N,\varepsilon}$ approaches to M_N if N increases and ε is small. This result corresponds to the estimates (8) with $m = 4$.

5   Conclusion

The accuracy of Lagrange interpolation for a function with large gradients in a boundary layer is investigated. It is shown that the application of Lagrange interpolation on a uniform mesh leads to the significant errors for small values of a parameter ε. The estimate of ε-uniform accuracy of Lagrange interpolation on Shishkin mesh is obtained. The numerical comparison of Lagrange interpolation on Shishkin mesh with the interpolation fitted to a boundary layer component is carried out.

Acknowledgements. Supported by Russian Foundation for Basic Research under Grants 13-01-00618, 15-01-06584.

References

1. Miller, J.J.H., O'Riordan, E., Shishkin, G.I.: Fitted Numerical Methods for Singular Perturbation Problems: Error Estimates in the Maximum Norm for Linear Problems in One and Two Dimensions, Revised edn. World Scientific, Singapore (2012)
2. Bakhvalov, N.S.: Numerical Methods. Nauka, Moskow (1975). (in Russian)
3. Zadorin, A.I.: Method of interpolation for a boundary layer problem. Sib. J. Numer. Math. **10**(3), 267–275 (2007). (in Russian)
4. Zadorin, A.I., Zadorin, N.A.: Spline interpolation on a uniform grid for functions with a boundary-layer component. Comput. Math. Math. Phys. **50**(2), 211–223 (2010)
5. Zadorin, A.I.: Spline interpolation of functions with a boundary layer component. Int. J. Numer. Anal. Model., Ser. B. **2**(2–3), 562–579 (2011)
6. Zadorin, A.I., Zadorin, N.A.: Interpolation formula for functions with a boundary layer component and its application to derivatives calculation. Siberian Electron. Math. Rep. **9**, 445–455 (2012)
7. Shishkin, G.I.: Grid Approximations of Singular Perturbation Elliptic and Parabolic Equations. UB RAS, Yekaterinburg (1992). (in Russian)
8. Lins, T.: The necessity of Shishkin decompositions. Appl. Math. Lett. **14**, 891–896 (2001)
9. Kornev, A.A., Chizhonkov, E.V.: Exercises on Numerical Methods. Part 2. Moscow State University, Moskow (2003). (in Russian)

Author Index

Printed in the United States
By Bookmasters